Numerical Analysis

Using MATLAB® and Excel®

Third Edition

Steven T. Karris

Orchard Publications
www.orchardpublications.com

Library of Congress Cataloging-in-Publication Data

Library of Congress Control Number: 2007922100

Copyright TX 5-589-152

ISBN-13: **978-1-934404-03-4**

ISBN-10: **1-934404-03-9**

Disclaimer

Preface

Numerical analysis is the branch of mathematics that is used to find approximations to difficult problems such as finding the roots of non–linear equations, integration involving complex expressions and solving differential equations for which analytical solutions do not exist. It is applied to a wide variety of disciplines such as business, all fields of engineering, computer science, education, geology, meteorology, and others.

Years ago, high–speed computers did not exist, and if they did, the largest corporations could only afford them; consequently, the manual computation required lots of time and hard work. But now that computers have become indispensable for research work in science, engineering and other fields, numerical analysis has become a much easier and more pleasant task.

This book is written primarily for students/readers who have a good background of high–school algebra, geometry, trigonometry, and the fundamentals of differential and integral calculus.[*] A prior knowledge of differential equations is desirable but not necessary; this topic is reviewed in Chapter 5.

One can use Fortran, Pascal, C, or Visual Basic or even a spreadsheet to solve a difficult problem. It is the opinion of this author that the best applications programs for solving engineering problems are 1) MATLAB which is capable of performing advanced mathematical and engineering computations, and 2) the Microsoft Excel spreadsheet since the versatility offered by spreadsheets have revolutionized the personal computer industry. We will assume that the reader has no prior knowledge of MATLAB and limited familiarity with Excel.

We intend to teach the student/reader how to use MATLAB via practical examples and for detailed explanations he/she will be referred to an Excel reference book or the MATLAB User's Guide. The MATLAB commands, functions, and statements used in this text can be executed with either MATLAB Student Version 12 or later. Our discussions are based on a PC with Windows XP platforms but if you have another platform such as Macintosh, please refer to the appropriate sections of the MATLAB's User Guide that also contains instructions for installation.

MATLAB is an acronym for MATrix LABoratory and it is a very large computer application which is divided to several special application fields referred to as toolboxes. In this book we will be using the toolboxes furnished with the Student Edition of MATLAB. As of this writing, the latest release is MATLAB Student Version Release 14 and includes SIMULINK which is a

---

[*] *These topics are discussed in **Mathematics for Business, Science, and Technology, Third Edition**, ISBN 0–9709511– 0–8. This text includes probability and other advanced topics which are supplemented by many practical applications using Microsoft Excel and MATLAB.*

software package used for modeling, simulating, and analyzing dynamic systems. SIMULINK is not discussed in this text; the interested reader may refer to Introduction to Simulink with Engineering Applications, ISBN 0-9744239-7-1. Additional information including purchasing the software may be obtained from The MathWorks, Inc., 3 Apple Hill Drive, Natick, MA 01760-2098. Phone: 508 647-7000, Fax: 508 647-7001, e-mail: info@mathwork.com and web site http://www.mathworks.com.

The author makes no claim to originality of content or of treatment, but has taken care to present definitions, statements of physical laws, theorems, and problems.

Chapter 1 is an introduction to MATLAB. The discussion is based on MATLAB Student Version 5 and it is also applicable to Version 6. Chapter 2 discusses root approximations by numerical methods. Chapter 3 is a review of sinusoids and complex numbers. Chapter 4 is an introduction to matrices and methods of solving simultaneous algebraic equations using Excel and MATLAB. Chapter 5 is an abbreviated, yet practical introduction to differential equations, state variables, state equations, eigenvalues and eigenvectors. Chapter 6 discusses the Taylor and Maclaurin series. Chapter 7 begins with finite differences and interpolation methods. It concludes with applications using MATLAB. Chapter 8 is an introduction to linear and parabolic regression. Chapters 9 and 10 discuss numerical methods for differentiation and integration respectively. Chapter 11 is a brief introduction to difference equations with a few practical applications. Chapters 12 is devoted to partial fraction expansion. Chapters 13, 14, and 15 discuss certain interesting functions that find wide application in science, engineering, and probability. This text concludes with Chapter 16 which discusses three popular optimization methods.

New to the Third Edition

This is an extensive revision of the first edition. The most notable changes are the inclusion of Fourier series, orthogonal functions and factorization methods, and the solutions to all end-of-chapter exercises. It is in response to many readers who expressed a desire to obtain the solutions in order to check their solutions to those of the author and thereby enhancing their knowledge. Another reason is that this text is written also for self-study by practicing engineers who need a review before taking more advanced courses such as digital image processing. The author has prepared more exercises and they are available with their solutions to those instructors who adopt this text for their class.

Another change is the addition of a rather comprehensive summary at the end of each chapter. Hopefully, this will be a valuable aid to instructors for preparation of view foils for presenting the material to their class.

The last major change is the improvement of the plots generated by the latest revisions of the MATLAB® Student Version, Release 14.

Orchard Publications
Fremont, California
www.orchardpublications.com
info@orchardpublications.com

Table of Contents

4 Matrices and Determinants — 4–1

5 Differential Equations, State Variables, and State Equations — 5–1

6   *Fourier, Taylor, and Maclaurin Series*   6-1

7   *Finite Differences and Interpolation*   7-1

16 Optimization Methods   16–1

A Difference Equations in Discrete–Time Systems   A–1

B Introduction to Simulink®   B–1

C Ill–Conditioned Matrices   C–1

References   R–1

Index   IN1

Chapter 1

Introduction to MATLAB

This chapter is an introduction of the basic MATLAB commands and functions, procedures for naming and saving the user generated files, comment lines, access to MATLAB's Editor/Debugger, finding the roots of a polynomial, and making plots. Several examples are provided with detailed explanations. Throughout this text, a left justified horizontal bar will denote the beginning of an example, and a right justified horizontal bar will denote the end of the example. These bars will not be shown whenever an example begins at the top of a page or at the bottom of a page. Also, when one example follows immediately after a previous example, the right justified bar will be omitted.

1.1 Command Window

To distinguish the screen displays from the user commands, important terms and MATLAB functions, we will use the following conventions:

Click: Click the left button of the mouse

`Courier Font`: Screen displays

Helvetica Font: User inputs at MATLAB's command window prompt EDU>>[*]

Helvetica Bold: MATLAB functions

Bold Italic: Important terms and facts, notes, and file names

When we first start MATLAB, we see the toolbar on top of the command screen and the prompt EDU>>. This prompt is displayed also after execution of a command; MATLAB now waits for a new command from the user. We can use the Editor/Debugger to write our program, save it, and return to the command screen to execute the program as explained below.

To use the Editor/Debugger:

1. From the File menu on the toolbar, we choose New and click on M–File. This takes us to the Editor Window where we can type our script (list of statements) for a new file, or open a previously saved file. We must save our program with a file name which starts with a letter. Important! MATLAB is case sensitive, that is, it distinguishes between upper– and lower–case letters. Thus, t and T are two different characters in MATLAB language. The files that we create are saved with the file name we use and the extension .m; for example, myfile01.m. It is a good

---

* *EDU>> is the MATLAB prompt in the Student Version.*

practice to save the script in a file name that is descriptive of our script content. For instance, if the script performs some matrix operations, we ought to name and save that file as matrices01.m or any other similar name. We should also use a separate disk to backup our files.

2. Once the script is written and saved as an m–file, we may exit the Editor/Debugger window by clicking on Exit Editor/Debugger of the File menu, and MATLAB returns to the command window.

3. To execute a program, we type the file name without the .m extension at the EDU>> prompt; then, we press <enter> and observe the execution and the values obtained from it. If we have saved our file in drive a or any other drive, we must make sure that it is added it to the desired directory in MATLAB's search path. The MATLAB User's Guide provides more information on this topic.

Henceforth, it will be understood that each input command is typed after the EDU>> prompt and followed by the <enter> key.

The command **help matlab iofun** will display input/output information. To get help with other MATLAB topics, we can type help followed by any topic from the displayed menu. For example, to get information on graphics, we type help matlab graphics. We can also get help from the Help pull-down menu. The MATLAB User's Guide contains numerous help topics.

To appreciate MATLAB's capabilities, we type **demo** and we see the MATLAB Demos menu. We can do this periodically to become familiar with them. Whenever we want to return to the command window, we click on the Close button.

When we are done and want to leave MATLAB, we type **quit** or **exit**. But if we want to clear all previous values, variables, and equations without exiting, we should use the **clear** command. This command erases everything; it is like exiting MATLAB and starting it again. The clc command clears the screen but MATLAB still remembers all values, variables and equations which we have already used. In other words, if we want MATLAB to retain all previously entered commands, but leave only the EDU>> prompt on the upper left of the screen, we can use the **clc** command.

All text after the % (percent) symbol is interpreted by MATLAB as a comment line and thus it is ignored during the execution of a program. A comment can be typed on the same line as the function or command or as a separate line. For instance, the statements

conv(p,q)   % performs multiplication of polynomials p and q

% The next statement performs partial fraction expansion of p(x) / q(x)

are both correct.

One of the most powerful features of MATLAB is the ability to do computations involving complex numbers. We can use either i, or j to denote the imaginary part of a complex number, such as $3 - 4i$ or $3 - 4j$. For example, the statement

z=3−4j

displays

```
z =
   3.0000 - 4.0000i
```

In the example above, a multiplication (*) sign between 4 and j was not necessary because the complex number consists of numerical constants. However, if the imaginary part is a function or variable such as cos(x), we must use the multiplication sign, that is, we must type cos(x)*j or j*cos(x).

1.2 Roots of Polynomials

In MATLAB, a polynomial is expressed as a row vector of the form $[a_n \ a_{n-1} \ a_2 \ a_1 \ a_0]$. The elements a_i of this vector are the coefficients of the polynomial in descending order. We must include terms whose coefficients are zero.

We can find the roots of any polynomial with the **roots(p)** function where **p** is a row vector containing the polynomial coefficients in descending order.

Example 1.1

Find the roots of the polynomial

$$p_1(x) = x^4 - 10x^3 + 35x^2 - 50x + 24 \tag{1.1}$$

Solution:

The roots are found with the following two statements. We have denoted the polynomial as **p1**, and the roots as **roots_ p1**.

```
p1=[1 −10 35 −50 24] % Specify the coefficients of p1(x)

p1 =
     1    -10     35    -50     24

roots_ p1=roots(p1) % Find the roots of p1(x)

roots_p1 =
    4.0000
    3.0000
    2.0000
    1.0000
```

We observe that MATLAB displays the polynomial coefficients as a row vector, and the roots as a column vector.

Example 1.2

Find the roots of the polynomial

$$p_2(x) = x^5 - 7x^4 + 16x^2 + 25x + 52 \qquad (1.2)$$

Solution:

There is no cube term; therefore, we must enter zero as its coefficient. The roots are found with the statements below where we have defined the polynomial as **p2**, and the roots of this polynomial as roots_ p2.

p2=[1 −7 0 16 25 52]

p2 =
         1   −7   0   16   25   52

roots_ p2=roots(p2)

roots_ p2 =
    6.5014
    2.7428
   -1.5711
   -0.3366 + 1.3202i
   -0.3366 - 1.3202i

The result indicates that this polynomial has three real roots, and two complex roots. Of course, complex roots always occur in complex conjugate* pairs.

1.3 Polynomial Construction from Known Roots

We can compute the coefficients of a polynomial from a given set of roots with the **poly(r)** function where r is a row vector containing the roots.

Example 1.3

It is known that the roots of a polynomial are $1, 2, 3,$ and 4. Compute the coefficients of this polynomial.

Solution:

We first define a row vector, say $r3$, with the given roots as elements of this vector; then, we find the coefficients with the poly(r) function as shown below.

---

* By definition, the conjugate of a complex number $A = a + jb$ is $A^* = a - jb$

```
r3=[1 2 3 4]  %  Specify the roots of the polynomial
r3 =
     1     2     3     4
poly_r3=poly(r3)  %  Find the polynomial coefficients
poly_r3 =
     1    -10    35    -50    24
```

We observe that these are the coefficients of the polynomial $p_1(x)$ of Example 1.1.

Example 1.4

It is known that the roots of a polynomial are -1, -2, -3, $4 + j5$, and $4 - j5$. Find the coefficients of this polynomial.

Solution:

We form a row vector, say $r4$, with the given roots, and we find the polynomial coefficients with the poly(r) function as shown below.

```
r4=[−1  −2  −3  4+5j  4−5j ]
r4 =
  Columns 1 through 4
  -1.0000    -2.0000    -3.0000    -4.0000 + 5.0000i
  Column 5
  -4.0000 - 5.0000i
poly_r4=poly(r4)
poly_r4 =
     1    14    100    340    499    246
```

Therefore, the polynomial is

$$p_4(x) = x^5 + 14x^4 + 100x^3 + 340x^2 + 499x + 246 \qquad (1.3)$$

1.4 Evaluation of a Polynomial at Specified Values

The **polyval(p,x)** function evaluates a polynomial $p(x)$ at some specified value of the independent variable x.

Example 1.5

Evaluate the polynomial

$$p_5(x) = x^6 - 3x^5 + 5x^3 - 4x^2 + 3x + 2 \qquad (1.4)$$

at $x = -3$.

Solution:

```
p5=[1 -3 0 5 -4 3 2]; % These are the coefficients
% The semicolon (;) after the right bracket suppresses the display of the row vector
% that contains the coefficients of p5.
%
val_minus3=polyval(p5, -3)% Evaluate p5 at x=-3. No semicolon is used here
% because we want the answer to be displayed

val_minus3 =
        1280
```

Other MATLAB functions used with polynomials are the following:

conv(a,b) – multiplies two polynomials **a** and **b**

[q,r]=deconv(c,d) –divides polynomial **c** by polynomial **d** and displays the quotient **q** and remainder **r**.

polyder(p) – produces the coefficients of the derivative of a polynomial **p**.

Example 1.6

Let

$$p_1 = x^5 - 3x^4 + 5x^2 + 7x + 9$$
$$p_2 = 2x^6 - 8x^4 + 4x^2 + 10x + 12 \qquad (1.5)$$

Compute the product $p_1 \cdot p_2$ with the **conv(a,b)** function.

Solution:

```
p1=[1 -3 0 5 7 9];
p2=[2 0 -8 0 4 10 12];
p1p2=conv(p1,p2)

p1p2 =
     2  -6  -8  34  18  -24  -74  -88  78  166  174  108
```

Therefore,

$$p_1 P p_2 = 2x^{11} - 6x^{10} - 8x^9 + 34x^8 + 18x^7 - 24x^6$$
$$-74x^5 - 88x^4 + 78x^3 + 166x^2 + 174x + 108$$

We can write MATLAB statements in one line if we separate them by commas or semicolons. *Commas will display the results whereas semicolons will suppress the display.*

Example 1.7

Let

$$p_3 = x^7 - 3x^5 + 5x^3 + 7x + 9$$

$$p_4 = 2x^6 - 8x^5 + 4x^2 + 10x + 12$$

(1.6)

Compute the quotient p_3/p_4 using the **deconv(p,q)** function.

Solution:

```
p3=[1  0 -3  0 5 7  9]; p4=[2 -8 0  0 4 10 12]; [q,r]=deconv(p3,p4)
q =
    0.5000
r =
    0     4    -3     0     3     2     3
```

Therefore, the quotient $q(x)$ and remainder $r(x)$ are

$$q(x) = 0.5 \qquad r(x) = 4x^5 - 3x^4 + 3x^2 + 2x + 3$$

Example 1.8

Let

$$p_5 = 2x^6 - 8x^4 + 4x^2 + 10x + 12$$

(1.7)

Compute the derivative dp_5/dx using the polyder(p) function.

Solution:

```
p5=[2  0 -8 0 4 10 12];
der_p5=polyder(p5)

der_p5 =
    12     0   -32     0     8    10
```

Therefore,

$$dp_5/dx = 12x^5 - 32x^3 + 4x^2 + 8x + 10$$

1.5 Rational Polynomials

Rational Polynomials are those which can be expressed in ratio form, that is, as

$$R(x) = \frac{Num(x)}{Den(x)} = \frac{b_n x^n + b_{n-1} x^{n-1} + b_{n-2} x^{n-2} + \ldots + b_1 x + b_0}{a_m x^m + a_{m-1} x^{m-1} + a_{m-2} x^{m-2} + \ldots + a_1 x + a_0} \tag{1.8}$$

where some of the terms in the numerator and/or denominator may be zero. We can find the roots of the numerator and denominator with the roots(p) function as before.

Example 1.9

Let

$$R(x) = \frac{p_{num}}{p_{den}} = \frac{x^5 - 3x^4 + 5x^2 + 7x + 9}{2x^6 - 8x^4 + 4x^2 + 10x + 12} \tag{1.9}$$

Express the numerator and denominator in factored form, using the roots(p) function.

Solution:

```
num=[1 -3 0 5 7 9]; den=[2 0 -8 0 4 10 12];% Do not display num and den coefficients
roots_num=roots(num), roots_den=roots(den)     % Display num and den roots

roots_num =
    2.4186 + 1.0712i     2.4186 - 1.0712i    -1.1633
   -0.3370 + 0.9961i    -0.3370 - 0.9961i

roots_den =
    1.6760 + 0.4922i     1.6760 - 0.4922i    -1.9304
   -0.2108 + 0.9870i    -0.2108 - 0.9870i    -1.0000
```

As expected, the complex roots occur in complex conjugate pairs.

For the numerator, we have the factored form

$$p_{num} = (x-2.4186 - j1.0712) \cdot (x-2.4186 + j1.0712) \cdot (x + 1.1633) \cdot$$
$$(x + 0.3370 - j0.9961) \cdot (x + 0.3370 + j0.9961)$$

and for the denominator, we have

$$p_{den} = (x-1.6760 - j0.4922) \cdot (x-1.6760 + j0.4922) \cdot (x + 1.9304) \cdot$$
$$(x + 0.2108 - j0.9870) \cdot (x + 0.2108 + j0.9870) \cdot (x + 1.0000)$$

We can also express the numerator and denominator of this rational function as a combination of linear and quadratic factors. We recall that in a quadratic equation of the form $x^2 + bx + c = 0$ whose roots are x_1 and x_2, the negative sum of the roots is equal to the coefficient b of the x term, that is, $-(x_1 + x_2) = b$, while the product of the roots is equal to the constant term c, that is, $x_1 \cdot x_2 = c$. Accordingly, we form the coefficient b by addition of the complex conjugate roots and this is done by inspection; then we multiply the complex conjugate roots to obtain the constant term c using MATLAB as indicated below.

```
(2.4186+1.0712i)*(2.4186 –1.0712i)    %  Form the product of the 1st set of complex conjugates

ans = 6.9971

(–0.3370+0.9961i)*(–0.3370–0.9961i) %  Form the product of the 2nd set of complex conjugates

ans = 1.1058

(1.6760+0.4922i)*(1.6760–0.4922i)

ans = 3.0512

(–0.2108+0.9870i)*(–0.2108–0.9870i)

ans = 1.0186
```

1.6 Using MATLAB to Make Plots

Quite often, we want to plot a set of ordered pairs. This is a very easy task with the MATLAB plot(x,y) command which plots y versus x. Here, x is the horizontal axis (abscissa) and y is the vertical axis (ordinate).

Example 1.10

Consider the electric circuit of Figure 1.1, where the radian frequency ω (radians/second) of the applied voltage was varied from 300 to 3000 in steps of 100 radians/second, while the amplitude was held constant. The ammeter readings were then recorded for each frequency. The magnitude of the impedance $|Z|$ was computed as $|Z| = |V/A|$ and the data were tabulated in Table 1.1.

Plot the magnitude of the impedance, that is, $|Z|$ versus radian frequency ω.

Solution:

We cannot type ω (omega) in the MATLAB command window, so we will use the English letter w instead.

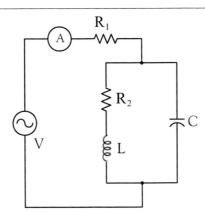

Figure 1.1. Electric circuit for Example 1.10

TABLE 1.1 Table for Example 1.10

ω (rads/s)	|Z| Ohms	ω (rads/s)	|Z| Ohms
300	39.339	1700	90.603
400	52.589	1800	81.088
500	71.184	1900	73.588
600	97.665	2000	67.513
700	140.437	2100	62.481
800	222.182	2200	58.240
900	436.056	2300	54.611
1000	1014.938	2400	51.428
1100	469.83	2500	48.717
1200	266.032	2600	46.286
1300	187.052	2700	44.122
1400	145.751	2800	42.182
1500	120.353	2900	40.432
1600	103.111	3000	38.845

If a statement, or a row vector is too long to fit in one line, it can be continued to the next line by typing three or more periods, then pressing <enter> to start a new line, and continue to enter data. This is illustrated below for the data of w and z. Also, as mentioned before, we use the semicolon (;) to suppress the display of numbers which we do not care to see on the screen.

The data are entered as follows:

```
w=[300 400 500 600 700 800 900 1000 1100 1200 1300 1400....  % Use 4 periods to continue
1500 1600 1700 1800 1900 2000 2100 2200 2300 2400 2500....
```

2600 2700 2800 2900 3000];   % Use semicolon to suppress display of these numbers
%
z=[39.339 52.789 71.104 97.665 140.437 222.182 436.056....
1014.938 469.830 266.032 187.052 145.751 120.353 103.111....
90.603 81.088 73.588 67.513 62.481 58.240 54.611 51.468....
48.717 46.286 44.122 42.182 40.432 38.845];

Of course, if we want to see the values of w or z or both, we simply type w or z, and we press <enter>.

To plot z (y – axis) versus w (x – axis), we use the **plot(x,y)** command. For this example, we use **plot(w,z)**. When this command is executed, MATLAB displays the plot on MATLAB's graph screen. This plot is shown in Figure 1.2.

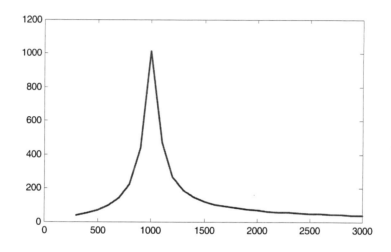

Figure 1.2. Plot of impedance |z| versus frequency ω for Example 1.10

This plot is referred to as the amplitude frequency response of the circuit.

To return to the command window, we press any key, or from the Window pull–down menu, we select MATLAB Command Window. To see the graph again, we click on the Window pull–down menu, and we select Figure.

We can make the above, or any plot, more presentable with the following commands:

grid on: This command adds grid lines to the plot. The **grid off** command removes the grid. The command **grid** toggles them, that is, changes from off to on or vice versa. The default[*] is off.

---

[*] *Default is a particular value for a variable or condition that is assigned automatically by an operating system, and remains in effect unless canceled or overridden by the operator.*

box off: This command removes the box (the solid lines which enclose the plot), and **box on** restores the box. The command **box** toggles them. The default is on.

title('string'): This command adds a line of the text string (label) at the top of the plot.

xlabel('string') and **ylabel('string')** are used to label the x– and y–axis respectively.

The amplitude frequency response is usually represented with the x–axis in a logarithmic scale. We can use the **semilogx(x,y)** command that is similar to the **plot(x,y)** command, except that the x–axis is represented as a log scale, and the y–axis as a linear scale. Likewise, the **semilogy(x,y)** command is similar to the **plot(x,y)** command, except that the y–axis is represented as a log scale, and the x–axis as a linear scale. The **loglog(x,y)** command uses logarithmic scales for both axes.

Throughout this text, it will be understood that log is the common (base 10) logarithm, and ln is the natural (base e) logarithm. We must remember, however, the function **log(x)** in MATLAB is the natural logarithm, whereas the common logarithm is expressed as **log10(x)**. Likewise, the logarithm to the base 2 is expressed as **log2(x)**.

Let us now redraw the plot with the above options, by adding the following statements:

```
semilogx(w,z); grid; % Replaces the plot(w,z) command
title('Magnitude of Impedance vs. Radian Frequency');
xlabel('w in rads/sec'); ylabel('|Z| in Ohms')
```

After execution of these commands, our plot is as shown in Figure 1.3.

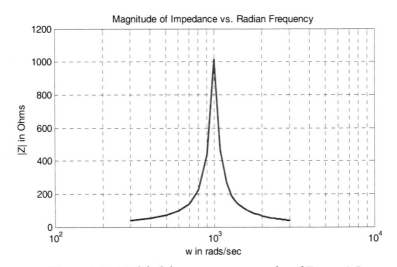

Figure 1.3. Modified frequency response plot of Figure 1.2.

If the y–axis represents power, voltage, or current, the x–axis of the frequency response is more often shown in a logarithmic scale, and the y–axis in dB (decibels) scale. A review of the decibel unit follows.

The ratio of any two values of the same quantity (power, voltage, or current) can be expressed in decibels (dB). Thus, we say that an amplifier has 10 dB power gain, or a transmission line has a power loss of 7 dB (or gain −7 dB). If the gain (or loss) is 0 dB the output is equal to the input. By definition,

$$dB = 10\log\left|\frac{P_{out}}{P_{in}}\right| \qquad (1.10)$$

Therefore,

10 dB represents a power ratio of 10

10n dB represents a power ratio of 10^n

It is very useful to remember that:

20 dB represents a power ratio of 100

30 dB represents a power ratio of 1, 000

60 dB represents a power ratio of 1, 000, 000

Also,

1 dB represents a power ratio of approximately 1.25

3 dB represents a power ratio of approximately 2

7 dB represents a power ratio of approximately 5

From these, we can estimate other values. For instance,

4 dB = 3 dB + 1 dB and since 3 dB ≅ power ratio of 2 and 1 dB ≅ power ratio of 1.25

then, 4 dB ≅ ratio of (2×1.25) = ratio of 2.5

Likewise, 27 dB = 20 dB + 7 dB and this is equivalent to a power ratio of approximately $100 \times 5 = 500$

Using the relations

$$y = \log x^2 = 2\log x$$

and

$$P = \frac{V^2}{Z} = I^2 Z$$

if we let Z = 1, the dB values for voltage and current ratios become

$$dB_v = 10\log\left|\frac{V_{out}}{V_{in}}\right|^2 = 20\log\left|\frac{V_{out}}{V_{in}}\right| \qquad (1.11)$$

and

$$dB_i = 10\log\left|\frac{I_{out}}{I_{in}}\right|^2 = 20\log\left|\frac{I_{out}}{I_{in}}\right| \qquad (1.12)$$

To display the voltage v in a dB scale on the $y-axis$, we add the relation **dB=20*log10(v)**, and we replace the **semilogx(w,z)** command with **semilogx(w,dB)**.

The command **gtext('string')** switches to the current Figure Window, and displays a cross–hair which can be moved around with the mouse. For instance, we can use the command **gtext('Impedance |Z| versus Frequency')**, and this will place a cross–hair in the Figure window. Then, using the mouse, we can move the cross–hair to the position where we want our label to begin, and we press <enter>.

The command **text(x,y,'string')** is similar to **gtext('string')**. It places a label on a plot in some specific location specified by x and y, and string is the label which we want to place at that location. We will illustrate its use with the following example which plots a 3–phase sinusoidal waveform.

The first line of the script below has the form

linspace(first_value, last_value, number_of_values)

This command specifies *the number of data points* but not the increments between data points. An alternate command uses the colon notation and has the format

x=first: increment: last

This format specifies *the increments between points* but not the number of data points.

The script for the 3–phase plot is as follows:

```
x=linspace(0, 2*pi, 60); %  pi is a built–in function in MATLAB;
%  we could have used x=0:0.02*pi:2*pi or x = (0: 0.02: 2)*pi instead;
y=sin(x); u=sin(x+2*pi/3); v=sin(x+4*pi/3);
plot(x,y,x,u,x,v); %  The x–axis must be specified for each function
grid on, box on,  %  turn grid and axes box on
text(0.75, 0.65, 'sin(x)');  text(2.85, 0.65, 'sin(x+2*pi/3)'); text(4.95, 0.65, 'sin(x+4*pi/3)')
```

These three waveforms are shown on the same plot of Figure 1.4.

In our previous examples, we did not specify line styles, markers, and colors for our plots. However, MATLAB allows us to specify various line types, plot symbols, and colors. These, or a combination of these, can be added with the plot(x,y,s) command, where s is a character string containing one or more characters shown on the three columns of Table 1.2.

MATLAB has no default color; it starts with blue and cycles through the first seven colors listed in Table 1.2 for each additional line in the plot. Also, there is no default marker; no markers are drawn unless they are selected. The default line is the solid line.

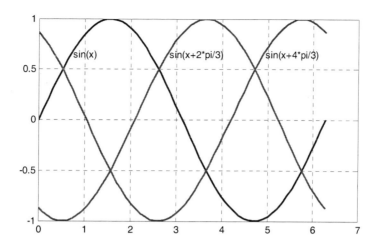

Figure 1.4. Three–phase waveforms

TABLE 1.2 Styles, colors, and markets used in MATLAB

Symbol	Color	Symbol	Marker	Symbol	Line Style
b	blue	.	point	–	solid line
g	green	o	circle	:	dotted line
r	red	x	x–mark	–.	dash–dot line
c	cyan	+	plus	––	dashed line
m	magenta	*	star		
y	yellow	s	square		
k	black	d	diamond		
w	white	/	triangle down		
		Ÿ	triangle up		
		<	triangle left		
		>	triangle right		
		p	pentagram		
		h	hexagram		

For example, the command **plot(x,y,'m*:')** plots a magenta dotted line with a star at each data point, and **plot(x,y,'rs')** plots a red square at each data point, but does not draw any line because no line was selected. If we want to connect the data points with a solid line, we must type **plot(x,y,'rs–')**. For additional information we can type **help plot** in MATLAB's command screen.

The plots which we have discussed thus far are two–dimensional, that is, they are drawn on two axes. MATLAB has also a three–dimensional (three–axes) capability and this is discussed next.

The command **plot3(x,y,z)** plots a line in 3–space through the points whose coordinates are the elements of x, y, and z, where x, y, and z are three vectors of the same length.

The general format is **plot3(x$_1$,y$_1$,z$_1$,s$_1$,x$_2$,y$_2$,z$_2$,s$_2$,x$_3$,y$_3$,z$_3$,s$_3$,...)** where **x$_n$**, **y$_n$**, and **z$_n$** are vectors or matrices, and **s$_n$** are strings specifying color, marker symbol, or line style. These strings are the same as those of the two–dimensional plots.

Example 1.11

Plot the function

$$z = -2x^3 + x + 3y^2 - 1 \qquad (1.13)$$

Solution:

We arbitrarily choose the interval (length) shown with the script below.

```
x= -10: 0.5: 10;          % Length of vector x
y= x;                      % Length of vector y must be same as x
z= -2.*x.^3+x+3.*y.^2-1;   % Vector z is function of both x and y*
plot3(x,y,z); grid
```

The three–dimensional plot is shown in Figure 1.5.

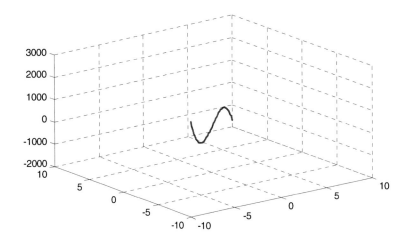

Figure 1.5. Three dimensional plot for Example 1.11

---

* This statement uses the so called dot multiplication, dot division, and dot exponentiation where these operations are preceded by a dot (period). These operations will be explained in Section 1.8, Page 1–19.

The command **plot3(x,y,z,'bd–')** will display the plot in blue diamonds, connected with a solid line.

In a three–dimensional plot, we can use the **zlabel('string')** command in addition to the **xlabel('string')** and **ylabel('string')**.

In a two–dimensional plot, we can set the limits of the x– and y– axes with the **axis([xmin xmax ymin ymax])** command. Likewise, in a three–dimensional plot we can set the limits of all three axes with the **axis([xmin xmax ymin ymax zmin zmax])** command. It must be placed after the **plot(x,y)** or **plot3(x,y,z)** commands, or on the same line without first executing the plot command. This must be done for each plot. The three–dimensional **text(x,y,z,'string')** command will place **string** beginning at the co–ordinate (x, y, z) on the plot.

For three–dimensional plots, **grid on** and **box off** are the default states.

The **mesh(x,y,z)** command displays a three–dimensional plot. Another command, **contour(Z,n)**, draws contour lines for **n** levels. We can also use the **mesh(x,y,z)** command with two vector arguments. These must be defined as length(x) = n and length(y) = m where [m, n] = size(Z). In this case, the vertices of the mesh lines are the triples $\{x(j), y(i), Z(i, j)\}$. We observe that x corresponds to the columns of Z, and y corresponds to the rows of Z.

To produce a mesh plot of a function of two variables, say $z = f(x, y)$, we must first generate the X and Y matrices which consist of repeated rows and columns over the range of the variables x and y. We can generate the matrices X and Y with the **[X,Y]=meshgrid(x,y)** function which creates the matrix X whose rows are copies of the vector **x**, and the matrix Y whose columns are copies of the vector **y**.

Example 1.12

The volume V of a right circular cone of radius r and height h is given by

$$V = \frac{1}{3}\pi r^2 h \tag{1.14}$$

Plot the volume of the cone as r and h vary on the intervals $0 \le r \le 4$ and $0 \le h \le 6$ meters.

Solution:

The volume of the cone is a function of both the radius r and the height h, that is, $V = f(r, h)$

The three–dimensional plot is created with the following MATLAB script where, as in the previous example, in the second line we have used the dot multiplication, division, and exponentiation. As mentioned in the footnote of the previous page, this topic will be explained in Section 1.8, Page 1–19.

```
[R,H]=meshgrid(0: 4, 0: 6);              % Creates R and H matrices from vectors r and h
V=(pi .* R .^ 2 .* H) ./ 3;  mesh(R, H, V)
xlabel('x–axis, radius r (meters)'); ylabel('y–axis, altitude h (meters)');
zlabel('z–axis, volume (cubic meters)'); title('Volume of Right Circular Cone'); box on
```

The three–dimensional plot of Figure 1.6, shows how the volume of the cone increases as the radius and height are increased.

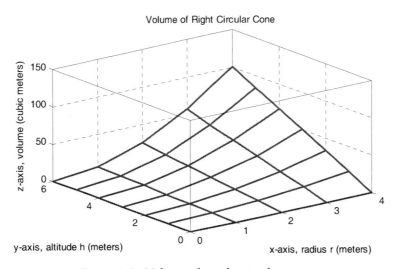

Figure 1.6. Volume of a right circular cone.

This, and the plot of Figure 1.5, are rudimentary; MATLAB can generate very sophisticated and impressive three–dimensional plots. The MATLAB User's manual contains more examples.

1.7 Subplots

MATLAB can display up to four windows of different plots on the Figure window using the command **subplot(m,n,p)**. This command divides the window into an m × n matrix of plotting areas and chooses the pth area to be active. No spaces or commas are required between the three integers m, n, and p. The possible combinations are shown in Figure 1.7.

We will illustrate the use of the **subplot(m,n,p)** command following the discussion on multiplication, division and exponentiation that follows.

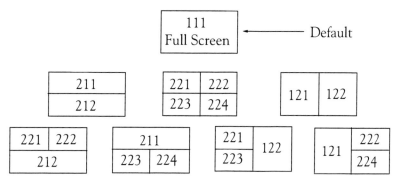

Figure 1.7. Possible subpot arrangements in MATLAB

1.8 Multiplication, Division and Exponentiation

MATLAB recognizes two types of multiplication, division, and exponentiation. These are the matrix multiplication, division, and exponentiation, and the element–by–element multiplication, division, and exponentiation. They are explained in the following paragraphs.

In Section 1.2, the arrays [a b c ...], such a those that contained the coefficients of polynomials, consisted of one row and multiple columns, and thus are called row vectors. If an array has one column and multiple rows, it is called a column vector. We recall that the elements of a row vector are separated by spaces. To distinguish between row and column vectors, the elements of a column vector must be separated by semicolons. An easier way to construct a column vector, is to write it first as a row vector, and then transpose it into a column vector. MATLAB uses the single quotation character (¢) to transpose a vector. Thus, a column vector can be written either as

b=[−1; 3; 6; 11]

or as

b=[−1 3 6 11]'

MATLAB produces the same display with either format as shown below.

b=[−1; 3; 6; 11]
```
b =
     -1
      3
      6
     11
```
b=[−1 3 6 11]'
```
b =
     -1
      3
```

```
      6
     11
```

We will now define Matrix Multiplication and Element–by–Element multiplication.

1. **Matrix Multiplication** (multiplication of row by column vectors)

   Let

   $$\mathbf{A} = [a_1 \quad a_2 \quad a_3 \quad \ldots \quad a_n]$$

   and

   $$\mathbf{B} = [b_1 \quad b_2 \quad b_3 \quad \ldots \quad b_n]'$$

be two vectors. We observe that $\mathbf{A}$ is defined as a row vector whereas $\mathbf{B}$ is defined as a column vector, as indicated by the transpose operator ('). Here, multiplication of the row vector $\mathbf{A}$ by the column vector $\mathbf{B}$, is performed with the matrix multiplication operator (*). Then,

$$\mathbf{A} * \mathbf{B} = [a_1 b_1 + a_2 b_2 + a_3 b_3 + \ldots + a_n b_n] = \text{single value} \qquad (B.15)$$

For example, if

$$\mathbf{A} = [1 \quad 2 \quad 3 \quad 4 \quad 5]$$

and

$$\mathbf{B} = [-2 \quad 6 \quad -3 \quad 8 \quad 7]'$$

the matrix multiplication $\mathbf{A} * \mathbf{B}$ produces the single value 68, that is,

$$\mathbf{A} * \mathbf{B} = 1 \times (-2) + 2 \times 6 + 3 \times (-3) + 4 \times 8 + 5 \times 7 = 68$$

and this is verified with the MATLAB script

```
A=[1 2 3 4 5]; B=[−2 6 −3 8 7]'; A*B    % Observe transpose operator (') in B
ans =
      68
```

Now, let us suppose that both $\mathbf{A}$ and $\mathbf{B}$ are row vectors, and we attempt to perform a row–by–row multiplication with the following MATLAB statements.

```
A=[1 2 3 4 5]; B=[−2 6 −3 8 7]; A*B        % No transpose operator (') here
```

When these statements are executed, MATLAB displays the following message:

```
??? Error using ==> *
Inner matrix dimensions must agree.
```

Here, because we have used the matrix multiplication operator (*) in **A*B**, MATLAB expects vector **B** to be a column vector, not a row vector. It recognizes that **B** is a row vector, and warns us that we cannot perform this multiplication using the matrix multiplication operator (*). Accordingly, we must perform this type of multiplication with a different operator. This operator is defined below.

2. **Element–by–Element Multiplication** (multiplication of a row vector by another row vector)

Let

$$\mathbf{C} = [c_1 \quad c_2 \quad c_3 \quad \ldots \quad c_n]$$

and

$$\mathbf{D} = [d_1 \quad d_2 \quad d_3 \quad \ldots \quad d_n]$$

be two row vectors. Here, multiplication of the row vector **C** by the row vector **D** is performed with the *dot multiplication operator* (.*). There is no space between the dot and the multiplication symbol. Thus,

$$\mathbf{C.*D} = [c_1 d_1 \quad c_2 d_2 \quad c_3 d_3 \quad \ldots \quad c_n d_n] \qquad (B.16)$$

This product is another row vector with the same number of elements, as the elements of **C** and **D**.

As an example, let

$$\mathbf{C} = [1 \quad 2 \quad 3 \quad 4 \quad 5]$$

and

$$\mathbf{D} = [-2 \quad 6 \quad -3 \quad 8 \quad 7]$$

Dot multiplication of these two row vectors produce the following result.

$$\mathbf{C.*D} = 1 \times (-2) \quad 2 \times 6 \quad 3 \times (-3) \quad 4 \times 8 \quad 5 \times 7 = -2 \quad 12 \quad -9 \quad 32 \quad 35$$

Check with MATLAB:

```
C=[1 2 3 4 5];     % Vectors C and D must have
D=[-2 6 -3 8 7];   % same number of elements
C.*D               % We observe that this is a dot multiplication
ans =
    -2     12     -9     32     35
```

Similarly, the division (/) and exponentiation (^) operators, are used for matrix division and exponentiation, whereas dot division (./) and dot exponentiation (. ^) are used for element–by–element division and exponentiation, as illustrated with the examples above.

We must remember that *no space is allowed between the dot (.) and the multiplication (*), division (/), and exponentiation (^) operators.*

Note: A dot (.) is never required with the plus (+) and minus (–) operators.

Example 1.13

Write the MATLAB script that produces a simple plot for the waveform defined as

$$y = f(t) = 3e^{-4t}\cos 5t - 2e^{-3t}\sin 2t + \frac{t^2}{t+1} \tag{1.17}$$

in the $0 \le t \le 5$ seconds interval.

Solution:

The MATLAB script for this example is as follows:

```
t=0: 0.01: 5;                        % Define t–axis in 0.01 increments
y=3 .* exp(–4 .* t) .* cos(5 .* t)–2 .* exp(–3 .* t) .* sin(2 .* t) + t .^2 ./ (t+1);
plot(t,y); grid; xlabel('t'); ylabel('y=f(t)'); title('Plot for Example 1.13')
```

Figure 1.8 shows the plot for this example.

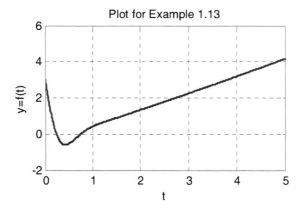

Figure 1.8. Plot for Example 1.13

Had we, in the example above, defined the time interval starting with a negative value equal to or less than -1, say as $-3 \le t \le 3$, MATLAB would have displayed the following message:

```
Warning: Divide by zero.
```

This is because the last term (the rational fraction) of the given expression, is divided by zero when $t = -1$. To avoid division by zero, we use the special MATLAB function **eps,** which is a number approximately equal to 2.2×10^{-16}. It will be used with the next example.

The command **axis([xmin xmax ymin ymax])** scales the current plot to the values specified by the arguments **xmin, xmax, ymin** and **ymax**. There are no commas between these four arguments. This command must be placed after the plot command and must be repeated for each plot.

The following example illustrates the use of the dot multiplication, division, and exponentiation, the eps number, the **axis([xmin xmax ymin ymax])** command, and also MATLAB's capability of displaying up to four windows of different plots.

Example 1.14

Plot the functions

$$y = \sin^2 x, \quad z = \cos^2 x, \quad w = \sin^2 x \cdot \cos^2 x, \quad v = \sin^2 x / \cos^2 x \qquad (1.18)$$

in the interval $0 \le x \le 2\pi$ using 100 data points. Use the **subplot** command to display these functions on four windows on the same graph.

Solution:

The MATLAB script to produce the four subplots is as follows:

```
x=linspace(0, 2*pi,100);          % Interval with 100 data points
y=(sin(x) .^ 2); z=(cos(x) .^ 2);
w=y .* z;
v=y ./ (z+eps);                   %  add eps to avoid division by zero
subplot(221);                     % upper left of four subplots
plot(x,y);  axis([0 2*pi 0 1]);
title('y=(sinx)^2');
subplot(222);                     % upper right of four subplots
plot(x,z);  axis([0 2*pi 0 1]);
title('z=(cosx)^2');
subplot(223);                     % lower left of four subplots
plot(x,w);  axis([0 2*pi 0 0.3]);
title('w=(sinx)^2*(cosx)^2');
subplot(224);                     % lower right of four subplots
plot(x,v);  axis([0 2*pi 0 400]);
title('v=(sinx)^2/(cosx)^2');
```

These subplots are shown in Figure 1.9.

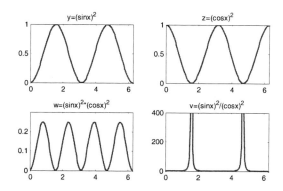

Figure 1.9. Subplots for the functions of Example 1.14

The next example illustrates MATLAB's capabilities with imaginary numbers. We will introduce the **real(z)** and **imag(z)** functions which display the real and imaginary parts of the complex quantity $z = x + iy$, the **abs(z)**, and the **angle(z)** functions that compute the absolute value (magnitude) and phase angle of the complex quantity $z = x + iy = r-\theta$. We will also use the **polar(theta,r)** function that produces a plot in polar coordinates, where **r** is the magnitude, **theta** is the angle in radians, and the **round(n)** function that rounds a number to its nearest integer.

Example 1.15

Consider the electric circuit of Figure 1.10.

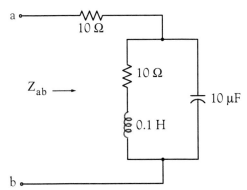

Figure 1.10. Electric circuit for Example 1.15

With the given values of resistance, inductance, and capacitance, the impedance Z_{ab} as a function of the radian frequency ω can be computed from the following expression.

$$Z_{ab} = Z = 10 + \frac{10^4 - j(10^6/w)}{10 + j(0.1w - 10^5/w)} \tag{1.19}$$

a. Plot Re{Z} (the real part of the impedance Z) versus frequency ω.

b. Plot Im{Z} (the imaginary part of the impedance Z) versus frequency ω.

c. Plot the impedance Z versus frequency ω in polar coordinates.

Solution:

The MATLAB script below computes the real and imaginary parts of Z_{ab} that is, for simplicity, denoted as z, and plots these as two separate graphs (parts a & b). It also produces a polar plot (part c).

```
w=0: 1: 2000;  %  Define interval with one radian interval
z=(10+(10 .^ 4 –j .* 10 .^ 6 ./ (w+eps)) ./ (10 + j .* (0.1 .* w –10.^5./ (w+eps))));
```

```
%
%  The first five statements (next two lines) compute and plot Re{z}
real_part=real(z);  plot(w,real_part);  grid;
xlabel('radian frequency w');  ylabel('Real part of Z');
%
%  The next five statements (next two lines) compute and plot Im{z}
imag_part=imag(z);  plot(w,imag_part);  grid;
xlabel('radian frequency w');  ylabel('Imaginary part of Z');
%  The last six statements (next six lines) below produce the polar plot of z
mag=abs(z);%  Computes |Z|
rndz=round(abs(z));%  Rounds |Z| to read polar plot easier
theta=angle(z);%  Computes the phase angle of impedance Z
polar(theta,rndz);%  Angle is the first argument
grid;
ylabel('Polar Plot of Z');
```

The real, imaginary, and polar plots are shown in Figures 1.11, 1.12, and 1.13 respectively.

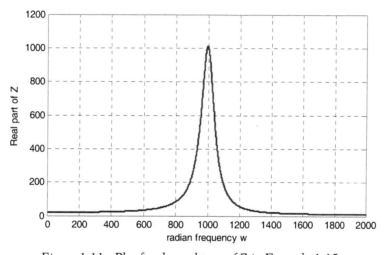

Figure 1.11. Plot for the real part of Z in Example 1.15

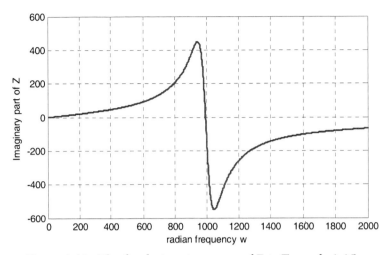

Figure 1.12. Plot for the imaginary part of Z in Example 1.15

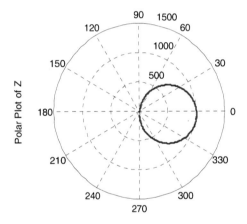

Figure 1.13. Polar plot of Z in Example 1.15

Example 1.15 clearly illustrates how powerful, fast, accurate, and flexible MATLAB is.

1.9 Script and Function Files

MATLAB recognizes two types of files: script files and function files. Both types are referred to as m–files since both require the .m extension.

A script file consists of two or more built–in functions such as those we have discussed thus far. Thus, the script for each of the examples we discussed earlier, make up a script file. Generally, a script file is one which was generated and saved as an m–file with an editor such as the MATLAB's

Editor/Debugger.

A function file is a user–defined function using MATLAB. We use function files for repetitive tasks. The first line of a function file must contain the word function, followed by the output argument, the equal sign (=), and the input argument enclosed in parentheses. *The function name and file name must be the same*, but the file name must have the extension .m. For example, the function file consisting of the two lines below

```
function y = myfunction(x)
y=x .^ 3 + cos(3 .* x)
```

is a function file and must be saved. To save it, from the File menu of the command window, we choose New and click on M–File. This takes us to the Editor Window where we type these two lines and we save it as myfunction.m.

We will use the following MATLAB functions with the next example.

The function **fzero(f,x)** tries to find a zero of a function of one variable, where f is a string containing the name of a real–valued function of a single real variable. MATLAB searches for a value near a point where the function f changes sign, and returns that value, or returns NaN if the search fails.

Important: We must remember that we use **roots(p)** to find the roots of polynomials only, such as those in Examples 1.1 and 1.2.

fplot(fcn,lims) – plots the function specified by the string **fcn** between the x–axis limits specified by **lims = [xmin xmax]**. Using **lims = [xmin xmax ymin ymax]** also controls the y–axis limits. The string **fcn** must be the name of an *m–file* function or a string with variable x .

NaN (Not–a–Number) is not a function; it is MATLAB's response to an undefined expression such as $0/0$, ∞/∞, or inability to produce a result as described on the next paragraph. We can avoid division by zero using the **eps** number, which we mentioned earlier.

Example 1.16

Find the zeros, maxima and minima of the function

$$f(x) = \frac{1}{(x-0.1)^2 + 0.01} + \frac{1}{(x-1.2)^2 + 0.04} - 10 \qquad (1.20)$$

in the interval $-1.5 \le x \le 1.5$

Solution:

We first plot this function to observe the approximate zeros, maxima, and minima using the following script:

```
x=−1.5: 0.01: 1.5;
y=1./ ((x−0.1).^ 2 + 0.01) −1./ ((x−1.2).^ 2 + 0.04) −10;
plot(x,y); grid
```

The plot is shown in Figure 1.14.

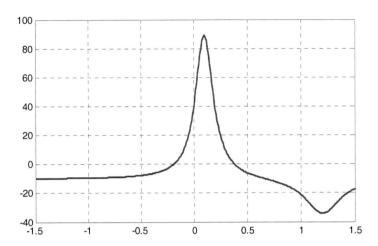

Figure 1.14. Plot for Example 1.16 using the plot command

The roots (zeros) of this function appear to be in the neighborhood of $x = -0.2$ and $x = 0.3$. The maximum occurs at approximately $x = 0.1$ where, approximately, $y_{max} = 90$, and the minimum occurs at approximately $x = 1.2$ where, approximately, $y_{min} = -34$.

Next, we define and save *f(x)* as the **funczero01.m** function m–file with the following script:

```
function y=funczero01(x)
% Finding the zeros of the function shown below
y=1/((x−0.1)^2+0.01)−1/((x−1.2)^2+0.04)−10;
```

To save this file, from the File drop menu on the Command Window, we choose New, and when the Editor Window appears, we type the script above and we save it as **funczero01.** MATLAB appends the extension **.m** to it.

Now, we can use the **fplot(fcn,lims)** command to plot f(x) as follows:

```
fplot('funczero01', [−1.5  1.5]); grid
```

This plot is shown in Figure 1.15. As expected, this plot is identical to the plot of Figure 1.14 which was obtained with the **plot(x,y)** command as shown in Figure 1.14.

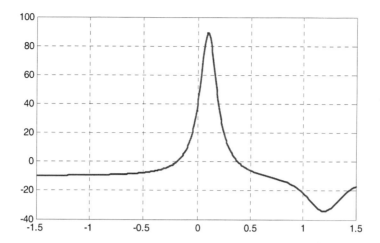

Figure 1.15. Plot for Example 1.16 using the fplot command

We will use the **fzero(f,x)** function to compute the roots of f(x) in Equation (1.20) more precisely. The MATLAB script below will accomplish this.

```
x1= fzero('funczero01', –0.2);
x2= fzero('funczero01', 0.3);
fprintf('The roots (zeros) of this function are r1= %3.4f', x1);
fprintf(' and r2= %3.4f \n', x2)
```
MATLAB displays the following:

```
The roots (zeros) of this function are r1= -0.1919 and r2= 0.3788
```

The earlier MATLAB versions included the function **fmin(f,x1,x2)** and with this function we could compute both a minimum of some function f(x) or a maximum of f(x) since a maximum of f(x) is equal to a minimum of –f(x). This can be visualized by flipping the plot of a function f(x) upside–down. This function is no longer used in MATLAB and thus we will compute the maxima and minima from the derivative of the given function.

From elementary calculus, we recall that the maxima or minima of a function y = f(x) can be found by setting the first derivative of a function equal to zero and solving for the independent variable x. For this example we use the **diff(x)** function which produces the approximate derivative of a function. Thus, we use the following MATLAB script:

```
syms x ymin zmin; ymin=1/((x–0.1)^2+0.01)–1/((x–1.2)^2+0.04)–10;...
zmin=diff(ymin)
```

```
zmin =
-1/((x-1/10)^2+1/100)^2*(2*x-1/5)+1/((x-6/5)^2+1/25)^2*(2*x-12/5)
```

When the command

```
solve(zmin)
```

is executed, MATLAB displays a very long expression which when copied at the command prompt and executed, produces the following:

```
ans =
   0.6585 + 0.3437i
ans =
   0.6585 - 0.3437i
ans =
    1.2012
```

The real value 1.2012 above is the value of x at which the function y has its minimum value as we observe also in the plot of Figure 1.15.

To find the value of y corresponding to this value of x, we substitute it into $f(x)$, that is,

x=1.2012; ymin=1 / ((x–0.1) ^ 2 + 0.01) –1 / ((x–1.2) ^ 2 + 0.04) –10

```
ymin = -34.1812
```

We can find the maximum value from $-f(x)$ whose plot is produced with the script

x=–1.5:0.01:1.5; ymax=–1./((x–0.1).^2+0.01)+1./((x–1.2).^2+0.04)+10; plot(x,ymax); grid

and the plot is shown in Figure 1.16.

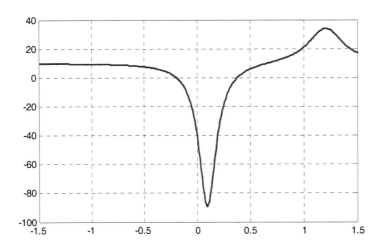

Figure 1.16. Plot of $-f(x)$ for Example 1.16

Next we compute the first derivative of $-f(x)$ and we solve for x to find the value where the maximum of **ymax** occurs. This is accomplished with the MATLAB script below.

syms x ymax zmax; ymax=–(1/((x–0.1)^2+0.01)–1/((x–1.2)^2+0.04)–10); zmax=diff(ymax)

```
zmax =
  1/((x-1/10)^2+1/100)^2*(2*x-1/5)-1/((x-6/5)^2+1/25)^2*(2*x-12/5)
```

solve(zmax)

When the command

solve(zmax)

is executed, MATLAB displays a very long expression which when copied at the command prompt and executed, produces the following:

```
ans =
   0.6585 + 0.3437i

ans =
   0.6585 - 0.3437i

ans =
   1.2012
ans =
   0.0999
```

From the values above we choose $x = 0.0999$ which is consistent with the plots of Figures 1.15 and 1.16. Accordingly, we execute the following script to obtain the value of ymin.

```
x=0.0999;            % Using this value find the corresponding value of ymax
ymax=1 / ((x–0.1) ^ 2 + 0.01) –1 / ((x–1.2) ^ 2 + 0.04) –10
```

```
ymax = 89.2000
```

1.10 Display Formats

MATLAB displays the results on the screen in integer format without decimals if the result is an integer number, or in short floating point format with four decimals if it a fractional number. The format displayed has nothing to do with the accuracy in the computations. MATLAB performs all computations with accuracy up to 16 decimal places.

The output format can changed with the format command. The available formats can be displayed with the **help format** command as follows:

help format

```
FORMAT Set output format.
All computations in MATLAB are done in double precision.
FORMAT may be used to switch between different output display
formats as follows:
```

```
FORMAT            Default. Same as SHORT.
FORMAT SHORT      Scaled fixed point format with 5 digits.
FORMAT LONG       Scaled fixed point format with 15 digits.
FORMAT SHORT E    Floating point format with 5 digits.
FORMAT LONG E     Floating point format with 15 digits.
FORMAT SHORT G    Best of fixed or floating point
                  format with 5 digits.
FORMAT LONG G     Best of fixed or floating point format
                  with 15 digits.
FORMAT HEX        Hexadecimal format.
FORMAT +          The symbols +, - and blank are printed
                  for positive, negative and zero elements.
                  Imaginary parts are ignored.
FORMAT BANK       Fixed format for dollars and cents.
FORMAT RAT        Approximation by ratio of small integers.

Spacing:

FORMAT COMPACT Suppress extra line-feeds.
FORMAT LOOSE   Puts the extra line-feeds back in.
```

Some examples with different format displays age given below.

format short 33.3335 Four decimal digits (default)

format long 33.33333333333334 16 digits

format short e 3.3333e+01 Four decimal digits plus exponent

format short g 33.333 Better of format short or format short e

format bank 33.33 two decimal digits

format + only + or – or zero are printed

format rat 100/3 rational approximation

1.11 Summary

- We can get help with MATLAB topics by typing help followed by any topic available. For example, the command **help matlab\iofun** will display input/output information, and **help matlab graphics** will display help on graphics.

- The MATLAB Demos menu displays MATLAB's capabilities. To access it, we type demo and we see the different topics. Whenever we want to return to the command window, we click on the Close button.

- We type **quit** or **exit** when we are done and want to leave MATLAB.

- We use the **clear** command if we want to clear all previous values, variables, and equations without exiting.

- The **clc** command clears the screen but MATLAB still remembers all values, variables and equations which we have already used.

- All text after the % (percent) symbol is interpreted by MATLAB as a comment line and thus it is ignored during the execution of a program. A comment can be typed on the same line as the function or command or as a separate line.

- For computations involving complex numbers we can use either i, or j to denote the imaginary part of the complex number.

- In MATLAB, a polynomial is expressed as a row vector of the form $[a_n \ a_{n-1} \ a_2 \ a_1 \ a_0]$. The elements a_i of this vector are the coefficients of the polynomial in descending order. We must include terms whose coefficients are zero.

- We find the roots of any polynomial with the **roots(p)** function where **p** is a row vector containing the polynomial coefficients in descending order.

- We can compute the coefficients of a polynomial from a given set of roots with the **poly(r)** function where **r** is a row vector containing the roots.

- The **polyval(p,x)** function evaluates a polynomial $p(x)$ at some specified value of the independent variable x.

- The **conv(a,b)** function multiplies the polynomials **a** and **b**.

- The **[q,r]=deconv(c,d)** function divides polynomial **c** by polynomial **d** and displays the quotient **q** and remainder **r**.

- The **polyder(p)** function produces the coefficients of the derivative of a polynomial **p**.

- We can write MATLAB statements in one line if we separate them by commas or semicolons. Commas will display the results whereas semicolons will suppress the display.

- Rational Polynomials are those which can be expressed in ratio form, that is, as

$$R(x) = \frac{Num(x)}{Den(x)} = \frac{b_n x^n + b_{n-1} x^{n-1} + b_{n-2} x^{n-2} + \dots + b_1 x + b_0}{a_m x^m + a_{m-1} x^{m-1} + a_{m-2} x^{m-2} + \dots + a_1 x + a_0}$$

where some of the terms in the numerator and/or denominator may be zero. Normally, we express the numerator and denominator of a rational function as a combination of linear and quadratic factors.

- We use the MATLAB command **plot(x,y)** to make two–dimensional plots. This command plots y versus x where **x** is the horizontal axis (abscissa), and **y** is the vertical axis (ordinate).

- If a statement, or a row vector is too long to fit in one line, it can be continued to the next line by typing three or more periods, then pressing <enter> to start a new line, and continue to enter data.

- We can make a two–dimensional plot more presentable with the commands **grid**, **box**, **title('string')**, **xlabel('string')**, and **ylabel('string')**. For a three–dimensional plot, we can also use the **zlabel('string')** command.

- The **semilogx(x,y)** command is similar to the **plot(x,y)** command, except that the x –axis is represented as a log scale, and the y –axis as a linear scale. Likewise, the **semilogy(x,y)** command is similar to the **plot(x,y)** command, except that the y –axis is represented as a log scale, and the x –axis as a linear scale. The **loglog(x,y)** command uses logarithmic scales for both axes.

- The function **log(x)** in MATLAB is the natural logarithm, whereas the common logarithm is expressed as **log10(x)**. Likewise, the logarithm to the base 2 is expressed as **log2(x)**.

- The ratio of any two values of the same quantity, typically power, is normally expressed in decibels (dB) and by definition,

$$dB = 10\log\left|\frac{P_{out}}{P_{in}}\right|$$

- The command **gtext('string')** switches to the current Figure Window, and displays a cross–hair which can be moved around with the mouse. The command **text(x,y,'string')** is similar to **gtext('string')**; it places a label on a plot in some specific location specified by **x** and **y**, and **string** is the label which we want to place at that location.

- The command **linspace(first_value, last_value, number_of_values)** specifies the number of data points but not the increments between data points. An alternate command uses the colon notation and has the format **x=first: increment: last**. This format specifies the increments between points but not the number of data points.

- MATLAB has no default color; it starts with blue and cycles through seven colors. Also, there is no default marker; no markers are drawn unless they are selected. The default line is the solid line.

- The **plot3(x,y,z)** command plots a line in 3–space through the points whose coordinates are the elements of x, y, and z, where **x**, **y**, and **z** are three vectors of the same length.

- In a two–dimensional plot, we can set the limits of the x– and y–axes with the **axis([xmin xmax ymin ymax])** command. Likewise, in a three–dimensional plot we can set the limits of all three axes with the **axis([xmin xmax ymin ymax zmin zmax])** command. It must be placed after the **plot(x,y)** or **plot3(x,y,z)** commands, or on the same line without first executing the plot command. This must be done for each plot. The three–dimensional **text(x,y,z,'string')** command will place string beginning at the co–ordinate (x, y, z) on the plot.

- The **mesh(x,y,z)** command displays a three–dimensional plot. Another command, **contour(Z,n)**, draws contour lines for **n** levels. We can also use the **mesh(x,y,z)** command with two vector arguments. These must be defined as $length(x) = n$ and $length(y) = m$ where $[m, n] = size(Z)$. In this case, the vertices of the mesh lines are the triples $\{x(j), y(i), Z(i, j)\}$. We observe that **x** corresponds to the columns of Z, and **y** corresponds to the rows of Z. To produce a mesh plot of a function of two variables, say $z = f(x, y)$, we must first generate the X and Y matrices which consist of repeated rows and columns over the range of the variables x and y. We can generate the matrices X and Y with the **[X,Y]=meshgrid(x,y)** function which creates the matrix X whose rows are copies of the vector **x**, and the matrix Y whose columns are copies of the vector **y**.

- MATLAB can display up to four windows of different plots on the Figure window using the command subplot(m,n,p). This command divides the window into an $m \times n$ matrix of plotting areas and chooses the pth area to be active.

- With MATLAB, matrix multiplication (multiplication of row by column vectors) is performed with the matrix multiplication operator (*), whereas element–by–element multiplication is performed with the dot multiplication operator (.*). Similarly, the division (/) and exponentiation ($\char`^$) operators, are used for matrix division and exponentiation, whereas dot division (./) and dot exponentiation (.$\char`^$) are used for element–by–element division and exponentiation.

- To avoid division by zero, we use the special MATLAB function **eps**, which is a number approximately equal to 2.2×10^{-16}.

- The command **axis([xmin xmax ymin ymax])** scales the current plot to the values specified by the arguments **xmin**, **xmax**, **ymin** and **ymax**. There are no commas between these four arguments. This command must be placed after the plot command and must be repeated for each plot.

- The **real(z)** and **imag(z)** functions display the real and imaginary parts of the complex quantity $z = x + iy$, and the **abs(z)**, and the **angle(z)** functions compute the absolute value (magnitude) and phase angle of the complex quantity $z = x + iy = r \text{-} \theta$. The **polar(theta,r)** function produces a plot in polar coordinates, where **r** is the magnitude, and **theta** is the angle in radians.

- MATLAB recognizes two types of files: script files and function files. Both types are referred to as m–files. A script file consists of two or more built–in functions. Generally, a script file is one which was generated and saved as an m–file with an editor such as the MATLAB's Editor/ Debugger. A function file is a user–defined function using MATLAB. We use function files for repetitive tasks. The first line of a function file must contain the word function, followed by the output argument, the equal sign (=), and the input argument enclosed in parentheses. The function name and file name must be the same, but the file name must have the extension .m.

- The MATLAB **fmin(f,x1,x2)** function minimizes a function of one variable. It attempts to return a value of x where f(x) is minimum in the interval $x_1 < x < x_2$. The string **f** contains the name of the function to be minimized.

- The MATLAB **fplot(fcn,lims)** command plots the function specified by the string fcn between the x–axis limits specified by **lims = [xmin xmax]**. Using **lims = [xmin xmax ymin ymax]** also controls the y–axis limits. The string **fcn** must be the name of an m–file function or a string with variable x.

- The MATLAB **fprintf(format,array)** command used above displays and prints both text and arrays. It uses specifiers to indicate where and in which format the values would be displayed and printed. Thus, if **%f** is used, the values will be displayed and printed in fixed decimal format, and if **%e** is used, the values will be displayed and printed in scientific notation format. With these commands only the real part of each parameter is processed.

- MATLAB displays the results on the screen in integer format without decimals if the result is an integer number, or in short floating point format with four decimals if it a fractional number. The format displayed has nothing to do with the accuracy in the computations. MATLAB performs all computations with accuracy up to 16 decimal places.

1.12 Exercises

1. Use MATLAB to compute the roots of the following polynomials:

   a. $p(x) = x^3 + 8x^2 + 10x + 4$

   b. $p(y) = y^5 + 7y^4 + 19y^3 + 25y^2 + 16y + 4$

2. Use MATLAB to derive the polynomials having the following roots:

   a. -6.5708 $-0.7146 + j0.3132$ $-0.7146 - j0.3132$

   b. Two roots at $x = -2.000$ and three roots at $x = -3.000$

3. Use MATLAB to evaluate the polynomials below at the specified values.

   a. $p(x) = x^3 + 8x^2 + 10x + 4$ at $x = 1.25$

   b. $p(y) = y^5 + 7y^4 + 19y^3 + 25y^2 + 16y + 4$ at $y = -3.75$

4. In the electric circuit below, the applied voltage V_S was kept constant and the voltage V_C across the capacitor was measured and recorded at several frequencies as shown on the table below.

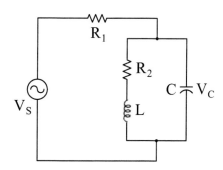

Capacitor voltage versus radian frequency						
ω	500	600	700	800	900	1000
V_C	88.9	98.5	103.0	104.9	105.3	104.8
ω	1100	1200	1300	1400	1500	1600
V_C	103.8	102.4	100.7	98.9	96.5	94.9

Plot V_C (in dB scale) versus ω (in common log scale) and label the axes appropriately.

1.13 Solutions to End–of–Chapter Exercises

Dear Reader:

The remaining pages on this chapter contain the solutions to the exercises.

You must, for your benefit, make an honest effort to find the solutions to the exercises without first looking at the solutions that follow. It is recommended that first you go through and work out those you feel that you know. For the exercises that you are uncertain, review this chapter and try again. Refer to the solutions as a last resort and rework those exercises at a later date.

You should follow this practice with the rest of the exercises of this book.

1.

a.

```
Px=[1  8  10  4]; roots(Px)

ans =
   -6.5708
   -0.7146 + 0.3132i
   -0.7146 - 0.3132i
```

b.

```
Py=[1  7  19  25  16  4]; roots(Py)

ans =
   -2.0000
   -2.0000
   -1.0000
   -1.0000 + 0.0000i
   -1.0000 - 0.0000i
```

2.

a.

```
r1=[-6.5708  -0.7146+0.3132j  -0.7146-0.3132j]; poly_r1=poly(r1)

poly_r1 =   1.0000     8.0000     9.9997     4.0000
```

$$p(x) = x^3 + 8x^2 + 10x + 4$$

b.

```
r2=[-2  -2  -3  -3  -3]; poly_r2=poly(r2)

poly_r2 =

       1     13     67    171    216    108
```

$$p(z) = z^5 + 13z^4 + 67z^3 + 171z^2 + 216z + 108$$

3.

a.

```
Pv=[1  8  10  4]; value=polyval(Pv, 1.25)

value = 30.9531
```

b.

```
Pw=[1  7  19  25  16  4]; value=polyval(Pw, -3.75)

value = -63.6904
```

4.

```
w=[5 6 7 8 9 10 11 12 13 14 15 16]*100;
Vc=[88.9 98.5 103 104.9 105.3 104.8 103.8 102.4 100.7 98.9 96.5 94.9];
dB=20*log10(Vc); semilogx(w,dB); grid; title('Magnitude of Vc vs. w');...
xlabel('w in rads/sec'); ylabel('|Vc| in volts')
```

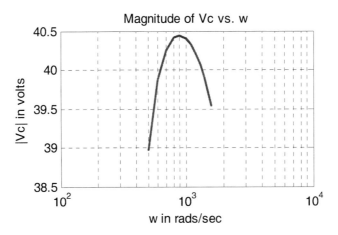

Numerical Analysis Using MATLAB® and Excel®, Third Edition
Copyright © Orchard Publications

Chapter 2

T his chapter is an introduction to Newton's and bisection methods for approximating roots of linear and non–linear equations. Several examples are presented to illustrate practical solutions using MATLAB and Excel spreadsheets.

2.1 Newton's Method for Root Approximation

Newton's (or Newton–Raphson) method can be used to approximate the roots of any linear or non–linear equation of any degree. This is an iterative (repetitive procedure) method and it is derived with the aid of Figure 2.1.

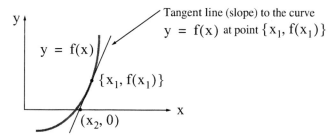

Figure 2.1. Newton's method for approximating real roots of a function

We assume that the slope is neither zero nor infinite. Then, the slope (first derivative) at $x = x_1$ is

$$f'(x_1) = \frac{y - f(x_1)}{x - x_1}$$

$$y - f(x_1) = f'(x_1)(x - x_1) \tag{2.1}$$

The slope crosses the x–axis at $x = x_2$ and $y = 0$. Since this point $[x_2, f(x_2)] = (x_2, 0)$ lies on the slope line, it satisfies (2.1). By substitution,

$$0 - f(x_1) = f'(x_1)(x_2 - x_1)$$

$$x_2 = x_1 - \frac{f(x_1)}{f'(x_1)} \tag{2.2}$$

and in general,

$$x_{n+1} = x_n - \frac{f(x_n)}{f'(x_n)} \tag{2.3}$$

Example 2.1

Use Newton's method to approximate the positive root of

$$f(x) = x^2 - 5 \qquad (2.4)$$

to four decimal places.

Solution:

As a first step, we plot the curve of (2.4) to find out where it crosses the $x-axis$. This can be done easily with a simple plot using MATLAB or a spreadsheet. We start with MATLAB and will discuss the steps for using a spreadsheet afterwards.

We will now introduce some new MATLAB functions and review some which are discussed in Chapter 1.

input('string'): It displays the text **string**, and waits for an input from the user. We must enclose the text in single quotation marks.

We recall that the **polyder(p)** function displays the row vector whose values are the coefficients of the first derivative of the polynomial **p**. The **polyval(p,x)** function evaluates the polynomial **p** at some value **x**. Therefore, we can compute the next iteration for approximating a root with Newton's method using these functions. Knowing the polynomial **p** and the first approximation x_0, we can use the following script for the next approximation x_1.

```
q=polyder(p)
x1=x0–polyval(p,x0)/polyval(q,x0)
```

We've used the **fprintf** command in Chapter 1; we will use it many more times. Therefore, let us review it again.

The following description was extracted from the **help fprintf** function.

*It formats the data in the real part of matrix A (and in any additional matrix arguments), under control of the specified format string, and writes it to the file associated with file identifier **fid** and contains C language conversion specifications. These specifications involve the character %, optional flags, optional width and precision fields, optional subtype specifier, and conversion characters **d, i, o, u, x, X, f, e, E, g, G, c,** and **s**. See the Language Reference Guide or a C manual for complete details. The special formats \n,\r,\t,\b,\f can be used to produce linefeed, carriage return, tab, backspace, and formfeed characters respectively. Use \\ to produce a backslash character and %% to produce the percent character.*

To apply Newton's method, we must start with a reasonable approximation of the root value. In all cases, this can best be done by plotting $f(x)$ versus x with the familiar statements below. The following two lines of script will display the graph of the given equation in the interval $-4 \le x \le 4$.

```
x=linspace(–4, 4, 100);      % Specifies 100 values between -4 and 4
y=x .^ 2 – 5; plot(x,y); grid  % The dot exponentiation is a must
```

We chose this interval because the given equation asks for the square root of 5; we expect this value to be a value between 2 and 3. For other functions, where the interval may not be so obvious, we can choose a larger interval, observe the $x-axis$ crossings, and then redefine the interval. The plot is shown in Figure 2.2.

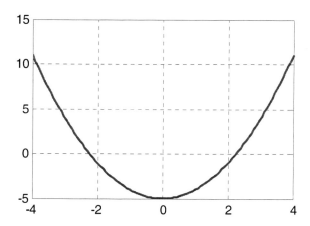

Figure 2.2. Plot for the curve of Example 2.1

As expected, the curve shows one crossing between $x = 2$ and $x = 3$, so we take $x_0 = 2$ as our first approximation, and we compute the next value x_1 as

$$x_1 = x_0 - \frac{f(x_0)}{f'(x_0)} = 2 - \frac{(2)^2 - 5}{2(2)} = 2 - \frac{(-1)}{4} = 2.25 \qquad (2.5)$$

The second approximation yields

$$x_2 = x_1 - \frac{f(x_1)}{f'(x_1)} = 2.25 - \frac{(2.25)^2 - 5}{2(2.25)} = 2.25 - \frac{0.0625}{4.5} = 2.2361 \qquad (2.6)$$

We will use the following MATLAB script to verify (2.5) and (2.6).

```
% Approximation of a root of a polynomial function p(x)
% Do not forget to enclose the coefficients in brackets [ ]
p=input('Enter coefficients of p(x) in descending order: ');
x0=input('Enter starting value: ');
q=polyder(p);                    % Calculates the derivative of p(x)
x1=x0–polyval(p,x0)/polyval(q,x0);
fprintf('\n');                   % Inserts a blank line
%
% The next function displays the value of x1 in decimal format as indicated
% by the specifier %9.6f, i.e., with 9 digits where 6 of these digits
% are to the right of the decimal point such as xxx.xxxxxx, and
```

```
% \n prints a blank line before printing x1
fprintf('The next approximation is: %9.6f \n', x1)
fprintf('\n');                          % Inserts another blank line
%
fprintf('Rerun the program using this value as your next....
approximation \n');
```

The following lines show MATLAB's inquiries and our responses (inputs) for the first two approximations.

```
Enter coefficients of P(x) in descending order:
[1  0 -5]
Enter starting value: 2
The next approximation is:   2.250000
Rerun the program using this value as your
next approximation
Enter polynomial coefficients in
descending order: [1 0 -5]
Enter starting value: 2.25
The next approximation is:   2.236111
```

We observe that this approximation is in close agreement with (2.6).

In Chapter 1 we discussed *script files* and *function files*. We recall that a function file is a user–defined function using MATLAB. We use function files for repetitive tasks. The first line of a function file must contain the word *function* followed by the output argument, the equal sign (=), and the input argument enclosed in parentheses. The function name and file name must be the same but the file name must have the extension *.m*. For example, the function file consisting of the two lines below

```
function y = myfunction(x)
y=x .^ 3 + cos(3 .* x)
```

is a function file and must be saved as *myfunction.m*

We will use the **while end** loop, whose general form is

while *expression*
    commands ...
end

where the *commands ...* in the second line are executed as long as all elements in *expression* of the first line are true.

We will also be using the following commands:

disp(x): Displays the array x without printing the array name. If x is a string, the text is displayed. For example, if v = 12, **disp(v)** displays 12, and **disp('volts')** displays volts.

sprintf(format,A): Formats the data in the real part of matrix **A** under control of the specified *format* string. For example,

```
sprintf('%d',round(pi))
```

```
ans =
3
```

where the format script **%d** specifies an integer. Likewise,

```
sprintf('%4.3f',pi)
```

```
ans =
3.142
```

where the format script **%4.3f** specifies a fixed format of 4 digits where 3 of these digits are allocated to the fractional part.

Example 2.2

Approximate one real root of the non–linear equation

$$f(x) = x^2 + 4x + 3 + \sin x - x\cos x \qquad (2.7)$$

to four decimal places using Newton's method.

Solution:

As a first step, we sketch the curve to find out where the curve crosses the $x-axis$. We generate the plot with the script below.

```
x=linspace(–pi, pi, 100); y=x .^ 2 + 4 .* x + 3 + sin(x) – x .* cos(x); plot(x,y); grid
```

The plot is shown in Figure 2.3.

The plot shows that one real root is approximately at $x = -1$, so we will use this value as our first approximation.

Next, we generate the function **funcnewt01** and we save it as an *m–file*. To save it, from the *File* menu of the command window, we choose *New* and click on *M-File*. This takes us to the *Editor Window* where we type the following three lines and we save it as *funcnewt01.m.*

```
function y=funcnewt01(x)
% Approximating roots with Newton's method
y=x .^ 2 + 4 .* x + 3 + sin(x) – x .* cos(x);
```

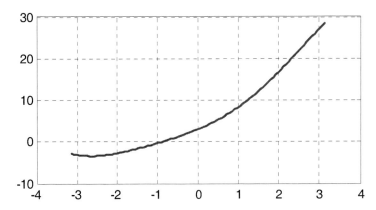

Figure 2.3. Plot for the equation of Example 2.2

We also need the first derivative of y; This is $y' = 2x + 4 + x \sin x$

The computation of the derivative for this example was a simple task; however, we can let MAT-LAB do the differentiation, just as a check, and to introduce the **diff(s)** function. This function performs differentiation of symbolic expressions. The **syms** function is used to define one or more symbolic expressions.

```
syms x
y = x^2+4*x+3+sin(x)–x*cos(x);    % Dot operations are not necessary with
                                  % symbolic expressions, but correct
                                  % answer will be displayed if they are used.
y1=diff(y)                        % Find the derivative of y

y1 =
2*x+4+x*sin(x)
```

Now, we generate the function **funcnewt02,** and we save it as m–file. To save it, from the *File* menu of the command window, we choose *New* and click on *M–File.* This takes us to the *Editor Window* where we type these two lines and we save it as *funcnewt02.m.*

```
function y=funcnewt02(x)
%  Finding roots by Newton's method
%  The following is the first derivative of the function defined as funcnewt02
y=2 .* x + 4 + x .* sin(x);
```

Our script for finding the next approximation with Newton's method follows.

```
x = input('Enter starting value: ');
fx = funcnewt01(x);
fprimex = funcnewt02(x);
xnext = x–fx/fprimex;
  x = xnext;
```

```
  fx = funcnewt01(x);
  fprimex = funcnewt02(x);
disp(sprintf('First approximation is x =  %9.6f \n', x))
while input('Next approximation? (<enter>=no,1=yes)');
  xnext=x−fx/fprimex;
  x=xnext;
  fx=funcnewt01(x);
  fprimex=funcnewt02(x);
disp(sprintf('Next approximation is x =  %9.6f \n', x))
end;
disp(sprintf('%9.6f \n', x))
```

MATLAB produces the following result with −1 as a starting value.

```
Enter starting value: −1
First approximation is: −0.894010
Next approximation? (<enter>=no,1=yes)1
−0.895225
Next approximation? (<enter>=no,1=yes) <enter>
```

We can also use the **fzero(f,x)** function. It was introduced in Chapter 1. This function tries to find a zero of a function of one variable. The string **f** contains the name of a real−valued function of a single real variable. As we recall, MATLAB searches for a value near a point where the function **f** changes sign and returns that value, or returns NaN if the search fails.

2.2 Approximations with Spreadsheets

In this section, we will go through several examples to illustrate the procedure of using a spreadsheet such as Excel[*] to approximate the real roots of linear and non−linear equations.

We recall that there is a standard procedure for finding the roots of a cubic equation; it is included here for convenience.

A cubic equation of the form

$$y^3 + py^2 + qy + r = 0 \tag{2.8}$$

can be reduced to the simpler form

$$x^3 + ax + b = 0 \tag{2.9}$$

where

---

* We will illustrate our examples with Excel, although others such as Lotus 1−2−3, and Quattro can also be used. Henceforth, all spreadsheet commands and formulas that we will be using, will be those of Excel.

$$x = y + \frac{p}{3} \qquad a = \frac{1}{3}(3q - p^2) \qquad b = \frac{1}{27}(2p^3 - 9pq + 27r) \qquad (2.10)$$

For the solution it is convenient to let

$$A = \sqrt[3]{\frac{-b}{2} + \sqrt{\frac{b^2}{4} + \frac{a^3}{27}}} \qquad B = -\sqrt[3]{\frac{-b}{2} + \sqrt{\frac{b^2}{4} + \frac{a^3}{27}}} \qquad (2.11)$$

Then, the values of x for which the cubic equation of (2.11) is equal to zero are

$$x_1 = A + B \qquad x_2 = -\frac{A+B}{2} + \frac{A-B}{2}\sqrt{-3} \qquad x_3 = -\frac{A+B}{2} - \frac{A-B}{2}\sqrt{-3} \qquad (2.12)$$

If the coefficients p, q, and r are real, then $\qquad\qquad (2.13)$

If $\dfrac{b^2}{4} + \dfrac{a^3}{27} > 0$ *one root will be real and the other two complex conjugates*

If $\dfrac{b^2}{4} + \dfrac{a^3}{27} < 0$ *the roots will be real and unequal*

If $\dfrac{b^2}{4} + \dfrac{a^3}{27} = 0$ *there will be three real roots with at least two equal*

While MATLAB handles complex numbers very well, spreadsheets do not. Therefore, unless we know that the roots are all real, we should not use a spreadsheet to find the roots of a cubic equation by substitution in the above formulas. However, we can use a spreadsheet to find the real root since in any cubic equation there is at least one real root. For real roots, we can use a spreadsheet to define a range of x values with small increments and compute the corresponding values of $y = f(x)$. Then, we can plot y versus x to observe the values of x that make $f(x) = 0$. This procedure is illustrated with the examples that follow.

Note: In our subsequent discussion we will omit the word *cell* and the key *<enter>*. Thus B3, C11, and so on will be understood to be cell B3, cell C11, and so on. Also, after an entry has been made, it will be understood that the *<enter>* key was pressed.

Example 2.3

Compute the roots of the polynomial

$$y = f(x) = x^3 - 7x^2 + 16x - 12 \qquad (2.14)$$

using Excel.

Solution:

We start with a blank worksheet. In an Excel worksheet, a *selected* cell is surrounded by a heavy border. We select a cell by moving the thick hollow white cross pointer to the desired cell and we *click*. For this example, we first select A1 and we type **x**. We observe that after pressing the <*enter*> key, the next cell moves downwards to A2; this becomes the next selected cell. We type 0.00 in A2. We observe that this value is displayed just as 0, that is, without decimals. Next, we type 0.05 in A3. We observe that this number is displayed exactly as it was typed.

We will enter more values in column A, and to make all values look uniform, we *click* on letter **A** on top of column A. We observe that the entire column is now highlighted, that is, the background on the monitor has changed from white to black. Next, from the *Tools* drop menu of the Menu bar, we choose *Options* and we click on the *Edit* tab. We *click* on the *Fixed Decimal* check box to place a check mark and we choose **2** as the number of decimal places. We repeat these steps for Column B and we choose **3** decimal places. Then, all numbers that we will type in Column A will be fixed numbers with two decimal places, and the numbers in Column B will be fixed with three decimal places.

To continue, we select A2, we *click* and holding the mouse left button down, we drag the mouse down to A3 so that both these two cells are highlighted; then we release the mouse button. When properly done, A2 will have a white background but A3 will have a black background. We will now use the *AutoFill*[*] feature to fill–in the other values of x in Column A. We will use values in **0.05** increments up to **5.00**. Column A now contains 100 values of x from 0.00 to 5.00 in increments of **0.05**.

Next, we select B1, and we type **f(x)**. In B2, we type the equation formula with the = sign in front of it, that is, we type

= A2^3-7*A2^2 + 16*A2-2

where A2 represents the first value of x = 0.00. We observe that B2 displays the value −12.000. This is the value of f(x) when x = 0.00 Next, we want to copy this formula to the range B3:B102 (the colon : means B3 through B102). With B2 still selected, we *click* on *Edit* on the main taskbar, and we *click* on *Copy*. We select the range B3:B102 with the mouse, we release the mouse button, and we observe that this range is now highlighted. We click on *Edit*, then on *Paste* and we observe that this range is now filled with the values of f(x). Alternately, we can use the *Copy* and *Paste* icons of the taskbar.

---

[*] *To use this feature, we highlight cells A2 and A3. We observe that on the lower right corner of A3, there is a small black square; this is called the fill handle. If it does not appear on the spreadsheet, we can make it visible by performing the sequential steps Tools>Options, select the Edit tab, and place a check mark on the Drag and Drop setting. Next, we point the mouse to the fill handle and we observe that the mouse pointer appears as a small cross. We click, hold down the mouse button, we drag it down to A102, and we release the mouse button. We observe that, as we drag the fill handle, a pop–up note shows the cell entry for the last value in the range.*

To plot $f(x)$ versus x, we click on the *Chart Wizard* icon of the Standard Toolbar, and on the *Chart type* column we click on **XY (Scatter)**. From the displayed charts, we choose the one on top of the right side (the smooth curves without connection points). Then, we *click* on *Next, Next, Next*, and *Finish*. A chart similar to the one on Figure 2.4 appears.

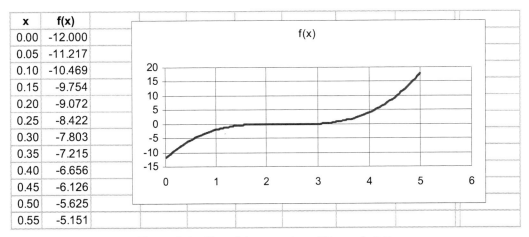

x	f(x)
0.00	-12.000
0.05	-11.217
0.10	-10.469
0.15	-9.754
0.20	-9.072
0.25	-8.422
0.30	-7.803
0.35	-7.215
0.40	-6.656
0.45	-6.126
0.50	-5.625
0.55	-5.151

Figure 2.4. Plot of the equation of Example 2.3.

We will modify this plot to make it more presentable, and to see more precisely the x – axis crossing(s), that is, the roots of $f(x)$. This is done with the following steps:

1. We *click* on the *Series 1* box to select it, and we delete it by pressing the *Delete* key.

2. We *click* anywhere inside the graph box. Then, we see it enclosed in six black square handles. From the *View* menu, we *click* on *Toolbars*, and we place a check mark on *Chart*. The *Chart* menu appears in two places, on the main taskbar and below it in a box where next to it is another small box with the hand icon. **Note:** The *Chart* menu appears on the main taskbar and on the box below it, only when the graph box is selected, that is, when it is enclosed in black square handles. From the *Chart* menu box (below the main taskbar), we select *Value (X) axis*, and we *click* on the small box next to it (the box with the hand icon). Then, on the *Format* axis menu, we *click* on the *Scale* tab and we make the following entries:

Minimum: **0.0**
Maximum: **5.0**
Major unit: **1.0**
Minor unit: **0.5**

We *click* on the *Number* tab, we select *Number* from the *Category* column, and we type **0** in the *Decimal places* box. We *click* on the *Font* tab, we select any font, *Regular style, Size 9*. We *click* on the *Patterns* tab to select it, and we *click* on *Low* on the *Tick* mark labels (lower right box). We *click* on OK to return to the graph.

3. From the *Chart* menu box we select *Value (Y) axis* and we *click* on the small box next to it (the

box with the hand icon). On the *Format* axis menu, we *click* on the *Scale* tab, and we make the following entries:

Minimum: −1.0
Maximum: 1.0
Major unit: 0.25
Minor unit: 0.05

We *click* on the *Number* tab, we select *Number* from the *Category* column, and we select **2** in the *Decimal places* box. We *click* on the *Font* tab, select any font, *Regular style, Size 9.* We *click* on the *Patterns* tab, and we *click* on *Outside* on the *Major tick* mark type (upper right box). We *click* on *OK* to return to the graph.

4. We *click* on *Chart* on the main taskbar, and on the *Chart Options.* We *click* on *Gridlines,* we place check marks on *Major gridlines* of both *Value (X) axis* and *Value (Y) axis.* Then, we *click* on the *Titles* tab and we make the following entries:

   Chart title: f(x) = the given equation (or whatever we wish)
   Value (X) axis: x (or whatever we wish)
   Value (Y) axis: y=f(x) (or whatever we wish)

5. Now, we will change the background of the plot area from gray to white. From the *Chart* menu box below the main task bar, we select *Plot Area* and we observe that the gray background of the plot area is surrounded by black square handles. We *click* on the box next to it (the box with the hand icon), and on the *Area* side of the *Patterns* tab, we *click* on the white square which is immediately below the gray box. The plot area on the chart now appears on white background.

6. To make the line of the curve f(x) thicker, we *click* at any point near it and we observe that several black square handles appear along the curve. *Series 1* appears on the *Chart* menu box. We *click* on the small box next to it, and on the *Patterns* tab. From the *Weight* selections we select the first of the thick lines.

7. Finally, to change *Chart Area* square corners to round, we select *Chart Area* from the *Chart* menu, and on the *Patterns* tab we place a check mark on the *Round corners* box.

The plot now resembles the one shown in Figure 2.5 where we have shown partial lists of x and $f(x)$. The given polynomial has two roots at $x = 2$, and the third root is $x = 3$.

We will follow the same procedure for generating the graphs of the other examples which follow; therefore, it is highly recommended that this file is saved with any name, say *poly01.xls* where *.xls* is the default extension for file names saved in Excel.

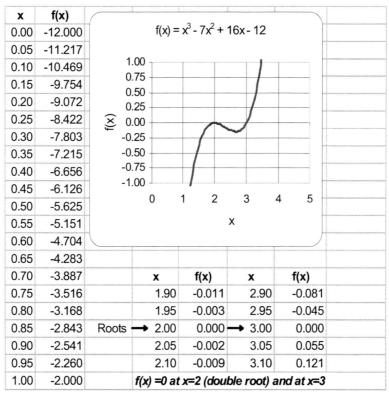

x	f(x)
0.00	-12.000
0.05	-11.217
0.10	-10.469
0.15	-9.754
0.20	-9.072
0.25	-8.422
0.30	-7.803
0.35	-7.215
0.40	-6.656
0.45	-6.126
0.50	-5.625
0.55	-5.151
0.60	-4.704
0.65	-4.283
0.70	-3.887
0.75	-3.516
0.80	-3.168
0.85	-2.843
0.90	-2.541
0.95	-2.260
1.00	-2.000

$f(x) = x^3 - 7x^2 + 16x - 12$

	x	f(x)	x	f(x)
	1.90	-0.011	2.90	-0.081
	1.95	-0.003	2.95	-0.045
Roots →	2.00	0.000 →	3.00	0.000
	2.05	-0.002	3.05	0.055
	2.10	-0.009	3.10	0.121

f(x) =0 at x=2 (double root) and at x=3

Figure 2.5. Modified plot of the equation of Example 2.3.

Example 2.4

Find a real root of the polynomial

$$y = f(x) = 3x^5 - 2x^3 + 6x - 8 \qquad (2.15)$$

using Excel.

Solution:

To save lots of unnecessary work, we invoke (open) the spreadsheet of the previous example, that is, *poly01.xls* (or any other file name that was assigned to it), and save it with another name such as *poly02.xls*. This is done by first opening the file *poly01.xls*, and from the *File* drop down menu, we choose the *Save as* option; then, we save it as *poly02.xls*, or any other name. When this is done, the spreadsheet of the previous example still exists as *poly01.xls*. Next, we perform the following steps:

1. For this example, the highest power of the polynomial is 5 (odd number), and since we know that complex roots occur in conjugate pairs, we expect that this polynomial will have at least one real root. Since we do not know where a real root is in the *x–axis* interval, we arbitrarily

choose the interval $-10 \leq x \leq 10$. Then, we enter -10 and -9 in A2 and A3 respectively. Using the *AutoFill* feature, we fill–in the range A4:A22, and we have the interval from -10 to 10 in increments of 1. We must now delete all rows starting with 23 and downward. We do this by highlighting the range A23:B102, and we press the *Delete* key. We observe that the chart has changed shape to conform to the new data.

Now we select *B2* where we enter the formula for the given equation, i.e.,

=3*A2^5–2*A2^3+6*A2–8

We copy this formula to B3:B22. Columns A and B now contain values of x and $f(x)$ respectively, and the plot shows that the curve crosses the x–axis somewhere between $x = 1$ and $x = 2$.

A part of the table is shown in Figure 2.6. Columns A (values of x), and B (values of $f(x)$), reveal some useful information.

x	f(x)
-10.00	-298068.000
-9.00	-175751.000
0.00	-8.000
1.00	-1.000
2.00	84.000
9.00	175735.000
10.00	298052.000

Sign Change

Figure 2.6. Partial table for Example 2.4

This table shows that $f(x)$ changes sign somewhere in the interval from $x = 1$ and $x = 2$. Let us then redefine our interval of the x values as $1 \leq x \leq 2$ in increments of **0.05**, to get better approximations. When this is done A1 contains **1.00**, A2 contains **1.05**, and so on. Our spreadsheet now shows that there is a sign change from B3 to B4, and thus we expect that a real root exists between $x = 1.05$ and $x = 1.10$. To obtain a good approximation of the real root in that interval, we perform Steps 2 through 4 below.

2. On the *View* menu, we *click* on *Toolbars* and place a check mark on *Chart*. We select the graph box by clicking inside it, and we observe the square handles surrounding it. The *Chart* menu on the main taskbar and the *Chart* menu box below it, are now displayed. From the *Chart* menu box (below the main taskbar) we select *Value (X)* axis, and we *click* on the small box next to it (the box with the hand). Next, on the *Format* axis menu, we *click* on the *Scale* tab and make the following entries:
Minimum: **1.0**
Maximum: **1.1**
Major unit: **0.02**
Minor unit: **0.01**

3. From the *Chart* menu we select *Value (Y) axis*, and we *click* on the small box next to it. Then, on the *Format axis* menu, we *click* on the *Scale* tab and make the following entries:

Minimum: −1.0
Maximum: 1.0
Major unit: 0.5
Minor unit: 0.1

4. We *click* on the Titles tab and make the following entries:

Chart title: f(x) = the given equation (or whatever we wish)
Value (X) axis: x (or whatever we wish)
Value (Y) axis: y=f(x) (or whatever we wish)

Our spreadsheet now should look like the one in Figure 2.7 and we see that one real root is approximately *1.06*.

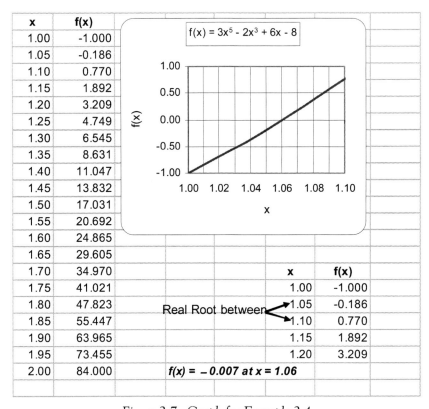

Figure 2.7. Graph for Example 2.4

Since no other roots are indicated on the plot, we suspect that the others are complex conjugates. We confirm this with MATLAB as follows:

p = [3 0 −2 0 6 −8]; roots_p=roots(p)

```
roots_p =
  -1.1415 + 0.8212i
  -1.1415 - 0.8212i
   0.6113 + 0.9476i
   0.6113 - 0.9476i
   1.0604
```

Example 2.5

Compute the real roots of the trigonometric function

$$y = f(x) = \cos 2x + \sin 2x + x - 1 \qquad (2.16)$$

using Excel.

Solution:

We invoke (open) the spreadsheet of one of the last two examples, that is, *poly01.xls* or *poly02.xls*, and save it with another name, such as *poly03.xls*.

Since we do not know where real roots (if any) are in the *x–axis* interval, we arbitrarily choose the interval $-1 \le x \le 6$. Then, we enter **–1.00** and **–0.90** in A2 and A3 respectively, Using the *Auto-Fill* feature, we fill–in the range A4:A72 and thus we have the interval from **–1** to **6** in increments of **0.10**. Next, we select B2 and we enter the formula for the given equation, i.e.,

=COS(2*A2)+SIN(2*A2)+A2–1

and we copy this formula to B3:B62.

There is a root at $x = 0$; this is found by substitution of zero into the given equation. We observe that Columns A and B contain the following sign changes (only a part of the table is shown):

x	f(x)	
1.20	0.138	↖ Sign Change
1.30	-0.041	↙
2.20	-0.059	↖ Sign Change
2.30	0.194	↙

We observe two sign changes. Therefore, we expect two more real roots, one in the $1.20 \le x \le 1.30$ interval and the other in the $2.20 \le x \le 2.30$ interval. If we redefine the $x - axis$ range as **1** to **2.5**, we will find that the other two roots are approximately $x = 1.30$ and $x = 2.24$.

Approximate values of these roots can also be observed on the plot of Figure 2.8 where the curve crosses the $x - axis$.

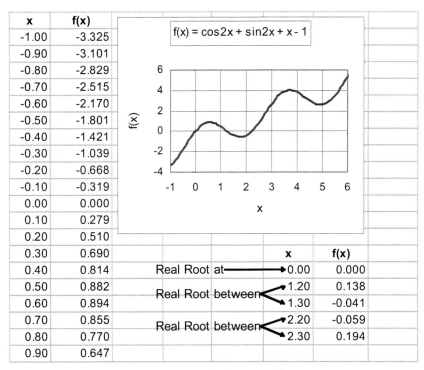

x	f(x)
-1.00	-3.325
-0.90	-3.101
-0.80	-2.829
-0.70	-2.515
-0.60	-2.170
-0.50	-1.801
-0.40	-1.421
-0.30	-1.039
-0.20	-0.668
-0.10	-0.319
0.00	0.000
0.10	0.279
0.20	0.510
0.30	0.690
0.40	0.814
0.50	0.882
0.60	0.894
0.70	0.855
0.80	0.770
0.90	0.647

$f(x) = \cos 2x + \sin 2x + x - 1$

	x	f(x)
Real Root at	0.00	0.000
Real Root between	1.20	0.138
	1.30	-0.041
Real Root between	2.20	-0.059
	2.30	0.194

Figure 2.8. Graph for Example 2.5

We can obtain more accurate approximations using Excel's **Goal Seek** feature. We use *Goal Seek* when we know the desired result of a single formula, but we do not know the input value which satisfies that result. Thus, if we have the function y = f(x), we can use *Goal Seek* to set the dependent variable y to the desired value (goal) and from it, find the value of the independent variable x which satisfies that goal. In the last three examples our goal was to find the values of x for which y = f(x) = 0.

To illustrate the *Goal Seek* feature, we will use it to find better approximations for the non–zero roots of Example 2.5. We do this with the following steps:

1. We copy range A24:B24 (or A25:B25) to two blank cells, say J1 and K1, so that J1 contains 1.20 and K1 contains 0.138 (or 1.30 and −0.041 if range A25:B25 was copied). We increase the accuracy of Columns J and K to 5 decimal places by clicking on *Format, Cells, Numbers* tab.

2. From the *Tools* drop menu, we *click* on *Goal Seek*, and when the *Goal Seek* dialog box appears, we make the following entries:

   Set cell: **K1**
   To value: **0**

   By changing cell: **J1**

3. When this is done properly, we will observe the changes in J1 and K1. These indicate that for

$x = 1.27647$, $y = f(x) = 0.00002$.

4. We repeat the above steps for the next root near $x = 2.20$, and we verify that for $x = 2.22515$, $y = f(x) = 0.00020$.

Another method of using the *Goal Seek* feature, is with a chart such as those we've created for the last three examples. We will illustrate the procedure with the chart of Example 2.5.

1. We point the mouse at the curve where it intersects the *x–axis*, near the $x = 1.30$ point. A square box appears and displays *Series 1, (1.30, –0.041)*. We observe that other points are also displayed as the mouse is moved at different points near the curve.

2. We *click* anywhere near the curve, and we observe that five handles (black square boxes) are displayed along different points on the curve. Next, we *click* on the handle near the $x = 1.30$ point, and when the cross symbol appears, we *drag* it towards the *x–axis* to change its value. The *Goal Seek* dialog box then appears where the *Set cell* shows B24. Then, in the *To value* box we enter 0, in the *By changing cell* we enter A24 and we *click* on OK. We observe now that A24 displays *1.28* and B24 displays *0.000*.

For repetitive tasks, such as finding the roots of polynomials, it is prudent to construct a template (model spreadsheet) with the appropriate formulas and then enter the coefficients of the polynomial to find its real roots[*]. This is illustrated with the next example.

Example 2.6

Construct a template (model spreadsheet), with Excel, which uses Newton's method to approximate a real root of any polynomial with real coefficients up to the seventh power; then, use it to compute a root of the polynomial

$$y = f(x) = x^7 - 6x^6 + 5x^5 - 4x^4 + 3x^3 - 2x^2 + x - 15 \qquad (2.17)$$

given that one real root lies in the $4 \le x \le 6$ interval.

Solution:

1. We begin with a blank spreadsheet and we make the entries shown in Figure 2.9.

---

[*] *There exists a numerical procedure, known as **Bairstow's method**, that we can use to find the complex roots of a polynomial with real coefficients. We will not discuss this method here; it can be found in advanced numerical analysis textbooks.*

	A	B	C	D	E	F	G	H
1	Spreadsheet for finding approximations of the real roots of polynomials							
2	up the 7th power by Newton's Method.							
3								
4	**Powers of x and corresponding coefficients of given polynomial p(x)**							
5	Enter coefficients of p(x) in Row 7							
6	x^7	x^6	x^5	x^4	x^3	x^2	x	**Constant**
7								
8								
9	**Coefficients of the derivative p'(x)**							
10	Enter coefficients of p'(x) in Row 12							
11		x^6	x^5	x^4	x^3	x^2	x	**Constant**
12								
13								
14	**Approximations: $x_{n+1} = x_n - p(x_n)/p'(x_n)$**							
15	**Initial (x_0)**	**1st (x_1)**	**2nd (x_2)**	**3rd (x_3)**	**4th (x_4)**	**5th (x_5)**	**6th (x_6)**	**7th (x_7)**
16								

Figure 2.9. Model spreadsheet for finding real roots of polynomials.

We save the spreadsheet of Figure 2.9 with a name, say *template.xls*. Then, we save it with a different name, say *Example_2_6.xls*, and in B16 we type the formula

=A16-($A$7*A16^7+$B$7*A16^6+$C$7*A16^5+$D$7*A16^4
+$E$7*A16^3+$F$7*A16^2+$G$7*A16^1+$H$7)/
($B$12*A16^6+$C$12*A16^5+$D$12*A16^4+$E$12*A16^3
+$F$12*A16^2+$G$12*A16^1+$H$12)

The use of the dollar sign ($) is explained in Paragraph 4 below.

The formula in B16 of Figure 2.10, is the familiar Newton's formula which also appears in Row 14. We observe that B16 now displays #DIV/0! (this is a warning that some value is being divided by zero), but this will change once we enter the polynomial coefficients, and the coefficients of the first derivative.

2. Since we are told that one real root is between 4 and 6, we take the average 5 and we enter it in A16. This value is our first (initial) approximation. We also enter the polynomial coefficients, and the coefficients of the first derivative in Rows 7 and 12 respectively.

3. Next, we copy B16 to C16:F16 and the spreadsheet now appears as shown in the spreadsheet of Figure 2.10. We observe that there is no change in the values of E16 and F16; therefore, we terminate the approximation steps there.

	A	B	C	D	E	F	G	H
1	Spreadsheet for finding approximations of the real roots of polynomials							
2	up the 7th power by Newton's Method.							
3								
4	**Powers of x and corresponding coefficients of given polynomial p(x)**							
5	Enter coefficients of p(x) in Row 7							
6	x^7	x^6	x^5	x^4	x^3	x^2	x	Constant
7	1	-6	5	-4	3	-2	1	-15
8								
9	**Coefficients of the derivative p'(x)**							
10	Enter coefficients of p'(x) in Row 12							
11		x^6	x^5	x^4	x^3	x^2	x	Constant
12		7	-36	25	-16	9	-4	1
13								
14	**Approximations: $x_{n+1} = x_n - p(x_n)/p'(x_n)$**							
15	Initial (x_0)	1st (x_1)	2nd (x_2)	3rd (x_3)	4th (x_4)	5th (x_5)	6th (x_6)	7th (x_7)
16	5.0	5.20409	5.16507	5.163194	5.163190	5.163190		

Figure 2.10. Spreadsheet for Example 2.6.

4. All cells in the formula of B16, except A16, have dollar signs ($) in front of the column letter, and in front of the row number. These cells are said to be *absolute*. The value of an absolute cell does not change when it is copied from one position to another. A cell that is not absolute is said to be *relative cell*. Thus, B16 is a relative cell, and $B$16 is an absolute cell. The contents of a relative cell changes when it is copied from one location to another. We can easily convert a relative cell to absolute or vice versa, by first placing the cursor in front, at the end, or between the letters and numbers of the cell, then, we press the function key *F4*. In this example, we made all cells, except A16, absolute so that the formula of B16 can be copied to C16, D16 and so on, without changing its value. The relative cell A16, when copied to the next column, changes to B16, when copied to the next column to the right, changes to C16, and so on.

We can now use this template with any other polynomial by just entering the coefficients of the new polynomial in row 7 and the coefficients of its derivative in Row 12; then, we observe the successive approximations in Row 16.

2.3 The Bisection Method for Root Approximation

The *Bisection* (or *interval halving*) method is an algorithm[*] for locating the real roots of a function.

---

[*] *This is a step–by–step problem–solving procedure, especially an established, recursive computational procedure for solving a problem in a finite number of steps.*

The objective is to find two values of x, say x_1 and x_2, so that $f(x_1)$ and $f(x_2)$ have opposite signs, that is, either $f(x_1) > 0$ and $f(x_2) < 0$, or $f(x_1) < 0$ and $f(x_2) > 0$. If any of these two conditions is satisfied, we can compute the midpoint x_m of the interval $x_1 \le x \le x_2$ with

$$x_m = \frac{x_1 + x_2}{2} \tag{2.18}$$

Knowing x_m, we can find $f(x_m)$. Then, the following decisions are made:

1. If $f(x_m)$ and $f(x_1)$ have the same sign, their product will be positive, that is, $f(x_m) \cdot f(x_1) > 0$. This indicates that x_m and x_1 are on the left side of the x–axis crossing as shown in Figure 2.11. In this case, we replace x_1 with x_m.

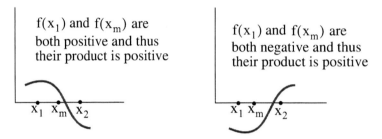

Figure 2.11. Sketches to illustrate the bisection method when $f(x_1)$ and $f(x_m)$ have same sign

2. If $f(x_m)$ and $f(x_1)$ have opposite signs, their product will be negative, that is, $f(x_m) \cdot f(x_1) < 0$. This indicates that x_m and x_2 are on the right side of the x–axis crossing as in Figure 2.12. In this case, we replace x_2 with x_m.

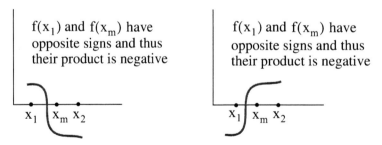

Figure 2.12. Sketches to illustrate the bisection method when $f(x_1)$ and $f(x_m)$ have opposite signs

After making the appropriate substitution, the above process is repeated until the root we are seeking has a specified tolerance. To terminate the iterations, we either:

a. specify a number of iterations

b. specify a tolerance on the error of $f(x)$

We will illustrate the *Bisection Method* with examples using both MATLAB and Excel.

Example 2.7

Use the Bisection Method with MATLAB to approximate one of the roots of

$$y = f(x) = 3x^5 - 2x^3 + 6x - 8 \qquad (2.19)$$

by

a. by specifying *16* iterations, and using a **for end** loop MATLAB program

b. by specifying *0.00001* tolerance for $f(x)$, and using a **while end** loop MATLAB program

Solution:

This is the same polynomial as in Example 2.4.

a. The **for end** loop allows a group of functions to be repeated a fixed and predetermined number of times. The syntax is:

```
for x = array
commands...
end
```

Before we write the program script, we must define a function assigned to the given polynomial and save it as a *function m–file*. We will define this function as *funcbisect01* and will save it as *funcbisect01.m*.

```
function y= funcbisect01(x);
y = 3 .* x .^ 5 – 2 .* x .^ 3 + 6 .* x – 8;
% We must not forget to type the semicolon at the end of the line above;
% otherwise our script will fill the screen with values of y
```

On the script below, the statement **for k = 1:16** says for $k = 1, k = 2, ..., k = 16$, evaluate all commands down to the **end** command. After the $k = 16$ iteration, the loop ends and any commands after the end are computed and displayed as commanded.

Let us also review the meaning of the **fprintf('%9.6f %13.6f \n', xm,fm)** line. Here, **%9.6f** and **%13.6f** are referred to as *format specifiers* or *format scripts*; the first specifies that the value of **xm** must be expressed in *decimal format* also called *fixed point format*, with a total of 9 digits, 6 of which will be to the right of the decimal point. Likewise, **fm** must be expressed in *decimal format* with a total of 13 digits, 6 of which will be to the right of the decimal point. Some other specifiers are **%e** for scientific format, **%s** for string format, and **%d** for integer format. For more information, we can type **help fprintf**. The special format \n specifies a linefeed, that is, it prints everything specified up to that point and starts a new line. We will discuss other special formats as they appear in subsequent examples.

The script for the first part of Example 2.7 is given below.

```
x1=1;  x2=2;                        % We know this interval from Example 2.4, Figure 2.6
disp('  xm          fm')           % xm is the average of x1 and x2, fm is f(xm)
disp('------------------------')    % insert line under xm and fm
for k=1:16;
    f1=funcbisect01(x1); f2=funcbisect01(x2);
xm=(x1+x2) / 2; fm=funcbisect01(xm);
fprintf('%9.6f %13.6f \n', xm,fm) % Prints xm and fm on same line;
if (f1*fm<0)
  x2=xm;
else
   x1=xm;
  end
end
```

When this program is executed, MATLAB displays the following:

```
xm                 fm
------------------------
1.500000         17.031250
1.250000          4.749023
1.125000          1.308441
1.062500          0.038318
1.031250         -0.506944
1.046875         -0.241184
1.054688         -0.103195
1.058594         -0.032885
1.060547          0.002604
1.059570         -0.015168
1.060059         -0.006289
1.060303         -0.001844
1.060425          0.000380
1.060364         -0.000732
1.060394         -0.000176
1.060410          0.000102
```

We observe that the values are displayed with 6 decimal places as we specified, but for the integer part unnecessary leading zeros are not displayed.

b. The **while end** loop evaluates a group of commands an indefinite number of times. The syntax is:

while expression
    commands...

end

The commands between **while** and **end** are executed as long as all elements in expression are *true*. The script should be written so that eventually a *false* condition is reached and the loop then terminates.

There is no need to create another function *m–file*; we will use the same as in part a. Now we type and execute the following **while end** loop program.

```
x1=1; x2=2; tol=0.00001;
disp('   xm          fm'); disp('------------------------')
while (abs(x1-x2)>2*tol);
  f1=funcbisect01(x1); f2=funcbisect01(x2); xm=(x1+x2)/2;
  fm=funcbisect01(xm);
  fprintf('%9.6f %13.6f \n', xm,fm);
  if (f1*fm<0);
    x2=xm;
  else
    x1=xm;
  end
end
```

When this program is executed, MATLAB displays the following:

```
xm                fm
------------------------
1.500000        17.031250
1.250000         4.749023
1.125000         1.308441
1.062500         0.038318
1.031250        -0.506944
1.046875        -0.241184
1.054688        -0.103195
1.058594        -0.032885
1.060547         0.002604
1.059570        -0.015168
1.060059        -0.006289
1.060303        -0.001844
1.060425         0.000380
1.060364        -0.000732
1.060394        -0.000176
1.060410         0.000102
1.060402        -0.000037
1.060406         0.000032
1.060404        -0.000003
```

Next, we will use an Excel spreadsheet to construct a template that approximates a real root of a function with the bisection method. This requires repeated use of the **IF** function which has the following syntax.

=IF(logical_test,value_if_true,value_if_false)

where

logical_test: any value or expression that can be evaluated to **true** or **false**.

value_if_true: the value that is returned if logical_test is **true**.

If logical_test is **true** and value_if_true is omitted, **true** is returned. Value_if_true can be another formula.

value_if_false is the value that is returned if logical_test is **false**. If *logical_test* is **false** and value_if_false is omitted, **false** is returned. Value_if_false can be another formula.

These statements may be clarified with the following examples.

=IF(C11>=1500,A15, B15):If the value in C11 is greater than or equal to *1500*, use the value in A15; otherwise use the value in B15.

=IF(D22<E22, 800, 1200):If the value in D22 is less than the value of E22, assign the number *800*; otherwise assign the number *1200*.

=IF(M8<>N17, K7*12, L8/24):If the value in M8 is not equal to the value in *N17*, use the value in *K7* multiplied by *12*; otherwise use the value in *L8* divided by 24.

Example 2.8

Use the bisection method with an Excel spreadsheet to approximate the value of $\sqrt{5}$ within *0.00001* accuracy.

Solution:

Finding the square root of 5 is equivalent to finding the roots of $x^2 - 5 = 0$. We expect the positive root to be in the $2 < x < 3$ interval so we assign $x_1 = 2$ and $x_2 = 3$. The average of these values is $x_m = 2.5$. We will create a template as we did in Example 2.6 so we can use it with any polynomial equation. We start with a blank spreadsheet and we make the entries in rows 1 through 12 as shown in Figure 2.13.

Now, we make the following entries in rows 13 and 14.

A13: 2
B13: 3
C13: =(A13+B13)/2

	A	B	C	D	E	F	G	H
1	Spreadsheet for finding approximations of the real roots							
2	of polynomials using the Bisection method							
3								
4	Equation:	$y = f(x) = x^2 - 5 = 0$						
5								
6	Powers of x and corresponding coefficients of given polynomial f(x)							
7	Enter coefficients of f(x) in Row 9							
8	x^7	x^6	x^5	x^4	x^3	x^2	x	Constant
9	0.00000	0.00000	0.00000	0.00000	0.00000	1.00000	0	-5
10								
11	x_1	x_2	x_m	$f(x_1)$	$f(x_m)$	$f(x_1)f(x_m)$		
12			$(x_1+x_2)/2$					

Figure 2.13. Partial spreadsheet for Example 2.8

D13: =$A$9*A13^7+$B$9*A13^6+$C$9*A13^5+$D$9*A13^4
    +$E$9*A13^3+$F$9*A13^2+$G$9*A13^1+$H$9*A13^0
E13: =$A$9*C13^7+$B$9*C13^6+$C$9*C13^5+$D$9*C13^4
    +$E$9*C13^3+$F$9*C13^2+$G$9*C13^1+$H$9*C13^0
F13: =D13*E13
A14: =IF(A14=A13, C13, B13)
B14: =IF(A14=A13, C13, B13)

We copy C13 into C14 and we verify that C14: =(A14+B14)/2

Next, we highlight D13:F13 and on the *Edit* menu we *click* on *Copy*. We place the cursor on D14 and from the *Edit* menu we *click* on *Paste*. We verify that the numbers on D14:F14 are as shown on the spreadsheet of Figure 2.14. Finally, we highlight A14:F14, from the *Edit* menu we *click* on *Copy*, we place the cursor on A15, and holding the mouse left button, we highlight the range A15:A30. Then, from the *Edit* menu, we click on *Paste* and we observe the values in A15:F30.

The square root of 5 accurate to six decimal places is shown on C30 in the spreadsheet of Figure 2.14.

	A	B	C	D	E	F	G	H
1	Spreadsheet for finding approximations of the real roots							
2	of polynomials using the Bisection method							
3								
4	Equation:	$y = f(x) = x^2 - 5 = 0$						
5								
6	Powers of x and corresponding coefficients of given polynomial f(x)							
7	Enter coefficients of f(x) in Row 9							
8	x^7	x^6	x^5	x^4	x^3	x^2	x	Constant
9	0.00000	0.00000	0.00000	0.00000	0.00000	1.00000	0	-5
10								
11	x_1	x_2	x_m	$f(x_1)$	$f(x_m)$	$f(x_1)f(x_m)$		
12			$(x_1+x_2)/2$					
13	2.00000	3.00000	2.50000	-1.00000	1.25000	-1.25000		
14	2.00000	2.50000	2.25000	-1.00000	0.06250	-0.06250		
15	2.00000	2.25000	2.12500	-1.00000	-0.48438	0.48438		
16	2.12500	2.25000	2.18750	-0.48438	-0.21484	0.10406		
17	2.18750	2.25000	2.21875	-0.21484	-0.07715	0.01657		
18	2.21875	2.25000	2.23438	-0.07715	-0.00757	0.00058		
19	2.23438	2.25000	2.24219	-0.00757	0.02740	-0.00021		
20	2.23438	2.24219	2.23828	-0.00757	0.00990	-0.00007		
21	2.23438	2.23828	2.23633	-0.00757	0.00116	-0.00001		
22	2.23438	2.23633	2.23535	-0.00757	-0.00320	0.00002		
23	2.23535	2.23633	2.23584	-0.00320	-0.00102	0.00000		
24	2.23584	2.23633	2.23608	-0.00102	0.00007	0.00000		
25	2.23584	2.23608	2.23596	-0.00102	-0.00047	0.00000		
26	2.23596	2.23608	2.23602	-0.00047	-0.00020	0.00000		
27	2.23602	2.23608	2.23605	-0.00020	-0.00006	0.00000		
28	2.23605	2.23608	2.23607	-0.00006	0.00000	0.00000		
29	2.23605	2.23607	2.23606	-0.00006	-0.00003	0.00000		
30	2.23606	2.23607	2.23606	-0.00003	-0.00001	0.00000		

Figure 2.14. Entire spreadsheet for Example 2.8

Numerical Analysis Using MATLAB® and Excel®, Third Edition
Copyright © Orchard Publications

2.4 Summary

- Newton's (or Newton–Raphson) method can be used to approximate the roots of any linear or non–linear equation of any degree. It uses the formula

$$x_{n+1} = x_n - \frac{f(x_n)}{f'(x_n)}$$

  To apply Newton's method, we must begin with a reasonable approximation of the root value. In all cases, this can best be done by plotting $f(x)$ versus x.

- We can use a spreadsheet to approximate the real roots of linear and non–linear equations but to approximate all roots (real and complex conjugates) it is advisable to use MATLAB.

- The MATLAB the **while end** loop evaluates a group of statements an indefinite number of times and thus can be effectively used for root approximation.

- For approximating real roots we can use Excel's Goal Seek feature. We use Goal Seek when we know the desired result of a single formula, but we do not know the input value which satisfies that result. Thus, if we have the function $y = f(x)$, we can use Goal Seek to set the dependent variable y to the desired value (goal) and from it, find the value of the independent variable x which satisfies that goal.

- For repetitive tasks, such as finding the roots of polynomials, it is prudent to construct a template (model spreadsheet) with the appropriate formulas and then enter the coefficients of the polynomial to find its real roots.

- The Bisection (or interval halving) method is an algorithm for locating the real roots of a function. The objective is to find two values of x, say x_1 and x_2, so that $f(x_1)$ and $f(x_2)$ have opposite signs, that is, either $f(x_1) > 0$ and $f(x_2) < 0$, or $f(x_1) < 0$ and $f(x_2) > 0$. If any of these two conditions is satisfied, we can compute the midpoint x_m of the interval $x_1 \leq x \leq x_2$ with

$$x_m = \frac{x_1 + x_2}{2}$$

- We can use the Bisection Method with MATLAB to approximate one of the roots by specifying a number of iterations using a **for end** or by specifying a tolerance using a **while end** loop program.

- We can use an Excel spreadsheet to construct a template that approximates a real root of a function with the bisection method. This requires repeated use of the IF function which has the =IF(logical_test,value_if_true,value_if_false)

2.5 Exercises

1. Use MATLAB to sketch the graph $y = f(x)$ for each of the following functions, and verify from the graph that $f(a)$ and $f(b)$, where a and b defined below, have opposite signs. Then, use Newton's method to estimate the root of $f(x) = 0$ that lies between a and b.

   a. $f_1(x) = x^4 + x - 3$ $a = 1$ $b = 2$

   b. $f_2(x) = \sqrt{2x + 1} - \sqrt{x + 4}$ $a = 2$ $b = 4$

   Hint: Start with $x_0 = (a + b)/2$

2. Repeat Exercise 1 above using the Bisection method.

3. Repeat Example 2.5 using MATLAB.

   Hint: Use the procedure of Example 2.2

Numerical Analysis Using MATLAB® and Excel®, Third Edition
Copyright © Orchard Publications

2.6 Solutions to End-of-Chapter Exercises

1.

  a.

    x=−2:0.05:2; f1x=x.^4+x−3; plot(x,f1x); grid

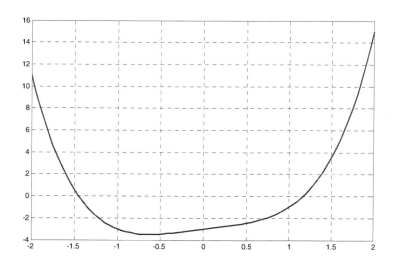

From the plot above we see that the positive root lies between $x = 1$ and $x = 1.25$ so we choose $a = 1$ and $b = 1.25$ so we take $x_0 = 1.1$ as our first approximation. We compute the next value x_1 as

$$x_1 = x_0 - \frac{f(x_0)}{f'(x_0)} = 1.1 - \frac{(1.1)^4 + 1.1 - 3}{4(1.1)^3 + 1} = 1.1 - \frac{(-0.436)}{6.324} = 1.169$$

The second approximation yields

$$x_2 = x_1 - \frac{f(x_1)}{f'(x_1)} = 1.169 - \frac{(1.169)^4 + 1.169 - 3}{4(1.169)^3 + 1} = 1.169 - \frac{0.0365}{7.39} = 1.164$$

Check with MATLAB:

pa=[1 0 0 1 −3]; roots(pa)

ans =

   −1.4526
    0.1443 + 1.3241i
    0.1443 − 1.3241i
    1.1640

b.

```
x=-5:0.05:5; f2x=sqrt(2.*x+1)-sqrt(x+4); plot(x,f2x); grid
```

Warning: Imaginary parts of complex X and/or Y arguments ignored.

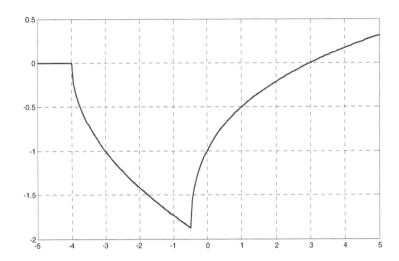

From the plot above we see that the positive root is very close to $x = 3$ and so we take $x_0 = 3$ as our first approximation. To compute the next value x_1 we first need to find the first derivative of $f_2(x)$. We rewrite it as

$$f_2(x) = \sqrt{2x+1} - \sqrt{x+4} = (2x+1)^{1/2} - (x+4)^{1/2}$$

Then,

$$\frac{d}{dx} \cdot f_2(x) = \frac{1}{2} \cdot (2x+1)^{-1/2} \cdot 2 - \frac{1}{2} \cdot (x+4)^{-1/2} \cdot 1 = \frac{1}{\sqrt{2x+1}} - \frac{1}{2\sqrt{x+4}}$$

and

$$x_1 = x_0 - \frac{f(x_0)}{f'(x_0)} = 3 - \frac{\sqrt{2 \times 3 + 1} - \sqrt{3+4}}{1/\sqrt{7} - 1/(2\sqrt{7})} = 3 - \frac{0}{1/(2\sqrt{7})} = 3$$

Thus, the real root is exactly $x = 3$. We also observe that since $f(x_0) = \sqrt{7} - \sqrt{7} = 0$, there was no need to find the first derivative $f'(x_0)$.

Check with MATLAB:

```
syms x; f2x=sqrt(2.*x+1)-sqrt(x+4); solve(f2x)

ans =

    3
```

2.

a. We will use the **for end** loop MATLAB program and specify 12 iterations. Before we write the program script, we must define a function assigned to the given polynomial and save it as a function m–file. We will define this function as **exercise2** and will save it as exercise2.m

```
function y= exercise2(x);
y = x .^ 4 +x – 3;
```

After saving this file as **exercise2.m**, we execute the following program:

```
x1=1;  x2=2;              % x1=a and x2=b
disp('   xm          fm')  % xm is the average of x1 and x2, fm is f(xm)
disp('------------------------')   % insert line under xm and fm
for k=1:12;
    f1=exercise2(x1); f2=exercise2(x2);
xm=(x1+x2) / 2; fm=exercise2(xm);
fprintf('%9.6f %13.6f \n', xm,fm)% Prints xm and fm on same line;
if (f1*fm<0)
  x2=xm;
 else
   x1=xm;
  end
end
```

MATLAB displays the following:

```
    xm                  fm
------------------------

 1.500000          3.562500
 1.250000          0.691406
 1.125000         -0.273193
 1.187500          0.176041
 1.156250         -0.056411
 1.171875          0.057803
 1.164063          0.000200
 1.160156         -0.028229
 1.162109         -0.014045
 1.163086         -0.006930
 1.163574         -0.003367
 1.163818         -0.001584
```

b. We will use the **while end** loop MATLAB program and specify a tolerance of **0.00001**.

We need to redefine the function *m–file* because the function in part (b) is not the same as in part a.

```
function y= exercise2(x);
y = sqrt(2.*x+1)–sqrt(x+4);
```

After saving this file as **exercise2.m**, we execute the following program:

```
x1=2.1; x2=4.3; tol=0.00001;    % If we specify x1=a=2 and x2=b=4, the program
% will not display any values because xm=(x1+x2)/2 = 3 = answer
disp('  xm         fm'); disp('------------------------')
while (abs(x1-x2)>2*tol);
  f1=exercise2(x1); f2=exercise2(x2); xm=(x1+x2)/2;
  fm=exercise2(xm);
  fprintf('%9.6f %13.6f \n', xm,fm);
  if (f1*fm<0);
 x2=xm;
   else
    x1=xm;
   end
end
```

When this program is executed, MATLAB displays the following:

```
    xm              fm
------------------------
3.200000        0.037013
2.650000       -0.068779
2.925000       -0.014289
3.062500        0.011733
2.993750       -0.001182
3.028125        0.005299
3.010938        0.002065
3.002344        0.000443
2.998047       -0.000369
3.000195        0.000037
2.999121       -0.000166
2.999658       -0.000065
2.999927       -0.000014
3.000061        0.000012
2.999994       -0.000001
3.000027        0.000005
3.000011        0.000002
```

3.

From Example 2.5,

$$y = f(x) = \cos 2x + \sin 2x + x - 1$$

We use the following script to plot this function.

```
x=−5:0.02:5; y=cos(2.*x)+sin(2.*x)+x−1; plot(x,y); grid
```

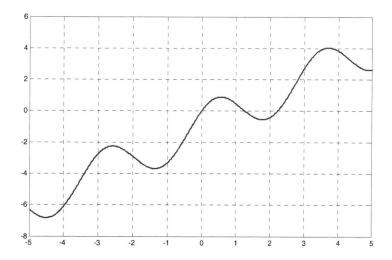

Let us find out what a symbolic solution gives.

```
syms x; y=cos(2*x)+sin(2*x)+x−1; solve(y)
```

```
ans =
   [0]
   [2]
```

The first value (0) is correct as it can be seen from the plot above and also verified by substitution of $x = 0$ into the given function. The second value (2) is not exactly correct as we can see from the plot. This is because when solving equations of periodic functions, there are an infinite number of solutions and MATLAB restricts its search for solutions to a limited range near zero and returns a non–unique subset of solutions.

To find a good approximation of the second root that lies between $x = 2$ and $x = 3$, we write and save the function files *exercise3* and *exercise3der* as defined below.

```
function y=exercise3(x)
% Finding roots by Newton's method using MATLAB
y=cos(2.*x)+sin(2.*x)+x−1;

function y=exercise3der(x)
```

```
% Finding roots by Newton's method
% The following is the first derivative of
% the function defined as exercise3
y=−2.*sin(2.*x)+2.*cos(2.*x)+1;
```

Now, we write and execute the following program and we find that the second root is x = 2.2295 and this is consistent with the value shown on the plot.

```
x = input('Enter starting value: ');
fx = exercise3(x);
fprimex = exercise3der(x);
xnext = x−fx/fprimex;
   x = xnext;
   fx = exercise3(x);
   fprimex = exercise3der(x);
disp(sprintf('First approximation is x =  %9.6f \n', x))
while input('Next approximation? (<enter>=no,1=yes)');
   xnext=x−fx/fprimex;
   x=xnext;
   fx=exercise3(x);
   fprimex=exercise3der(x);
disp(sprintf('Next approximation is x =  %9.6f \n', x))
end;
disp(sprintf('%9.6f \n', x))

Enter starting value: 3

First approximation is x = 2.229485
```

Chapter 3

T his chapter is an introduction to alternating current waveforms. The characteristics of sinusoids are discussed and the frequency, phase angle, and period are defined. Voltage and current relationships are expressed in sinusoidal terms. Phasors which are rotating vectors in terms of complex numbers, are also introduced and their relationships to sinusoids are derived.

3.1 Alternating Voltages and Currents

The waveforms shown in Figure 3.1 may represent alternating currents or voltages.

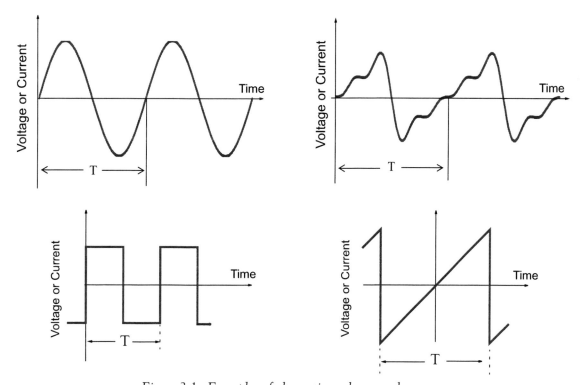

Figure 3.1. Examples of alternating voltages and currents

Thus an *alternating current* (AC) is defined as a periodic current whose average value over a period is zero. Stated differently, an alternating current alternates between positive and negative values at regularly recurring intervals of time. Also, the average of the positive and negative values over a period is zero.

As shown in Figure 3.1, the *period* T of an alternating current or voltage is the smallest value of time which separates recurring values of the alternating waveform.

Unless otherwise stated, our subsequent discussion will be restricted to sine or cosine waveforms and these are referred to as *sinusoids*. Two main reasons for studying sinusoids are: (1) many physical phenomena such as electric machinery produce (nearly) sinusoidal voltages and currents and (2) by Fourier analysis, any *periodic* waveform which is not a sinusoid, such as the square and sawtooth waveforms on the previous page, can be represented by a sum of sinusoids.

3.2 Characteristics of Sinusoids

Consider the sine waveform shown in Figure 3.2, where $f(t)$ may represent either a voltage or a current function, and let $f(t) = A\sin t$ where A is the amplitude of this function. A sinusoid (sine or cosine function) can be constructed graphically from the *unit circle*, which is a circle with radius of one unit, that is, $A = 1$ as shown, or any other unit. Thus, if we let the *phasor* (rotating vector) travel around the unit circle with an angular velocity ω, the $\cos\omega t$ and $\sin\omega t$ functions are generated from the projections of the phasor on the horizontal and vertical axis respectively. We observe that when the phasor has completed a *cycle* (one revolution), it has traveled 2π radians or $360°$ degrees, and then repeats itself to form another cycle.

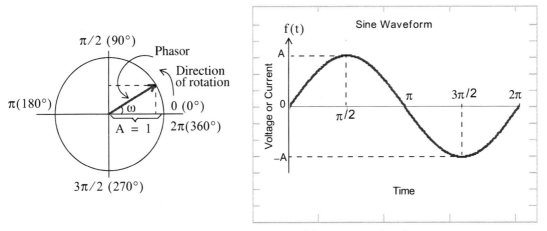

Figure 3.2. Generation of a sinusoid by rotation of a phasor

At the completion of one cycle, $t = T$ (one period), and since ω is the angular velocity, commonly known as *angular* or *radian frequency*, then

$$\boxed{\omega T = 2\pi} \qquad \text{or} \qquad \boxed{T = \frac{2\pi}{\omega}} \tag{3.1}$$

The term *frequency* in Hertz, denoted as Hz, is used to express the number of cycles per second. Thus, if it takes one second to complete one cycle (one revolution around the unit circle), we say that the frequency is 1 Hz or one cycle per second.

The frequency is denoted by the letter f and in terms of the period T and (3.1) we have

$$f = \frac{1}{T} \qquad \text{or} \qquad \omega = 2\pi f \qquad (3.2)$$

The frequency f is often referred to as the *cyclic frequency* to distinguish it from the *radian frequency* ω.

Since the cosine and sine functions are usually known in terms of degrees or radians, it is convenient to plot sinusoids versus ωt (radians) rather that time t. For example, $v(t) = V_{max}\cos\omega t$, and $i(t) = I_{max}\sin\omega t$ are plotted as shown in Figure 3.3.

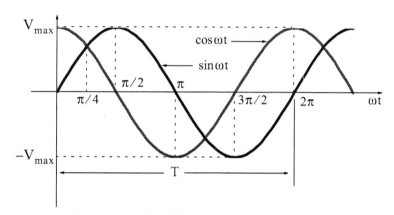

Figure 3.3. Plot of the cosine and sine functions

By comparing the sinusoidal waveforms of Figure 3.3, we see that the cosine function will be the same as the sine function if the latter is shifted to the left by $\pi/2$ radians, or $90°$. Thus, we say that *the cosine function leads (is ahead of) the sine function by $\pi/2$ radians or $90°$*. Likewise, if we shift the cosine function to the right by $\pi/2$ radians or $90°$, we obtain the sine waveform; in this case, we say that *the sine function lags (is behind) the cosine function by $\pi/2$ radians or $90°$*.

Another common expression is that *the cosine and sine functions are out-of-phase by $90°$*, or *there is a phase angle of $90°$ between the cosine and sine functions*. It is possible, of course, that two sinusoids are out-of-phase by a phase angle other than $90°$. Figure 3.4 shows three sinusoids which are *out-of-phase*. If the phase angle between them is $0°$ degrees, the two sinusoids are said to be *in-phase*.

We must remember that when we say that one sinusoid *leads* or *lags* another sinusoid, these are of the same frequency. Obviously, two sinusoids of different frequencies can never be in phase.

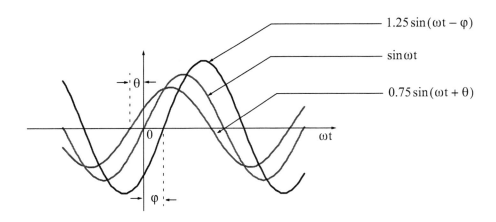

Figure 3.4. Out-of-phase sinusoids

It is convenient to express the phase angle in degrees rather than in radians in a sinusoidal function. For example, it is acceptable to express

$$v(t) = 100\sin(2000\pi t - \pi/6)$$

as

$$v(t) = 100\sin(2000\pi t - 30°)$$

since the subtraction inside the parentheses needs not to be performed.

When two sinusoids are to be compared in terms of their phase difference, these must first be written either both as cosine functions, or both as sine functions, and should also be written with positive amplitudes. We should remember also that *a negative amplitude implies* 180° *phase shift.*

Example 3.1

Find the phase difference between the sinusoids

$$i_1 = 120\cos(100\pi t - 30°)$$

and

$$i_2 = -6\sin(100\pi t - 30°)$$

Solution:

We recall that the minus (–) sign indicates a ±180° phase shift, and that the sine function lags the cosine by 90°. Then,

$$-\sin x = \sin(x \pm 180°) \ \text{and} \ \sin x = \cos(x - 90°)$$

and

$$i_2 = 6\sin(100\pi t - 210^\circ) = 6\sin(100\pi t + 150^\circ)$$

$$= 6\cos(100\pi t + 150^\circ - 90^\circ) = 6\cos(100\pi t + 60^\circ)$$

and comparing i_2 with i_1, we see that i_2 *leads* i_1 by 90°, or i_1 *lags* i_2 by 90°.

In our subsequent discussion, we will be using several trigonometric identities, derivatives and integrals involving trigonometric functions. We, therefore, provide the following relations and formulas for quick reference. Let us also review the definition of a radian and its relationship to degrees with the aid of Figure 3.5.

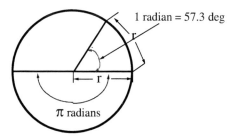

Figure 3.5. Definition of radian

As shown in Figure 3.5, the *radian* is a circular angle subtended by an arc equal in length to the radius of the circle, whose radius is r units in length. The circumference of a circle is $2\pi r$ units; therefore, there are 2π or 6.283... radians in 360° degrees. Then,

$$1 \text{ radian } = \frac{360^\circ}{2\pi} \approx 57.3^\circ \qquad (3.3)$$

The angular velocity is expressed in radians per second, and it is denoted by the symbol ω. Then, a *rotating vector* that completes n revolutions per second, has an angular velocity $\omega = 2\pi n$ radians per second.

Some useful trigonometric relations are given below for quick reference.

$$\cos 0^\circ = \cos 360^\circ = \cos 2\pi = 1 \qquad (3.4)$$

$$\cos 30^\circ = \cos\frac{\pi}{6} = \frac{\sqrt{3}}{2} = 0.866 \qquad (3.5)$$

$$\cos 45^\circ = \cos\frac{\pi}{4} = \frac{\sqrt{2}}{2} = 0.707 \qquad (3.6)$$

$$\cos 60^\circ = \cos\frac{\pi}{3} = \frac{1}{2} = 0.5 \qquad (3.7)$$

$$\cos 90° = \cos\frac{\pi}{2} = 0 \tag{3.8}$$

$$\cos 120° = \cos\frac{2\pi}{3} = \frac{-1}{2} = -0.5 \tag{3.9}$$

$$\cos 150° = \cos\frac{5\pi}{6} = \frac{-\sqrt{3}}{2} = -0.866 \tag{3.10}$$

$$\cos 180° = \cos\pi = -1 \tag{3.11}$$

$$\cos 210° = \cos\frac{7\pi}{6} = \frac{-\sqrt{3}}{2} = -0.866 \tag{3.12}$$

$$\cos 225° = \cos\frac{5\pi}{4} = \frac{-\sqrt{2}}{2} = -0.707 \tag{3.13}$$

$$\cos 240° = \cos\frac{4\pi}{3} = \frac{-1}{2} = -0.5 \tag{3.14}$$

$$\cos 270° = \cos\frac{3\pi}{2} = 0 \tag{3.15}$$

$$\cos 300° = \cos\frac{5\pi}{3} = 0.5 \tag{3.16}$$

$$\cos 330° = \cos\frac{11\pi}{6} = 0.866 \tag{3.17}$$

$$\sin 0° = \sin 360° = \sin 2\pi = 0 \tag{3.18}$$

$$\sin 30° = \sin\frac{\pi}{6} = \frac{1}{2} = 0.5 \tag{3.19}$$

$$\sin 45° = \sin\frac{\pi}{4} = \frac{\sqrt{2}}{2} = 0.707 \tag{3.20}$$

$$\sin 60° = \sin\frac{\pi}{3} = \frac{1}{2} = 0.866 \tag{3.21}$$

$$\sin 90° = \sin\frac{\pi}{2} = 1 \tag{3.22}$$

$$\sin 120° = \sin\frac{2\pi}{3} = \frac{\sqrt{3}}{2} = 0.866 \tag{3.23}$$

$$\sin 150° = \sin\frac{5\pi}{6} = \frac{1}{2} = 0.5 \tag{3.24}$$

$$\sin 180° = \sin\pi = 0 \tag{3.25}$$

$$\sin 210° = \sin\frac{7\pi}{6} = \frac{-1}{2} = -0.5 \tag{3.26}$$

$$\sin 225° = \sin\frac{5\pi}{4} = \frac{-\sqrt{2}}{2} = -0.707 \tag{3.27}$$

$$\sin 240° = \sin\frac{4\pi}{3} = \frac{-\sqrt{3}}{2} = -0.866 \tag{3.28}$$

$$\sin 270° = \sin\frac{3\pi}{2} = -1 \tag{3.29}$$

$$\sin 300° = \sin\frac{5\pi}{3} = \frac{-\sqrt{3}}{2} = -0.866 \tag{3.30}$$

$$\sin 330° = \sin\frac{11\pi}{6} = \frac{-1}{2} = -0.5 \tag{3.31}$$

$$\cos(-\theta) = \cos\theta \tag{3.32}$$

$$\cos(90° + \theta) = -\sin\theta \tag{3.33}$$

$$\cos(180° - \theta) = -\cos\theta \tag{3.34}$$

$$\sin(-\theta) = -\sin\theta \tag{3.35}$$

$$\sin(90° + \theta) = \cos\theta \tag{3.36}$$

$$\sin(180° - \theta) = \sin\theta \tag{3.37}$$

$$\tan\theta = \frac{\sin\theta}{\cos\theta} \tag{3.38}$$

$$\cot\theta = \frac{\cos\theta}{\sin\theta} = \frac{1}{\tan\theta} \tag{3.39}$$

$$\sec\theta = \frac{1}{\cos\theta} \tag{3.40}$$

$$\csc\theta = \frac{1}{\sin\theta} \tag{3.41}$$

$$\tan(90° + \theta) = -\cot\theta \tag{3.42}$$

$$\tan(180° - \theta) = -\tan\theta \tag{3.43}$$

$$\cos(\theta + \phi) = \cos\theta\cos\phi - \sin\theta\sin\phi \tag{3.44}$$

$$\cos(\theta - \phi) = \cos\theta\cos\phi + \sin\theta\sin\phi \tag{3.45}$$

$$\sin(\theta + \phi) = \sin\theta\cos\phi + \cos\theta\sin\phi \tag{3.46}$$

$$\sin(\theta - \phi) = \sin\theta\cos\phi - \cos\theta\sin\phi \tag{3.47}$$

$$\tan(\theta + \phi) = \frac{\tan\theta + \tan\phi}{1 - \tan\theta\tan\phi} \tag{3.48}$$

$$\tan(\theta - \phi) = \frac{\tan\theta - \tan\phi}{1 + \tan\theta\tan\phi} \tag{3.49}$$

$$\cos^2\theta + \sin^2\theta = 1 \tag{3.50}$$

$$\cos 2\theta = \cos^2\theta - \sin^2\theta \tag{3.51}$$

$$\sin 2\theta = 2\sin\theta\cos\theta \tag{3.52}$$

$$\tan 2\theta = \frac{2\tan\theta}{1 - \tan^2\theta} \tag{3.53}$$

$$\cos^2\theta = \frac{1}{2}(1 + \cos 2\theta) \tag{3.54}$$

$$\sin^2\theta = \frac{1}{2}(1 - \cos 2\theta) \tag{3.55}$$

$$\cos\theta\cos\phi = \frac{1}{2}\cos(\theta + \phi) + \frac{1}{2}\cos(\theta - \phi) \tag{3.56}$$

$$\cos\theta\sin\phi = \frac{1}{2}\sin(\theta + \phi) - \frac{1}{2}\sin(\theta - \phi) \tag{3.57}$$

$$\sin\theta\cos\phi = \frac{1}{2}\sin(\theta + \phi) + \frac{1}{2}\sin(\theta - \phi) \tag{3.58}$$

$$\sin\theta\sin\phi = \frac{1}{2}\cos(\theta - \phi) - \frac{1}{2}\cos(\theta + \phi) \tag{3.59}$$

Let Figure 3.6 be any triangle.

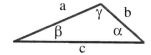

Figure 3.6. General triangle

Then,

by the *law of sines*,

$$\frac{a}{\sin\alpha} = \frac{b}{\sin\beta} = \frac{c}{\sin\gamma} \qquad (3.60)$$

by the *law of cosines*,

$$a^2 = b^2 + c^2 - 2bc\cos\alpha \qquad (3.61)$$

$$b^2 = a^2 + c^2 - 2ac\cos\beta \qquad (3.62)$$

$$c^2 = a^2 + b^2 - 2ab\cos\gamma \qquad (3.63)$$

and by the *law of tangents*,

$$\frac{a-b}{a+b} = \frac{\tan\frac{1}{2}(\alpha-\beta)}{\tan\frac{1}{2}(\alpha+\beta)} \qquad \frac{b-c}{b+c} = \frac{\tan\frac{1}{2}(\beta-\gamma)}{\tan\frac{1}{2}(\beta+\gamma)} \qquad \frac{c-a}{c+a} = \frac{\tan\frac{1}{2}(\gamma-\alpha)}{\tan\frac{1}{2}(\gamma+\alpha)} \qquad (3.64)$$

The following differential and integral trigonometric and exponential functions, are used extensively in engineering.

$$\frac{d}{dx}(\sin v) = \cos v \frac{dv}{dx} \qquad (3.65)$$

$$\frac{d}{dx}(\cos v) = -\sin v \frac{dv}{dx} \qquad (3.66)$$

$$\frac{d}{dx}(e^v) = e^v \frac{dv}{dx} \qquad (3.67)$$

$$\int \sin ax\, dx = -\frac{1}{a}\cos ax + c \qquad (3.68)$$

$$\int \cos ax\, dx = \frac{1}{a}\sin ax + c \qquad (3.69)$$

$$\int e^{ax}\, dx = \frac{1}{a}e^{ax} + c \qquad (3.70)$$

3.3 Inverse Trigonometric Functions

The notation $\cos^{-1}y$ or $\arccos y$ is used to denote an angle whose cosine is y. Thus, if $y = \cos x$, then $x = \cos^{-1}y$. Similarly, if $w = \sin v$, then $v = \sin^{-1}w$, and if $z = \tan u$, then $u = \tan^{-1}z$. These are called *Inverse Trigonometric Functions*.

Example 3.2

Find the angle θ if $\cos^{-1}0.5 = \theta$

Solution:

Here, we want to find the angle θ given that its cosine is 0.5. From (3.7), $\cos 60° = 0.5$. Therefore, $\theta = 60°$

3.4 Phasors

In the language of mathematics, the square root of minus one is denoted as i, that is, $i = \sqrt{-1}$. In the electrical engineering field, we denote i as j to avoid confusion with current i. Essentially, j is an operator that produces a $90°$ counterclockwise rotation to any vector to which it is applied as a multiplying factor. Thus, if it is given that a vector **A** has the direction along the right side of the x-axis as shown in Figure 3.7, multiplication of this vector by the operator j will result in a new vector $j\mathbf{A}$ whose magnitude remains the same, but it has been rotated counterclockwise by $90°$. Also, another multiplication of the new vector $j\mathbf{A}$ by j will produce another $90°$ counterclockwise direction. In this case, the vector **A** has rotated $180°$ and its new value now is $-\mathbf{A}$. When this vector is rotated by another $90°$ for a total of $270°$, its value becomes $j(-\mathbf{A}) = -j\mathbf{A}$. A fourth $90°$ rotation returns the vector to its original position, and thus its value is again **A**. Therefore, we conclude that $j^2 = -1$, $j^3 = -j$, $j^4 = 1$, and the rotating vector **A** is referred to as a *phasor*.

Note: In our subsequent discussion, we will designate the x-axis (abscissa) as the *real axis*, and the y-axis (ordinate) as the *imaginary axis* with the understanding that the "imaginary" axis is just as "real" as the real axis. In other words, the imaginary axis is just as important as the real axis.[*]

An *imaginary number* is the product of a real number, say r, by the operator j. Thus, r is a real number and jr is an imaginary number.

---

[*] *A more appropriate nomenclature for the real and imaginary axes would be the axis of the cosines and the axis of the sines respectively.*

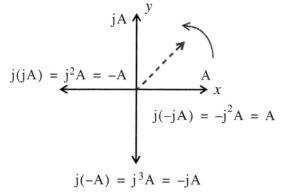

$$j(jA) = j^2A = -A \qquad j(-jA) = -j^2A = A$$

$$j(-A) = j^3A = -jA$$

Figure 3.7. The j operator

A *complex number* is the sum (or difference) of a real number and an imaginary number. For example, the number $A = a + jb$ where a and b are both real numbers, is a complex number. Then, $a = \text{Re}\{A\}$ and $b = \text{Im}\{A\}$ where $\text{Re}\{A\}$ denotes real part of A, and $b = \text{Im}\{A\}$ the imaginary part of A. When written as $A = a + jb$, it is said to be expressed in *rectangular form*.

Since in engineering we use complex quantities as phasors, henceforth any complex number will be referred to as a phasor.

By definition, two phasors A *and* B where $A = a + jb$ and $B = c + jd$, are equal if and only if their real parts are equal and also their imaginary parts are equal. Thus, $A = B$ if and only if $a = c$ and $b = d$.

3.5 Addition and Subtraction of Phasors

The sum of two phasors has a real component equal to the sum of the real components, and an imaginary component equal to the sum of the imaginary components. For subtraction, we change the signs of the components of the subtrahend and we perform addition. Thus, if $A = a + jb$ and $B = c + jd$, then

$$A + B = (a + c) + j(b + d)$$

and

$$A - B = (a - c) + j(b - d)$$

Example 3.3

It is given that $A = 3 + j4$, and $B = 4 - j2$. Find $A + B$ and $A - B$

Solution:

$$A + B = (3 + j4) + (4 - j2) = (3 + 4) + j(4 - 2) = 7 + j2$$

$$A - B = (3 + j4) - (4 - j2) = (3 - 4) + j(4 + 2) = -1 + j6$$

3.6 Multiplication of Phasors

Phasors are multiplied using the rules of elementary algebra, and making use of the fact that $j^2 = -1$. Thus, if $A = a + jb$ and $B = c + jd$, then

$$A \cdot B = (a + jb) \cdot (c + jd) = ac + jad + jbc + j^2bd$$

and since $j^2 = -1$, it follows that

$$A \cdot B = ac + jad + jbc - bd = (ac - bd) + j(ad + bc) \qquad (3.71)$$

Example 3.4

It is given that $A = 3 + j4$ and $B = 4 - j2$. Find $A \cdot B$

Solution:

$$A \cdot B = (3 + j4) \cdot (4 - j2) = 12 - j6 + j16 - j^2 8 = 20 + j10$$

The *conjugate* of a phasor, denoted as A^*, is another phasor with the same real component, and with an imaginary component of opposite sign. Thus, if $A = a + jb$, then $A^* = a - jb$.

Example 3.5

It is given that $A = 3 + j5$. Find A^*

Solution:

The conjugate of the phasor A has the same real component, but the imaginary component has opposite sign. Then, $A^* = 3 - j5$

If a phasor A is multiplied by its conjugate, the result is a real number. Thus, if $A = a + jb$, then

$$A \cdot A^* = (a + jb)(a - jb) = a^2 - jab + jab - j^2b^2 = a^2 + b^2$$

Example 3.6

It is given that $A = 3 + j5$. Find $A \cdot A*$

Solution:

$$A \cdot A* = (3 + j5)(3 - j5) = 3^2 + 5^2 = 9 + 25 = 34$$

3.7 Division of Phasors

When performing division of phasors, it is desirable to obtain the quotient separated into a real part and an imaginary part. This procedure is called *rationalization of the quotient*, and it is done by multiplying the denominator by its conjugate. Thus, if $A = a + jb$ and $B = c + jd$, then,

$$\frac{A}{B} = \frac{a + jb}{c + jd} = \frac{(a + jb)(c - jd)}{(c + jd)(c - jd)} = \frac{A}{B} \cdot \frac{B*}{B*} = \frac{(ac + bd) + j(bc - ad)}{c^2 + d^2}$$

$$= \frac{(ac + bd)}{c^2 + d^2} + j\frac{(bc - ad)}{c^2 + d^2}$$

(3.72)

In (3.72), we multiplied both the numerator and denominator by the conjugate of the denominator to eliminate the j operator from the denominator of the quotient. Using this procedure, we see that the quotient is easily separated into a real and an imaginary part.

Example 3.7

It is given that $A = 3 + j4$, and $B = 4 + j3$. Find A/B

Solution:

Using the procedure of (3.72), we get

$$\frac{A}{B} = \frac{3 + j4}{4 + j3} = \frac{(3 + j4)(4 - j3)}{(4 + j3)(4 - j3)} = \frac{12 - j9 + j16 + 12}{4^2 + 3^2} = \frac{24 + j7}{25} = \frac{24}{25} + j\frac{7}{25} = 0.96 + j0.28$$

3.8 Exponential and Polar Forms of Phasors

The relations

$$\boxed{e^{j\theta} = \cos\theta + j\sin\theta}$$

(3.73)

and

$$e^{-j\theta} = \cos\theta - j\sin\theta \tag{3.74}$$

are known as the *Euler's identities.*

Multiplying (3.73) by the *real* positive constant C we get:

$$Ce^{j\theta} = C\cos\theta + jC\sin\theta \tag{3.75}$$

This expression represents a phasor, say $a + jb$, and thus

$$Ce^{j\theta} = a + jb \tag{3.76}$$

Equating real and imaginary parts in (3.75) and (3.76), we get

$$a = C\cos\theta \quad \text{and} \quad b = C\sin\theta \tag{3.77}$$

Squaring and adding the expressions in (3.77), we get

$$a^2 + b^2 = (C\cos\theta)^2 + (C\sin\theta)^2 = C^2(\cos^2\theta + \sin^2\theta) = C^2$$

Then,

$$C^2 = a^2 + b^2$$

or

$$C = \sqrt{a^2 + b^2} \tag{3.78}$$

Also, from (3.77)

$$\frac{b}{a} = \frac{C\sin\theta}{C\cos\theta} = \tan\theta$$

or

$$\theta = \tan^{-1}\left(\frac{b}{a}\right) \tag{3.79}$$

Therefore, to convert a phasor from rectangular to exponential form, we use the expression

$$a + jb = \sqrt{a^2 + b^2}\, e^{j\left(\tan^{-1}\frac{b}{a}\right)} \tag{3.80}$$

To convert a phasor from exponential to rectangular form, we use the expressions

$$Ce^{j\theta} = C\cos\theta + jC\sin\theta$$
$$Ce^{-j\theta} = C\cos\theta - jC\sin\theta \tag{3.81}$$

The *polar form* is essentially the same as the exponential form but the notation is different, that is,

$$\boxed{Ce^{j\theta} = C\angle\theta}$$

(3.82)

where the left side of (3.82) is the exponential form, and the right side is the polar form.

We must remember that *the phase angle* θ *is always measured with respect to the positive real axis, and rotates in the counterclockwise direction.*

In Examples 3.8 and 3.9 below, we will verify the results with the following MATLAB co-ordinate transformation functions:

[theta,r] = cart2pol(x,y) – transforms from Cartesian to polar co–ordinates.

[x,y] = pol2cart(theta,r) – transforms from polar to Cartesian co–ordinates

---

Example 3.8

Convert the following phasors to exponential and polar forms:

a. $3 + j4$ b. $-1 + j2$ c. $-2 - j$ d. $4 - j3$

Solution:

a. The real and imaginary components of this phasor are shown in Figure 3.8.

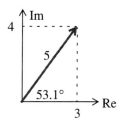

Figure 3.8. The components of $3 + j4$

Then,

$$3 + j4 = \sqrt{3^2 + 4^2} \cdot e^{j\tan^-(4/3)} = 5e^{j53.1°} = 5\angle 53.1°$$

Check with MATLAB:

```
x=3+j*4; magx=abs(x); thetax=angle(x)*180/pi; disp(magx); disp(thetax)
```

    5

    53.1301

or

```
x = 3; y = 4; [theta,r] = cart2pol(x,y), deg = theta*180/pi

theta =

   0.9273

r =

   5

deg =

   53.1301
```

We can also verify the result with Simulink®[*] as shown in the model of Figure 3.9. The K value for the Gain block has been specified as $180/\pi$ to convert radians into degrees.

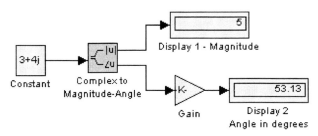

Figure 3.9. Simulink model for Example 3.8 (a)

b. The real and imaginary components of this phasor are shown in Figure 3.10.

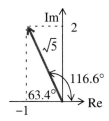

Figure 3.10. The components of $-1 + j2$

Then,

$$-1 + j2 = \sqrt{1^2 + 2^2}\, e^{j\tan^{-}(2/-1)} = \sqrt{5}\, e^{j116.6^\circ} = \sqrt{5}\angle 116.6^\circ = 2.236\angle 116.6^\circ$$

Check with MATLAB:

```
y=−1+j*2; magy=abs(y); thetay=angle(y)*180/pi;  disp(magy); disp(thetay)
```

---

* *The reader who is not familiar with Simulink may skip this model and all others without loss of continuity. For an introduction to Simulink, please refer to "Introduction to Simulink with Engineering Applications", ISBN 0-9744239-7-1. A brief introduction to Simulink is provided in Appendix B.*

```
2.2361

116.5651
```

or

x = −1; y = 2; [theta,r] = cart2pol(x,y), deg = theta*180/pi

```
theta =

    2.0344

r =

    2.2361

deg =

   116.5651
```

Check with the Simulink model of Figure 3.11:

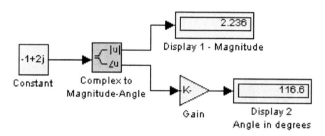

Figure 3.11. Simulink model for Example 3.8 (b)

c. The real and imaginary components of this phasor are shown in Figure 3.12.

Then,

$$-2-j1 = \sqrt{2^2+1^2} \cdot e^{j\tan^{-}(-1/-2)} = \sqrt{5}e^{j206.6^\circ} = \sqrt{5}\angle 206.6^\circ = \sqrt{5}e^{j(-153.4)^\circ} = 2.236\angle -153.4^\circ$$

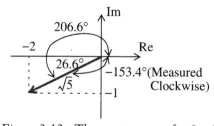

Figure 3.12. The components of − 2 − j

Check with MATLAB:

v=−2−j*1; magv=abs(v); thetav=angle(v)*180/pi; disp(magv); disp(thetav)

```
   2.2361

  -153.4349
```

or

x = −2; y = −1; [theta,r] = cart2pol(x,y), deg = theta*180/pi

```
theta =

   -2.6779

r =

   2.2361

deg =

  -153.4349
```

Check with the Simulink model of Figure 3.13:

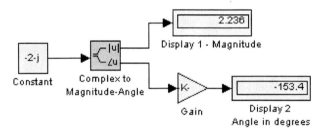

Figure 3.13. Simulink model for Example 3.8 (c)

d. The real and imaginary components of this phasor are shown in Figure 3.14.

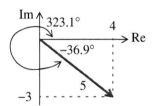

Figure 3.14. The components of $4 - j3$

Then,

$$4 - j3 = \sqrt{4^2 + 3^2} \cdot e^{j\tan^{-}(-3/4)} = 5e^{j323.1°} = 5\angle 323.1° = 5e^{-j36.9°} = 5\angle -36.9°$$

Check with MATLAB:

w=4−j*3; magw=abs(w); thetaw=angle(w)*180/pi; disp(magw); disp(thetaw)

```
     5
-36.8699
```

or

x = 4; y = −3; [theta,r] = cart2pol(x,y), deg = theta*180/pi

```
theta =
   -0.6435
r =
        5
deg =
  -36.8699
```

Check with the Simulink model of Figure 3.15:

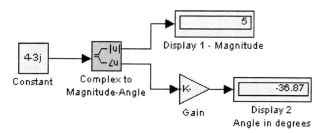

Figure 3.15. Simulink model for Example 3.8 (d)

Example 3.9

Express the phasor $-2\angle 30°$ in exponential and in rectangular forms.

Solution:

We recall that $-1 = j^2$. Since each j rotates a vector by $90°$ counterclockwise, then $-2\angle 30°$ is the same as $2\angle 30°$ rotated counterclockwise by $180°$. Therefore,

$$-2\angle 30° = 2\angle(30° + 180°) = 2\angle 210° = 2\angle{-150°}$$

The components of this phasor are shown in Figure 3.16.

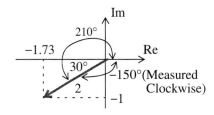

Figure 3.16. The components of $2\angle{-150°}$

Then,

$$2\angle{-150°} = 2e^{-j150°} = 2(\cos 150° - j\sin 150°) = 2(-0.866 - j0.5) = -1.73 - j$$

Check with MATLAB:

r = –2; theta = 30/pi; [x,y] = pol2cart(theta*180/pi,r)

x =

   –1.7578

y =

   –0.9541

Check with the Simulink model of Figure 3.17:

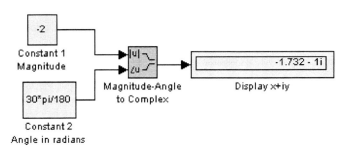

Figure 3.17. Simulink model for Example 3.9

Note: The rectangular form is most useful when we add or subtract phasors; however, the exponential and polar forms are most convenient when we multiply or divide phasors.

To multiply two phasors in exponential (or polar) form, we multiply the magnitudes and we add the phase angles, that is, if

$$A = M\angle\theta \quad \text{and} \quad B = N\angle\phi$$

then,

$$AB = MN\angle(\theta + \phi) = Me^{j\theta}Ne^{j\phi} = MNe^{j(\theta + \phi)} \qquad (3.83)$$

Example 3.10

Multiply $A = 12.58\angle 74.3°$ by $B = 7.22\angle -118.7°$

Solution:

Multiplication in polar form yields

$$AB = (12.58 \times 7.22)\angle[74.3° + (-118.7°)] = 90.83\angle -44.4°$$

and multiplication in exponential form yields

$$AB = (12.58e^{j74.3°})(7.22e^{-j118.7°}) = 90.83e^{j(74.3° - 118.7°)} = 90.83e^{-j44.4°}$$

Check with MATLAB:

r1=12.58; r2=7.22; deg1=74.3; deg2=–118.7; r=r1*r2, deg=deg1+deg2

```
    r =

       90.8276

    deg =

       -44.4000
```

Check with the Simulink model of Figure 3.18[*]:

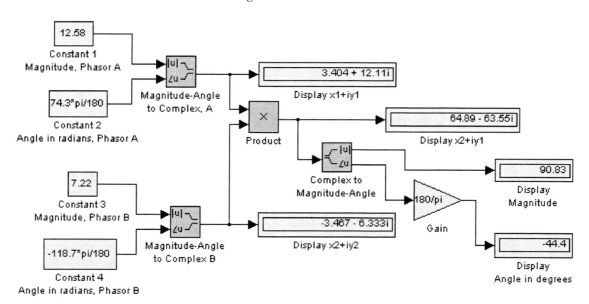

Figure 3.18. Simulink model for Example 3.10

---

[*] *It would certainly be a waste of time to use Simulink for such an application. It can be done faster with just MATLAB. The intent here is to introduce relevant Simulink blocks for more complicated models.*

To divide one phasor by another when both are expressed in exponential or polar form, we divide the magnitude of the dividend by the magnitude of the divisor, and we subtract the phase angle of the divisor from the phase angle of the dividend, that is, if

$$A = M\angle\theta \quad \text{and} \quad B = N\angle\phi$$

then,

$$\frac{A}{B} = \frac{M}{N}\angle(\theta - \phi) = \frac{Me^{j\theta}}{Ne^{j\phi}} = \frac{M}{N}e^{j(\theta - \phi)}$$ (3.84)

Example 3.11

Divide $A = 12.58\angle74.3°$ by $B = 7.22\angle-118.7°$

Solution:

Division in polar form yields

$$\frac{A}{B} = \frac{12.58\angle74.3°}{7.22\angle-118.7°} = 1.74\angle[74.3° - (-118.7°)] = 1.74\angle193° = 1.74\angle-167°$$

Division in exponential form yields

$$\frac{A}{B} = \frac{12.58e^{j74.3°}}{7.22e^{-j118.7°}} = 1.74e^{j74.3°}e^{j118.7°} = 1.74e^{j193°} = 1.74e^{-j167°}$$

Check with MATLAB:

r1=12.58; r2=7.22; deg1=74.3; deg2=−118.7; r=r1/r2, deg=deg1−deg2

```
r =

    1.7424

deg =

   193
```

Check with the Simulink model of Figure 3.19[*]:

---

* *Same comment as on the footnote of the previous page.*

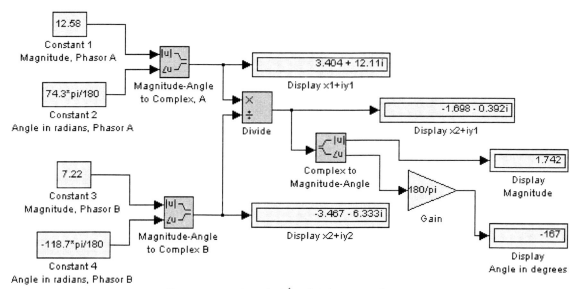

Figure 3.19. Simulink model for Example 3.11

3.9 Summary

- An alternating current (or voltage) alternates between positive and negative values at regularly recurring intervals of time.

- The period T of an alternating current or voltage is the smallest value of time which separates recurring values of the alternating waveform.

- Sine and cosine waveforms and these are referred to as sinusoids.

- The angular velocity ω is commonly known as angular or radian frequency and $\omega T = 2\pi$

- The term *frequency* in Hertz, denoted as Hz, is used to express the number of cycles per second. The frequency is denoted by the letter f and in terms of the period T, $f = 1/T$. The frequency f is often referred to as the cyclic frequency to distinguish it from the radian frequency ω.

- The cosine function leads (is ahead of) the sine function by $\pi/2$ radians or $90°$, and the sine function lags (is behind) the cosine function by $\pi/2$ radians or $90°$. Alternately, we say that the cosine and sine functions are out-of-phase by $90°$, or there is a phase angle of $90°$ between the cosine and sine functions.

- Two (or more) sinusoids can be out-of-phase by a phase angle other than $90°$.

- It is important to remember that when we say that one sinusoid leads or lags another sinusoid, these are of the same frequency since two sinusoids of different frequencies can never be in phase.

- It is customary to express the phase angle in degrees rather than in radians in a sinusoidal function. For example, we write $v(t) = 100\sin(2000\pi t - \pi/6)$ as $v(t) = 100\sin(2000\pi t - 30°)$

- When two sinusoids are to be compared in terms of their phase difference, these must first be written either both as cosine functions, or both as sine functions, and should also be written with positive amplitudes.

- A negative amplitude implies $180°$ phase shift.

- The radian is a circular angle subtended by an arc equal in length to the radius of the circle, whose radius is r units in length. The circumference of a circle is $2\pi r$.

- The notation $\cos^{-1}y$ or $\arccos y$ is used to denote an angle whose cosine is y. Thus, if $y = \cos x$, then $x = \cos^{-1}y$. These are called Inverse Trigonometric Functions.

- A phasor is a rotating vector expressed as a complex number where j is an operator that rotates a vector by $90°$ in a counterclockwise direction.

- Two phasors A *and* B where $A = a + jb$ and $B = c + jd$, are equal if and only if their real parts are equal and also their imaginary parts are equal. Thus, $A = B$ if and only if $a = c$ and $b = d$.

- The sum of two phasors has a real component equal to the sum of the real components, and an imaginary component equal to the sum of the imaginary components. For subtraction, we change the signs of the components of the subtrahend and we perform addition. Thus, if $A = a + jb$ and $B = c + jd$, then $A + B = (a + c) + j(b + d)$ and $A - B = (a - c) + j(b - d)$

- Phasors are multiplied using the rules of elementary algebra. If $A = a + jb$ and $B = c + jd$, then $A \cdot B = ac + jad + jbc - bd = (ac - bd) + j(ad + bc)$

- The conjugate of a phasor, denoted as A^*, is another phasor with the same real component, and with an imaginary component of opposite sign. Thus, if $A = a + jb$, then $A^* = a - jb$.

- When performing division of phasors, it is desirable to obtain the quotient separated into a real part and an imaginary part. This is achieved by multiplying the denominator by its conjugate. Thus, if $A = a + jb$ and $B = c + jd$, then,

$$\frac{A}{B} = \frac{a + jb}{c + jd} = \frac{(a + jb)(c - jd)}{(c + jd)(c - jd)} = \frac{(ac + bd) + j(bc - ad)}{c^2 + d^2} = \frac{(ac + bd)}{c^2 + d^2} + j\frac{(bc - ad)}{c^2 + d^2}$$

- The relations $e^{j\theta} = \cos\theta + j\sin\theta$ and $e^{-j\theta} = \cos\theta - j\sin\theta$ are known as the Euler's identities.

- To convert a phasor from rectangular to exponential form, we use the expression

$$a + jb = \sqrt{a^2 + b^2}\, e^{j\left(\tan^{-1}\frac{b}{a}\right)}$$

- To convert a phasor from exponential to rectangular form, we use the expressions

$$Ce^{j\theta} = C\cos\theta + jC\sin\theta$$

$$Ce^{-j\theta} = C\cos\theta - jC\sin\theta$$

- The polar form is essentially the same as the exponential form but the notation is different, that is,

$$Ce^{j\theta} = C\angle\theta$$

and it is important to remember that the phase angle θ is always measured with respect to the positive real axis, and rotates in the counterclockwise direction.

- The rectangular form is most useful when we add or subtract phasors; however, the exponential and polar forms are most convenient when we multiply or divide phasors.

- To multiply two phasors in exponential (or polar) form, we multiply the magnitudes and we add the phase angles, that is, if

$$A = M\angle\theta \text{ and } B = N\angle\phi$$

then,

$$AB = MN\angle(\theta + \phi) = Me^{j\theta}Ne^{j\phi} = MNe^{j(\theta + \phi)}$$

- To divide one phasor by another when both are expressed in exponential or polar form, we divide the magnitude of the dividend by the magnitude of the divisor, and we subtract the phase angle of the divisor from the phase angle of the dividend, that is, if

$$A = M\angle\theta \text{ and } B = N\angle\phi$$

then,

$$\frac{A}{B} = \frac{M}{N}\angle(\theta - \phi) = \frac{Me^{j\theta}}{Ne^{j\phi}} = \frac{M}{N}e^{j(\theta - \phi)}$$

3.10 Exercises

1. Perform the following operations, and check your answers with MATLAB.

   a. $(2-j4)+(3+j4)$ b. $(-3+j5)-(1+j6)$ c. $(2-j3)-(2-j3)^*$ d. $(3-j2)\cdot(3-j2)^*$

   e. $(2-j4)\cdot(3+j5)$ f. $(3-j2)\cdot(-2-j3)$ g. $(2-j4)\cdot(3+j5)\cdot(3-j2)\cdot(-2-j3)$

2. Perform the following operations, and check your answers with MATLAB.

   a. $\dfrac{22+j6}{3+j2}$ b. $\dfrac{8+j6}{-3-j}$ c. $\dfrac{120}{4-j10}$ d. $\dfrac{(3-j2)}{(3-j2)^*}$

3. Any phasor A can be expressed as

$$A = a+jb = r(\cos\theta + j\sin\theta) = re^{j\theta}$$

Using the identities $(re^{j\theta}) = r^n e^{jn\theta}$ or $\sqrt[n]{re^{j\theta}} = \sqrt[n]{r}\, e^{j\theta/n}$, compute:

   a. $\sqrt[6]{12+j5}$ b. $\sqrt[4]{100\sqrt{2}(1-j)}$

Check your answers with MATLAB

4. Compute the exponential and polar forms of

   a. $\dfrac{9+j5}{-4-j2}$ b. $\dfrac{-8+j3}{-2+j4}$

Check your answers with MATLAB.

5. Compute the rectangular form of

   a. $\dfrac{4\angle 30°}{5\angle -150°}$ b. $\dfrac{e^{j60°}}{-2e^{-j30°}}$

Check your answers with MATLAB

6. Find the real and imaginary components of $\dfrac{9-j4}{-5+jx}$

3.11 Solutions to End–of–Chapter Exercises

1.

   a. $(2 - j4) + (3 + j4) = 5 + 0 = 5$

   b. $(-3 + j5) - (1 + j6) = -4 - j$

   c. $(2 - j3) - (2 - j3)^* = (2 - j3) - (2 + j3) = 0 - j6$

   d. $(3 - j2) \cdot (3 - j2)^* = (3 - j2) \cdot (3 + j2) = 9 + j6 - j6 + 4 = 13$

   e. $(2 - j4) \cdot (3 + j5) = 6 + j10 - j12 + 20 = 26 - j2$

   f. $(3 - j2) \cdot (-2 - j3) = -6 - j9 + j4 - 6 = -12 - j5$

   g.
$$\begin{aligned}
(2 - j4) \cdot (3 + j5) \cdot (3 - j2) \cdot (-2 - j3) &= (6 + j10 - j12 + 20) \cdot (-6 - j9 + j4 - 6) \\
&= (26 - j2) \cdot (-12 - j5) \\
&= -312 - j130 + j24 - 10 = -322 - j106
\end{aligned}$$

Check with MATLAB:

```
(2–4j)+(3+4j), (–3+5j)–(1+6j), (2–3j)–(2+3j), (3–2j)*(3+2j),...
(2–4j)*(3+5j), (3–2j)*(–2–3j), (2–4j)*(3+5j)*(3–2j)*(–2–3j)
ans =

     5
ans =

  -4.0000 - 1.0000i
ans =

   0 - 6.0000i
ans =

     13
ans =

  26.0000 - 2.0000i
ans =

 -12.0000 - 5.0000i
ans =

 -3.2200e+002 - 1.0600e+002i
```

2.

a. $\dfrac{22+j6}{3+j2} = \dfrac{22+j6}{3+j2} \cdot \dfrac{3-j2}{3-j2} = \dfrac{66-j44+j18+12}{3^2+2^2} = \dfrac{78-j26}{13} = 6-j2$

b. $\dfrac{8+j6}{-3-j} = \dfrac{8+j6}{-3-j} \cdot \dfrac{-3+j}{-3+j} = \dfrac{-24+j8-j18-6}{3^2+1^2} = \dfrac{-30-j10}{10} = -3-j$

c. $\dfrac{120}{4-j10} = \dfrac{120}{4-j10} \cdot \dfrac{4+j10}{4+j10} = \dfrac{480+j1200}{4^2+10^2} = \dfrac{480}{116} + j \cdot \dfrac{1200}{116} = \dfrac{120}{29} + j \cdot \dfrac{300}{29}$

d. $\dfrac{(3-j2)}{(3-j2)^*} = \dfrac{(3-j2)}{(3+j2)} \cdot \dfrac{(3-j2)}{(3-j2)} = \dfrac{9-j6-j6-4}{3^2+2^2} = \dfrac{5-j12}{13} = \dfrac{5}{13} - j \cdot \dfrac{12}{13}$

Check with MATLAB:

22+6j)/(3+2j), (8+6j)/(–3–j), 120/(4–10j), (3–2j)/(3+2j)

ans =

    6 - 2i

ans =

    -3 - 1i

ans =

    120/29 + 300/29i

ans =

    5/13 - 12/13i

3.

a.

$\sqrt[6]{12+j5} = \sqrt[6]{13e^{j0.395}} = \sqrt[6]{13} \cdot e^{j0.3948/6} = 13^{1/6} \cdot e^{j0.0658}$
$= 1.5334(\cos 0.0658 + j\sin 0.0658) = 1.53 + j0.10$

b.

$\sqrt[4]{100\sqrt{2}(1-j)} = \sqrt[4]{100\sqrt{2} \cdot \sqrt{2}e^{-j\pi/4}} = (100\sqrt{2} \cdot \sqrt{2}e^{-j\pi/4})^{1/4} = (100\sqrt{2})^{1/4} \cdot \sqrt{2}^{1/4} e^{-j\pi/16}$
$= (3.4485 \times 1.0905)(\cos(\pi/16) - j\sin(\pi/16)) = 3.6883 - j0.7337$

Check with MATLAB:

(12+5j)^(1/6), (100*sqrt(2)*(1–j))^(1/4)

```
ans =

    1.5301 + 0.1008i

ans =

    3.6883 - 0.7337i
```

4.

a.

$$\frac{9+j5}{-4-j2} = \frac{\sqrt{9^2+5^2} \cdot e^{j\tan^{-1}(5/9)}}{\sqrt{4^2+2^2} \cdot e^{j\tan^{-1}(-2/-4)}} = \frac{\sqrt{106} \cdot e^{j0.5071}}{\sqrt{20} \cdot e^{j3.6052}} = 2.3022e^{-j3.0981}$$

$$= 2.3022e^{-j177.5081°} = 2.3022\angle-177.5081°$$

b.

$$\frac{-8+j3}{-2+j4} = \frac{\sqrt{8^2+3^2} \cdot e^{j\tan^{-1}(3/-8)}}{\sqrt{2^2+4^2} \cdot e^{j\tan^{-1}(4/-2)}} = \frac{\sqrt{73}}{\sqrt{20}} \cdot \frac{e^{-j0.3588}}{e^{-j1.1071}} = 1.9105e^{j0.7483}$$

$$= 1.9105e^{j42.8744°} = 1.9105\angle42.8744°$$

Check with MATLAB:

x=(9+5j)/(−4−2j); abs(x), angle(x)*180/pi,...

y=(−8+3j)/(−2+4j); abs(y), angle(y)*180/pi

```
ans =

    2.3022

ans =

  -177.5104

ans =

    1.9105

ans =

    42.8789
```

5.

a.

$$\frac{4\angle30°}{5\angle-150°} = (4/5)\angle180° = -0.8$$

b. $$\frac{e^{j60°}}{-2e^{-j30°}} = -0.5e^{j90°} = -0.5(\cos90° + j\sin90°) = -0.5(0+j) = -j0.5$$

Check with MATLAB:

4*(cos(pi/6)+sin(pi/6)*j)/(5*(cos(–5*pi/6)+sin(–5*pi/6)*j)),...

exp(pi*j/3)/(–2*exp(–pi*j/6))

```
ans =

  -0.8000 - 0.0000i

ans =

  -0.0000 - 0.5000i
```

6.

$$\frac{9-j4}{-5+jx} = \frac{9-j4}{-5+jx} \cdot \frac{-5-jx}{-5-jx} = \frac{-45-j9x+j20-4x}{5^2+x^2} = \frac{-4x-45}{x^2+25} + j\frac{-9x+20}{x^2+25}$$

NOTES

Chapter 4

Matrices and Determinants

T his chapter is an introduction to matrices and matrix operations. Determinants, Cramer's rule, and Gauss's elimination method are introduced. Some definitions and examples are not applicable to subsequent material presented in this text, but are included for subject continuity, and reference to more advance topics in matrix theory. These are denoted with a dagger (†) and may be skipped.

4.1 Matrix Definition

A *matrix* is a rectangular array of numbers such as those shown below.

$$\begin{bmatrix} 2 & 3 & 7 \\ 1 & -1 & 5 \end{bmatrix} \quad \text{or} \quad \begin{bmatrix} 1 & 3 & 1 \\ -2 & 1 & -5 \\ 4 & -7 & 6 \end{bmatrix}$$

In general form, a matrix A is denoted as

$$A = \begin{bmatrix} a_{11} & a_{12} & a_{13} & \cdots & a_{1n} \\ a_{21} & a_{22} & a_{23} & \cdots & a_{2n} \\ a_{31} & a_{32} & a_{33} & \cdots & a_{3n} \\ \cdots & \cdots & \cdots & \cdots & \cdots \\ a_{m1} & a_{m2} & a_{m3} & \cdots & a_{mn} \end{bmatrix} \qquad (4.1)$$

The numbers a_{ij} are the *elements* of the matrix where the index i indicates the row, and j indicates the column in which each element is positioned. Thus, a_{43} indicates the element positioned in the fourth row and third column.

A matrix of m rows and n columns is said to be of $m \times n$ *order matrix*.

If $m = n$, the matrix is said to be a *square matrix of order* m (or n). Thus, if a matrix has five rows and five columns, it is said to be a square matrix of order 5.

In a square matrix, the elements a_{11}, a_{22}, a_{33}, ..., a_{nn} are called the *main diagonal elements*. Alternately, we say that the matrix elements a_{11}, a_{22}, a_{33}, ..., a_{nn} , are located on the *main diagonal*.

Numerical Analysis Using MATLAB® and Excel®, Third Edition 4-1
Copyright © Orchard Publications

† The sum of the diagonal elements of a square matrix A is called the *trace*[*] of A.

† A matrix in which every element is zero, is called a *zero matrix*.

4.2 Matrix Operations

Two matrices $A = \begin{bmatrix} a_{ij} \end{bmatrix}$ and $B = \begin{bmatrix} b_{ij} \end{bmatrix}$ are equal, that is, $A = B$, if and only if

$$a_{ij} = b_{ij} \qquad i = 1, 2, 3, ..., m \qquad j = 1, 2, 3, ..., n \qquad (4.2)$$

Two matrices are said to be *conformable for addition (subtraction)*, if they are of the same order $m \times n$.

If $A = \begin{bmatrix} a_{ij} \end{bmatrix}$ and $B = \begin{bmatrix} b_{ij} \end{bmatrix}$ are conformable for addition (subtraction), their sum (difference) will be another matrix C with the same order as A and B, where each element of C is the sum (difference) of the corresponding elements of A and B, that is,

$$C = A \pm B = [a_{ij} \pm b_{ij}] \qquad (4.3)$$

Example 4.1

Compute $A + B$ and $A - B$ given that

$$A = \begin{bmatrix} 1 & 2 & 3 \\ 0 & 1 & 4 \end{bmatrix} \text{ and } B = \begin{bmatrix} 2 & 3 & 0 \\ -1 & 2 & 5 \end{bmatrix}$$

Solution:

$$A + B = \begin{bmatrix} 1+2 & 2+3 & 3+0 \\ 0-1 & 1+2 & 4+5 \end{bmatrix} = \begin{bmatrix} 3 & 5 & 3 \\ -1 & 3 & 9 \end{bmatrix}$$

and

$$A - B = \begin{bmatrix} 1-2 & 2-3 & 3-0 \\ 0+1 & 1-2 & 4-5 \end{bmatrix} = \begin{bmatrix} -1 & -1 & 3 \\ 1 & -1 & -1 \end{bmatrix}$$

Check with MATLAB:

```
A=[1 2 3; 0 1 4]; B=[2 3 0; -1 2 5];   % Define matrices A and B
A+B                                     % Add A and B
```

---

[*] *Henceforth, all paragraphs and topics preceded by a dagger (†) may be skipped. These are discussed in matrix theory textbooks.*

```
ans =
     3     5     3
    -1     3     9
```

A–B   % Subtract B from A

```
ans =
    -1    -1     3
     1    -1    -1
```

If k is any scalar (a positive or negative number), and not $[k]$ which is a 1×1 matrix, then multiplication of a matrix A by the scalar k, is the multiplication of every element of A by k.

Example 4.2

Multiply the matrix

$$A = \begin{bmatrix} 1 & -2 \\ 2 & 3 \end{bmatrix}$$

by (a) $k_1 = 5$ and (b) $k_2 = -3 + j2$

Solution:

a.

$$k_1 \cdot A = 5 \times \begin{bmatrix} 1 & -2 \\ 2 & 3 \end{bmatrix} = \begin{bmatrix} 5 \times 1 & 5 \times (-2) \\ 5 \times 2 & 5 \times 3 \end{bmatrix} = \begin{bmatrix} 5 & -10 \\ 10 & 15 \end{bmatrix}$$

b.

$$k_2 \cdot A = (-3 + j2) \times \begin{bmatrix} 1 & -2 \\ 2 & 3 \end{bmatrix} = \begin{bmatrix} (-3 + j2) \times 1 & (-3 + j2) \times (-2) \\ (-3 + j2) \times 2 & (-3 + j2) \times 3 \end{bmatrix} = \begin{bmatrix} -3 + j2 & 6 - j4 \\ -6 + j4 & -9 + j6 \end{bmatrix}$$

Check with MATLAB:

```
k1=5; k2=(-3 + 2*j);        % Define scalars k1 and k2
A=[1 -2; 2 3];              % Define matrix A
k1*A                        % Multiply matrix A by constant k1

ans =
     5    -10
    10     15

k2*A                        %Multiply matrix A by constant k2
```

```
ans =
   -3.0000+ 2.0000i    6.0000- 4.0000i
   -6.0000+ 4.0000i   -9.0000+ 6.0000i
```

Two matrices A and B are said to be *conformable for multiplication* $A \cdot B$ in that order, only when the number of columns of matrix A is equal to the number of rows of matrix B. That is, the product $A \cdot B$ (but not $B \cdot A$) is conformable for multiplication only if A is an $m \times p$ and matrix B is an $p \times n$ matrix. The product $A \cdot B$ will then be an $m \times n$ matrix. A convenient way to determine if two matrices are conformable for multiplication is to write the dimensions of the two matrices side–by–side as shown below.

Shows that A and B are conformable for multiplication

Indicates the dimension of the product $A \cdot B$

For the product $B \cdot A$ we have:

Here, B and A are not conformable for multiplication

$$B \quad A$$
$$p \times n \quad m \times p$$

For matrix multiplication, the operation is row by column. Thus, to obtain the product $A \cdot B$, we multiply each element of a row of A by the corresponding element of a column of B; then, we add these products.

Example 4.3

Given that

$$C = \begin{bmatrix} 2 & 3 & 4 \end{bmatrix} \text{ and } D = \begin{bmatrix} 1 \\ -1 \\ 2 \end{bmatrix}$$

compute the products $C \cdot D$ and $D \cdot C$

Solution:

The dimensions of matrices C and D are respectively 1×3 3×1; therefore the product $C \cdot D$ is feasible, and will result in a 1×1, that is,

$$C \cdot D = \begin{bmatrix} 2 & 3 & 4 \end{bmatrix} \begin{bmatrix} 1 \\ -1 \\ 2 \end{bmatrix} = \begin{bmatrix} (2) \cdot (1) + (3) \cdot (-1) + (4) \cdot (2) \end{bmatrix} = \begin{bmatrix} 7 \end{bmatrix}$$

The dimensions for D and C are respectively 3×1 1×3 and therefore, the product $D \cdot C$ is also feasible. Multiplication of these will produce a 3×3 matrix as follows.

$$D \cdot C = \begin{bmatrix} 1 \\ -1 \\ 2 \end{bmatrix} \begin{bmatrix} 2 & 3 & 4 \end{bmatrix} = \begin{bmatrix} (1) \cdot (2) & (1) \cdot (3) & (1) \cdot (4) \\ (-1) \cdot (2) & (-1) \cdot (3) & (-1) \cdot (4) \\ (2) \cdot (2) & (2) \cdot (3) & (2) \cdot (4) \end{bmatrix} = \begin{bmatrix} 2 & 3 & 4 \\ -2 & -3 & -4 \\ 4 & 6 & 8 \end{bmatrix}$$

Check with MATLAB:

```
C=[2 3 4]; D=[1; -1; 2];        % Define matrices C and D
C*D                             % Multiply C by D

ans =
      7

D*C                             % Multiply D by C

ans =
      2       3       4
     -2      -3      -4
      4       6       8
```

Division of one matrix by another, is not defined. However, an equivalent operation exists, and it will become apparent later in this chapter, when we discuss the inverse of a matrix.

4.3 Special Forms of Matrices

† A square matrix is said to be *upper triangular* when all the elements below the diagonal are zero. The matrix A below is an upper triangular matrix.

$$A = \begin{bmatrix} a_{11} & a_{12} & a_{13} & \cdots & a_{1n} \\ 0 & a_{22} & a_{23} & \cdots & a_{2n} \\ 0 & 0 & \ddots & \cdots & \cdots \\ \cdots & \cdots & 0 & \ddots & \cdots \\ 0 & 0 & 0 & \cdots & a_{mn} \end{bmatrix} \qquad (4.4)$$

In an upper triangular matrix, not all elements above the diagonal need to be non–zero. For applications, refer to Chapter 14.

† A square matrix is said to be *lower triangular*, when all the elements above the diagonal are zero. The matrix B below is a lower triangular matrix. For applications, refer to Chapter 14.

$$B = \begin{bmatrix} a_{11} & 0 & 0 & \dots & 0 \\ a_{21} & a_{22} & 0 & \dots & 0 \\ \dots & \dots & \ddots & 0 & 0 \\ \dots & \dots & \dots & \ddots & 0 \\ a_{m1} & a_{m2} & a_{m3} & \dots & a_{mn} \end{bmatrix} \tag{4.5}$$

In a lower triangular matrix, not all elements below the diagonal need to be non–zero.

† A square matrix is said to be *diagonal*, if all elements are zero, except those in the diagonal. The matrix C below is a diagonal matrix.

$$C = \begin{bmatrix} a_{11} & 0 & 0 & \dots & 0 \\ 0 & a_{22} & 0 & \dots & 0 \\ 0 & 0 & \ddots & 0 & 0 \\ 0 & 0 & 0 & \ddots & 0 \\ 0 & 0 & 0 & \dots & a_{nn} \end{bmatrix} \tag{4.6}$$

† A diagonal matrix is called a *scalar matrix*, if $a_{11} = a_{22} = a_{33} = \dots = a_{nn} = k$ where k is a scalar. The matrix D below is a scalar matrix with $k = 4$.

$$D = \begin{bmatrix} 4 & 0 & 0 & 0 \\ 0 & 4 & 0 & 0 \\ 0 & 0 & 4 & 0 \\ 0 & 0 & 0 & 4 \end{bmatrix} \tag{4.7}$$

A scalar matrix with $k = 1$, is called an *identity matrix* I. Shown below are 2×2, 3×3, and 4×4 identity matrices.

$$\begin{bmatrix} 1 & 0 \\ 0 & 1 \end{bmatrix} \quad \begin{bmatrix} 1 & 0 & 0 \\ 0 & 1 & 0 \\ 0 & 0 & 1 \end{bmatrix} \quad \begin{bmatrix} 1 & 0 & 0 & 0 \\ 0 & 1 & 0 & 0 \\ 0 & 0 & 1 & 0 \\ 0 & 0 & 0 & 1 \end{bmatrix} \tag{4.8}$$

The MATLAB **eye(n)** function displays an $n \times n$ identity matrix. For example, eye(4)% Display a 4 by 4 identity matrix

ans =

1	0	0	0
0	1	0	0
0	0	1	0
0	0	0	1

Likewise, the **eye(size(A))** function, produces an identity matrix whose size is the same as matrix A. For example, let A be defined as

A=[1 3 1; –2 1 –5; 4 –7 6]   % Define matrix A

A =

1	3	1
-2	1	-5
4	-7	6

then,

eye(size(A))

displays

ans =

1	0	0
0	1	0
0	0	1

† The *transpose of a matrix* A, denoted as A^T, is the matrix that is obtained when the rows and columns of matrix A are interchanged. For example, if

$$A = \begin{bmatrix} 1 & 2 & 3 \\ 4 & 5 & 6 \end{bmatrix} \text{ then } A^T = \begin{bmatrix} 1 & 4 \\ 2 & 5 \\ 3 & 6 \end{bmatrix} \qquad (4.9)$$

In MATLAB we use the apostrophe (') symbol to denote and obtain the transpose of a matrix. Thus, for the above example,

A=[1 2 3; 4 5 6] % Define matrix A

A =

1	2	3
4	5	6

A'% Display the transpose of A

```
ans =
     1     4
     2     5
     3     6
```

† A *symmetric matrix* A, is one such that $A^T = A$, that is, the transpose of a matrix A is the same as A. An example of a symmetric matrix is shown below.

$$A = \begin{bmatrix} 1 & 2 & 3 \\ 2 & 4 & -5 \\ 3 & -5 & 6 \end{bmatrix} \qquad A^T = \begin{bmatrix} 1 & 2 & 3 \\ 2 & 4 & -5 \\ 3 & -5 & 6 \end{bmatrix} = A \qquad (4.10)$$

† If a matrix A has complex numbers as elements, the matrix obtained from A by replacing each element by its conjugate, is called the *conjugate of* A, and it is denoted as A*.

An example is shown below.

$$A = \begin{bmatrix} 1 + j2 & j \\ 3 & 2 - j3 \end{bmatrix} \qquad A* = \begin{bmatrix} 1 - j2 & -j \\ 3 & 2 + j3 \end{bmatrix}$$

† MATLAB has two built-in functions which compute the complex conjugate of a number. The first, **conj(x)**, computes the complex conjugate of any complex number, and the second, **conj(A)**, computes the conjugate of a matrix A. Using MATLAB with the matrix A defined as above, we obtain

A = [1+2j j; 3 2–3j]   % Define and display matrix A

```
A =
   1.0000 + 2.0000i        0 + 1.0000i
   3.0000              2.0000 - 3.0000i
```

conj_A=conj(A)   % Compute and display the conjugate of A

```
conj_A =
   1.0000 - 2.0000i        0 - 1.0000i
   3.0000              2.0000 + 3.0000i
```

† A square matrix A such that $A^T = -A$, is called *skew-symmetric*. For example,

$$A = \begin{bmatrix} 0 & 2 & -3 \\ -2 & 0 & -4 \\ 3 & 4 & 0 \end{bmatrix} \qquad A^T = \begin{bmatrix} 0 & -2 & 3 \\ 2 & 0 & 4 \\ -3 & -4 & 0 \end{bmatrix} = -A$$

Therefore, matrix A above is skew symmetric.

† A square matrix A such that $A^{T*} = A$, is called *Hermitian*. For example,

$$A = \begin{bmatrix} 1 & 1-j & 2 \\ 1+j & 3 & j \\ 2 & -j & 0 \end{bmatrix} \quad A^T = \begin{bmatrix} 1 & 1+j & 2 \\ 1-j & 3 & -j \\ 2 & j & 0 \end{bmatrix} \quad A^{T*} = \begin{bmatrix} 1 & 1+j & 2 \\ 1-j & 3 & -j \\ 2 & j & 0 \end{bmatrix} = A$$

Therefore, matrix A above is Hermitian.

† A square matrix A such that $A^{T*} = -A$, is called *skew–Hermitian*. For example,

$$A = \begin{bmatrix} j & 1-j & 2 \\ -1-j & 3j & j \\ -2 & j & 0 \end{bmatrix} \quad A^T = \begin{bmatrix} j & -1-j & -2 \\ 1-j & 3j & j \\ 2 & j & 0 \end{bmatrix} \quad A^{T*} = \begin{bmatrix} -j & -1+j & -2 \\ 1+j & -3j & -j \\ 2 & -j & 0 \end{bmatrix} = -A$$

Therefore, matrix A above is skew–Hermitian.

4.4 Determinants

Let matrix A be defined as the square matrix

$$A = \begin{bmatrix} a_{11} & a_{12} & a_{13} & \cdots & a_{1n} \\ a_{21} & a_{22} & a_{23} & \cdots & a_{2n} \\ a_{31} & a_{32} & a_{33} & \cdots & a_{3n} \\ \cdots & \cdots & \cdots & \cdots & \cdots \\ a_{n1} & a_{n2} & a_{n3} & \cdots & a_{nn} \end{bmatrix} \tag{4.11}$$

then, the *determinant of* A, denoted as $\det A$, is defined as

$$\det A = a_{11}a_{22}a_{33}\cdots a_{nn} + a_{12}a_{23}a_{34}\cdots a_{n1} + a_{13}a_{24}a_{35}\cdots a_{n2} + \cdots \tag{4.12}$$
$$-a_{n1}\cdots a_{22}a_{13}\cdots -a_{n2}\cdots a_{23}a_{14} - a_{n3}\cdots a_{24}a_{15} - \cdots$$

The determinant of a square matrix of order n is referred to as *determinant of order* n.

Let A be a *determinant of order* 2, that is,

$$A = \begin{bmatrix} a_{11} & a_{12} \\ a_{21} & a_{22} \end{bmatrix} \tag{4.13}$$

Then,

$$\det A = a_{11}a_{22} - a_{21}a_{12} \qquad (4.14)$$

Example 4.4

Given that

$$A = \begin{bmatrix} 1 & 2 \\ 3 & 4 \end{bmatrix} \text{ and } B = \begin{bmatrix} 2 & -1 \\ 2 & 0 \end{bmatrix}$$

compute $\det A$ and $\det B$.

Solution:

$$\det A = 1 \cdot 4 - 3 \cdot 2 = 4 - 6 = -2$$
$$\det A = 2 \cdot 0 - 2 \cdot (-1) = 0 - (-2) = 2$$

Check with MATLAB:

```
A=[1  2; 3  4]; B=[2  -1; 2  0];    % Define matrices A and B
det(A)                               % Compute the determinant of A

ans =
    -2

det(B)                               % Compute the determinant of B

ans =
     2
```

While MATLAB has the built–in function **det(A)** for computing the determinant of a matrix **A**, this function is not included in the MATLAB Run–Time Function Library List that is used with the Simulink **Embedded MATLAB Function** block.[*] The MATLAB user–defined function file below can be used to compute the determinant of a 2×2 matrix.

```
% This file computes the determinant of a 2x2 matrix
% It must be saved as function (user defined) file
% det2x2.m in the current Work Directory. Make sure
% that his directory is added to MATLAB's search
% path accessed from the Editor Window as File>Set Path>
% Add Folder. It is highly recommended that this
% function file is created in MATLAB's Editor Window.
%
function y=det2x2(A);
```

---

[*] *For an example using this block, please refer to Introduction to Simulink with Engineering Applications, ISBN 0–9744239–7–1, Page 16–3.*

```
y=A(1,1)*A(2,2)–A(1,2)*A(2,1);
%
% To run this program, define the 2x2 matrix in
% MATLAB's Command Window as A=[....] and then
% type det2x2(A) at the command prompt.
```

Let A be a matrix of order 3, that is,

$$A = \begin{bmatrix} a_{11} & a_{12} & a_{13} \\ a_{21} & a_{22} & a_{23} \\ a_{31} & a_{32} & a_{33} \end{bmatrix} \tag{4.15}$$

then, detA is found from

$$\det A = a_{11}a_{22}a_{33} + a_{12}a_{23}a_{31} + a_{11}a_{22}a_{33}$$
$$-a_{11}a_{22}a_{33} - a_{11}a_{22}a_{33} - a_{11}a_{22}a_{33} \tag{4.16}$$

A convenient method to evaluate the determinant of order 3, is to write the first two columns to the right of the 3×3 matrix, and add the products formed by the diagonals from upper left to lower right; then subtract the products formed by the diagonals from lower left to upper right as shown on the diagram of the next page. When this is done properly, we obtain (4.16) above.

This method works only with second and third order determinants. To evaluate higher order determinants, we must first compute the *cofactors*; these will be defined shortly.

Example 4.5

Compute detA and detB given that

$$A = \begin{bmatrix} 2 & 3 & 5 \\ 1 & 0 & 1 \\ 2 & 1 & 0 \end{bmatrix} \text{ and } B = \begin{bmatrix} 2 & -3 & -4 \\ 1 & 0 & -2 \\ 0 & -5 & -6 \end{bmatrix}$$

Solution:

$$\det A = \begin{matrix} 2 & 3 & 5 & 2 & 3 \\ 1 & 0 & 1 & 1 & 0 \\ 2 & 1 & 0 & 2 & 1 \end{matrix}$$

or

$$\det A = (2 \times 0 \times 0) + (3 \times 1 \times 1) + (5 \times 1 \times 1)$$
$$- (2 \times 0 \times 5) - (1 \times 1 \times 2) - (0 \times 1 \times 3) = 11 - 2 = 9$$

Likewise,

$$\det B = \begin{matrix} 2 & -3 & -4 & 2 & -3 \\ 1 & 0 & -2 & 1 & -2 \\ 0 & -5 & -6 & 2 & -6 \end{matrix}$$

or

$$\det B = [2 \times 0 \times (-6)] + [(-3) \times (-2) \times 0] + [(-4) \times 1 \times (-5)]$$
$$- [0 \times 0 \times (-4)] - [(-5) \times (-2) \times 2] - [(-6) \times 1 \times (-3)] = 20 - 38 = -18$$

Check with MATLAB:

A=[2 3 5; 1 0 1; 2 1 0]; det(A)   % Define matrix A and compute detA

ans =
      9

B=[2 −3 −4; 1 0 −2; 0 −5 −6]; det(B)   % Define matrix B and compute detB

ans =
    −18

The MATLAB user–defined function file below can be used to compute the determinant of a 3×3 matrix.

```
% This file computes the determinant of a 3x3 matrix
% It must be saved as function (user defined) file
% det3x3.m in the current Work Directory. Make sure
% that his directory is added to MATLAB's search
% path accessed from the Editor Window as File>Set Path>
% Add Folder. It is highly recommended that this
% function file is created in MATLAB's Editor Window.
%
function y=det3x3(A);
y=A(1,1)*A(2,2)*A(3,3)+A(1,2)*A(2,3)*A(3,1)+A(1,3)*A(2,1)*A(3,2)...
   −A(3,1)*A(2,2)*A(1,3)−A(3,2)*A(2,3)*A(1,1)−A(3,3)*A(2,1)*A(1,2);
%
% To run this program, define the 3x3 matrix in
% MATLAB's Command Window as A=[....] and then
% type det3x3(A) at the command prompt.
```

4.5 Minors and Cofactors

Let matrix A be defined as the square matrix of order n as shown below.

$$A = \begin{bmatrix} a_{11} & a_{12} & a_{13} & \cdots & a_{1n} \\ a_{21} & a_{22} & a_{23} & \cdots & a_{2n} \\ a_{31} & a_{32} & a_{33} & \cdots & a_{3n} \\ \cdots & \cdots & \cdots & \cdots & \cdots \\ a_{n1} & a_{n2} & a_{n3} & \cdots & a_{nn} \end{bmatrix} \qquad (4.17)$$

If we remove the elements of its ith row, and jth column, the determinant of the remaining n − 1 square matrix is called the *minor of determinant* A, and it is denoted as $\begin{bmatrix} M_{ij} \end{bmatrix}$.

The signed minor $(-1)^{i+j}\begin{bmatrix} M_{ij} \end{bmatrix}$ is called the *cofactor* of a_{ij} and it is denoted as α_{ij}.

Example 4.6

Given that

$$A = \begin{bmatrix} a_{11} & a_{12} & a_{13} \\ a_{21} & a_{22} & a_{23} \\ a_{31} & a_{32} & a_{33} \end{bmatrix} \qquad (4.18)$$

compute the minors $\begin{bmatrix} M_{11} \end{bmatrix}$, $\begin{bmatrix} M_{12} \end{bmatrix}$, $\begin{bmatrix} M_{13} \end{bmatrix}$ and the cofactors α_{11}, α_{12} and α_{13}.

Solution:

$$\begin{bmatrix} M_{11} \end{bmatrix} = \begin{bmatrix} a_{22} & a_{23} \\ a_{32} & a_{33} \end{bmatrix} \qquad \begin{bmatrix} M_{12} \end{bmatrix} = \begin{bmatrix} a_{21} & a_{23} \\ a_{31} & a_{33} \end{bmatrix} \qquad \begin{bmatrix} M_{11} \end{bmatrix} = \begin{bmatrix} a_{21} & a_{22} \\ a_{31} & a_{32} \end{bmatrix}$$

and

$$\alpha_{11} = (-1)^{1+1}\begin{bmatrix} M_{11} \end{bmatrix} = \begin{bmatrix} M_{11} \end{bmatrix} \qquad \alpha_{12} = (-1)^{1+2}\begin{bmatrix} M_{12} \end{bmatrix} = -\begin{bmatrix} M_{12} \end{bmatrix} \qquad \alpha_{13} = \begin{bmatrix} M_{13} \end{bmatrix} = (-1)^{1+3}\begin{bmatrix} M_{13} \end{bmatrix}$$

The remaining minors

$$\begin{bmatrix} M_{21} \end{bmatrix}, \quad \begin{bmatrix} M_{22} \end{bmatrix}, \quad \begin{bmatrix} M_{23} \end{bmatrix}, \quad \begin{bmatrix} M_{31} \end{bmatrix}, \quad \begin{bmatrix} M_{32} \end{bmatrix}, \quad \begin{bmatrix} M_{33} \end{bmatrix}$$

and cofactors

$$\alpha_{21}, \alpha_{22}, \alpha_{23}, \alpha_{31}, \alpha_{32}, \text{ and } \alpha_{33}$$

are defined similarly.

Example 4.7

Given that

$$A = \begin{bmatrix} 1 & 2 & -3 \\ 2 & -4 & 2 \\ -1 & 2 & -6 \end{bmatrix} \tag{4.19}$$

compute its cofactors.

Solution:

$$\alpha_{11} = (-1)^{1+1} \begin{bmatrix} -4 & 2 \\ 2 & -6 \end{bmatrix} = 20 \qquad \alpha_{12} = (-1)^{1+2} \begin{bmatrix} 2 & 2 \\ -1 & -6 \end{bmatrix} = 10 \tag{4.20}$$

$$\alpha_{13} = (-1)^{1+3} \begin{bmatrix} 2 & -4 \\ -1 & 2 \end{bmatrix} = 0 \qquad \alpha_{21} = (-1)^{2+1} \begin{bmatrix} 2 & -3 \\ 2 & -6 \end{bmatrix} = 6 \tag{4.21}$$

$$\alpha_{22} = (-1)^{2+2} \begin{bmatrix} 1 & -3 \\ -1 & -6 \end{bmatrix} = -9 \qquad \alpha_{23} = (-1)^{2+3} \begin{bmatrix} 1 & 2 \\ -1 & 2 \end{bmatrix} = -4 \tag{4.22}$$

$$\alpha_{31} = (-1)^{3+1} \begin{bmatrix} 2 & -3 \\ -4 & 2 \end{bmatrix} = -8, \qquad \alpha_{32} = (-1)^{3+2} \begin{bmatrix} 1 & -3 \\ 2 & 2 \end{bmatrix} = -8 \tag{4.23}$$

$$\alpha_{33} = (-1)^{3+3} \begin{bmatrix} 1 & 2 \\ 2 & -4 \end{bmatrix} = -8 \tag{4.24}$$

It is useful to remember that the signs of the cofactors follow the pattern

$$\begin{matrix} + & - & + & - & + \\ - & + & - & + & - \\ + & - & + & - & + \\ - & + & - & + & - \\ + & - & + & - & + \end{matrix}$$

that is, the cofactors on the diagonals have the same sign as their minors.

Let A be a square matrix of any size; the value of the determinant of A is the sum of the products obtained by multiplying each element of *any* row or *any* column by its cofactor.

Example 4.8

Compute the determinant of A using the elements of the first row.

$$A = \begin{bmatrix} 1 & 2 & -3 \\ 2 & -4 & 2 \\ -1 & 2 & -6 \end{bmatrix} \tag{4.25}$$

Solution:

$$\det A = 1 \begin{bmatrix} -4 & 2 \\ 2 & -6 \end{bmatrix} - 2 \begin{bmatrix} 2 & 2 \\ -1 & -6 \end{bmatrix} - 3 \begin{bmatrix} 2 & -4 \\ -1 & 2 \end{bmatrix} = 1 \times 20 - 2 \times (-10) - 3 \times 0 = 40$$

Check with MATLAB:

A=[1 2 −3; 2 −4 2; −1 2 −6]; det(A)   % Define matrix A and compute detA

ans =
        40

The MATLAB user–defined function file below can be used to compute the determinant of a 4×4 matrix.

We must use the above procedure to find the determinant of a matrix A of order *4* or higher. Thus, a fourth–order determinant can first be expressed as the sum of the products of the elements of its first row by its cofactor as shown below.

$$A = \begin{bmatrix} a_{11} & a_{12} & a_{13} & a_{14} \\ a_{21} & a_{22} & a_{23} & a_{24} \\ a_{31} & a_{32} & a_{33} & a_{34} \\ a_{41} & a_{42} & a_{43} & a_{44} \end{bmatrix} = a_{11} \begin{bmatrix} a_{22} & a_{23} & a_{24} \\ a_{32} & a_{33} & a_{34} \\ a_{42} & a_{43} & a_{44} \end{bmatrix} - a_{21} \begin{bmatrix} a_{12} & a_{13} & a_{14} \\ a_{32} & a_{33} & a_{34} \\ a_{42} & a_{43} & a_{44} \end{bmatrix} \tag{4.26}$$

$$+ a_{31} \begin{bmatrix} a_{12} & a_{13} & a_{14} \\ a_{22} & a_{23} & a_{24} \\ a_{42} & a_{43} & a_{44} \end{bmatrix} - a_{41} \begin{bmatrix} a_{12} & a_{13} & a_{14} \\ a_{22} & a_{23} & a_{24} \\ a_{32} & a_{33} & a_{34} \end{bmatrix}$$

Determinants of order five or higher can be evaluated similarly.

Example 4.9

Compute the value of the determinant

$$A = \begin{bmatrix} 2 & -1 & 0 & -3 \\ -1 & 1 & 0 & -1 \\ 4 & 0 & 3 & -2 \\ -3 & 0 & 0 & 1 \end{bmatrix} \qquad (4.27)$$

Solution:

Using the above procedure, we will multiply each element of the first column by its cofactor. Then,

$$A = 2\underbrace{\begin{bmatrix} 1 & 0 & -1 \\ 0 & 3 & -2 \\ 0 & 0 & 1 \end{bmatrix}}_{[a]} \quad -(-1)\underbrace{\begin{bmatrix} -1 & 0 & -3 \\ 0 & 3 & -2 \\ 0 & 0 & 1 \end{bmatrix}}_{[b]} \quad +4\underbrace{\begin{bmatrix} -1 & 0 & -3 \\ 1 & 0 & -1 \\ 0 & 0 & 1 \end{bmatrix}}_{[c]} \quad -(-3)\underbrace{\begin{bmatrix} -1 & 0 & -3 \\ 1 & 0 & -1 \\ 0 & 3 & -2 \end{bmatrix}}_{[d]}$$

Next, using the procedure of Example 4.5 or Example 4.8, we find

$$[a] = 6 , \ [b] = -3 , \ [c] = 0 , \ [d] = -36$$

and thus

$$\det A = [a] + [b] + [c] + [d] = 6 - 3 + 0 - 36 = -33$$

We can verify our answer with MATLAB as follows:

```
A=[ 2  -1  0  -3; -1  1  0  -1; 4  0  3  -2;  -3  0  0  1]; delta = det(A)

delta =
   -33
```

The MATLAB user–defined function file below can be used to compute the determinant of a $n \times n$ matrix.

```
% This file computes the determinant of a nxn matrix
% It must be saved as function (user defined) file
% detnxn.m in the current Work Directory. Make sure
% that his directory is added to MATLAB's search
% path accessed from the Editor Window as File>Set Path>
% Add Folder. It is highly recommended that this
% function file is created in MATLAB's Editor Window.
%
function y=detnxn(A);
% The following statement initializes y
y=0;
% The following statement defines the size of the matrix A
[n,n]=size(A);
% MATLAB allows us to use the user-defined functions to be recursively
% called on themselves so we can call det2x2(A) for a 2x2 matrix,
% and det3x3(A) for a 3x3 matrix.
```

```
if n==2
    y=det2x2(A);
    return
end
%
if n==3
    y=det3x3(A);
    return
end
% For 4x4 or higher order matrices we use the following:
% (We can define n and matrix A in Command Window
for i=1:n
    y=y+(−1)^(i+1)*A(1,i)*detnxn(A(2:n, [1:(i−1) (i+1):n]));
end
%
% To run this program, define the nxn matrix in
% MATLAB's Command Window as A=[....] and then
% type detnxn(A) at the command prompt.
```

Some useful properties of determinants are given below.

Property 1:

If all elements of one row or one column are zero, the determinant is zero. An example of this is the determinant of the cofactor [c] above.

Property 2:

If all the elements of one row or column are m times the corresponding elements of another row or column, the determinant is zero. For example, if

$$A = \begin{bmatrix} 2 & 4 & 1 \\ 3 & 6 & 1 \\ 1 & 2 & 1 \end{bmatrix} \tag{4.28}$$

then,

$$\det A = \begin{vmatrix} 2 & 4 & 1 \\ 3 & 6 & 1 \\ 1 & 2 & 1 \end{vmatrix} \begin{matrix} 2 & 4 \\ 3 & 6 \\ 1 & 2 \end{matrix} = 12 + 4 + 6 - 6 - 4 - 12 = 0 \tag{4.29}$$

Here, $\det A$ is zero because the second column in A is 2 times the first column.

Check with MATLAB:

```
A=[2 4 1; 3 6 1; 1 2 1]; det(A)

ans =
     0
```

Property 3:

If two rows or two columns of a matrix are identical, the determinant is zero. This follows from Property 2 with $m = 1$.

4.6 Cramer's Rule

Let us consider the systems of the three equations below

$$a_{11}x + a_{12}y + a_{13}z = A$$
$$a_{21}x + a_{22}y + a_{23}z = B \qquad (4.30)$$
$$a_{31}x + a_{32}y + a_{33}z = C$$

and let

$$\Delta = \begin{vmatrix} a_{11} & a_{12} & a_{13} \\ a_{21} & a_{22} & a_{23} \\ a_{31} & a_{32} & a_{33} \end{vmatrix} \quad D_1 = \begin{vmatrix} A & a_{11} & a_{13} \\ B & a_{21} & a_{23} \\ C & a_{31} & a_{33} \end{vmatrix} \quad D_2 = \begin{vmatrix} a_{11} & A & a_{13} \\ a_{21} & B & a_{23} \\ a_{31} & C & a_{33} \end{vmatrix} \quad D_3 = \begin{vmatrix} a_{11} & a_{12} & A \\ a_{21} & a_{22} & B \\ a_{31} & a_{32} & C \end{vmatrix}$$

Cramer's rule states that the unknowns x, y, and z can be found from the relations

$$x = \frac{D_1}{\Delta} \qquad y = \frac{D_2}{\Delta} \qquad z = \frac{D_3}{\Delta} \qquad (4.31)$$

provided that the determinant Δ (delta) is not zero.

We observe that the numerators of (4.31) are determinants that are formed from Δ by the substitution of the known values A, B, and C, for the coefficients of the desired unknown.

Cramer's rule applies to systems of two or more equations.

If (4.30) is a homogeneous set of equations, that is, if $A = B = C = 0$, then, D_1, D_2, and D_3 are all zero as we found in Property 1 above. Then, $x = y = z = 0$ also.

Example 4.10

Use Cramer's rule to find v_1, v_2, and v_3 if

$$2v_1 - 5 - v_2 + 3v_3 = 0$$
$$-2v_3 - 3v_2 - 4v_1 = 8 \qquad (4.32)$$
$$v_2 + 3v_1 - 4 - v_3 = 0$$

and verify your answers with MATLAB.

Solution:

Rearranging the unknowns v, and transferring known values to the right side, we obtain

$$2v_1 - v_2 + 3v_3 = 5$$
$$-4v_1 - 3v_2 - 2v_3 = 8 \qquad\qquad (4.33)$$
$$3v_1 + v_2 - v_3 = 4$$

Now, by Cramer's rule,

$$\Delta = \begin{vmatrix} 2 & -1 & 3 \\ -4 & -3 & -2 \\ 3 & 1 & -1 \end{vmatrix} \begin{matrix} 2 & -1 \\ -4 & -3 \\ 3 & 1 \end{matrix} = 6 + 6 - 12 + 27 + 4 + 4 = 35$$

$$D_1 = \begin{vmatrix} 5 & -1 & 3 \\ 8 & -3 & -2 \\ 4 & 1 & -1 \end{vmatrix} \begin{matrix} 5 & -1 \\ 8 & -3 \\ 4 & 1 \end{matrix} = 15 + 8 + 24 + 36 + 10 - 8 = 85$$

$$D_2 = \begin{vmatrix} 2 & 5 & 3 \\ -4 & 8 & -2 \\ 3 & 4 & -1 \end{vmatrix} \begin{matrix} 2 & 5 \\ -4 & 8 \\ 3 & 4 \end{matrix} = -16 - 30 - 48 - 72 + 16 - 20 = -170$$

$$D_3 = \begin{vmatrix} 2 & -1 & 5 \\ -4 & -3 & 8 \\ 3 & 1 & 4 \end{vmatrix} \begin{matrix} 2 & -1 \\ -4 & -3 \\ 3 & 1 \end{matrix} = -24 - 24 - 20 + 45 - 16 - 16 = -55$$

Therefore, using (4.31) we obtain

$$x_1 = \frac{D_1}{\Delta} = \frac{85}{35} = \frac{17}{7} \qquad x_2 = \frac{D_2}{\Delta} = -\frac{170}{35} = -\frac{34}{7} \qquad x_3 = \frac{D_3}{\Delta} = -\frac{55}{35} = -\frac{11}{7} \qquad (4.34)$$

We will verify with MATLAB as follows.

```
% The following script will compute and display the values of v₁, v₂ and v₃.
format rat                      % Express answers in ratio form
B=[2 -1 3; -4 -3 -2; 3 1 -1];   % The elements of the determinant D
delta=det(B);                   % Compute the determinant D of B
d1=[5 -1 3; 8 -3 -2; 4 1 -1];   % The elements of D₁
detd1=det(d1);                  % Compute the determinant of D₁
d2=[2 5 3; -4 8 -2; 3 4 -1];    % The elements of D₂
detd2=det(d2);                  % Compute the determinant of D₂
d3=[2 -1 5; -4 -3 8; 3 1 4];    % The elements of D₃
detd3=det(d3);                  % Compute he determinant of D₃
```

```
v1=detd1/delta;                    % Compute the value of v₁
v2=detd2/delta;                    % Compute the value of v₂
v3=detd3/delta;                    % Compute the value of v₃
%
disp('v1=');disp(v1);              % Display the value of v₁
disp('v2=');disp(v2);              % Display the value of v₂
disp('v3=');disp(v3);              % Display the value of v₃
```

```
v1=
    17/7
v2=
   -34/7
v3=
   -11/7
```

These are the same values as in (4.34)

4.7 Gaussian Elimination Method

We can find the unknowns in a system of two or more equations also by the *Gaussian elimination method*. With this method, the objective is to eliminate one unknown at a time. This can be done by multiplying the terms of any of the equations of the system by a number such that we can add (or subtract) this equation to another equation in the system so that one of the unknowns will be eliminated. Then, by substitution to another equation with two unknowns, we can find the second unknown. Subsequently, substitution of the two values found can be made into an equation with three unknowns from which we can find the value of the third unknown. This procedure is repeated until all unknowns are found. This method is best illustrated with the following example which consists of the same equations as the previous example.

Example 4.11

Use the Gaussian elimination method to find v_1, v_2, and v_3 of

$$2v_1 - v_2 + 3v_3 = 5$$
$$-4v_1 - 3v_2 - 2v_3 = 8 \tag{4.35}$$
$$3v_1 + v_2 - v_3 = 4$$

Solution:

As a first step, we add the first equation of (4.35) with the third to eliminate the unknown v_2 and we obtain the following equation.

$$5v_1 + 2v_3 = 9 \qquad (4.36)$$

Next, we multiply the third equation of (4.35) by 3, and we add it with the second to eliminate v_2. Then, we obtain the following equation.

$$5v_1 - 5v_3 = 20 \qquad (4.37)$$

Subtraction of (4.37) from (4.36) yields

$$7v_3 = -11 \ \text{ or } \ v_3 = -\frac{11}{7} \qquad (4.38)$$

Now, we can find the unknown v_1 from either (4.36) or (4.37). By substitution of (4.38) into (4.36) we obtain

$$5v_1 + 2 \cdot \left(-\frac{11}{7}\right) = 9 \ \text{ or } \ v_1 = \frac{17}{7} \qquad (4.39)$$

Finally, we can find the last unknown v_2 from any of the three equations of (4.35). By substitution into the first equation we obtain

$$v_2 = 2v_1 + 3v_3 - 5 = \frac{34}{7} - \frac{33}{7} - \frac{35}{7} = -\frac{34}{7} \qquad (4.40)$$

These are the same values as those we found in Example 4.10.

The Gaussian elimination method works well if the coefficients of the unknowns are small integers, as in Example 4.11. However, it becomes impractical if the coefficients are large or fractional numbers.

The Gaussian elimination is further discussed in Chapter 14 in conjunction with the LU factorization method.

4.8 The Adjoint of a Matrix

Let us assume that A is an n square matrix and α_{ij} is the cofactor of a_{ij}. Then *the adjoint of* A, denoted as adjA, is defined as the n square matrix shown on the next page.

$$adjA = \begin{bmatrix} \alpha_{11} & \alpha_{21} & \alpha_{31} & \cdots & \alpha_{n1} \\ \alpha_{12} & \alpha_{22} & \alpha_{32} & \cdots & \alpha_{n2} \\ \alpha_{13} & \alpha_{23} & \alpha_{33} & \cdots & \alpha_{n3} \\ \cdots & \cdots & \cdots & \cdots & \cdots \\ \alpha_{1n} & \alpha_{2n} & \alpha_{3n} & \cdots & \alpha_{nn} \end{bmatrix} \qquad (4.41)$$

We observe that the cofactors of the elements of the ith row (column) of A, are the elements of the ith column (row) of $adjA$.

Example 4.12

Compute $adjA$ given that

$$A = \begin{bmatrix} 1 & 2 & 3 \\ 1 & 3 & 4 \\ 1 & 4 & 3 \end{bmatrix} \qquad (4.42)$$

Solution:

$$adjA = \begin{bmatrix} \begin{vmatrix} 3 & 4 \\ 4 & 3 \end{vmatrix} & -\begin{vmatrix} 2 & 3 \\ 4 & 3 \end{vmatrix} & \begin{vmatrix} 2 & 3 \\ 3 & 4 \end{vmatrix} \\ -\begin{vmatrix} 1 & 4 \\ 1 & 3 \end{vmatrix} & \begin{vmatrix} 1 & 3 \\ 1 & 3 \end{vmatrix} & -\begin{vmatrix} 2 & 3 \\ 3 & 4 \end{vmatrix} \\ \begin{vmatrix} 1 & 3 \\ 1 & 4 \end{vmatrix} & -\begin{vmatrix} 1 & 2 \\ 1 & 4 \end{vmatrix} & \begin{vmatrix} 1 & 2 \\ 1 & 3 \end{vmatrix} \end{bmatrix} = \begin{bmatrix} -7 & 6 & -1 \\ 1 & 0 & -1 \\ 1 & -2 & 1 \end{bmatrix}$$

4.9 Singular and Non–Singular Matrices

An n square matrix A is called *singular* if $detA = 0$; if $detA \neq 0$, A is called *non–singular*. If an n square matrix A is nearly singular, that is, if the determinant of that matrix is very small, the matrix is said to be *ill–conditioned*. This topic is discussed in Appendix C.

Example 4.13

Given that

$$A = \begin{bmatrix} 1 & 2 & 3 \\ 2 & 3 & 4 \\ 3 & 5 & 7 \end{bmatrix} \qquad (4.43)$$

determine whether this matrix is singular or non–singular.

Solution:

$$\det A = \begin{vmatrix} 1 & 2 & 3 \\ 2 & 3 & 4 \\ 3 & 5 & 7 \end{vmatrix} \begin{matrix} 1 & 2 \\ 2 & 3 \\ 3 & 5 \end{matrix} = 21 + 24 + 30 - 27 - 20 - 28 = 0$$

Therefore, matrix A is singular.

4.10 The Inverse of a Matrix

If A and B are n square matrices such that $AB = BA = I$, where I is the identity matrix, B is called the *inverse* of A, denoted as $B = A^{-1}$, and likewise, A is called the *inverse* of B, that is, $A = B^{-1}$

If a matrix A is non–singular, we can compute its inverse from the relation

$$\boxed{A^{-1} = \frac{1}{\det A} \mathrm{adj} A} \qquad (4.44)$$

Example 4.14

Given that

$$A = \begin{bmatrix} 1 & 2 & 3 \\ 1 & 3 & 4 \\ 1 & 4 & 3 \end{bmatrix} \qquad (4.45)$$

compute its inverse, that is, find A^{-1}

Solution:

Here, $\det A = 9 + 8 + 12 - 9 - 16 - 6 = -2$, and since this is a non–zero value, it is possible to compute the inverse of A using (4.44).
From Example 4.12,

$$\text{adj}A = \begin{bmatrix} -7 & 6 & -1 \\ 1 & 0 & -1 \\ 1 & -2 & 1 \end{bmatrix}$$

Then,

$$A^{-1} = \frac{1}{\det A}\text{adj}A = \frac{1}{-2}\begin{bmatrix} -7 & 6 & -1 \\ 1 & 0 & -1 \\ 1 & -2 & 1 \end{bmatrix} = \begin{bmatrix} 3.5 & -3 & 0.5 \\ -0.5 & 0 & 0.5 \\ -0.5 & 1 & -0.5 \end{bmatrix} \tag{4.46}$$

Check with MATLAB:

A=[1 2 3; 1 3 4; 1 4 3], invA=inv(A)   % Define matrix A and compute its inverse

```
A =
      1       2       3
      1       3       4
      1       4       3

invA =
      3.5000    -3.0000     0.5000
     -0.5000          0     0.5000
     -0.5000     1.0000    -0.5000
```

Multiplication of a matrix A by its inverse A^{-1} produces the identity matrix I, that is,

$$AA^{-1} = I \quad \text{or} \quad A^{-1}A = I \tag{4.47}$$

Example 4.15

Prove the validity of (4.47) for

$$A = \begin{bmatrix} 4 & 3 \\ 2 & 2 \end{bmatrix}$$

Proof:

$$\det A = 8 - 6 = 2 \quad \text{and} \quad \text{adj}A = \begin{bmatrix} 2 & -3 \\ -2 & 4 \end{bmatrix}$$

Then,

$$A^{-1} = \frac{1}{\det A}\text{adj}A = \frac{1}{2}\begin{bmatrix} 2 & -3 \\ -2 & 4 \end{bmatrix} = \begin{bmatrix} 1 & -3/2 \\ -1 & 2 \end{bmatrix}$$

and

$$AA^{-1} = \begin{bmatrix} 4 & 3 \\ 2 & 2 \end{bmatrix} \begin{bmatrix} 1 & -3/2 \\ -1 & 2 \end{bmatrix} = \begin{bmatrix} 4-3 & -6+6 \\ 2-2 & -3+4 \end{bmatrix} = \begin{bmatrix} 1 & 0 \\ 0 & 1 \end{bmatrix} = I$$

4.11 Solution of Simultaneous Equations with Matrices

Consider the relation

$$AX = B \tag{4.48}$$

where A and B are matrices whose elements are known, and X is a matrix (a column vector) whose elements are the unknowns. We assume that A and X are conformable for multiplication. Multiplication of both sides of (4.48) by A^{-1} yields:

$$A^{-1}AX = A^{-1}B = IX = A^{-1}B \tag{4.49}$$

or

$$\boxed{X = A^{-1}B} \tag{4.50}$$

Therefore, we can use (4.50) to solve any set of simultaneous equations that have solutions. We will refer to this method as the *inverse matrix method of solution* of simultaneous equations.

Example 4.16

Given the system of equations

$$\begin{cases} 2x_1 + 3x_2 + x_3 = 9 \\ x_1 + 2x_2 + 3x_3 = 6 \\ 3x_1 + x_2 + 2x_3 = 8 \end{cases} \tag{4.51}$$

compute the unknowns $x_1, x_2,$ and x_3 using the inverse matrix method.

Solution:

In matrix form, the given set of equations is $AX = B$ where

$$A = \begin{bmatrix} 2 & 3 & 1 \\ 1 & 2 & 3 \\ 3 & 1 & 2 \end{bmatrix}, \quad X = \begin{bmatrix} x_1 \\ x_2 \\ x_3 \end{bmatrix}, \quad B = \begin{bmatrix} 9 \\ 6 \\ 8 \end{bmatrix} \tag{4.52}$$

Then,

$$X = A^{-1}B \tag{4.53}$$

or

$$\begin{bmatrix} x_1 \\ x_2 \\ x_3 \end{bmatrix} = \begin{bmatrix} 2 & 3 & 1 \\ 1 & 2 & 3 \\ 3 & 1 & 2 \end{bmatrix}^{-1} \begin{bmatrix} 9 \\ 6 \\ 8 \end{bmatrix} \tag{4.54}$$

Next, we find the determinant $\det A$, and the adjoint $\text{adj}A$.

$$\det A = 18 \quad \text{and} \quad \text{adj}A = \begin{bmatrix} 1 & -5 & 7 \\ 7 & 1 & -5 \\ -5 & 7 & 1 \end{bmatrix}$$

Therefore,

$$A^{-1} = \frac{1}{\det A}\, \text{adj}A = \frac{1}{18} \begin{bmatrix} 1 & -5 & 7 \\ 7 & 1 & -5 \\ -5 & 7 & 1 \end{bmatrix}$$

and by (4.53) we obtain the solution as follows.

$$X = \begin{bmatrix} x_1 \\ x_2 \\ x_3 \end{bmatrix} = \frac{1}{18} \begin{bmatrix} 1 & -5 & 7 \\ 7 & 1 & -5 \\ -5 & 7 & 1 \end{bmatrix} \begin{bmatrix} 9 \\ 6 \\ 8 \end{bmatrix} = \frac{1}{18} \begin{bmatrix} 35 \\ 29 \\ 5 \end{bmatrix} = \begin{bmatrix} 35/18 \\ 29/18 \\ 5/18 \end{bmatrix} = \begin{bmatrix} 1.94 \\ 1.61 \\ 0.28 \end{bmatrix} \tag{4.55}$$

To verify our results, we could use the MATLAB **inv(A)** function, and multiply A^{-1} by B. However, it is easier to use the *matrix left division* operation $X = A \setminus B$; this is MATLAB's solution of $A^{-1}B$ for the matrix equation $A \cdot X = B$, where matrix X is the same size as matrix B. For this example,

```
A=[2 3 1; 1 2 3; 3 1 2]; B=[9 6 8]'; X=A \ B    % Observe that B is a column vector
X =
    1.9444
    1.6111
    0.2778
```

As stated earlier, while MATLAB has the built–in function **det(A)** for computing the determinant of a matrix **A**, this function is not included in the MATLAB Run–Time Function Library List that is used with the Simulink **Embedded MATLAB Function** block. The MATLAB user–defined function file below can be used to compute the determinant of a 2×2 matrix. A user–defined function to compute the inverse of an $n \times n$ is presented in Chapter 14.

Example 4.17

For the electric circuit of Figure 4.1, the mesh equations are

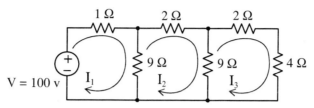

Figure 4.1. Circuit for Example 4.17

$$10I_1 - 9I_2 \qquad = 100$$
$$-9I_1 + 20I_2 - 9I_3 = \quad 0 \qquad (4.56)$$
$$-9I_2 + 15I_3 = \quad 0$$

Use the inverse matrix method to compute the values of the currents I_1, I_2, and I_3.

Solution:

For this example, the matrix equation is $RI = V$ or $I = R^{-1}V$, where

$$R = \begin{bmatrix} 10 & -9 & 0 \\ -9 & 20 & -9 \\ 0 & -9 & 15 \end{bmatrix}, \quad V = \begin{bmatrix} 100 \\ 0 \\ 0 \end{bmatrix} \quad \text{and} \quad I = \begin{bmatrix} I_1 \\ I_2 \\ I_3 \end{bmatrix}$$

The next step is to find R^{-1}. This is found from the relation

$$R^{-1} = \frac{1}{\det R} \, adjR \qquad (4.57)$$

Therefore, we find the determinant and the adjoint of R. For this example, we find that

$$\det R = 975, \quad adjR = \begin{bmatrix} 219 & 135 & 81 \\ 135 & 150 & 90 \\ 81 & 90 & 119 \end{bmatrix} \qquad (4.58)$$

Then,

$$R^{-1} = \frac{1}{\det R} adjR = \frac{1}{975} \begin{bmatrix} 219 & 135 & 81 \\ 135 & 150 & 90 \\ 81 & 90 & 119 \end{bmatrix}$$

and

$$I = \begin{bmatrix} I_1 \\ I_2 \\ I_3 \end{bmatrix} = \frac{1}{975} \begin{bmatrix} 219 & 135 & 81 \\ 135 & 150 & 90 \\ 81 & 90 & 119 \end{bmatrix} \begin{bmatrix} 100 \\ 0 \\ 0 \end{bmatrix} = \frac{100}{975} \begin{bmatrix} 219 \\ 135 \\ 81 \end{bmatrix} = \begin{bmatrix} 22.46 \\ 13.85 \\ 8.31 \end{bmatrix}$$

Check with MATLAB:

R=[10 −9 0; −9 20 −9; 0 −9 15]; V=[100 0 0]'; I=R\V

```
I =
   22.4615
   13.8462
    8.3077
```

We can also use subscripts to address the individual elements of the matrix. Accordingly, the above script could also have been written as:

R(1,1)=10; R(1,2)=−9;   % No need to make entry for A(1,3) since it is zero.
R(2,1)=−9; R(2,2)=20; R(2,3)=−9; R(3,2)=−9; R(3,3)=15; V=[100 0 0]'; I=R\V

```
I =
   22.4615
   13.8462
    8.3077
```

Spreadsheets also have the capability of solving simultaneous equations using the inverse matrix method. For instance, we can use Microsoft Excel's MINVERSE (Matrix Inversion) and MMULT (Matrix Multiplication) functions, to obtain the values of the three currents in Example 4.17.

The procedure is as follows:

1. We start with a blank spreadsheet and in a block of cells, say B3:D5, we enter the elements of matrix R as shown in Figure 4.2. Then, we enter the elements of matrix V in G3:G5.

	A	B	C	D	E	F	G	H
1	Spreadsheet for Matrix Inversion and Matrix Multiplication							
2								
3			10	-9	0			100
4		R=	-9	20	-9		V=	0
5			0	-9	15			0
6								
7			0.225	0.138	0.083			22.462
8		R^{-1}=	0.138	0.154	0.092		I=	13.846
9			0.083	0.092	0.122			8.3077
10								

Figure 4.2. Solution of Example 4.17 with a spreadsheet

2. Next, we compute and display the inverse of R, that is, R^{-1}. We choose B7:D9 for the elements of this inverted matrix. We format this block for number display with three decimal places. With this range highlighted and making sure that the cell marker is in B7, we type the formula

=MININVERSE(B3:D5)

and we press the *Crtl–Shift–Enter* keys simultaneously. We observe that R^{-1} appears in these cells.

3. Now, we choose the block of cells G7:G9 for the values of the current I. As before, we highlight them, and with the cell marker positioned in G7, we type the formula

=MMULT(B7:D9,G3:G5)

and we press the *Crtl–Shift–Enter* keys simultaneously. The values of *I* then appear in G7:G9.

Example 4.18

For the phasor circuit of Figure 4.3, the current I_X can be found from the relation

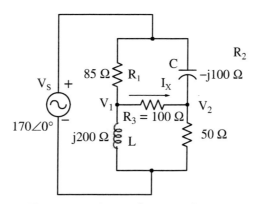

Figure 4.3. Circuit for Example 4.18

$$I_X = \frac{V_1 - V_2}{R_3} \qquad (4.59)$$

and the voltages V_1 and V_2 can be computed from the nodal equations

$$\frac{V_1 - 170\angle 0°}{85} + \frac{V_1 - V_2}{100} + \frac{V_1 - 0}{j200} = 0 \qquad (4.60)$$

$$\frac{V_2 - 170\angle 0°}{-j100} + \frac{V_2 - V_1}{100} + \frac{V_2 - 0}{50} = 0 \qquad (4.61)$$

Compute, and express the current I_x in both rectangular and polar forms by first simplifying like

---

terms, collecting, and then writing the above relations in matrix form as $YV = I$, where Y = admittance, V = voltage, and I = current.

Solution:

The elements of the Y matrix are the coefficients of V_1 and V_2. Simplifying and rearranging the nodal equations of (4.60) and (4.61), we obtain

$$(0.0218 - j0.005)V_1 - 0.01V_2 = 2$$

$$-0.01V_1 + (0.03 + j0.01)V_2 = j1.7$$

Next, we write (4.62) in matrix form as

$$\underbrace{\begin{bmatrix} 0.0218 - j0.005 & -0.01 \\ -0.01 & 0.03 + j0.01 \end{bmatrix}}_{Y} \underbrace{\begin{bmatrix} V_1 \\ V_2 \end{bmatrix}}_{V} = \underbrace{\begin{bmatrix} 2 \\ j1.7 \end{bmatrix}}_{I} \qquad (4.62)$$

where the matrices Y, V, and I are as indicatedin (4.63).

We will use MATLAB to compute the voltages V_1 and V_2, and to do all other computations. The script is shown below.

```
Y=[0.0218-0.005j -0.01; -0.01 0.03+0.01j]; I=[2; 1.7j]; V=Y\I;    % Define Y, I, and find V
fprintf('\n');                              % Insert a line
disp(' V1     V2'); disp(' -----------------');    % Display V1 and V2 with dash line underneath
fprintf('%9.3f %9.3f\n',V(1),V(2))          % Display values of V1 and V2 in tabular form
fprintf('\n')% Insert another line

     V1            V2
   -----------------
   104.905    53.416
```

Next, we find I_X from

```
R3=100; IX=(V(1)-V(2))/R3               % Compute the value of IX

IX =
   0.5149 - 0.0590i
```

and this is the rectangular form of I_X. For the polar form we use

```
magIX=abs(IX)                           % Compute the magnitude of IX

magIX =
```

```
    0.5183
```

thetaIX=angle(IX)*180/pi   % Compute angle theta in degrees

```
thetaIX =
   -6.5326
```

Therefore, in polar form $I_X = 0.518\angle{-6.53°}$

Spreadsheets have limited capabilities with complex numbers, and thus we cannot use them to compute matrices that include complex numbers in their elements.

4.12 Summary

- A matrix is a rectangular array of numbers whose general form is

$$A = \begin{bmatrix} a_{11} & a_{12} & a_{13} & \cdots & a_{1n} \\ a_{21} & a_{22} & a_{23} & \cdots & a_{2n} \\ a_{31} & a_{32} & a_{33} & \cdots & a_{3n} \\ \cdots & \cdots & \cdots & \cdots & \cdots \\ a_{m1} & a_{m2} & a_{m3} & \cdots & a_{mn} \end{bmatrix}$$

The numbers a_{ij} are the elements of the matrix where the index i indicates the row, and j indicates the column in which each element is positioned. A matrix of m rows and n columns is said to be of $m \times n$ order matrix. If $m = n$, the matrix is said to be a square matrix of order m.

- Two matrices $A = \begin{bmatrix} a_{ij} \end{bmatrix}$ and $B = \begin{bmatrix} b_{ij} \end{bmatrix}$ are equal, that is, $A = B$, if and only if

$$a_{ij} = b_{ij} \qquad i = 1, 2, 3, ..., m \qquad j = 1, 2, 3, ..., n$$

- Two matrices are said to be *conformable for addition* (*subtraction*), if they are of the same order $m \times n$. If $A = \begin{bmatrix} a_{ij} \end{bmatrix}$ and $B = \begin{bmatrix} b_{ij} \end{bmatrix}$ are conformable for addition (subtraction), their sum (difference) will be another matrix C with the same order as A and B, where each element of C is the sum (difference) of the corresponding elements of A and B, i.e., $C = A \pm B = [a_{ij} \pm b_{ij}]$

- If k is any scalar (a positive or negative number), and not [k] which is a 1×1 matrix, then multiplication of a matrix A by the scalar k, is the multiplication of every element of A by k.

- Two matrices A and B are said to be *conformable for multiplication* $A \cdot B$ in that order, only when the number of columns of matrix A is equal to the number of rows of matrix B. That is, the product $A \cdot B$ (but not $B \cdot A$) is conformable for multiplication only if A is an $m \times p$ and matrix B is an $p \times n$ matrix. The product $A \cdot B$ will then be an $m \times n$ matrix.

- For matrix multiplication, the operation is row by column. Thus, to obtain the product $A \cdot B$, we multiply each element of a row of A by the corresponding element of a column of B; then, we add these products.

- Division of one matrix by another, is not defined.

- A scalar matrix is a square matrix where $a_{11} = a_{22} = a_{33} = ... = a_{nn} = k$ and k is a scalar. A scalar matrix with $k = 1$, is called an identity matrix I.

- The MATLAB **eye(n)** function displays an $n \times n$ identity matrix and the **eye(size(A))** function displays an identity matrix whose size is the same as matrix A.

- The transpose of a matrix A, denoted as A^T, is the matrix that is obtained when the rows and columns of matrix A are interchanged.

- The determinant of a square matrix A where

$$A = \begin{bmatrix} a_{11} & a_{12} & a_{13} & \cdots & a_{1n} \\ a_{21} & a_{22} & a_{23} & \cdots & a_{2n} \\ a_{31} & a_{32} & a_{33} & \cdots & a_{3n} \\ \cdots & \cdots & \cdots & \cdots & \cdots \\ a_{n1} & a_{n2} & a_{n3} & \cdots & a_{nn} \end{bmatrix}$$

is denoted as $\det A$ and it is defined as

$$\det A = a_{11}a_{22}a_{33}...a_{nn} + a_{12}a_{23}a_{34}...a_{n1} + a_{13}a_{24}a_{35}...a_{n2} + ...$$
$$-a_{n1}...a_{22}a_{13}...-a_{n2}...a_{23}a_{14} - a_{n3}...a_{24}a_{15} - ...$$

- If from a matrix A be defined as

$$A = \begin{bmatrix} a_{11} & a_{12} & a_{13} & \cdots & a_{1n} \\ a_{21} & a_{22} & a_{23} & \cdots & a_{2n} \\ a_{31} & a_{32} & a_{33} & \cdots & a_{3n} \\ \cdots & \cdots & \cdots & \cdots & \cdots \\ a_{n1} & a_{n2} & a_{n3} & \cdots & a_{nn} \end{bmatrix}$$

we remove the elements of its ith row, and jth column, the determinant of the remaining $n - 1$ square matrix is called the *minor of determinant* A, and it is denoted as $\left[M_{ij} \right]$.

- The signed minor $(-1)^{i+j} \left[M_{ij} \right]$ is called the *cofactor* of a_{ij} and it is denoted as α_{ij}.

- Let A be a square matrix of any size; the value of the determinant of A is the sum of the products obtained by multiplying each element of *any* row or *any* column by its cofactor. We must use this procedure to find the determinant of a matrix A of order 4 or higher.

- Some useful properties of determinants are:

  a. If all elements of one row or one column are zero, the determinant is zero.

  b. If all the elements of one row or column are m times the corresponding elements of another row or column, the determinant is zero.

  c. If two rows or two columns of a matrix are identical, the determinant is zero.

- Cramer's rule states that if a system of equations is defined as

$$a_{11}x + a_{12}y + a_{13}z = A$$

$$a_{21}x + a_{22}y + a_{23}z = B$$

$$a_{31}x + a_{32}y + a_{33}z = C$$

and we let

$$\Delta = \begin{vmatrix} a_{11} & a_{12} & a_{13} \\ a_{21} & a_{22} & a_{23} \\ a_{31} & a_{32} & a_{33} \end{vmatrix} \quad D_1 = \begin{vmatrix} A & a_{11} & a_{13} \\ B & a_{21} & a_{23} \\ C & a_{31} & a_{33} \end{vmatrix} \quad D_2 = \begin{vmatrix} a_{11} & A & a_{13} \\ a_{21} & B & a_{23} \\ a_{31} & C & a_{33} \end{vmatrix} \quad D_3 = \begin{vmatrix} a_{11} & a_{12} & A \\ a_{21} & a_{22} & B \\ a_{31} & a_{32} & C \end{vmatrix}$$

the unknowns x, y, and z can be found from the relations

$$x = \frac{D_1}{\Delta} \qquad y = \frac{D_2}{\Delta} \qquad z = \frac{D_3}{\Delta}$$

provided that the determinant Δ (delta) is not zero.

- We can find the unknowns in a system of two or more equations also by the *Gaussian elimination method*. With this method, the objective is to eliminate one unknown at a time. This can be done by multiplying the terms of any of the equations of the system by a number such that we can add (or subtract) this equation to another equation in the system so that one of the unknowns will be eliminated. Then, by substitution to another equation with two unknowns, we can find the second unknown. Subsequently, substitution of the two values found can be made into an equation with three unknowns from which we can find the value of the third unknown. This procedure is repeated until all unknowns are found.

- If A is an n square matrix and α_{ij} is the cofactor of a_{ij}, the adjoint of A, denoted as $\mathrm{adj}A$, is defined as the n square matrix below.

$$adjA = \begin{bmatrix} \alpha_{11} & \alpha_{21} & \alpha_{31} & \cdots & \alpha_{n1} \\ \alpha_{12} & \alpha_{22} & \alpha_{32} & \cdots & \alpha_{n2} \\ \alpha_{13} & \alpha_{23} & \alpha_{33} & \cdots & \alpha_{n3} \\ \cdots & \cdots & \cdots & \cdots & \cdots \\ \alpha_{1n} & \alpha_{2n} & \alpha_{3n} & \cdots & \alpha_{nn} \end{bmatrix}$$

- An n square matrix A is called singular if $detA = 0$; if $detA \neq 0$, A is called non-singular.

- If A and B are n square matrices such that $AB = BA = I$, where I is the identity matrix, B is called the *inverse* of A, denoted as $B = A^{-1}$, and likewise, A is called the *inverse* of B, that is, $A = B^{-1}$

- If a matrix A is non-singular, we can compute its inverse from the relation

$$A^{-1} = \frac{1}{detA} adjA$$

- Multiplication of a matrix A by its inverse A^{-1} produces the identity matrix I, that is,

$$AA^{-1} = I \quad or \quad A^{-1}A = I$$

- If A and B are matrices whose elements are known, X is a matrix (a column vector) whose elements are the unknowns and A and X are conformable for multiplication, we can use the relation $X = A^{-1}B$ to solve any set of simultaneous equations that have solutions. We refer to this method as the inverse matrix method of solution of simultaneous equations.

- The matrix left division operation is defined as $X = A \backslash B$; this is MATLAB's solution of $A^{-1}B$ for the matrix equation $A \cdot X = B$, where matrix X is the same size as matrix B.

- We can use Microsoft Excel's MINVERSE (Matrix Inversion) and MMULT (Matrix Multiplication) functions, to solve any set of simultaneous equations that have solutions. However, we cannot use them to compute matrices that include complex numbers in their elements.

4.13 Exercises

For Exercises 1 through 3 below, the matrices A, B, C and D are defined as:

$$A = \begin{bmatrix} 1 & -1 & -4 \\ 5 & 7 & -2 \\ 3 & -5 & 6 \end{bmatrix} \quad B = \begin{bmatrix} 5 & 9 & -3 \\ -2 & 8 & 2 \\ 7 & -4 & 6 \end{bmatrix} \quad C = \begin{bmatrix} 4 & 6 \\ -3 & 8 \\ 5 & -2 \end{bmatrix} \quad D = \begin{bmatrix} 1 & -2 & 3 \\ -3 & 6 & -4 \end{bmatrix}$$

1. Perform the following computations, if possible. Verify your answers with Excel or MATLAB.

   a. $A + B$ b. $A + C$ c. $B + D$ d. $C + D$ e. $A - B$ f. $A - C$ g. $B - D$ h. $C - D$

2. Perform the following computations, if possible. Verify your answers with Excel or MATLAB.

   a. $A \cdot B$ b. $A \cdot C$ c. $B \cdot D$ d. $C \cdot D$ e. $B \cdot A$ f. $C \cdot A$ g. $D \cdot A$ h. $\dot{D} \cdot C$

3. Perform the following computations, if possible. Verify your answers with Excel or MATLAB.

   a. $\det A$ b. $\det B$ c. $\det C$ d. $\det D$ e. $\det(A \cdot B)$ f. $\det(A \cdot C)$

4. Solve the following system of equations using Cramer's rule. Verify your answers with Excel or MATLAB.

$$x_1 - 2x_2 + x_3 = -4$$
$$-2x_1 + 3x_2 + x_3 = 9$$
$$3x_1 + 4x_2 - 5x_3 = 0$$

5. Repeat Exercise 4 using the Gaussian elimination method.

6. Use the MATLAB **det(A)** function to find the unknowns of the system of equations below.

$$-x_1 + 2x_2 - 3x_3 + 5x_4 = 14$$
$$x_1 + 3x_2 + 2x_3 - x_4 = 9$$
$$3x_1 - 3x_2 + 2x_3 + 4x_4 = 19$$
$$4x_1 + 2x_2 + 5x_3 + x_4 = 27$$

7. Solve the following system of equations using the inverse matrix method. Verify your answers with Excel or MATLAB.

$$\begin{bmatrix} 1 & 3 & 4 \\ 3 & 1 & -2 \\ 2 & 3 & 5 \end{bmatrix} \cdot \begin{bmatrix} x_1 \\ x_2 \\ x_3 \end{bmatrix} = \begin{bmatrix} -3 \\ -2 \\ 0 \end{bmatrix}$$

8. Use Excel to find the unknowns for the system

$$\begin{bmatrix} 2 & 4 & 3 & -2 \\ 2 & -4 & 1 & 3 \\ -1 & 3 & -4 & 2 \\ 2 & -2 & 2 & 1 \end{bmatrix} \cdot \begin{bmatrix} x_1 \\ x_2 \\ x_3 \\ x_4 \end{bmatrix} = \begin{bmatrix} 1 \\ 10 \\ -14 \\ 7 \end{bmatrix}$$

Verify your answers with the MATLAB left division operation.

4.14 Solutions to End-of-Chapter Exercises

1.

a. $A + B = \begin{bmatrix} 1+5 & -1+9 & -4-3 \\ 5-2 & 7+8 & -2+2 \\ 3+7 & -5-4 & 6+6 \end{bmatrix} = \begin{bmatrix} 6 & 8 & -7 \\ 3 & 15 & 0 \\ 10 & -9 & 12 \end{bmatrix}$ b. $A + C$ not conformable for addition

c. $B + D$ not conformable for addition   d. $C + D$ not conformable for addition

e. $A - B = \begin{bmatrix} 1-5 & -1-9 & -4+3 \\ 5+2 & 7-8 & -2-2 \\ 3-7 & -5+4 & 6-6 \end{bmatrix} = \begin{bmatrix} -4 & -10 & -1 \\ 7 & -1 & -4 \\ -4 & -1 & 0 \end{bmatrix}$ f. $A - C$ not conformable for subtraction

g. $B - D$ not conformable for subtraction   h. $C - D$ not conformable for subtraction

2.

a.
$A \cdot B = \begin{bmatrix} 1 \times 5 + (-1) \times (-2) + (-4) \times 7 & 1 \times 9 + (-1) \times 8 + (-4) \times (-4) & 1 \times (-3) + (-1) \times 2 + (-4) \times 6 \\ 5 \times 5 + 7 \times (-2) + (-2) \times 7 & 5 \times 9 + 7 \times 8 + (-2) \times (-4) & 5 \times (-3) + 7 \times 2 + (-2) \times 6 \\ 3 \times 5 + (-5) \times (-2) + 6 \times 7 & 3 \times 9 + (-5) \times 8 + 6 \times (-4) & 3 \times (-3) + (-5) \times 2 + 6 \times 6 \end{bmatrix}$

$= \begin{bmatrix} -21 & 17 & -29 \\ -3 & 109 & -13 \\ 67 & -37 & 17 \end{bmatrix}$

Check with MATLAB:

A=[1 −1 −4; 5 7 −2; 3 −5 6]; B=[5 9 −3; −2 8 2; 7 −4 6]; A*B

ans =

```
   -21      17     -29
    -3     109     -13
    67     -37      17
```

b. $\cdot C = \begin{bmatrix} 1 \times 4 + (-1) \times (-3) + (-4) \times 5 & 1 \times 6 + (-1) \times 8 + (-4) \times (-2) \\ 5 \times 4 + 7 \times (-3) + (-2) \times 5 & 5 \times 6 + 7 \times 8 + (-2) \times (-2) \\ 3 \times 4 + (-5) \times (-3) + 6 \times 5 & 3 \times 6 + (-5) \times 8 + 6 \times (-2) \end{bmatrix} = \begin{bmatrix} -13 & 6 \\ -11 & 90 \\ 57 & -34 \end{bmatrix}$

c. $B \cdot D$ not conformable for multiplication

d. $\cdot D = \begin{bmatrix} 4 \times 1 + 6 \times (-3) & 4 \times (-2) + 6 \times 6 & 4 \times 3 + 6 \times (-4) \\ (-3) \times 1 + 8 \times (-3) & (-3) \times (-2) + 8 \times 6 & (-3) \times 3 + 8 \times (-4) \\ 5 \times 1 + (-2) \times (-3) & 5 \times (-2) + (-2) \times 6 & 5 \times 3 + (-2) \times (-4) \end{bmatrix} = \begin{bmatrix} -14 & 28 & -1 \\ -27 & 54 & -4 \\ 11 & -22 & 23 \end{bmatrix}$

e.

$$B \cdot A = \begin{bmatrix} 5 \times 1 + 9 \times 5 + (-3) \times 3 & (-2) \times 1 + 8 \times 5 + 2 \times 3 & 7 \times 1 + (-4) \times 5 + 6 \times 3 \\ 5 \times (-1) + 9 \times 7 + (-3) \times (-5) & (-2) \times (-1) + 8 \times 7 + 2 \times (-5) & 7 \times (-1) + (-4) \times 7 + 6 \times (-5) \\ 5 \times (-4) + 9 \times (-2) + (-3) \times 6 & (-2) \times (-4) + 8 \times (-2) + 2 \times 6 & 7 \times (-4) + (-4) \times (-2) + 6 \times 6 \end{bmatrix}$$

$$= \begin{bmatrix} 41 & 73 & -56 \\ 44 & 48 & 4 \\ 5 & -65 & 16 \end{bmatrix}$$

f. $C \cdot A$ not conformable for multiplication

g.

$$D \cdot A = \begin{bmatrix} 1 \times 1 + (-2) \times 5 + 3 \times 3 & 1 \times (-1) + (-2) \times 7 + 3 \times (-5) & 1 \times (-4) + (-2) \times (-2) + 3 \times 6 \\ (-3) \times 1 + 6 \times 5 + (-4) \times 3 & (-3) \times (-1) + 6 \times 7 + (-4) \times (-5) & (-3) \times (-4) + 6 \times (-2) + (-4) \times 6 \end{bmatrix}$$

$$= \begin{bmatrix} 0 & -30 & 18 \\ 15 & 65 & -24 \end{bmatrix}$$

h. $D \cdot C = \begin{bmatrix} 1 \times 4 + (-2) \times (-3) + 3 \times 5 & 1 \times 6 + (-2) \times 8 + 3 \times (-2) \\ (-3) \times 4 + 6 \times (-3) + (-4) \times 5 & (-3) \times 6 + 6 \times 8 + (-4) \times (-2) \end{bmatrix} = \begin{bmatrix} 25 & -16 \\ -50 & 38 \end{bmatrix}$

3.

a.

$$detA = \begin{matrix} 1 & -1 & -4 & 1 & -1 \\ 5 & 7 & -2 & 5 & 7 \\ 3 & -5 & 6 & 3 & -5 \end{matrix}$$

$$= 1 \times 7 \times 6 + (-1) \times (-2) \times 3 + (-4) \times 5 \times (-5) - [3 \times 7 \times (-4) + (-5) \times (-2) \times 1 + 6 \times 5 \times (-1)]$$

$$= 42 + 6 + 100 - (-84) - 10 - (-30) = 252$$

b.

$$detB = \begin{matrix} 5 & 9 & -3 & 5 & 9 \\ -2 & 8 & 2 & -2 & 8 \\ 7 & -4 & 6 & 7 & -4 \end{matrix}$$

$$= 5 \times 8 \times 6 + 9 \times 2 \times 7 + (-3) \times (-2) \times (-4) - [7 \times 8 \times (-3) + (-4) \times 2 \times 5 + 6 \times (-2) \times 9]$$

$$= 240 + 126 - 24 - (-168) + 40 - (-108) = 658$$

c. detC does not exist; matrix must be square

d. detD does not exist; matrix must be square

e. et($A \cdot B$) = detA · detl and from parts (a) and (b), det($A \cdot B$) = 252×658 = 165816

f. det($A \cdot C$) does not exist because detC does not exist

4.

$$\Delta = \begin{matrix} 1 & -2 & 1 & 1 & -2 \\ -2 & 3 & 1 & -2 & 3 \\ 3 & 4 & -5 & 3 & 4 \end{matrix}$$

$= 1 \times 3 \times (-5) + (-2) \times 1 \times 3 + 1 \times (-2) \times 4 - [3 \times 3 \times 1 + 4 \times 1 \times 1 + (-5) \times (-2) \times (-2)]$

$= -15 - 6 - 8 - 9 - 4 + 20 = -22$

$$D_1 = \begin{matrix} -4 & -2 & 1 & 4 & -2 \\ 9 & 3 & 1 & 9 & 3 \\ 0 & 4 & -5 & 0 & 4 \end{matrix}$$

$= -4 \times 3 \times (-5) + (-2) \times 1 \times 0 + 1 \times 9 \times 4 - [0 \times 3 \times 1 + 4 \times 1 \times 4 + (-5) \times 9 \times (-2)]$

$= 60 + 0 + 36 - 0 + 16 - 90 = 22$

$$D_2 = \begin{matrix} 1 & -4 & 1 & 1 & -4 \\ -2 & 9 & 1 & -2 & 9 \\ 3 & 0 & -5 & 3 & 0 \end{matrix}$$

$= 1 \times 9 \times (-5) + (-4) \times 1 \times 3 + 1 \times (-2) \times 0 - [3 \times 9 \times 1 + 0 \times 1 \times 1 + (-5) \times (-2) \times (-4)]$

$= -45 - 12 - 0 - 27 - 0 + 40 = -44$

$$D_3 = \begin{matrix} 1 & -2 & -4 & 1 & -2 \\ -2 & 3 & 9 & -2 & 3 \\ 3 & 4 & 0 & 3 & 4 \end{matrix}$$

$= 1 \times 3 \times 0 + (-2) \times 9 \times 3 + (-4) \times (-2) \times 4 - [3 \times 3 \times (-4) + 4 \times 9 \times 1 + 0 \times (-2) \times (-2)]$

$= 0 - 54 + 32 + 36 - 36 - 0 = -22$

$$x_1 = \frac{D_1}{\Delta} = \frac{22}{-22} = -1 \qquad x_2 = \frac{D_2}{\Delta} = \frac{-44}{-22} = 2 \qquad x_3 = \frac{D_3}{\Delta} = \frac{-22}{-22} = 1$$

5.

$$x_1 - 2x_2 + x_3 = -4 \quad (1)$$

$$-2x_1 + 3x_2 + x_3 = 9 \quad (2)$$

$$3x_1 + 4x_2 - 5x_3 = 0 \quad (3)$$

Multiplication of (1) by 2 yields

$$2x_1 - 4x_2 + 2x_3 = -8 \quad (4)$$

Addition of (2) and (4) yields

$$-x_2 + 3x_3 = 1 \quad (5)$$

Multiplication of (1) by –3 yields

$$-3x_1 + 6x_2 - 3x_3 = 12 \quad (6)$$

Addition of (3) and (6) yields

$$10x_2 - 8x_3 = 12 \quad (7)$$

Multiplication of (5) by 10 yields

$$-10x_2 + 30x_3 = 10 \quad (8)$$

Addition of (7) and (8) yields

$$22x_3 = 22 \quad (9)$$

or

$$x_3 = 1 \quad (10)$$

Substitution of (10) into (7) yields

$$10x_2 - 8 = 12 \quad (11)$$

or

$$x_2 = 2 \quad (12)$$

and substitution of (10) and (12) into (1) yields

$$x_1 - 4 + 1 = -4 \quad (13)$$

or

$$x_1 = -1 \quad (14)$$

6.

```
Delta=[-1 2 -3 5; 1 3 2 -1; 3 -3 2 4; 4 2 5 1];
D1=[14 2 -3 5; 9 3 2 -1; 19 -3 2 4; 27 2 5 1];
D2=[-1 14 -3 5; 1 9 2 -1; 3 19 2 4; 4 27 5 1];
D3=[-1 2 14 5; 1 3 9 -1; 3 -3 19 4; 4 2 27 1];
D4=[-1 2 -3 14; 1 3 2 9; 3 -3 2 19; 4 2 5 27];
x1=det(D1)/det(Delta), x2=det(D2)/det(Delta),...
x3=det(D3)/det(Delta), x4=det(D4)/det(Delta)
x1=1      x2=2      x3=3      x4=4
```

7.

$$\det A = \begin{matrix} 1 & 3 & 4 & 1 & 3 \\ 3 & 1 & -2 & 3 & 1 \\ 2 & 3 & 5 & 2 & 3 \end{matrix}$$

$$= 1 \times 1 \times 5 + 3 \times (-2) \times 2 + 4 \times 3 \times 3 - [2 \times 1 \times 4 + 3 \times (-2) \times 1 + 5 \times 3 \times 3]$$

$$= 5 - 12 + 36 - 8 + 6 - 45 = -18$$

$$\text{adj}A = \begin{bmatrix} 11 & -3 & -10 \\ -19 & -3 & 14 \\ 7 & 3 & -8 \end{bmatrix}$$

$$A^{-1} = \frac{1}{\det A} \cdot \text{adj}A = \frac{1}{-18} \cdot \begin{bmatrix} 11 & -3 & -10 \\ -19 & -3 & 14 \\ 7 & 3 & -8 \end{bmatrix} = \begin{bmatrix} -11/18 & 3/18 & 10/18 \\ 19/18 & 3/18 & -14/18 \\ -7/18 & -3/18 & 8/18 \end{bmatrix}$$

$$X = \begin{bmatrix} x_1 \\ x_2 \\ x_3 \end{bmatrix} = \begin{bmatrix} -11/18 & 3/18 & 10/18 \\ 19/18 & 3/18 & -14/18 \\ -7/18 & -3/18 & 8/18 \end{bmatrix} \begin{bmatrix} -3 \\ -2 \\ 0 \end{bmatrix} = \begin{bmatrix} 33/18 - 6/18 + 0 \\ -57/18 - 6/18 + 0 \\ 21/18 + 6/18 + 0 \end{bmatrix} = \begin{bmatrix} 27/18 \\ -63/18 \\ 27/18 \end{bmatrix} = \begin{bmatrix} 1.50 \\ -3.50 \\ 1.50 \end{bmatrix}$$

1	Spreadsheet for Matrix Inversion and					
2	Matrix Multiplication - Exercise 7					
3						
4		1.00	3.00	4.00		-3.00
5	A=	3.00	1.00	-2.00	B=	-2.00
6		2.00	3.00	5.00		0.00
7						
8		-0.61	0.17	0.56		1.50
9	A⁻¹	1.06	0.17	-0.78	X=	-3.50
10		-0.39	-0.17	0.44		1.50

8.

	A	B	C	D	E	F	G	H
1	Spreadsheet for Matrix Inversion and							
2	Matrix Multiplication - Exercise 8							
3								
4		2.00	4.00	3.00	-2.00			1.00
5	A=	2.00	-4.00	1.00	3.00		B=	10.00
6		-1.00	3.00	-4.00	2.00			-14.00
7		2.00	-2.00	2.00	1.00			7.00
8								
9		-1.58	-4.08	1.17	6.75			-11.50
10	A⁻¹	0.58	1.08	-0.17	-1.75		X=	1.50
11		1.50	3.50	-1.00	-5.50			12.00
12		1.33	3.33	-0.67	-5.00			9.00

A=[2 4 3 –2; 2 –4 1 3; –1 3 –4 2; 2 –2 2 1];

B=[1 10 –14 7]'; A\B

ans =

  –11.5000

    1.5000

   12.0000

    9.0000

NOTES:

Chapter 5

Differential Equations, State Variables, and State Equations

This chapter is a review of ordinary differential equations and an introduction to state variables and state equations. Solutions of differential equations with numerical methods is discussed in Chapter 9.

5.1 Simple Differential Equations

In this section we present two simple examples to show the importance of differential equations in engineering applications.

Example 5.1

The current and voltage in a capacitor are related by

$$i_C(t) = C\frac{dv_C}{dt} \tag{5.1}$$

where $i_C(t)$ is the current through the capacitor, $v_C(t)$ is the voltage across the capacitor, and the constant C is the capacitance in farads (F). For this example $C = 1$ F and the capacitor is being charged by a constant current I. Find the voltage v_C across this capacitor as a function of time given that the voltage at some reference time $t = 0$ is V_0.

Solution:

It is given that the current, as a function of time, is constant, that is,

$$i_C(t) = I = constant \tag{5.2}$$

By substitution of (5.2) into (5.1) we obtain

$$\frac{dv_C}{dt} = I$$

and by separation of the variables,

$$dv_C = Idt \tag{5.3}$$

Integrating both sides of (5.3) we obtain

$$v_C(t) = It + k \tag{5.4}$$

where k represents the constants of integration of both sides.

We can find the value of the constant k by making use of the initial condition, i.e., at $t = 0$, $v_C = V_0$ and (5.4) then becomes

$$V_0 = 0 + k \tag{5.5}$$

or $k = V_0$, and by substitution into (5.4),

$$v_C(t) = It + V_0 \tag{5.6}$$

This example shows that *when a capacitor is charged with a constant current, a linear voltage is produced across the terminals of the capacitor.*

Example 5.2

Find the current $i_L(t)$ through an inductor whose slope at the coordinate (t, i_L) is $\cos t$ and the current i_L passes through the point $(\pi/2, 1)$.

Solution:

We are given that

$$\frac{di_L}{dt} = \cos t \tag{5.7}$$

By separating the variables we obtain

$$di_L = \cos t \, dt \tag{5.8}$$

and integrating both sides we obtain

$$i_L(t) = \sin t + k \tag{5.9}$$

where k represents the constants of integration of both sides.

We find the value of the constant k by making use of the initial condition. For this example, $\omega = 1$ and thus at $\omega t = t = \pi/2$, $i_L = 1$. With these values (5.9) becomes

$$1 = \sin\frac{\pi}{2} + k \tag{5.10}$$

or $k = 0$, and by substitution into (5.9),

$$i_L(t) = \sin t \tag{5.11}$$

5.2 Classification

Differential equations are classified by:

1. *Type* - Ordinary or Partial

2. *Order* - The highest order derivative which is included in the differential equation

3. *Degree* - The exponent of the highest power of the highest order derivative after the differential equation has been cleared of any fractions or radicals in the dependent variable and its derivatives

For example, the differential equation

$$\left(\frac{d^4y}{dx^4}\right)^2 + 5\left(\frac{d^3y}{dx^3}\right)^4 + 6\left(\frac{d^2y}{dx^2}\right)^6 + 3\left(\frac{dy}{dx}\right)^8 + \frac{y^2}{x^3+1} = ye^{-2x}$$

is an ordinary differential equation of order 4 and degree 2.

If the dependent variable y is a function of only a single variable x, that is, if $y = f(x)$, the differential equation which relates y and x is said to be an *ordinary differential equation* and it is abbreviated as ODE.

The differential equation

$$\frac{d^2y}{dt^2} + 3\frac{dy}{dt} + 2 = 5\cos 4t$$

is an ODE with constant coefficients.

The differential equation

$$x^2\frac{d^2y}{dt^2} + x\frac{dy}{dt} + (x^2 - n^2) = 0$$

is an ODE with variable coefficients.

If the dependent variable y is a function of two or more variables such as $y = f(x, t)$, where x and t are independent variables, the differential equation that relates y, x, and t is said to be a *partial differential equation* and it is abbreviated as *PDE*.

An example of a partial differential equation is the well-known *one-dimensional wave equation* shown below.

$$\frac{\partial^2 y}{\partial t^2} = a^2\frac{\partial^2 y}{\partial x^2}$$

Most engineering problems are solved with ordinary differential equations with constant coefficients; however, partial differential equations provide often quick solutions to some practical applications as illustrated with the following three examples.

Example 5.3

The equivalent resistance R_T of three resistors R_1, R_2, and R_3 in parallel is obtained from

$$\frac{1}{R_T} = \frac{1}{R_1} + \frac{1}{R_2} + \frac{1}{R_3}$$

Given that initially $R_1 = 5 \ \Omega$, $R_2 = 20 \ \Omega$, and $R_3 = 4 \ \Omega$, compute the change in R_T if R_2 is increased by 10% and R_3 is decreased by 5% while R_1 does not change.

Solution:

The initial value of the equivalent resistance is $R_T = 5 \parallel 20 \parallel 4 = 2 \ \Omega$.

Now, we treat R_2 and R_3 as constants and differentiating R_T with respect to R_1 we obtain

$$-\frac{1}{R_T^2}\frac{\partial R_T}{\partial R_1} = -\frac{1}{R_1^2} \quad \text{or} \quad \frac{\partial R_T}{\partial R_1} = \left(\frac{R_T}{R_1}\right)^2$$

Similarly,

$$\frac{\partial R_T}{\partial R_2} = \left(\frac{R_T}{R_2}\right)^2 \quad \text{and} \quad \frac{\partial R_T}{\partial R_3} = \left(\frac{R_T}{R_3}\right)^2$$

and the total differential dR_T is

$$R_T = \frac{\partial R_T}{\partial R_1}dR_1 + \frac{\partial R_T}{\partial R_2}dR_2 + \frac{\partial R_T}{\partial R_3}dR_3 = \left(\frac{R_T}{R_1}\right)^2 dR_1 + \left(\frac{R_T}{R_2}\right)^2 dR_2 + \left(\frac{R_T}{R_3}\right)^2 dF$$

By substitution of the given numerical values we obtain

$$dR_T = \left(\frac{2}{5}\right)^2(0) + \left(\frac{2}{20}\right)^2(2) + \left(\frac{2}{4}\right)^2(-0.2) = 0.02 - 0.05 = -0.03$$

Therefore, the eequivalent resistance decreases by 3%.

Example 5.4

In a series RC electric circuit that is excited by a sinusoidal voltage, the magnitude of the impedance Z is computed from $Z = \sqrt{R^2 + X_C^2}$. Initially, $R = 4 \ \Omega$ and $X_C = 3 \ \Omega$. Find the change in the impedance Z if the resistance R is increased by 0.25 Ω (6.25%) and the capacitive reactance X_C is decreased by 0.125 Ω (−4.167%).

Solution:

We will first find the partial derivatives $\frac{\partial Z}{\partial R}$ and $\frac{\partial Z}{\partial X_C}$; then we compute the change in impedance from the total differential dZ. Thus,

$$\frac{\partial Z}{\partial R} = \frac{R}{\sqrt{R^2 + X_C^2}} \quad \text{and} \quad \frac{\partial Z}{\partial X_C} = \frac{X_C}{\sqrt{R^2 + X_C^2}}$$

and

$$dZ = \frac{\partial Z}{\partial R} dR + \frac{\partial Z}{\partial X_C} dX_C = \frac{R\, dR + X_C\, dX_C}{\sqrt{R^2 + X_C^2}}$$

and by substitution of the given values

$$dZ = \frac{4\,(0.25) + 3\,(-0.125)}{\sqrt{4^2 + 3^2}} = \frac{1 - 0.375}{5} = 0.125$$

Therefore, if R increases by 6.25% and X_C decreases by 4.167%, the impedance Z increases by 4.167%.

Example 5.5

A light bulb is rated at 120 volts and 75 watts. If the voltage decreases by 5 volts and the resistance of the bulb is increased by 8 Ω, by how much will the power change?

Solution:

At $V = 120$ volts and $P = 75$ watts, the bulb resistance is

$$R = \frac{V^2}{P} = \frac{120^2}{75} = 192\ \Omega$$

and since

$$P = \frac{V^2}{R} \quad \text{then} \quad \frac{\partial P}{\partial V} = \frac{2V}{R} \quad \text{and} \quad \frac{\partial P}{\partial R} = -\frac{V^2}{R^2}$$

and the total differential is

$$dP = \frac{\partial P}{\partial V} dV + \frac{\partial P}{\partial R} dR = \frac{2V}{R}dV - \frac{V^2}{R^2}dR = \frac{2(120)}{192}(-5) - \frac{120^2}{192^2}(8) = -9.375$$

That is, the power will decrease by 9.375 watts.

5.3 Solutions of Ordinary Differential Equations (ODE)

A function $y = f(x)$ is a solution of a differential equation if the latter is satisfied when y and its derivatives are replaced throughout by $f(x)$ and its corresponding derivatives. Also, the initial conditions must be satisfied.

For example a solution of the differential equation

$$\frac{d^2 y}{dx^2} + y = 0$$

is

$$y = k_1 \sin x + k_2 \cos x$$

since y and its second derivative satisfy the given differential equation.

Any linear, time-invariant system can be described by an ODE which has the form

$$
\boxed{
\begin{array}{c}
a_n \dfrac{d^n y}{dt^n} + a_{n-1} \dfrac{d^{n-1} y}{dt^{n-1}} + \ldots + a_1 \dfrac{dy}{dt} + a_0 y \\[2mm]
= \underbrace{b_m \dfrac{d^m x}{dt^m} + b_{m-1} \dfrac{d^{m-1} x}{dt^{n-1}} + \ldots + b_1 \dfrac{dx}{dt} + b_0 x}_{\text{Excitation (Forcing) Function } x(t)} \\[2mm]
\text{NON–HOMOGENEOUS DIFFERENTIAL EQUATION}
\end{array}
}
\tag{5.12}
$$

If the excitation in (B12) is not zero, that is, if $x(t) \neq 0$, the ODE is called a *non-homogeneous ODE*. If $x(t) = 0$, it reduces to:

$$
\boxed{
\begin{array}{c}
a_n \dfrac{d^n y}{dt^n} + a_{n-1} \dfrac{d^{n-1} y}{dt^{n-1}} + \ldots + a_1 \dfrac{dy}{dt} + a_0 y = 0 \\[2mm]
\text{HOMOGENEOUS DIFFERENTIAL EQUATION}
\end{array}
}
\tag{5.13}
$$

The differential equation of (5.13) above is called a *homogeneous ODE* and has n different linearly independent solutions denoted as $y_1(t), y_2(t), y_3(t), \ldots, y_n(t)$.

We will now prove that the *most general solution* of (5.13) is:

$$y_H(t) = k_1 y_1(t) + k_2 y_2(t) + k_3 y_3(t) + \ldots + k_n y_n(t) \tag{5.14}$$

where the subscript H on the left side is used to emphasize that this is the form of the solution of the homogeneous ODE and $k_1, k_2, k_3, \ldots, k_n$ are arbitrary constants.

Proof:

Let us assume that $y_1(t)$ is a solution of (5.13); then by substitution,

$$a_n \frac{d^n y_1}{dt^n} + a_{n-1} \frac{d^{n-1} y_1}{dt^{n-1}} + \dots + a_1 \frac{dy_1}{dt} + a_0 y_1 = 0 \qquad (5.15)$$

A solution of the form $k_1 y_1(t)$ will also satisfy (5.13) since

$$a_n \frac{d^n}{dt^n}(k_1 y_1) + a_{n-1} \frac{d^{n-1}}{dt^{n-1}}(k_1 y_1) + \dots + a_1 \frac{d}{dt}(k_1 y_1) + a_0(k_1 y_1)$$

$$= k_1 \left(a_n \frac{d^n y_1}{dt^n} + a_{n-1} \frac{d^{n-1} y_1}{dt^{n-1}} + \dots + a_1 \frac{dy_1}{dt} + a_0 y_1 \right) = 0 \qquad (5.16)$$

If $y = y_1(t)$ and $y = y_2(t)$ are any two solutions, then $y = y_1(t) + y_2(t)$ will also be a solution since

$$a_n \frac{d^n y_1}{dt^n} + a_{n-1} \frac{d^{n-1} y_1}{dt^{n-1}} + \dots + a_1 \frac{dy_1}{dt} + a_0 y_1 = 0$$

and

$$a_n \frac{d^n y_2}{dt^n} + a_{n-1} \frac{d^{n-1} y_2}{dt^{n-1}} + \dots + a_1 \frac{d y_2}{dt} + a_0 y_2 = 0$$

Therefore,

$$a_n \frac{d^n}{dt^n}(y_1 + y_2) + a_{n-1} \frac{d^{n-1}}{dt^{n-1}}(y_1 + y_2) + \dots + a_1 \frac{d}{dt}(y_1 + y_2) + a_0(y_1 + y_2) \qquad (5.17)$$

$$= a_n \frac{d^n}{dt^n} y_1 + a_{n-1} \frac{d^{n-1}}{dt^{n-1}} y_1 + \dots + a_1 \frac{d}{dt} y_1 + a_0 y_1$$

$$+ a_n \frac{d^n}{dt^n} y_2 + a_{n-1} \frac{d^{n-1}}{dt^{n-1}} y_2 + \dots + a_1 \frac{d}{dt} y_2 + a_0 y_2 = 0$$

In general, if

$$y = k_1 y_1(t), k_2 y_1(t), k_3 y_3(t), \dots, k_n y_n(t)$$

are the n solutions of the homogeneous ODE of (5.13), the linear combination

$$y = k_1 y_1(t) + k_2 y_1(t) + k_3 y_3(t) + \dots + k_n y_n(t)$$

is also a solution.

In our subsequent discussion, the solution of the homogeneous ODE, i.e., the complementary solution, will be referred to as the *natural response*, and will be denoted as $y_N(t)$ or simply y_N. The particular solution of a non-homogeneous ODE will be referred to as the *forced response*, and will be denoted as $y_F(t)$ or simply y_F. Accordingly, we express the total solution of the non-homogeneous ODE of (5.12) as:

$$\boxed{y(t) = y_{\underset{\text{Response}}{\text{Natural}}} + y_{\underset{\text{Response}}{\text{Forced}}} = y_N + y_F} \tag{5.18}$$

The natural response y_N contains arbitrary constants and these can be evaluated from the given initial conditions. The forced response y_F, however, contains no arbitrary constants. It is imperative to remember that the arbitrary constants of the natural response must be evaluated from the total response.

5.4 Solution of the Homogeneous ODE

Let the solutions of the homogeneous ODE

$$a_n \frac{d^n y}{dt^n} + a_{n-1} \frac{d^{n-1} y}{dt^{n-1}} + \ldots + a_1 \frac{dy}{dt} + a_0 y = 0 \tag{5.19}$$

be of the form

$$y = ke^{st} \tag{5.20}$$

Then, by substitution of (5.20) into (5.19) we obtain

$$a_n ks^n e^{st} + a_{n-1} ks^{n-1} e^{st} + \ldots + a_1 kse^{st} + a_0 ke^{st} = 0$$

or

$$(a_n s^n + a_{n-1} s^{n-1} + \ldots + a_1 s + a_0) ke^{st} = 0 \tag{5.21}$$

We observe that (5.21) can be satisfied when

$$(a_n s^n + a_{n-1} s^{n-1} + \ldots + a_1 s + a_0) = 0 \quad \text{or} \quad k = 0 \quad \text{or} \quad s = -\infty \tag{5.22}$$

but the only meaningful solution is the quantity enclosed in parentheses since the latter two yield trivial (meaningless) solutions. We, therefore, accept the expression inside the parentheses as the only meaningful solution and this is referred to as the *characteristic (auxiliary) equation*, that is,

$$\boxed{\underbrace{a_n s^n + a_{n-1} s^{n-1} + \ldots + a_1 s + a_0 = 0}_{\text{Characteristic Equation}}} \tag{5.23}$$

Since the characteristic equation is an algebraic equation of an *nth-power* polynomial, its solutions are $s_1, s_2, s_3, \ldots, s_n$, and thus the solutions of the homogeneous ODE are:

$$y_1 = k_1 e^{s_1 t}, \quad y_2 = k_2 e^{s_2 t}, \quad y_3 = k_3 e^{s_3 t}, \quad ..., \quad y_n = k_n e^{s_n t} \tag{5.24}$$

Case I – Distinct Roots

If the roots of the characteristic equation are *distinct* (different from each another), the n solutions of (5.23) are independent and the most general solution is:

$$\boxed{\begin{array}{c} y_N = k_1 e^{s_1 t} + k_2 e^{s_2 t} + ... + k_n e^{s_n t} \\ \text{FOR DISTINCT ROOTS} \end{array}} \tag{5.25}$$

Case II – Repeated Roots

If two or more roots of the characteristic equation are *repeated* (same roots), then some of the terms of (5.24) are not independent and therefore (5.25) does not represent the most general solution. If, for example, $s_1 = s_2$, then,

$$k_1 e^{s_1 t} + k_2 e^{s_2 t} = k_1 e^{s_1 t} + k_2 e^{s_1 t} = (k_1 + k_2)e^{s_1 t} = k_3 e^{s_1 t}$$

and we see that one term of (5.25) is lost. In this case, we express one of the terms of (5.25), say $k_2 e^{s_1 t}$ as $k_2 t e^{s_1 t}$. These two represent two independent solutions and therefore the most general solution has the form:

$$y_N = (k_1 + k_2 t)e^{s_1 t} + k_3 e^{s_3 t} + ... + k_n e^{s_n t} \tag{5.26}$$

If there are m equal roots the most general solution has the form:

$$\boxed{\begin{array}{c} y_N = (k_1 + k_2 t + ... + k_m t^{m-1})e^{s_1 t} + k_{n-i} e^{s_2 t} + ... + k_n e^{s_n t} \\ \text{FOR M EQUAL ROOTS} \end{array}} \tag{5.27}$$

Case III – Complex Roots

If the characteristic equation contains complex roots, these occur as complex conjugate pairs. Thus, if one root is $s_1 = -\alpha + j\beta$ where α and β are real numbers, then another root is $s_1 = -\alpha - j\beta$. Then,

$$\boxed{\begin{aligned} k_1 e^{s_1 t} + k_2 e^{s_2 t} &= k_1 e^{-\alpha t + j\beta t} + k_2 e^{-\alpha t - j\beta t} = e^{-\alpha t}(k_1 e^{j\beta t} + k_2 e^{-j\beta t}) \\ &= e^{-\alpha t}(k_1 \cos\beta t + jk_1 \sin\beta t + k_2 \cos\beta t - jk_2 \sin\beta t) \\ &= e^{-\alpha t}[(k_1 + k_2)\cos\beta t + j(k_1 - k_2)\sin\beta t] \\ &= e^{-\alpha t}(k_3 \cos\beta t + k_4 \sin\beta t) = e^{-\alpha t}k_5 \cos(\beta t + \varphi) \\ \text{FOR TWO COMPLEX CONJUGATE ROOTS} \end{aligned}} \tag{5.28}$$

If (5.28) is to be a real function of time, the constants k_1 and k_2 must be complex conjugates. The other constants k_3, k_4, k_5, and the phase angle φ are real constants.

The forced response can be found by

a. *The Method of Undetermined Coefficients* or

b. *The Method of Variation of Parameters*

We will study the Method of Undetermined Coefficients first.

5.5 Using the Method of Undetermined Coefficients for the Forced Response

For simplicity, we will only consider ODEs of order 2. Higher order ODEs are discussed in differential equations textbooks.

Consider the non-homogeneous ODE

$$a\frac{d^2y}{dt^2} + b\frac{d}{dt}y + cy = f(x) \tag{5.29}$$

where a, b, and c are real constants.

We have learned that the total (complete) solution consists of the summation of the natural and forced responses.

For the natural response, if y_1 and y_2 are any two solutions of (5.29), the linear combination $y_3 = k_1y_1 + k_2y_2$, where k_1 and k_2 are arbitrary constants, is also a solution, that is, if we know the two solutions, we can obtain the most general solution by forming the linear combination of y_1 and y_2. To be certain that there exist no other solutions, we examine the Wronskian Determinant defined below.

$$W(y_1, y_2) \equiv \begin{bmatrix} y_1 & y_2 \\ \dfrac{d}{dx}y_1 & \dfrac{d}{dx}y_2 \end{bmatrix} = y_1\frac{d}{dx}y_2 - y_2\frac{d}{dx}y_1 \neq 0 \tag{5.30}$$

WRONSKIAN DETERMINANT

If (5.30) is true, we can be assured that all solutions of (5.29) are indeed the linear combination of y_1 and y_2.

The forced response is obtained by observation of the right side of the given ODE as it is illustrated by the examples that follow.

Example 5.6

Find the total solution of the ODE

$$\frac{d^2y}{dt^2} + 4\frac{dy}{dt} + 3y = 0 \qquad (5.31)$$

subject to the initial conditions $y(0) = 3$ and $y'(0) = 4$ where $y' = dy/dt$

Solution:

This is a homogeneous ODE and its total solution is just the natural response found from the characteristic equation $s^2 + 4s + 3 = 0$ whose roots are $s_1 = -1$ and $s_2 = -3$. The total response is:

$$y(t) = y_N(t) = k_1 e^{-t} + k_2 e^{-3t} \qquad (5.32)$$

The constants k_1 and k_2 are evaluated from the given initial conditions. For this example,

$$y(0) = 3 = k_1 e^0 + k_2 e^0$$

or

$$k_1 + k_2 = 3 \qquad (5.33)$$

Also,

$$y'(0) = 4 = \frac{dy}{dt}\bigg|_{t=0} = -k_1 e^{-t} - 3k_2 e^{-3t}\bigg|_{t=0}$$

or

$$-k_1 - 3k_2 = 4 \qquad (5.34)$$

Simultaneous solution of (5.33) and (5.34) yields $k_1 = 6.5$ and $k_2 = -3.5$. By substitution into (5.32), we obtain

$$y(t) = y_N(t) = 6.5e^{-t} - 3.5e^{-3t} \qquad (5.35)$$

Check with MATLAB:

```
y=dsolve('D2y+4*Dy+3*y=0', 'y(0)=3', 'Dy(0)=4')

y =
(-7/2*exp(-3*t)*exp(t)+13/2)/exp(t)

pretty(y)
        - 7/2 exp(-3 t) exp(t) + 13/2
        -----------------------------
                  exp(t)
```

The function $y = f(t)$, of relation (5.35), shown in Figure 5.1, was plotted with the use of the MATLAB script y=dsolve('D2y+4*Dy+3*y=0', 'y(0)=3', 'Dy(0)=4'); ezplot(y,[0 5])

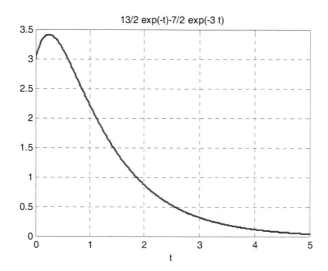

$$13/2\ exp(-t)-7/2\ exp(-3\ t)$$

Figure 5.1. Plot for the function $y = f(t)$ *of Example 5.6.*

Example 5.7

Find the total solution of the ODE

$$\frac{d^2 y}{dt^2} + 4\frac{dy}{dt} + 3y = 3e^{-2t} \tag{5.36}$$

subject to the initial conditions $y(0) = 1$ and $y'(0) = -1$

Solution:

The left side of (5.36) is the same as that of Example 5.6. Therefore,

$$y_N(t) = k_1 e^{-t} + k_2 e^{-3t} \tag{5.37}$$

(We must remember that the constants k_1 and k_2 must be evaluated from the total response).

To find the forced response, we assume a solution of the form

$$y_F = Ae^{-2t} \tag{5.38}$$

We can find out whether our assumption is correct by substituting (5.38) into the given ODE of (5.36). Then,

$$4Ae^{-2t} - 8Ae^{-2t} + 3Ae^{-2t} = 3e^{-2t} \tag{5.39}$$

from which $A = -3$ and the total solution is

$$y(t) = y_N + y_F = k_1 e^{-t} + k_2 e^{-3t} - 3e^{-2t} \qquad (5.40)$$

The constants k_1 and k_2 are evaluated from the given initial conditions. For this example,

$$y(0) = 1 = k_1 e^0 + k_2 e^0 - 3e^0$$

or

$$k_1 + k_2 = 4 \qquad (5.41)$$

Also,

$$y'(0) = -1 = \left.\frac{dy}{dt}\right|_{t=0} = \left.-k_1 e^{-t} - 3k_2 e^{-3t} + 6e^{-2t}\right|_{t=0}$$

or

$$-k_1 - 3k_2 = -7 \qquad (5.42)$$

Simultaneous solution of (5.41) and (5.42) yields $k_1 = 2.5$ and $k_2 = 1.5$. By substitution into (5.40), we obtain

$$y(t) = y_N + y_F = 2.5e^{-t} + 1.5e^{-3t} - 3e^{-2t} \qquad (5.43)$$

Check with MATLAB:

y=dsolve('D2y+4*Dy+3*y=3*exp(−2*t)', 'y(0)=1', 'Dy(0)=−1')

```
y =
(-3*exp(-2*t)*exp(t)+3/2*exp(-3*t)*exp(t)+5/2)/exp(t)
```

pretty(y)

```
   -3 exp(-2 t) exp(t) + 3/2 exp(-3 t) exp(t) + 5/2
   -------------------------------------------------
                        exp(t)
```

The plot is shown in Figure 5.2 was produced with the MATLAB script

y=dsolve('D2y+4*Dy+3*y=3*exp(−2*t)', 'y(0)=1', 'Dy(0)=−1'); ezplot(y,[0 8])

Example 5.8

Find the total solution of the ODE

$$\frac{d^2 y}{dt^2} + 6\frac{dy}{dt} + 9y = 0 \qquad (5.44)$$

subject to the initial conditions $y(0) = -1$ and $y'(0) = 1$

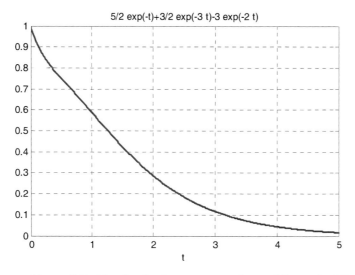

Figure 5.2. Plot for the function $y = f(t)$ *of Example 5.7.*

Solution:

This is a homogeneous ODE and therefore its total solution is just the natural response found from the characteristic equation $s^2 + 6s + 9 = 0$ whose roots are $s_1 = s_2 = -3$ (repeated roots). Thus, the total response is

$$y(t) = y_N = k_1 e^{-3t} + k_2 t e^{-3t} \tag{5.45}$$

Next, we evaluate the constants k_1 and k_2 from the given initial conditions. For this example,

$$y(0) = -1 = k_1 e^0 + k_2(0)e^0$$

or

$$k_1 = -1 \tag{5.46}$$

Also,

$$y'(0) = 1 = \left.\frac{dy}{dt}\right|_{t=0} = -3k_1 e^{-3t} + k_2 e^{-3t} - 3k_2 t e^{-3t}\Big|_{t=0}$$

or

$$-3k_1 + k_2 = 1 \tag{5.47}$$

From (5.46) and (5.47) we obtain $k_1 = -1$ and $k_2 = -2$. By substitution into (5.45),

$$y(t) = -e^{-3t} - 2te^{-3t} \tag{5.48}$$

Check with MATLAB:

```
y=dsolve('D2y+6*Dy+9*y=0', 'y(0)=-1', 'Dy(0)=1')
```

y =
-exp(-3*t)-2*exp(-3*t)*t

The plot shown in Figure 5.3 was produced with the MATLAB script

y=dsolve('D2y+6*Dy+9*y=0', 'y(0)=−1', 'Dy(0)=1'); ezplot(y,[0 3])

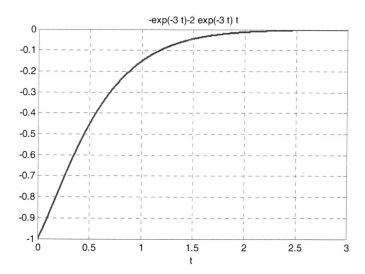

Figure 5.3. Plot for the function y = f(t) *of Example 5.8.*

Example 5.9

Find the total solution of the ODE

$$\frac{d^2y}{dt^2} + 5\frac{dy}{dt} + 6y = 3e^{-2t} \tag{5.49}$$

Solution:

No initial conditions are given; therefore, we will express the solution in terms of the constants k_1 and k_2. By inspection, the roots of the characteristic equation of (5.49) are $s_1 = -2$ and $s_2 = -3$ and thus the natural response has the form

$$y_N = k_1 e^{-2t} + k_2 e^{-3t} \tag{5.50}$$

Next, we find the forced response by assuming a solution of the form

$$y_F = Ae^{-2t} \tag{5.51}$$

We can find out whether our assumption is correct by substitution of (5.51) into the given ODE of (5.49). Then,

$$4Ae^{-2t} - 10Ae^{-2t} + 6Ae^{-2t} = 3e^{-2t} \quad\quad (5.52)$$

but the sum of the three terms on the left side of (5.52) is zero whereas the right side can never be zero unless we let $t \to \infty$ and this produces a meaningless result.

The problem here is that the right side of the given ODE of (5.49) has the same form as one of the terms of the natural response of (5.50), namely the term $k_1 e^{-2t}$.

To work around this problem, we assume that the forced response has the form

$$y_F = Ate^{-2t} \quad\quad (5.53)$$

that is, we multiply (5.51) by t in order to eliminate the duplication of terms in the total response. Then, by substitution of (5.53) into (5.49) and equating like terms, we find that $A = 3$. Therefore, the total response is

$$y(t) = y_N + y_F = k_1 e^{-2t} + k_2 e^{-3t} + 3te^{-2t} \quad\quad (5.54)$$

Check with MATLAB:

```
y=dsolve('D2y+5*Dy+6*y=3*exp(-2*t)')
```

```
y =
-3*exp(-2*t)+3*t*exp(-2*t)+C1*exp(-3*t)+C2*exp(-2*t)
```

We observe that the first and last terms of the displayed expression above have the same form and thus they can be combined to form a single term `C3*exp(-2*t)`.

Example 5.10

Find the total solution of the ODE

$$\frac{d^2y}{dt^2} + 5\frac{dy}{dt} + 6y = 4\cos 5t \quad\quad (5.55)$$

Solution:

No initial conditions are given; therefore, we will express solution in terms of the constants k_1 and k_2. We observe that the left side of (5.55) is the same of that of Example 5.9. Therefore, the natural response is the same, that is, it has the form

$$y_N = k_1 e^{-2t} + k_2 e^{-3t} \quad\quad (5.56)$$

Next, to find the forced response and we assume a solution of the form

$$y_F = A\cos 5t \qquad (5.57)$$

We can find out whether our assumption is correct by substitution of the assumed solution of (5.57) into the given ODE of (5.56). Then,

$$-25A\cos 5t - 25A\sin 5t + 6A\cos 5t = -19A\cos 5t - 25A\sin 5t = 4\cos 5t$$

but this relation is invalid since by equating cosine and sine terms, we find that $A = -4/19$ and also $A = 0$. This inconsistency is a result of our failure to recognize that the derivatives of $A\cos 5t$ produce new terms of the form $B\sin 5t$ and these terms must be included in the forced response. Accordingly, we let

$$y_F = k_3\sin 5t + k_4\cos 5t \qquad (5.58)$$

and by substitution into (5.55) we obtain

$$-25k_3\sin 5t - 25k_4\cos 5t + 25k_3\cos 5t - 25k_4\sin 5t + 6k_3\sin 5t + 6k_4\cos 5t = 4\cos 5t$$

Collecting like terms and equating sine and cosine terms, we obtain the following set of equations

$$\begin{aligned} 19k_3 + 25k_4 &= 0 \\ 25k_3 - 19k_4 &= 4 \end{aligned} \qquad (5.59)$$

We use MATLAB to solve (5.59)

format rat; [k3 k4]=solve(19*x+25*y, 25*x–19*y–4)

```
k3 =
50/493
k4 =
-38/493
```

Therefore, the total solution is

$$y(t) = y_N + y_F(t) = k_1 e^{-2t} + k_2 e^{-3t} + \frac{50}{493}\sin 5t + \frac{-38}{493}\cos 5t \qquad (5.60)$$

Check with MATLAB:

y=dsolve('D2y+5*Dy+6*y=4*cos(5*t)'); y=simple(y)

```
y =
-38/493*cos(5*t)+50/493*sin(5*t)+C1*exp(-3*t)+C2*exp(-2*t)
```

In most engineering problems the right side of the non–homogeneous ODE consists of elementary functions such as k (constant), x^n where n is a positive integer, e^{kx}, $\cos kx$, $\sin kx$, and linear combinations of these. Table 5.1 summarizes the forms of the forced response for a second order

ODE with constant coefficients.

TABLE 5.1 Form of the forced response for 2nd order differential equations

Forced Response of the ODE $a\dfrac{d^2y}{dt^2} + b\dfrac{dy}{dt} + cy = f(t)$	
$f(t)$	**Form of Forced Response $y_F(t)$**
k (constant)	K (constant)
kt^n (n = positive integer)	$K_0 t^n + K_1 t^{n-1} + \ldots + K_{n-1}t + K_n$
ke^{rt} (r = real or complex)	Ke^{rt}
$k\cos\alpha t$ or $k\sin\alpha t$ (α = constant)	$K_1\cos\alpha t + K_2\sin\alpha t$
$kt^n e^{rt}\cos\alpha t$ or $kt^n e^{rt}\sin\alpha t$	$(K_0 t^n + K_1 t^{n-1} + \ldots + K_{n-1}t + K_n)e^{rt}\cos\alpha t$ $+ (K_0 t^n + K_1 t^{n-1} + \ldots + K_{n-1}t + K_n)e^{rt}\sin\alpha t$

We must remember that if $f(t)$ is the sum of several terms, the most general form of the forced response $y_F(t)$ is the linear combination of these terms. Also, if a term in $y_F(t)$ is a duplicate of a term in the natural response $y_N(t)$, we must multiply $y_F(t)$ by the lowest power of t that will eliminate the duplication.

Example 5.11

Find the total solution of the ODE

$$\frac{d^2y}{dt^2} + 4\frac{dy}{dt} + 4y = te^{-2t} - e^{-2t} \tag{5.61}$$

Solution:

No initial conditions are given; therefore we will express solution in terms of the constants k_1 and k_2. The roots of the characteristic equation are equal, that is, $s_1 = s_2 = -2$, and thus the natural response has the form

$$y_N = k_1 e^{-2t} + k_2 te^{-2t} \tag{5.62}$$

To find the forced response (particular solution), we refer to the table of the previous page and from the last row we choose the term $kt^n e^{rt}\cos\alpha t$. This term with $n = 1$, $r = -2$, and $\alpha = 0$, reduces to kte^{-2t}. Therefore the forced response will have the form

$$y_F = (k_3 t + k_4)e^{-2t} \tag{5.63}$$

But the terms e^{-2t} and te^{-2t} are also present in (5.61); therefore, we multiply (5.62) by t^2 to obtain a suitable form for the forced response which now is

$$y_F = (k_3 t^3 + k_4 t^2)e^{-2t} \tag{5.64}$$

Now, we need to evaluate the constants k_3 and k_4. This is done by substituting (5.64) into the given ODE of (5.61) and equating with the right side. We use MATLAB do the computations as shown below.

```
syms t k3 k4                    % Define symbolic variables
f0=(k3*t^3+k4*t^2)*exp(-2*t);   % Forced response (5.64)
f1=diff(f0); f1=simple(f1)      % Compute and simplify first derivative

f1 =
-t*exp(-2*t)*(-3*k3*t-2*k4+2*k3*t^2+2*k4*t)

f2=diff(f0,2); f2=simple(f2)    % Compute and simplify second derivative

f2 =
2*exp(-2*t)*(3*k3*t+k4-6*k3*t^2-4*k4*t+2*k3*t^3+2*k4*t^2)

f=f2+4*f1+4*f0; f=simple(f)     % Form and simplify the left side of the given ODE

f = 2*(3*k3*t+k4)*exp(-2*t)
```

Finally, we equate f above with the right side of the given ODE, that is

$$2(3k_3 t + k_4)e^{-2t} = te^{-2t} - e^{-2t} \tag{5.65}$$

and we find $k_3 = 1/6$ and $k_4 = -1/2$. By substitution of these values into (5.64) and combining the forced response with the natural response, we obtain the total solution

$$y(t) = k_1 e^{-2t} + k_2 te^{-2t} + \frac{1}{6}t^3 e^{-2t} - \frac{1}{2}t^2 e^{-2t} \tag{5.66}$$

We verify this solution with MATLAB as follows:

```
z=dsolve('D2y+4*Dy+4*y=t*exp(-2*t)-exp(-2*t)')

z =
1/6*exp(-2*t)*t^3-1/2*exp(-2*t)*t^2
+C1*exp(-2*t)+C2*t*exp(-2*t)
```

5.6 Using the Method of Variation of Parameters for the Forced Response

In certain non–homogeneous ODEs, the right side $f(t)$ cannot be determined by the method of undetermined coefficients. For these ODEs we must use the method of variation of parameters. This method will work with all linear equations including those with variable coefficients such as

$$\frac{d^2 y}{dt^2} + \alpha(t)\frac{dy}{dt} + \beta(t)y = f(t) \tag{5.67}$$

provided that the general form of the natural response is known.

Our discussion will be restricted to second order ODEs with constant coefficients.

The method of variation of parameters replaces the constants k_1 and k_2 by two variables u_1 and u_2 that satisfy the following three relations:

$$\boxed{y = u_1 y_1 + u_2 y_2} \tag{5.68}$$

$$\boxed{\frac{du_1}{dt} y_1 + \frac{du_2}{dt} y_2 = 0} \tag{5.69}$$

$$\boxed{\frac{du_1}{dt} \cdot \frac{dy_1}{dt} + \frac{du_2}{dt} \cdot \frac{dy_2}{dt} = f(t)} \tag{5.70}$$

Simultaneous solution of (5.68) and (5.69) will yield the values of du_1/dt and du_2/dt; then, integration of these will produce u_1 and u_2, which when substituted into (5.67) will yield the total solution.

Example 5.12

Find the total solution of

$$\frac{d^2 y}{dt^2} + 4\frac{dy}{dt} + 3y = 12 \tag{5.71}$$

in terms of the constants k_1 and k_2 by the

a. method of undetermined coefficients

b. method of variation of parameters

Solution:

With either method, we must first find the natural response. The characteristic equation yields the roots $s_1 = -1$ and $s_2 = -3$. Therefore, the natural response is

$$y_N = k_1 e^{-t} + k_2 e^{-3t} \qquad (5.72)$$

a. Using the method of undetermined coefficients we let $y_F = k_3$ (a constant). Then, by substitution into (5.71) we obtain $k_3 = 4$ and thus the total solution is

$$y(t) = y_N + y_F = k_1 e^{-t} + k_2 e^{-3t} + 4 \qquad (5.73)$$

b. With the method of variation of parameters we begin with the natural response found above as (5.72) and we let the solutions y_1 and y_2 be represented as

$$y_1 = e^{-t} \text{ and } y_2 = e^{-3t} \qquad (5.74)$$

Then by (5.68), the total solution is

$$y = u_1 y_1 + u_2 y_2$$

or

$$y = u_1 e^{-t} + u_2 e^{-3t} \qquad (5.75)$$

Also, from (5.69),

$$\frac{du_1}{dt} y_1 + \frac{du_2}{dt} y_2 = 0$$

or

$$\frac{du_1}{dt} e^{-t} + \frac{du_2}{dt} e^{-3t} = 0 \qquad (5.76)$$

and from (5.70),

$$\frac{du_1}{dt} \cdot \frac{dy_1}{dt} + \frac{du_2}{dt} \cdot \frac{dy_2}{dt} = f(t)$$

or

$$\frac{du_1}{dt}(-e^{-t}) + \frac{du_2}{dt}(-3e^{-3t}) = 12 \qquad (5.77)$$

Next, we find du_1/dt and du_2/dt by Cramer's rule as follows:

$$\frac{du_1}{dt} = \frac{\begin{vmatrix} 0 & e^{-3t} \\ 12 & -3e^{-3t} \end{vmatrix}}{\begin{vmatrix} e^{-t} & e^{-3t} \\ -e^{-t} & -3e^{-3t} \end{vmatrix}} = \frac{-12e^{-3t}}{-3e^{-4t} + e^{-4t}} = \frac{-12e^{-3t}}{-2e^{-4t}} = 6e^{t} \qquad (5.78)$$

and

$$\frac{du_2}{dt} = \frac{\begin{vmatrix} e^{-t} & 0 \\ -e^{-t} & 12 \end{vmatrix}}{-2e^{-4t}} = \frac{12e^{-t}}{-2e^{-4t}} = -6e^{3t} \tag{5.79}$$

Now, integration of (5.78) and (5.79) and substitution into (5.75) yields

$$u_1 = 6\int e^t dt = 6e^t + k_1 \qquad u_2 = -6\int e^{3t} dt = -2e^{3t} + k_2 \tag{5.80}$$

$$y = u_1 e^{-t} + u_2 e^{-3t} = (6e^t + k_1)e^{-t} + (-2e^{3t} + k_2)e^{-3t}$$
$$= 6 + k_1 e^{-t} - 2 + k_2 e^{-3t} = k_1 e^{-t} + k_2 e^{-3t} + 4 \tag{5.81}$$

We observe that the last expression in (5.81) is the same as (5.73) of part (a).

Check with MATLAB:

```
y=dsolve('D2y+4*Dy+3*y=12')

y =
  (4*exp(t)+C1*exp(-3*t)*exp(t)+C2)/exp(t)
```

Example 5.13

Find the total solution of

$$\frac{d^2 y}{dt^2} + 4y = \tan 2t \tag{5.82}$$

in terms of the constants k_1 and k_2 by any method.

Solution:

This ODE cannot be solved by the method of undetermined coefficients; therefore, we will use the method of variation of parameters.

The characteristic equation is $s^2 + 4 = 0$ from which $s = \pm j2$ and thus the natural response is

$$y_N = k_1 e^{j2t} + k_2 e^{-j2t} \tag{5.83}$$

We let

$$y_1 = \cos 2t \text{ and } y_2 = \sin 2t \tag{5.84}$$

Then, by (5.68) the solution is

$$y = u_1 y_1 + u_2 y_2 = u_1 \cos 2t + u_2 \sin 2t \tag{5.85}$$

Also, from (5.69),

$$\frac{du_1}{dt}y_1 + \frac{du_2}{dt}y_2 = 0$$

or

$$\frac{du_1}{dt}\cos 2t + \frac{du_2}{dt}\sin 2t = 0 \tag{5.86}$$

and from (5.70),

$$\frac{du_1}{dt}\cdot\frac{dy_1}{dt} + \frac{du_2}{dt}\cdot\frac{dy_2}{dt} = f(t) = \frac{du_1}{dt}(-2\sin 2t) + \frac{du_2}{dt}(2\cos 2t) = \tan 2t \tag{5.87}$$

Next, we find du_1/dt and du_2/dt by Cramer's rule as follows:

$$\frac{du_1}{dt} = \frac{\begin{vmatrix} 0 & \sin 2t \\ \tan 2t & 2\cos 2t \end{vmatrix}}{\begin{vmatrix} \cos 2t & \sin 2t \\ -2\sin 2t & 2\cos 2t \end{vmatrix}} = \frac{-\dfrac{\sin^2 2t}{\cos 2t}}{2\cos^2 2t + 2\sin^2 2t} = \frac{-\sin^2 2t}{2\cos 2t} \tag{5.88}$$

and

$$\frac{du_2}{dt} = \frac{\begin{vmatrix} \cos 2t & 0 \\ -2\sin 2t & \tan 2t \end{vmatrix}}{2} = \frac{\sin 2t}{2} \tag{5.89}$$

Now, integration of (5.88) and (5.89) and substitution into (5.85) yields

$$u_1 = -\frac{1}{2}\int\frac{\sin^2 2t}{\cos 2t}dt = \frac{\sin 2t}{4} - \frac{1}{4}\ln(\sec 2t + \tan 2t) + k_1 \tag{5.90}$$

$$u_2 = \frac{1}{2}\int\sin 2t\,dt = -\frac{\cos 2t}{4} + k_2 \tag{5.91}$$

$$y = u_1 y_1 + u_2 y_2 = \frac{\sin 2t\cos 2t}{4} - \frac{1}{4}\cos 2t\ln(\sec 2t + \tan 2t) + k_1\cos 2t - \frac{\sin 2t\cos 2t}{4} + k_2\sin 2t$$

$$= -\frac{1}{4}\cos 2t\ln(\sec 2t + \tan 2t) + k_1\cos 2t + k_2\sin 2t \tag{5.92}$$

Check with MATLAB:

```
y=dsolve('D2y+4*y=tan(2*t)')

y =
-1/4*cos(2*t)*log((1+sin(2*t))/cos(2*t))+C1*cos(2*t)+C2*sin(2*t)
```

5.7 Expressing Differential Equations in State Equation Form

A first order differential equation with constant coefficients has the form

$$a_1 \frac{dy}{dt} + a_0 \, y(t) = x(t) \tag{5.93}$$

In a second order differential equation the highest order is a second derivative.

An *nth-order* differential equation can be resolved to n first-order simultaneous differential equations with a set of auxiliary variables called *state variables*. The resulting first-order differential equations are called *state space equations*, or simply *state equations*. The state variable method offers the advantage that it can also be used with non-linear and time-varying systems. However, our discussion will be limited to linear, time-invariant systems.

State equations can also be solved with numerical methods such as Taylor series and Runge-Kutta methods; these will be discussed in Chapter 9. The state variable method is best illustrated through several examples presented in this chapter.

Example 5.14

A system is described by the integro-differential equation

$$Ri + L\frac{di}{dt} + \frac{1}{C}\int_{-\infty}^{t} i \, dt = e^{j\omega t} \tag{5.94}$$

Differentiating both sides and dividing by L we obtain

$$\frac{d^2 i}{dt^2} + \frac{R}{L}\frac{di}{dt} + \frac{1}{LC}i = \frac{1}{L}j\omega e^{j\omega t} \tag{5.95}$$

or

$$\frac{d^2 i}{dt^2} = -\frac{R}{L}\frac{di}{dt} - \frac{1}{LC}i + \frac{1}{L}j\omega e^{j\omega t} \tag{5.96}$$

Next, we define two state variables x_1 and x_2 such that

$$x_1 = i \tag{5.97}$$

and

$$x_2 = \frac{di}{dt} = \frac{dx_1}{dt} = \dot{x}_1 \tag{5.98}$$

Then,

$$\dot{x}_2 = d^2 i / dt^2 \tag{5.99}$$

where $\dot{x}_k$ denotes the derivative of the state variable x_k.

From (5.96) through (5.99), we obtain the state equations

$$\dot{x}_1 = x_2$$
$$\dot{x}_2 = -\frac{R}{L}x_2 - \frac{1}{LC}x_1 + \frac{1}{L}j\omega e^{j\omega t} \qquad (5.100)$$

It is convenient and customary to express the state equations in matrix form. Thus, we write the state equations of (5.100) as

$$\begin{bmatrix} \dot{x}_1 \\ \dot{x}_2 \end{bmatrix} = \begin{bmatrix} 0 & 1 \\ -\dfrac{1}{LC} & -\dfrac{R}{L} \end{bmatrix} \begin{bmatrix} x_1 \\ x_2 \end{bmatrix} + \begin{bmatrix} 0 \\ \dfrac{1}{L}j\omega e^{j\omega t} \end{bmatrix} u \qquad (5.101)$$

We usually write (5.101) in a compact form as

$$\dot{x} = Ax + bu \qquad (5.102)$$

where

$$\dot{x} = \begin{bmatrix} \dot{x}_1 \\ \dot{x}_2 \end{bmatrix}, \quad A = \begin{bmatrix} 0 & 1 \\ -\dfrac{1}{LC} & -\dfrac{R}{L} \end{bmatrix}, \quad x = \begin{bmatrix} x_1 \\ x_2 \end{bmatrix}, \quad b = \begin{bmatrix} 0 \\ \dfrac{1}{L}j\omega e^{j\omega t} \end{bmatrix}, \quad and \quad u = any \ input \qquad (5.103)$$

The output $y(t)$ is expressed by the state equation

$$y = Cx + du \qquad (5.104)$$

where C is another matrix, and d is a column vector. Therefore, the state representation of a system can be described by the pair of the of the state space equations

$$\boxed{\begin{aligned} \dot{x} &= Ax + bu \\ y &= Cx + du \end{aligned}} \qquad (5.105)$$

The state space equations of (5.105) can be realized with the block diagram of Figure 5.1.

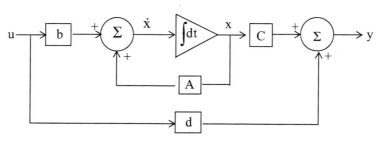

Figure 5.4. Block diagram for the realization of the state equations of (5.105)

We will learn how to solve the matrix equations of (5.105) in the subsequent sections.

Example 5.15

A fourth–order system is described by the differential equation

$$\frac{d^4y}{dt^4} + a_3\frac{d^3y}{dt^3} + a_2\frac{d^2y}{dt^2} + a_1\frac{dy}{dt} + a_0\, y(t) = u(t) \tag{5.106}$$

where $y(t)$ is the output and $u(t)$ is any input. Express (5.106) as a set of state equations.

Solution:

The differential equation of (5.106) is of fourth–order; therefore, we must define four state variables that will be used with the resulting four first–order state equations.

We denote the state variables as x_1, x_2, x_3, and x_4, and we relate them to the terms of the given differential equation as

$$x_1 = y(t) \qquad x_2 = \frac{dy}{dt} \qquad x_3 = \frac{d^2y}{dt^2} \qquad x_4 = \frac{d^3y}{dt^3} \tag{5.107}$$

We observe that

$$\dot{x}_1 = x_2$$
$$\dot{x}_2 = x_3$$
$$\dot{x}_3 = x_4 \tag{5.108}$$
$$\frac{d^4y}{dt^4} = \dot{x}_4 = -a_0x_1 - a_1x_2 - a_2x_3 - a_3x_4 + u(t)$$

and in matrix form

$$\begin{bmatrix} \dot{x}_1 \\ \dot{x}_2 \\ \dot{x}_3 \\ \dot{x}_4 \end{bmatrix} = \begin{bmatrix} 0 & 1 & 0 & 0 \\ 0 & 0 & 1 & 0 \\ 0 & 0 & 0 & 1 \\ -a_0 & -a_1 & -a_2 & -a_3 \end{bmatrix} \begin{bmatrix} x_1 \\ x_2 \\ x_3 \\ x_4 \end{bmatrix} + \begin{bmatrix} 0 \\ 0 \\ 0 \\ 1 \end{bmatrix} u(t) \tag{5.109}$$

In compact form, (5.109) is written as

$$\dot{x} = Ax + bu \tag{5.110}$$

where

$$\dot{x} = \begin{bmatrix} \dot{x}_1 \\ \dot{x}_2 \\ \dot{x}_3 \\ \dot{x}_4 \end{bmatrix}, \quad A = \begin{bmatrix} 0 & 1 & 0 & 0 \\ 0 & 0 & 1 & 0 \\ 0 & 0 & 0 & 1 \\ -a_0 & -a_1 & -a_2 & -a_3 \end{bmatrix}, \quad x = \begin{bmatrix} x_1 \\ x_2 \\ x_3 \\ x_4 \end{bmatrix}, \quad b = \begin{bmatrix} 0 \\ 0 \\ 0 \\ 1 \end{bmatrix}, \quad \text{and } u = u(t)$$

5.8 Solution of Single State Equations

Let us consider the state equations

$$\begin{aligned} \dot{x} &= \alpha x + \beta u \\ y &= k_1 x + k_2 u \end{aligned} \tag{5.111}$$

where α, β, k_1, and k_2 are scalar constants, and the initial condition, if non–zero, is denoted as

$$x_0 = x(t_0) \tag{5.112}$$

We will now prove that the solution of the first state equation in (5.111) is

$$x(t) = e^{\alpha(t-t_0)} x_0 + e^{\alpha t} \int_{t_0}^{t} e^{-\alpha \tau} \beta u(\tau) d\tau \tag{5.113}$$

Proof:

First, we must show that (5.113) satisfies the initial condition of (5.112). This is done by substitution of $t = t_0$ in (5.113). Then,

$$x(t_0) = e^{\alpha(t_0-t_0)} x_0 + e^{\alpha t} \int_{t_0}^{t_0} e^{-\alpha \tau} \beta u(\tau) d\tau \tag{5.114}$$

The first term in the right side of (5.114) reduces to x_0 since

$$e^{\alpha(t_0-t_0)} x_0 = e^0 x_0 = x_0 \tag{5.115}$$

The second term of (5.114) is zero since the upper and lower limits of integration are the same. Therefore, (5.114) reduces to $x(t_0) = x_0$ and thus the initial condition is satisfied.

Next, we must prove that (5.113) satisfies also the first equation in (5.111). To prove this, we differentiate (5.113) with respect to t and we obtain

$$\dot{x}(t) = \frac{d}{dt}(e^{\alpha(t-t_0)} x_0) + \frac{d}{dt}\left\{ e^{\alpha t} \int_{t_0}^{t} e^{-\alpha \tau} \beta u(\tau) d\tau \right\}$$

$$\dot{x}(t) = \alpha e^{\alpha(t-t_0)}x_0 + \alpha e^{\alpha t}\int_{t_0}^{t} e^{-\alpha\tau}\beta u(\tau)d\tau + e^{\alpha t}[e^{-\alpha\tau}\beta u(\tau)]\Big|_{\tau=t}$$

$$= \alpha\left[e^{\alpha(t-t_0)}x_0 + e^{\alpha t}\int_{t_0}^{t} e^{-\alpha\tau}\beta u(\tau)d\tau\right] + e^{\alpha t}e^{-\alpha t}\beta u(t)$$

$$\dot{x}(t) = \alpha\left[e^{\alpha(t-t_0)}x_0 + \int_{t_0}^{t} e^{\alpha(t-\tau)}\beta u(\tau)d\tau\right] + \beta u(t) \tag{5.116}$$

We observe that the bracketed terms of (5.116) are the same as the right side of the assumed solution of (5.113). Therefore,

$$\dot{x} = \alpha x + \beta u$$

and this is the same as the first equation of (5.111). The second equation of (5.111) is an algebraic equation whose coefficients are scalar constants.

In summary, if α and β are scalar constants, the solution of

$$\dot{x} = \alpha x + \beta u \tag{5.117}$$

with initial condition

$$x_0 = x(t_0) \tag{5.118}$$

is obtained from the relation

$$x(t) = e^{\alpha(t-t_0)}x_0 + e^{\alpha t}\int_{t_0}^{t} e^{-\alpha\tau}\beta u(\tau)d\tau \tag{5.119}$$

5.9 The State Transition Matrix

Let us again consider the state equations pair

$$\begin{aligned}\dot{x} &= Ax + bu \\ y &= Cx + du\end{aligned} \tag{5.120}$$

where for two or more simultaneous differential equations A and C are 2×2 or higher order matrices, and b and d are column vectors with two or more rows. In this section we will introduce the *state transition matrix* e^{At}, and we will prove that the solution of the matrix differential equation

$$\dot{x} = Ax + bu \tag{5.121}$$

with initial conditions

$$x(t_0) = x_0 \tag{5.122}$$

is obtained from the relation

$$x(t) = e^{A(t-t_0)}x_0 + e^{At}\int_{t_0}^t e^{-A\tau}bu(\tau)d\tau \tag{5.123}$$

Proof:

Let A be any $n \times n$ matrix whose elements are constants. Then, another $n \times n$ matrix denoted as $\varphi(t)$, is said to be the state transition matrix of (5.34), if it is related to the matrix A as the matrix power series

$$\boxed{\varphi(t) \equiv e^{At} = I + At + \frac{1}{2!}A^2t^2 + \frac{1}{3!}A^3t^3 + \ldots + \frac{1}{n!}A^nt^n} \tag{5.124}$$

where I is the $n \times n$ identity matrix.

From (5.124), we find that

$$\varphi(0) = e^{A0} = I + A0 + \ldots = I \tag{5.125}$$

Differentiation of (5.124) with respect to t yields

$$\varphi'(t) = \frac{d}{dt}e^{At} = 0 + A \cdot 1 + A^2t + \ldots = A + A^2t + \ldots \tag{5.126}$$

and by comparison with (5.124) we obtain

$$\frac{d}{dt}e^{At} = Ae^{At} \tag{5.127}$$

To prove that (5.123) is the solution of the first equation of (5.120), we must prove that it satisfies both the initial condition and the matrix differential equation. The initial condition is satisfied from the relation

$$x(t_0) = e^{A(t_0-t_0)}x_0 + e^{At_0}\int_{t_0}^{t_0} e^{-A\tau}bu(\tau)d\tau = e^{A0}x_0 + 0 = Ix_0 = x_0 \tag{5.128}$$

where we have used (5.125) for the initial condition. The integral is zero since the upper and lower limits of integration are the same.

To prove that the first equation of (5.120) is also satisfied, we differentiate the assumed solution

$$x(t) = e^{A(t-t_0)}x_0 + e^{At}\int_{t_0}^t e^{-A\tau}bu(\tau)d\tau$$

with respect to t and we use (5.127), that is,

$$\frac{d}{dt}e^{At} = Ae^{At}$$

Then,

$$\dot{x}(t) = Ae^{A(t-t_0)}x_0 + Ae^{At}\int_{t_0}^t e^{-A\tau}bu(\tau)d\tau + e^{At}e^{-At}bu(t)$$

or

$$\dot{x}(t) = A\left[e^{A(t-t_0)}x_0 + e^{At}\int_{t_0}^t e^{-A\tau}bu(\tau)d\tau\right] + e^{At}e^{-At}bu(t) \qquad (5.129)$$

We recognize the bracketed terms in (5.129) as $x(t)$, and the last term as $bu(t)$. Thus, the expression (5.129) reduces to

$$\dot{x}(t) = Ax + bu$$

In summary, if A is an $n \times n$ matrix whose elements are constants, $n \geq 2$, and b is a column vector with *n* elements, the solution of

$$\dot{x}(t) = Ax + bu \qquad (5.130)$$

with initial condition

$$x_0 = x(t_0) \qquad (5.131)$$

is

$$\boxed{x(t) = e^{A(t-t_0)}x_0 + e^{At}\int_{t_0}^t e^{-A\tau}bu(\tau)d\tau} \qquad (5.132)$$

Therefore, the solution of second or higher order systems using the state variable method, entails the computation of the state transition matrix e^{At}, and integration of (5.132).

5.10 Computation of the State Transition Matrix e^{At}

Let A be an $n \times n$ matrix, and I be the $n \times n$ identity matrix. By definition, the *eigenvalues* λ_i, $i = 1, 2, ..., n$ of A are the roots of the nth order polynomial

$$\boxed{\det[A - \lambda I] = 0} \qquad (5.133)$$

We recall that expansion of a determinant produces a polynomial. The roots of the polynomial of (5.133) can be real (unequal or equal), or complex numbers.

Evaluation of the state transition matrix e^{At} is based on the *Cayley–Hamilton theorem*. This theorem states that a matrix can be expressed as an $(n-1)$th degree polynomial in terms of the matrix A as

$$\boxed{e^{At} = a_0 I + a_1 A + a_2 A^2 + ... + a_{n-1}A^{n-1}} \qquad (5.134)$$

where the coefficients a_i are functions of the eigenvalues λ. We accept (5.134) without proving it. The proof can be found in Linear Algebra and Matrix Theory textbooks.

Since the coefficients a_i are functions of the eigenvalues λ, we must consider the following cases:

Case I: Distinct Eigenvalues (Real or Complex)

If $\lambda_1 \neq \lambda_2 \neq \lambda_3 \neq \dots \neq \lambda_n$, that is, if all eigenvalues of a given matrix A are distinct, the coefficients a_i are found from the simultaneous solution of the following system of equations:

$$
\begin{aligned}
a_0 + a_1\lambda_1 + a_2\lambda_1^2 + \dots + a_{n-1}\lambda_1^{n-1} &= e^{\lambda_1 t} \\
a_0 + a_1\lambda_2 + a_2\lambda_2^2 + \dots + a_{n-1}\lambda_2^{n-1} &= e^{\lambda_2 t} \\
&\dots \\
a_0 + a_1\lambda_n + a_2\lambda_n^2 + \dots + a_{n-1}\lambda_n^{n-1} &= e^{\lambda_n t}
\end{aligned}
\tag{5.135}
$$

Example 5.16

Compute the state transition matrix e^{At} given that $A = \begin{bmatrix} -2 & 1 \\ 0 & -1 \end{bmatrix}$

Solution:

We must first find the eigenvalues λ of the given matrix A. These are found from the expansion of

$$\det[A - \lambda I] = 0$$

For this example,

$$\det[A - \lambda I] = \det\left\{\begin{bmatrix} -2 & 1 \\ 0 & -1 \end{bmatrix} - \lambda \begin{bmatrix} 1 & 0 \\ 0 & 1 \end{bmatrix}\right\} = \det\begin{bmatrix} -2-\lambda & 1 \\ 0 & -1-\lambda \end{bmatrix} = 0$$

$$= (-2-\lambda)(-1-\lambda) = 0$$

or

$$(\lambda + 1)(\lambda + 2) = 0$$

Therefore,

$$\lambda_1 = -1 \quad \text{and} \quad \lambda_2 = -2 \tag{5.136}$$

Next, we must find the coefficients a_i of (5.134). Since A is a 2×2 matrix, we only need to consider the first two terms of that relation, that is,

$$e^{At} = a_0 I + a_1 A \tag{5.137}$$

The coefficients a_0 and a_1 are found from (5.135). For this example,

$$a_0 + a_1 \lambda_1 = e^{\lambda_1 t}$$

$$a_0 + a_1 \lambda_2 = e^{\lambda_2 t}$$

or

$$a_0 + a_1(-1) = e^{-t}$$

$$a_0 + a_1(-2) = e^{-2t} \tag{5.138}$$

Simultaneous solution of (5.138) yields

$$a_0 = 2e^{-t} - e^{-2t}$$

$$a_1 = e^{-t} - e^{-2t} \tag{5.139}$$

and by substitution into (5.137),

$$e^{At} = (2e^{-t} - e^{-2t}) \begin{bmatrix} 1 & 0 \\ 0 & 1 \end{bmatrix} + (e^{-t} - e^{-2t}) \begin{bmatrix} -2 & 1 \\ 0 & -1 \end{bmatrix}$$

or

$$e^{At} = \begin{bmatrix} e^{-2t} & e^{-t} - e^{-2t} \\ 0 & e^{-t} \end{bmatrix} \tag{5.140}$$

In summary, we compute the state transition matrix e^{At} for a given matrix A using the following procedure:

1. We find the eigenvalues λ from $\det[A - \lambda I] = 0$. We can write $[A - \lambda I]$ at once by subtracting λ from each of the main diagonal elements of A. If the dimension of A is a 2×2 matrix, it will yield two eigenvalues; if it is a 3×3 matrix, it will yield three eigenvalues, and so on. If the eigenvalues are distinct, we perform steps 2 through 4 below; otherwise we refer to Case II.

2. If the dimension of A is a 2×2 matrix, we use only the first 2 terms of the right side of the state transition matrix

$$e^{At} = a_0 I + a_1 A + a_2 A^2 + \ldots + a_{n-1} A^{n-1} \tag{5.141}$$

If A matrix is a 3×3 matrix, we use the first 3 terms, and so on.

3. We obtain the a_i coefficients from

$$a_0 + a_1\lambda_1 + a_2\lambda_1^2 + \ldots + a_{n-1}\lambda_1^{n-1} = e^{\lambda_1 t}$$

$$a_0 + a_1\lambda_2 + a_2\lambda_2^2 + \ldots + a_{n-1}\lambda_2^{n-1} = e^{\lambda_2 t}$$

$$\ldots$$

$$a_0 + a_1\lambda_n + a_2\lambda_n^2 + \ldots + a_{n-1}\lambda_n^{n-1} = e^{\lambda_n t}$$

We use as many equations as the number of the eigenvalues, and we solve for the coefficients a_i.

4. We substitute the a_i coefficients into the state transition matrix of (5.141), and we simplify.

Example 5.17

Compute the state transition matrix e^{At} given that

$$A = \begin{bmatrix} 5 & 7 & -5 \\ 0 & 4 & -1 \\ 2 & 8 & -3 \end{bmatrix} \tag{5.142}$$

Solution:

1. We first compute the eigenvalues from $\det[A - \lambda I] = 0$. We obtain $[A - \lambda I]$ at once, by subtracting λ from each of the main diagonal elements of A. Then,

$$\det[A - \lambda I] = \det\begin{bmatrix} 5-\lambda & 7 & -5 \\ 0 & 4-\lambda & -1 \\ 2 & 8 & -3-\lambda \end{bmatrix} = 0 \tag{5.143}$$

and expansion of this determinant yields the polynomial

$$\lambda^3 - 6\lambda^2 + 11\lambda - 6 = 0 \tag{5.144}$$

We will use MATLAB **roots(p)** function to obtain the roots of (5.144).

```
p=[1 −6 11 −6]; r=roots(p); fprintf(' \n'); fprintf('lambda1 = %5.2f \t', r(1));...
fprintf('lambda2 = %5.2f \t', r(2)); fprintf('lambda3 = %5.2f', r(3))
```

```
lambda1 = 3.00    lambda2 = 2.00    lambda3 = 1.00
```

and thus the eigenvalues are

$$\lambda_1 = 1 \qquad \lambda_2 = 2 \qquad \lambda_3 = 3 \tag{5.145}$$

2. Since A is a 3×3 matrix, we need to use the first 3 terms of (5.134), that is,

$$e^{At} = a_0 I + a_1 A + a_2 A^2 \tag{5.146}$$

3. We obtain the coefficients a_0, a_1, and a_2 from

$$a_0 + a_1 \lambda_1 + a_2 \lambda_1^2 = e^{\lambda_1 t}$$

$$a_0 + a_1 \lambda_2 + a_2 \lambda_2^2 = e^{\lambda_2 t}$$

$$a_0 + a_1 \lambda_3 + a_2 \lambda_3^2 = e^{\lambda_3 t}$$

or

$$a_0 + a_1 + a_2 = e^t$$

$$a_0 + 2a_1 + 4a_2 = e^{2t} \tag{5.147}$$

$$a_0 + 3a_1 + 9a_2 = e^{3t}$$

We will use the following MATLAB script for the solution of (5.147).

```
B=sym('[1 1 1; 1 2 4; 1 3 9]'); b=sym('[exp(t); exp(2*t); exp(3*t)]'); a=B\b; fprintf(' \n');...
disp('a0 = '); disp(a(1)); disp('a1 = '); disp(a(2)); disp('a2 = '); disp(a(3))

a0 =
3*exp(t)-3*exp(2*t)+exp(3*t)
a1 =
-5/2*exp(t)+4*exp(2*t)-3/2*exp(3*t)
a2 =
1/2*exp(t)-exp(2*t)+1/2*exp(3*t)
```

Thus,

$$a_0 = 3e^t - 3e^{2t} + e^{3t}$$

$$a_1 = -\frac{5}{2}e^t + 4e^{2t} - \frac{3}{2}e^{3t} \tag{5.148}$$

$$a_2 = \frac{1}{2}e^t - e^{2t} + \frac{1}{2}e^{3t}$$

4. We also use MATLAB to perform the substitution into the state transition matrix, and to perform the matrix multiplications. The script is shown below.

```
syms t; a0 = 3*exp(t)+exp(3*t)-3*exp(2*t); a1 = -5/2*exp(t)-3/2*exp(3*t)+4*exp(2*t);...
a2 = 1/2*exp(t)+1/2*exp(3*t)-exp(2*t);...
A = [5 7 -5; 0 4 -1; 2 8 -3]; eAt=a0*eye(3)+a1*A+a2*A^2
```

```
eAt =
  [-2*exp(t)+2*exp(2*t)+exp(3*t),-6*exp(t)+5*exp(2*t)+exp(3*t),
  4*exp(t)-3*exp(2*t)-exp(3*t)]
  [-exp(t)+2*exp(2*t)-exp(3*t),-3*exp(t)+5*exp(2*t)-exp(3*t),
  2*exp(t)-3*exp(2*t)+exp(3*t)]
  [-3*exp(t)+4*exp(2*t)-exp(3*t),-9*exp(t)+10*exp(2*t)-exp(3*t),
  6*exp(t)-6*exp(2*t)+exp(3*t)]
```

Thus,

$$e^{At} = \begin{bmatrix} -2e^t + 2e^{2t} + e^{3t} & -6e^t + 5e^{2t} + e^{3t} & 4e^t - 3e^{2t} - e^{3t} \\ -e^t + 2e^{2t} - e^{3t} & -3e^t + 5e^{2t} - e^{3t} & 2e^t - 3e^{2t} + e^{3t} \\ -3e^t + 4e^{2t} - e^{3t} & -9e^t + 10e^{2t} - e^{3t} & 6e^t - 6e^{2t} + e^{3t} \end{bmatrix}$$

Case II: Multiple Eigenvalues

In this case, we will assume that the polynomial of

$$\det[A - \lambda I] = 0 \tag{5.149}$$

has n roots, and m of these roots are equal. In other words, the roots are

$$\lambda_1 = \lambda_2 = \lambda_3 \ldots = \lambda_m, \ \lambda_{m+1}, \ \lambda_n \tag{5.150}$$

The coefficients a_i of the state transition matrix

$$e^{At} = a_0 I + a_1 A + a_2 A^2 + \ldots + a_{n-1} A^{n-1} \tag{5.151}$$

are found from the simultaneous solution of the system of equations of (5.152) below.

Example 5.18

Compute the state transition matrix e^{At} given that

$$A = \begin{bmatrix} -1 & 0 \\ 2 & -1 \end{bmatrix}$$

Solution:

1. We first find the eigenvalues λ of the matrix A and these are found from the polynomial of $\det[A - \lambda I] = 0$. For this example,

$$a_0 + a_1\lambda_1 + a_2\lambda_1^2 + \ldots + a_{n-1}\lambda_1^{n-1} = e^{\lambda_1 t}$$

$$\frac{d}{d\lambda_1}(a_0 + a_1\lambda_1 + a_2\lambda_1^2 + \ldots + a_{n-1}\lambda_1^{n-1}) = \frac{d}{d\lambda_1}e^{\lambda_1 t}$$

$$\frac{d^2}{d\lambda_1^2}(a_0 + a_1\lambda_1 + a_2\lambda_1^2 + \ldots + a_{n-1}\lambda_1^{n-1}) = \frac{d^2}{d\lambda_1^2}e^{\lambda_1 t}$$

$$\ldots$$

$$\frac{d^{m-1}}{d\lambda_1^{m-1}}(a_0 + a_1\lambda_1 + a_2\lambda_1^2 + \ldots + a_{n-1}\lambda_1^{n-1}) = \frac{d^{m-1}}{d\lambda_1^{m-1}}e^{\lambda_1 t}$$

$$a_0 + a_1\lambda_{m+1} + a_2\lambda_{m+1}^2 + \ldots + a_{n-1}\lambda_{m+1}^{n-1} = e^{\lambda_{m+1} t}$$

$$\ldots$$

$$a_0 + a_1\lambda_n + a_2\lambda_n^2 + \ldots + a_{n-1}\lambda_n^{n-1} = e^{\lambda_n t}$$

$$(5.152)$$

$$\det[A - \lambda I] = \det\begin{bmatrix} -1-\lambda & 0 \\ 2 & -1-\lambda \end{bmatrix} = 0$$

$$= (-1-\lambda)(-1-\lambda) = 0$$

$$= (\lambda + 1)^2 = 0$$

and thus,

$$\lambda_1 = \lambda_2 = -1$$

2. Since A is a 2×2 matrix, we only need the first two terms of the state transition matrix, that is,

$$e^{At} = a_0 I + a_1 A \qquad (5.153)$$

3. We find a_0 and a_1 from (5.152). For this example,

$$a_0 + a_1\lambda_1 = e^{\lambda_1 t}$$

$$\frac{d}{d\lambda_1}(a_0 + a_1\lambda_1) = \frac{d}{d\lambda_1}e^{\lambda_1 t}$$

or

$$a_0 + a_1\lambda_1 = e^{\lambda_1 t}$$

$$a_1 = te^{\lambda_1 t}$$

and by substitution with $\lambda_1 = \lambda_2 = -1$, we obtain

$$a_0 - a_1 = e^{-t}$$

$$a_1 = te^{-t}$$

Simultaneous solution of the last two equations yields

$$a_0 = e^{-t} + te^{-t}$$

$$a_1 = te^{-t}$$

(5.154)

4. By substitution of (5.154) into (5.153), we obtain

$$e^{At} = (e^{-t} + te^{-t}) \begin{bmatrix} 1 & 0 \\ 0 & 1 \end{bmatrix} + te^{-t} \begin{bmatrix} -1 & 0 \\ 2 & -1 \end{bmatrix} = e^{At} = \begin{bmatrix} e^{-t} & 0 \\ 2te^{-t} & e^{-t} \end{bmatrix}$$

(5.155)

We can use the MATLAB **eig(x)** function to find the eigenvalues of an $n \times n$ matrix. To find out how it is used, we invoke the **help eig** command.

We will first use MATLAB to verify the values of the eigenvalues found in Examples 5.16 through 5.18, and we will briefly discuss eigenvectors on the next section.

For Example 5.16:

A= [–2 1; 0 –1]; lambda=eig(A)

lambda =
    –2
    –1

For Example 5.17:

B = [5 7 –5; 0 4 –1; 2 8 –3]; lambda=eig(B)

lambda =
    1.0000
    3.0000
    2.0000

For Example 5.18:

C = [–1 0; 2 –1]; lambda=eig(C)

lambda =
    –1
    –1

5.11 Eigenvectors

Consider the relation

$$AX = \lambda X \qquad (5.156)$$

where A is an $n \times n$ matrix, X is a column vector, and λ is a scalar number. We can express this relation in matrix form as

$$
\begin{bmatrix}
a_{11} & a_{12} & \cdots & a_{1n} \\
a_{21} & a_{22} & \cdots & a_{2n} \\
\cdots & \cdots & \cdots & \cdots \\
a_{n1} & a_{n2} & \cdots & a_{nn}
\end{bmatrix}
\begin{bmatrix}
x_1 \\ x_2 \\ \cdots \\ x_n
\end{bmatrix}
= \lambda
\begin{bmatrix}
x_1 \\ x_2 \\ \cdots \\ x_n
\end{bmatrix}
\qquad (5.157)
$$

We write (5.157) as

$$(A - \lambda I)X = 0 \qquad (5.158)$$

or

$$
\begin{bmatrix}
(a_{11} - \lambda)x_1 & a_{12}x_2 & \cdots & a_{1n}x_n \\
a_{21}x_1 & (a_{22} - \lambda)x_2 & \cdots & a_{2n}x_n \\
\cdots & \cdots & \cdots & \cdots \\
a_{n1}x_1 & a_{n2}x_2 & \cdots & (a_{nn} - \lambda)x_n
\end{bmatrix}
= 0
\qquad (5.159)
$$

The equations of (5.159) will have non–trivial solutions if and only if its determinant is zero[*], that is, if

$$
\det
\begin{bmatrix}
(a_{11} - \lambda) & a_{12} & \cdots & a_{1n} \\
a_{21} & (a_{22} - \lambda) & \cdots & a_{2n} \\
\cdots & \cdots & \cdots & \cdots \\
a_{n1} & a_{n2} & \cdots & (a_{nn} - \lambda)
\end{bmatrix}
= 0
\qquad (5.160)
$$

Expansion of the determinant of (5.160) results in a polynomial equation of degree n in λ, and it is called the *characteristic equation*.

We can express (5.160) in a compact form as

$$\det(A - \lambda I) = 0 \qquad (5.161)$$

As we know, the roots λ of the characteristic equation are the eigenvalues of the matrix A, and corresponding to each eigenvalue λ, there is a non–trivial solution of the column vector X, i.e.,

---

[*]. *This is because we want the vector X in (5.158) to be a non–zero vector and the product (A–λI)X to be zero.*

$X \neq 0$. This vector X is called *eigenvector*. Obviously, there is a different eigenvector for each eigenvalue. Eigenvectors are generally expressed as *unit eigenvectors*, that is, they are normalized to unit length. This is done by dividing each component of the eigenvector by the square root of the sum of the squares of their components, so that the sum of the squares of their components is equal to unity.

In many engineering applications the unit eigenvectors are chosen such that $X \cdot X^T = I$ where X^T is the transpose of the eigenvector X, and I is the identity matrix.

Two vectors X and Y are said to be *orthogonal* if their inner (dot) product is zero. A set of eigenvectors constitutes an *orthonormal basis* if the set is normalized (expressed as unit eigenvectors) and these vector are mutually orthogonal. An orthonormal basis can be formed with the *Gram–Schmidt Orthogonalization Procedure*; it is discussed in Chapter 14.

The example which follows, illustrates the relationships between a matrix A, its eigenvalues, and eigenvectors.

Example 5.19

Given the matrix

$$A = \begin{bmatrix} 5 & 7 & -5 \\ 0 & 4 & -1 \\ 2 & 8 & -3 \end{bmatrix}$$

a. Find the eigenvalues of A

b. Find eigenvectors corresponding to each eigenvalue of A

c. Form a set of unit eigenvectors using the eigenvectors of part (b).

Solution:

a. This is the same matrix as in Example 5.17, where we found the eigenvalues to be

$$\lambda_1 = 1 \qquad \lambda_2 = 2 \qquad \lambda_3 = 3$$

b. We begin with

$$AX = \lambda X$$

and we let

$$X = \begin{bmatrix} x_1 \\ x_2 \\ x_3 \end{bmatrix}$$

Then,

$$\begin{bmatrix} 5 & 7 & -5 \\ 0 & 4 & -1 \\ 2 & 8 & -3 \end{bmatrix} \begin{bmatrix} x_1 \\ x_2 \\ x_3 \end{bmatrix} = \lambda \begin{bmatrix} x_1 \\ x_2 \\ x_3 \end{bmatrix} \qquad (5.162)$$

or

$$\begin{bmatrix} 5x_1 & 7x_2 & -5x_3 \\ 0 & 4x_2 & -x_3 \\ 2x_1 & 8x_2 & -3x_3 \end{bmatrix} = \begin{bmatrix} \lambda x_1 \\ \lambda x_2 \\ \lambda x_3 \end{bmatrix} \qquad (5.163)$$

Equating corresponding rows and rearranging, we obtain

$$\begin{bmatrix} (5-\lambda)x_1 & 7x_2 & -5x_3 \\ 0 & (4-\lambda)x_2 & -x_3 \\ 2x_1 & 8x_2 & -(3-\lambda)x_3 \end{bmatrix} = \begin{bmatrix} 0 \\ 0 \\ 0 \end{bmatrix} \qquad (5.164)$$

For $\lambda = 1$, (5.164) reduces to

$$4x_1 + 7x_2 - 5x_3 = 0$$
$$3x_2 - x_3 = 0 \qquad (5.165)$$
$$2x_1 + 8x_2 - 4x_3 = 0$$

By Crame's rule, or MATLAB, we obtain the indeterminate values

$$x_1 = 0/0 \qquad x_2 = 0/0 \qquad x_3 = 0/0 \qquad (5.166)$$

Since the unknowns x_1, x_2, and x_3 are scalars, we can assume that one of these, say x_2, is known, and solve x_1 and x_3 in terms of x_2. Then, we obtain $x_1 = 2x_2$, and $x_3 = 3x_2$.

Therefore, an eigenvector for $\lambda = 1$ is

$$X_{\lambda=1} = \begin{bmatrix} x_1 \\ x_2 \\ x_3 \end{bmatrix} = \begin{bmatrix} 2x_2 \\ x_2 \\ 3x_2 \end{bmatrix} = x_2 \begin{bmatrix} 2 \\ 1 \\ 3 \end{bmatrix} = \begin{bmatrix} 2 \\ 1 \\ 3 \end{bmatrix} \qquad (5.167)$$

since any eigenvector is a scalar multiple of the last vector in (5.167).

Similarly, for $\lambda = 2$, we obtain $x_1 = x_2$, and $x_3 = 2x_2$. Then, an eigenvector for $\lambda = 2$ is

$$X_{\lambda\,=\,2} = \begin{bmatrix} x_1 \\ x_2 \\ x_3 \end{bmatrix} = \begin{bmatrix} x_2 \\ x_2 \\ 2x_2 \end{bmatrix} = x_2\begin{bmatrix} 1 \\ 1 \\ 2 \end{bmatrix} = \begin{bmatrix} 1 \\ 1 \\ 2 \end{bmatrix} \qquad (5.168)$$

Finally, for $\lambda = 3$, we obtain $x_1 = -x_2$, and $x_3 = x_2$. Then, an eigenvector for $\lambda = 3$ is

$$X_{\lambda\,=\,3} = \begin{bmatrix} x_1 \\ x_2 \\ x_3 \end{bmatrix} = \begin{bmatrix} -x_2 \\ x_2 \\ x_2 \end{bmatrix} = x_2\begin{bmatrix} -1 \\ 1 \\ 1 \end{bmatrix} = \begin{bmatrix} -1 \\ 1 \\ 1 \end{bmatrix} \qquad (5.169)$$

c. We find the unit eigenvectors by dividing the components of each vector by the square root of the sum of the squares of the components. These are:

$$\sqrt{2^2 + 1^2 + 3^2} = \sqrt{14}$$

$$\sqrt{1^2 + 1^2 + 2^2} = \sqrt{6}$$

$$\sqrt{(-1)^2 + 1^2 + 1^2} = \sqrt{3}$$

The unit eigenvectors are

$$\text{Unit } X_{\lambda\,=\,1} = \begin{bmatrix} \dfrac{2}{\sqrt{14}} \\[2mm] \dfrac{1}{\sqrt{14}} \\[2mm] \dfrac{3}{\sqrt{14}} \end{bmatrix} \qquad \text{Unit } X_{\lambda\,=\,2} = \begin{bmatrix} \dfrac{1}{\sqrt{6}} \\[2mm] \dfrac{1}{\sqrt{6}} \\[2mm] \dfrac{2}{\sqrt{6}} \end{bmatrix} \qquad \text{Unit } X_{\lambda\,=\,3} = \begin{bmatrix} \dfrac{-1}{\sqrt{3}} \\[2mm] \dfrac{1}{\sqrt{3}} \\[2mm] \dfrac{1}{\sqrt{3}} \end{bmatrix} \qquad (5.170)$$

We observe that for the first unit eigenvector the sum of the squares is unity, that is,

$$\left(\frac{2}{\sqrt{14}}\right)^2 + \left(\frac{1}{\sqrt{14}}\right)^2 + \left(\frac{3}{\sqrt{14}}\right)^2 = \frac{4}{14} + \frac{1}{14} + \frac{9}{14} = 1 \qquad (5.171)$$

and the same is true for the other two unit eigenvectors in (5.170).

5.12 Summary

- Differential equations are classified by:

  Type – Ordinary or Partial

  Order – The highest order derivative which is included in the differential equation

  Degree – The exponent of the highest power of the highest order derivative after the differential equation has been cleared of any fractions or radicals in the dependent variable and its derivatives

- If the dependent variable y is a function of only a single variable x, that is, if $y = f(x)$, the differential equation which relates y and x is said to be an ordinary differential equation and it is abbreviated as ODE.

- If the dependent variable y is a function of two or more variables such as $y = f(x, t)$, where x and t are independent variables, the differential equation that relates y, x, and t is said to be a partial differential equation and it is abbreviated as PDE.

- A function $y = f(x)$ is a solution of a differential equation if the latter is satisfied when y and its derivatives are replaced throughout by $f(x)$ and its corresponding derivatives. Also, the initial conditions must be satisfied.

- The ODE

$$a_n \frac{d^n y}{dt^n} + a_{n-1} \frac{d^{n-1} y}{dt^{n-1}} + \ldots + a_1 \frac{dy}{dt} + a_0 y = b_m \frac{d^m x}{dt^m} + b_{m-1} \frac{d^{m-1} x}{dt^{n-1}} + \ldots + b_1 \frac{dx}{dt} + b_0 x$$

  is a non–homogeneous differential equation if the right side, known as forcing function, is not zero. If the forcing function is zero, the differential equation is referred to as homogeneous differential equation.

- The most general solution of an **homogeneous ODE is the linear combination**

$$y_H(t) = k_1 y_1(t) + k_2 y_2(t) + k_3 y_3(t) + \ldots + k_n y_n(t)$$

  where the subscript H is used to denote homogeneous and $k_1, k_2, k_3, \ldots, k_n$ are arbitrary constants.

- Generally, in engineering the solution of the homogeneous ODE, also known as the complementary solution, is referred to as the **natural response**, and is denoted as $y_N(t)$ or simply y_N. The particular solution of a non–homogeneous ODE is be referred to as the **forced response**, and is denoted as $y_F(t)$ or simply y_F. The total solution of the non–homogeneous ODE is the summation of the natural and forces responses, that is,

$$y(t) = y_{\text{Natural}} + y_{\text{Forced}} = y_N + y_F$$
$$ \text{Response} \quad \text{Response}$$

The natural response y_N contains arbitrary constants and these can be evaluated from the given initial conditions. The forced response y_F, however, contains no arbitrary constants. It is imperative to remember that the arbitrary constants of the natural response must be evaluated from the total response.

- For an nth order homogeneous differential equation the solutions are

$$y_1 = k_1 e^{s_1 t}, \quad y_2 = k_2 e^{s_2 t}, \quad y_3 = k_3 e^{s_3 t}, \quad ..., \quad y_n = k_n e^{s_n t}$$

where $s_1, s_2, ..., s_n$ are the solutions of the characteristic equation

$$a_n s^n + a_{n-1} s^{n-1} + ... + a_1 s + a_0 = 0$$

and $a_n, a_{n-1}, ..., a_1, a_0$ are the constant coefficients of the ODE

- If the roots of the characteristic equation are distinct, the n solutions of the natural response are independent and the most general solution is:

$$y_N = k_1 e^{s_1 t} + k_2 e^{s_2 t} + ... + k_n e^{s_n t}$$

- If the solution of the characteristic equation contains m equal roots, the most general solution has the form:

$$y_N = (k_1 + k_2 t + ... + k_m t^{m-1}) e^{s_1 t} + k_{n-i} e^{s_2 t} + ... + k_n e^{s_n t}$$

- If the characteristic equation contains complex roots, these occur as complex conjugate pairs. Thus, if one root is $s_1 = -\alpha + j\beta$ where α and β are real numbers, then another root is $s_2 = -\alpha - j\beta$. Then, for two complex conjugate roots we evaluate the constants from the expressions

$$k_1 e^{s_1 t} + k_2 e^{s_2 t} = e^{-\alpha t}(k_3 \cos\beta t + k_4 \sin\beta t) = e^{-\alpha t} k_5 \cos(\beta t + \varphi)$$

- The forced response of a non–homogeneous ODE can be found by the method of undetermined coefficients or the method of variation of parameters.

- With the method of undetermined coefficients, the forced response is a function similar to the right side of the non–homogeneous ODE. The form of the forced response for second order non–homogeneous ODEs is given in Table 5.1.

- In certain non–homogeneous ODEs, the right side $f(t)$ cannot be determined by the method of undetermined coefficients. For these ODEs we must use the method of variation of parameters. This method will work with all linear equations including those with variable coefficients provided that the general form of the natural response is known.

- For second order ODEs with constant coefficients, the method of variation of parameters replaces the constants k_1 and k_2 by two variables u_1 and u_2 that satisfy the following three relations:

$$y = u_1 y_1 + u_2 y_2$$

$$\frac{du_1}{dt} y_1 + \frac{du_2}{dt} y_2 = 0$$

$$\frac{du_1}{dt} \cdot \frac{dy_1}{dt} + \frac{du_2}{dt} \cdot \frac{dy_2}{dt} = f(t)$$

Simultaneous solution of last two expressions above will yield the values of du_1/dt and du_2/dt; then, integration of these will produce u_1 and u_2, which when substituted into the first will yield the total solution.

- An nth–order differential equation can be resolved to n first–order simultaneous differential equations with a set of auxiliary variables called state variables. The resulting first–order differential equations are called state space equations, or simply state equations.

- The state representation of a system can be described by the pair of the of the state space equations

$$\dot{x} = Ax + bu$$
$$y = Cx + du$$

- In a system of state equations of the form

$$\dot{x} = \alpha x + \beta u$$
$$y = k_1 x + k_2 u$$

where α, β, k_1, and k_2 are scalar constants, and the initial condition, if non–zero is denoted as $x_0 = x(t_0)$, the solution of the first state equation above is

$$x(t) = e^{\alpha(t - t_0)} x_0 + e^{\alpha t} \int_{t_0}^{t} e^{-\alpha \tau} \beta u(\tau) d\tau$$

- In a system of state equations of the form

$$\dot{x} = Ax + bu$$
$$y = Cx + du$$

where for two or more simultaneous differential equations A and C are 2×2 or higher order matrices, and b and d are column vectors with two or more rows, the solution of the matrix differential equation $\dot{x} = Ax + bu$ with initial conditions $x(t_0) = x_0$ is obtained from the relation

$$x(t) = e^{A(t - t_0)}x_0 + e^{At}\int_{t_0}^{t} e^{-A\tau}bu(\tau)d\tau$$

where the state transition matrix e^{At} is defined as the matrix power series

$$\varphi(t) \equiv e^{At} = I + At + \frac{1}{2!}A^2t^2 + \frac{1}{3!}A^3t^3 + ... + \frac{1}{n!}A^nt^n$$

and I is the $n \times n$ identity matrix.

- If A is an $n \times n$ matrix, and I be the $n \times n$ identity matrix, the eigenvalues λ_i, $i = 1, 2, ..., n$ of A are the roots of the nth order polynomial

$$\det[A - \lambda I] = 0$$

- Evaluation of the state transition matrix e^{At} is based on the *Cayley–Hamilton theorem*. This theorem states that a matrix can be expressed as an $(n - 1)$th degree polynomial in terms of the matrix A as

$$e^{At} = a_0 I + a_1 A + a_2 A^2 + ... + a_{n-1}A^{n-1}$$

where the coefficients a_i are functions of the eigenvalues λ.

- If $\lambda_1 \neq \lambda_2 \neq \lambda_3 \neq ... \neq \lambda_n$, that is, if all eigenvalues of a given matrix A are distinct, the coefficients a_i are found from the simultaneous solution of the following system of equations:

$$a_0 + a_1\lambda_1 + a_2\lambda_1^2 + ... + a_{n-1}\lambda_1^{n-1} = e^{\lambda_1 t}$$
$$a_0 + a_1\lambda_2 + a_2\lambda_2^2 + ... + a_{n-1}\lambda_2^{n-1} = e^{\lambda_2 t}$$
$$...$$
$$a_0 + a_1\lambda_n + a_2\lambda_n^2 + ... + a_{n-1}\lambda_n^{n-1} = e^{\lambda_n t}$$

- If the polynomial of $\det[A - \lambda I] = 0$ has n roots, and m of these roots are equal, that is, if $\lambda_1 = \lambda_2 = \lambda_3 ... = \lambda_m$, λ_{m+1}, λ_n, the coefficients a_i of the state transition matrix

$$e^{At} = a_0 I + a_1 A + a_2 A^2 + ... + a_{n-1}A^{n-1}$$

are found from the simultaneous solution of the system of equations below.

$$a_0 + a_1\lambda_1 + a_2\lambda_1^2 + \ldots + a_{n-1}\lambda_1^{n-1} = e^{\lambda_1 t}$$

$$\frac{d}{d\lambda_1}(a_0 + a_1\lambda_1 + a_2\lambda_1^2 + \ldots + a_{n-1}\lambda_1^{n-1}) = \frac{d}{d\lambda_1}e^{\lambda_1 t}$$

$$\frac{d^2}{d\lambda_1^2}(a_0 + a_1\lambda_1 + a_2\lambda_1^2 + \ldots + a_{n-1}\lambda_1^{n-1}) = \frac{d^2}{d\lambda_1^2}e^{\lambda_1 t}$$

$$\ldots$$

$$\frac{d^{m-1}}{d\lambda_1^{m-1}}(a_0 + a_1\lambda_1 + a_2\lambda_1^2 + \ldots + a_{n-1}\lambda_1^{n-1}) = \frac{d^{m-1}}{d\lambda_1^{m-1}}e^{\lambda_1 t}$$

$$a_0 + a_1\lambda_{m+1} + a_2\lambda_{m+1}^2 + \ldots + a_{n-1}\lambda_{m+1}^{n-1} = e^{\lambda_{m+1} t}$$

$$\ldots$$

$$a_0 + a_1\lambda_n + a_2\lambda_n^2 + \ldots + a_{n-1}\lambda_n^{n-1} = e^{\lambda_n t}$$

- We can use the MATLAB **eig(x)** function to find the eigenvalues of an $n \times n$ matrix.

- If A is an $n \times n$ matrix, X is a non–zero column vector, and λ is a scalar number, the vector X is called *eigenvector*. Obviously, there is a different eigenvector for each eigenvalue. Eigenvectors are generally expressed as *unit eigenvectors*, that is, they are normalized to unit length. This is done by dividing each component of the eigenvector by the square root of the sum of the squares of their components, so that the sum of the squares of their components is equal to unity.

5.13 Exercises

Solve the following ODEs by any method and verify your answers with MATLAB.

1. $\dfrac{d^2y}{dt^2} + 4\dfrac{dy}{dt} + 3y = t - 1$

2. $\dfrac{d^2y}{dt^2} + 4\dfrac{dy}{dt} + 3y = 4e^{-t}$

3. $\dfrac{d^2y}{dt^2} + 2\dfrac{dy}{dt} + y = \cos^2 t$ Hint: Use $\cos^2 t = \dfrac{1}{2}(\cos 2t + 1)$

4. $\dfrac{d^2y}{dt^2} + y = \sec t$

5. Express the integro–differential equation below as a matrix of state equations where $k_1, k_2,$ and k_3 are constants.

$$\frac{dv^2}{dt^2} + k_3\frac{dv}{dt} + k_2 v + k_1 \int_0^t v\,dt = \sin 3t + \cos 3t$$

6. Express the matrix of the state equations below as a single differential equation, and let $x(y) = y(t)$.

$$\begin{bmatrix} \dot{x}_1 \\ \dot{x}_2 \\ \dot{x}_3 \\ \dot{x}_4 \end{bmatrix} = \begin{bmatrix} 0 & 1 & 0 & 0 \\ 0 & 0 & 1 & 0 \\ 0 & 0 & 0 & 1 \\ -1 & -2 & -3 & -4 \end{bmatrix} \cdot \begin{bmatrix} x_1 \\ x_2 \\ x_3 \\ x_4 \end{bmatrix} + \begin{bmatrix} 0 \\ 0 \\ 0 \\ 1 \end{bmatrix} u(t)$$

7. Compute the eigenvalues of the matrices A, B, and C below.

$$A = \begin{bmatrix} 1 & 2 \\ 3 & -1 \end{bmatrix} \qquad B = \begin{bmatrix} a & 0 \\ -a & b \end{bmatrix} \qquad C = \begin{bmatrix} 0 & 1 & 0 \\ 0 & 0 & 1 \\ -6 & -11 & -6 \end{bmatrix}$$

Hint: One of the eigenvalues of matrix C is -1.

8. Compute e^{At} given that

$$A = \begin{bmatrix} 0 & 1 & 0 \\ 0 & 0 & 1 \\ -6 & -11 & -6 \end{bmatrix}$$

5.14 Solutions to End-of-Chapter Exercises

1. The characteristic equation of the homogeneous part is $s^2 + 4s + 3 = 0$ from which $s_1 = -1$ and $s_2 = -3$. Thus $y_N = k_1 e^{-t} + k_2 e^{-3t}$. For the forced response, we refer to Table 5.1 and we assume a solution of the form $y_F = k_3 t + k_4$ and the total solution is

$$y = k_1 e^{-t} + k_2 e^{-3t} + k_3 t + k_4$$

The first and second derivatives of y are

$$dy/dt = -k_1 e^{-t} - 3k_2 e^{-3t} + k_3$$

$$d^2 y/dt^2 = k_1 e^{-t} + 9k_2 e^{-3t}$$

and by substitution into the given ODE

$$k_1 e^{-t} + 9k_2 e^{-3t} + 4(-k_1 e^{-t} - 3k_2 e^{-3t} + k_3) + 3(k_1 e^{-t} + k_2 e^{-3t} + k_3 t + k_4) = t - 1$$

Equating like terms we obtain

$$4k_3 + 3k_3 t + 3k_4 = t - 1$$
$$3k_3 t = t$$
$$4k_3 + 3k_4 = -1$$

and simultaneous solution of the last two yields $k_3 = 1/3$ and $k_4 = -7/9$. Therefore,

$$y = k_1 e^{-t} + k_2 e^{-3t} + \frac{1}{3}t - \frac{7}{9}$$

Check with MATLAB:

```
y=dsolve('D2y+4*Dy+3*y=t-1'); y=simple(y)
y =
 -7/9+1/3*t+C1/exp(t)+C2/exp(t)^3
```

2. The characteristic equation of the homogeneous part is the same as for Exercise 1 and thus $y_N = k_1 e^{-t} + k_2 e^{-3t}$. For the forced response, we refer to Table 5.1 and we assume a solution of the form $y_F = k_3 t e^{-t}$ where we multiplied e^{-t} by t to avoid the duplication with $k_1 e^{-t}$. By substitution of this assumed solution into the given ODE and using MATLAB to find the first and second derivatives we obtain:

$$y = k_1 e^{-t} + k_2 e^{-3t} + k_3 t e^{-t}$$

We will use MATLAB to find the first and second derivatives of this expression.

```
syms t k3                          % Define symbolic variables
y0=k3*t*exp(-t);                   % Assumed form of total solution
y1=diff(y0); f1=simple(y1)         % Compute and simplify first derivative
```

```
f1 =
-k3*exp(-t)*(-1+t)
```

Thus, the first derivative of y_F is

$$dy_F/dt = k_3 e^{-t} - k_3 t e^{-t}$$

```
y2=diff(y0,2); f2=simple(y2)       % Compute and simplify second derivative
```

```
f2 =
k3*exp(-t)*(-2+t)
```

and the second derivative of y is

$$d^2 y_F/dt^2 = -2k_3 e^{-t} + k_3 t e^{-t}$$

```
f=y2+4*y1+3*y0; f=simple(f)        % Form and simplify the left side of the given ODE
```

```
f =
2*k3/exp(t)
```

and by substitution into the given ODE

$$2k_3 e^{-t} = 4e^{-t}$$

or $k_3 = 2$. Therefore,

$$y = k_1 e^{-t} + k_2 e^{-3t} + 2t e^{-t}$$

Check with MATLAB:

```
y=dsolve('D2y+4*Dy+3*y=4*exp(-t)'); y=simple(y)
```

```
2*t/exp(t)-1/exp(t)+C1/exp(t)+C2/exp(t)^3
```

We observe that the second and third terms of the displayed expression above have the same form and thus they can be combined to form a single term C3/exp(t).

3. The characteristic equation yields two equal roots $s_1 = s_2 = -1$ and thus the natural response has the form

$$y_N = k_1 e^{-t} + k_2 t e^{-t}$$

For the forced response we assume a solution of the form

$$y_F = k_3 \cos 2t + k_4 \sin 2t + k_5$$

We will use MATLAB to find the first and second derivatives of this expression.

```
syms t k1 k2 k3 k4 k5              % Define symbolic variables
y0=k3*cos(2*t)+k4*sin(2*t)+k5;     % Assumed form of total solution
y1=diff(y0); f1=simple(y1)         % Compute and simplify first derivative

f1 =
-2*k3*sin(2*t)+2*k4*cos(2*t)
```

Thus, the first derivative of y_F is

$$dy_F / dt = -2k_3 \sin 2t + 2k_4 \cos 2t$$

```
y2=diff(y0,2); f2=simple(y2)       % Compute and simplify second derivative

f2 =
-4*k3*cos(2*t)-4*k4*sin(2*t)
```

and the second derivative of y is

$$d^2 y_F / dt^2 = -4k_3 \cos 2t - 4k_4 \sin 2t$$

```
f=y2+2*y1+y0; f=simple(f)          % Form and simplify the left side of the given ODE

f =
-3*k3*cos(2*t)-3*k4*sin(2*t)-4*k3*sin(2*t)+4*k4*cos(2*t)+k5
```

Simplifying this expression and equating with the right side of the given ODE we obtain:

$$(-3k_3 + 4k_4) \cos 2t - (4k_3 + 3k_4) \sin 2t + k_5 = \frac{\cos 2t}{2} + \frac{1}{2}$$

Equating like terms and solving for the k terms we obtain

$$-3k_3 + 4k_4 = 1/2$$
$$-4k_3 - 3k_4 = 0$$
$$k_5 = 1/2$$

Simultaneous solution of the first two equations above yields $k_3 = -3/50$ and $k_4 = 4/50$. Therefore, the forced response is

$$y_F = (-3/50) \cos 2t + (4/50) \sin 2t + 1/2$$

and the total response is

$$y = k_1 e^{-t} + k_2 t e^{-t} + \frac{1}{2} - \frac{3 \cos 2t - 4 \sin 2t}{50}$$

Check with MATLAB:

```
y=dsolve('D2y+2*Dy+y=cos(2*t)/2+1/2'); f=simple(y)

f =
-3/50*cos(2*t)+2/25*sin(2*t)+1/2+C1*exp(-t)+C2*exp(-t)*t
```

4. It is very difficult, if not impossible, to assume a solution for the forced response of this ODE. Therefore, we will use the method of variation of parameters.

The characteristic equation is $s^2 + 1 = 0$ from which $s = \pm j$ and thus the natural response is

$$y_N = k_1 e^{jt} + k_2 e^{-jt}$$

We let

$$y_1 = \cos t \ \text{ and } \ y_2 = \sin t$$

Then, by (5.68) the solution is

$$y = u_1 y_1 + u_2 y_2 = u_1 \cos t + u_2 \sin t \quad (1)$$

Also, from (5.69),

$$\frac{du_1}{dt} y_1 + \frac{du_2}{dt} y_2 = 0$$

or

$$\frac{du_1}{dt} \cos t + \frac{du_2}{dt} \sin t = 0$$

and from (5.70),

$$\frac{du_1}{dt} \cdot \frac{dy_1}{dt} + \frac{du_2}{dt} \cdot \frac{dy_2}{dt} = f(t) = \frac{du_1}{dt}(-\sin t) + \frac{du_2}{dt}(\cos t) = \sec t$$

Next, we find du_1/dt and du_2/dt by Cramer's rule as follows:

$$\frac{du_1}{dt} = \frac{\begin{vmatrix} 0 & \sin t \\ \sec t & \cos t \end{vmatrix}}{\begin{vmatrix} \cos t & \sin t \\ -\sin t & \cos t \end{vmatrix}} = \frac{-\dfrac{\sin t}{\cos t}}{\cos^2 t + \sin^2 t} = \frac{-\tan t}{1} = -\tan t \quad (2)$$

and

$$\frac{du_2}{dt} = \frac{\begin{vmatrix} \cos t & 0 \\ -\sin t & \sec t \end{vmatrix}}{1} = \frac{1}{1} = 1 \quad (3)$$

Integration of (2) and (3) above and substitution into (1) yields

$$u_1 = \int (-\tan t) dt = -(-\ln \cos t) + k_1 = \ln \cos t + k_1$$

$$u_2 = \int dt = t + k_2$$

$$y = u_1 y_1 + u_2 y_2 = (\ln\cos t + k_1)\cos t + (t + k_2)\sin t$$
$$= k_1 \cos t + k_2 \sin t + t \sin t + \cos t(\ln\cos t)$$

Check with MATLAB:

y=dsolve('D2y+y=sec(t)'); f=simple(y)

```
f =
sin(t)*t+log(cos(t))*cos(t)+C1*sin(t)+C2*cos(t)
```

5. Differentiating the given integro–differential equation with respect to t we obtain

$$\frac{dv^3}{dt^3} + k_3\frac{dv^2}{dt^2} + k_2\frac{dv}{dt} + k_1 v = 3\cos 3t - 3\sin 3t = 3(\cos 3t - \sin 3t)$$

or

$$\frac{dv^3}{dt^3} = -k_3\frac{dv^2}{dt^2} - k_2\frac{dv}{dt} - k_1 v + 3(\cos 3t - \sin 3t) \quad (1)$$

We let

$$v = x_1 \qquad \frac{dv}{dt} = x_2 = \dot{x}_1 \qquad \frac{dv^2}{dt^2} = x_3 = \dot{x}_2$$

Then,

$$\frac{dv^3}{dt^3} = \dot{x}_3$$

and by substitution into (1)

$$\dot{x}_3 = -k_1 x_1 - k_2 x_2 - k_3 x_3 + 3(\cos 3t - \sin 3t)$$

Thus, the state equations are

$$\dot{x}_1 = x_2$$
$$\dot{x}_2 = x_3$$
$$\dot{x}_3 = -k_1 x_1 - k_2 x_2 - k_3 x_3 + 3(\cos 3t - \sin 3t)$$

and in matrix form

$$\begin{bmatrix} \dot{x}_1 \\ \dot{x}_2 \\ \dot{x}_3 \end{bmatrix} = \begin{bmatrix} 0 & 1 & 0 \\ 0 & 0 & 1 \\ -k_1 & -k_2 & -k_3 \end{bmatrix} \cdot \begin{bmatrix} x_1 \\ x_2 \\ x_3 \end{bmatrix} + \begin{bmatrix} 0 \\ 0 \\ 1 \end{bmatrix} \cdot 3(\cos 3t - \sin 3t)$$

6. Expansion of the given matrix yields

$$\dot{x}_1 = x_2 \quad \dot{x}_2 = x_3 \quad \dot{x}_3 = x_2 \quad \dot{x}_4 = -x_1 - 2x_2 - 3x_3 - 4x_4 + u(t)$$

Letting $x = y$ we obtain

$$\frac{dy^4}{dt^4} + 4\frac{dy^3}{dt^3} + 3\frac{dy^2}{dt^2} + 2\frac{dy}{dt} + y = u(t)$$

7.

a.

$$A = \begin{bmatrix} 1 & 2 \\ 3 & -1 \end{bmatrix} \quad \det(A - \lambda I) = \det\left(\begin{bmatrix} 1 & 2 \\ 3 & -1 \end{bmatrix} - \lambda\begin{bmatrix} 1 & 0 \\ 0 & 1 \end{bmatrix}\right) = \det\begin{bmatrix} 1-\lambda & 2 \\ 3 & -1-\lambda \end{bmatrix} = 0$$

$(1-\lambda)(-1-\lambda) - 6 = 0$, $-1 - \lambda + \lambda + \lambda^2 - 6 = 0$, $\lambda^2 = 7$, and thus $\lambda_1 = \sqrt{7}$ $\lambda_2 = -\sqrt{7}$

b.

$$B = \begin{bmatrix} a & 0 \\ -a & b \end{bmatrix} \quad \det(B - \lambda I) = \det\left(\begin{bmatrix} a & 0 \\ -a & b \end{bmatrix} - \lambda\begin{bmatrix} 1 & 0 \\ 0 & 1 \end{bmatrix}\right) = \det\begin{bmatrix} a-\lambda & 0 \\ -a & b-\lambda \end{bmatrix} = 0$$

$(a-\lambda)(b-\lambda) = 0$, and thus $\lambda_1 = a$ $\lambda_2 = b$

c.

$$C = \begin{bmatrix} 0 & 1 & 0 \\ 0 & 0 & 1 \\ -6 & -11 & -6 \end{bmatrix} \quad \det(C - \lambda I) = \det\left(\begin{bmatrix} 0 & 1 & 0 \\ 0 & 0 & 1 \\ -6 & -11 & -6 \end{bmatrix} - \lambda\begin{bmatrix} 1 & 0 & 0 \\ 0 & 1 & 0 \\ 0 & 0 & 1 \end{bmatrix}\right)$$

$$= \det\begin{bmatrix} -\lambda & 1 & 0 \\ 0 & -\lambda & 1 \\ -6 & -11 & -6-\lambda \end{bmatrix} = 0$$

$\lambda^2(-6-\lambda) - 6 - (-11)(-\lambda) = \lambda^3 + 6\lambda^2 + 11\lambda + 6 = 0$ and it is given that $\lambda_1 = -1$. Then,

$$\frac{\lambda^3 + 6\lambda^2 + 11\lambda + 6}{(\lambda + 1)} = \lambda^2 + 5\lambda + 6 \Rightarrow (\lambda + 1)(\lambda + 2)(\lambda + 3) = 0$$

and thus $\lambda_1 = -1$ $\lambda_2 = -2$ $\lambda_1 = -3$

8.

a. Matrix A is the same as Matrix C in Exercise 7. Then,

$$\lambda_1 = -1 \qquad \lambda_2 = -2 \qquad \lambda_1 = -3$$

and since A is a 3×3 matrix the state transition matrix is

$$e^{At} = a_0 I + a_1 A + a_2 A^2 \quad (1)$$

Then,

$$a_0 + a_1 \lambda_1 + a_2 \lambda_1^2 = e^{\lambda_1 t} \Rightarrow a_0 - a_1 + a_2 = e^{-t}$$

$$a_0 + a_1 \lambda_2 + a_2 \lambda_2^2 = e^{\lambda_2 t} \Rightarrow a_0 - 2a_1 + 4a_2 = e^{-2t}$$

$$a_0 + a_1 \lambda_3 + a_2 \lambda_3^2 = e^{\lambda_3 t} \Rightarrow a_0 - 3a_1 + 9a_2 = e^{-3t}$$

```
syms t; A=[1 –1 1; 1 –2 4; 1 –3 9];...
a=sym('[exp(–t); exp(–2*t); exp(–3*t)]'); x=A\a; fprintf(' \n');...
disp('a0 = '); disp(x(1)); disp('a1 = '); disp(x(2)); disp('a2 = '); disp(x(3))
```

```
a0 =
3*exp(-t)-3*exp(-2*t)+exp(-3*t)
a1 =
5/2*exp(-t)-4*exp(-2*t)+3/2*exp(-3*t)
a2 =
1/2*exp(-t)-exp(-2*t)+1/2*exp(-3*t)
```
Thus,

$$a_0 = 3e^{-t} - 3e^{-2t} + 3e^{-3t}$$

$$a_1 = 2.5e^{-t} - 4e^{-2t} + 1.5e^{-3t}$$

$$a_2 = 0.5e^{-t} - e^{-2t} + 0.5e^{-3t}$$

Now, we compute e^{At} of (1) with the following MATLAB code:

```
syms t; a0=3*exp(–t)–3*exp(–2*t)+exp(–3*t); a1=5/2*exp(–t)–4*exp(–2*t)+3/2*exp(–3*t);...
a2=1/2*exp(–t)–exp(–2*t)+1/2*exp(–3*t); A=[0 1 0; 0 0 1; –6 –11 –6]; fprintf(' \n');...
eAt=a0*eye(3)+a1*A+a2*A^2
```

```
eAt =
[3*exp(-t)-3*exp(-2*t)+exp(-3*t), 5/2*exp(-t)-4*exp(-2*t)+3/
2*exp(-3*t), 1/2*exp(-t)-exp(-2*t)+1/2*exp(-3*t)]
[-3*exp(-t)+6*exp(-2*t)-3*exp(-3*t), -5/2*exp(-t)+8*exp(-
2*t)-9/2*exp(-3*t), -1/2*exp(-t)+2*exp(-2*t)-3/2*exp(-3*t)]
[3*exp(-t)-12*exp(-2*t)+9*exp(-3*t), 5/2*exp(-t)-16*exp(-
2*t)+27/2*exp(-3*t), 1/2*exp(-t)-4*exp(-2*t)+9/2*exp(-
3*t)]
```

Then,

$$e^{At} = \begin{bmatrix} 3e^{-t} - 3e^{-2t} + e^{-3t} & 2.5e^{-t} - 4e^{-2t} + 1.5e^{-3t} & 0.5e^{-t} - e^{-2t} + 0.5e^{-3t} \\ -3e^{-t} + 6e^{-2t} - 3e^{-3t} & -2.5e^{-t} + 8e^{-2t} - 4.5e^{-3t} & -0.5e^{-t} + 2e^{-2t} - 1.5e^{-3t} \\ 3e^{-t} - 12e^{-2t} + 9e^{-3t} & 2.5e^{-t} - 16e^{-2t} + 13.5e^{-3t} & 0.5e^{-t} - 4e^{-2t} + 4.5e^{-3t} \end{bmatrix}$$

Chapter 6

Fourier, Taylor, and Maclaurin Series

This chapter is an introduction to Fourier and power series. We begin with the definition of sinusoids that are harmonically related and the procedure for determining the coefficients of the trigonometric form of the series. Then, we discuss the different types of symmetry and how they can be used to predict the terms that may be present. Several examples are presented to illustrate the approach. The alternate trigonometric and the exponential forms are also presented. We conclude with a discussion on power series expansion with the Taylor and Maclaurin series.

6.1 Wave Analysis

The French mathematician Fourier found that any *periodic* waveform, that is, a waveform that repeats itself after some time, can be expressed as a series of harmonically related sinusoids, i.e., sinusoids whose frequencies are multiples of a *fundamental* frequency (or first harmonic). For example, a series of sinusoids with frequencies 1 MHz, 2 MHz, 3 MHz, and so on, contains the fundamental frequency of 1 MHz, a second harmonic of 2 MHz, a third harmonic of 3 MHz, and so on. In general, any periodic waveform $f(t)$ can be expressed as

$$f(t) = \frac{1}{2}a_0 + a_1\cos\omega t + a_2\cos 2\omega t + a_3\cos 3\omega t + a_4\cos 4\omega t + \ldots$$
$$+ b_1\sin\omega t + b_2\sin 2\omega t + b_3\sin 3\omega t + b_4\sin 4\omega t + \ldots \qquad (6.1)$$

or

$$f(t) = \frac{1}{2}a_0 + \sum_{n=1}^{\infty}(a_n\cos n\omega t + b_n\sin n\omega t) \qquad (6.2)$$

where the first term $a_0/2$ is a constant, and represents the DC (average) component of $f(t)$. Thus, if $f(t)$ represents some voltage $v(t)$, or current $i(t)$, the term $a_0/2$ is the average value of $v(t)$ or $i(t)$.

The terms with the coefficients a_1 and b_1 together, represent the fundamental frequency component ω [*]. Likewise, the terms with the coefficients a_2 and b_2 together, represent the second harmonic component 2ω, and so on.

---

[*] We recall that $k_1\cos\omega t + k_2\sin\omega t = k\cos(\omega t + \theta)$ where θ is a constant.

Since any periodic waveform f(t) can be expressed as a Fourier series, it follows that the sum of the DC, the fundamental, the second harmonic, and so on, must produce the waveform f(t). Generally, the sum of two or more sinusoids of different frequencies produce a waveform that is not a sinusoid as shown in Figure 6.1.

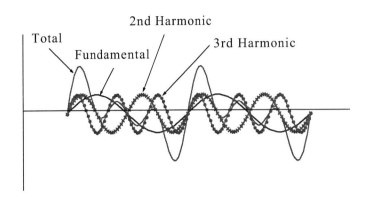

Figure 6.1. Summation of a fundamental, second and third harmonic

6.2 Evaluation of the Coefficients

Evaluations of a_i and b_i coefficients of (6.1) is not a difficult task because the sine and cosine are *orthogonal functions*, that is, the product of the sine and cosine functions under the integral evaluated from 0 to 2π is zero. This will be shown shortly.

Let us consider the functions $\sin mt$ and $\cos mt$ where m and n are any integers, and for convenience, we have assumed that $\omega = 1$. Then,

$$\int_0^{2\pi} \sin mt \, dt = 0 \tag{6.3}$$

$$\int_0^{2\pi} \cos mt \, dt = 0 \tag{6.4}$$

$$\int_0^{2\pi} (\sin mt)(\cos nt) \, dt = 0 \tag{6.5}$$

The integrals of (6.3) and (6.4) are zero since the net area over the *0* to 2π area is zero. The integral of (6.5) is also is zero since

$$\sin x \cos y = \frac{1}{2}[\sin(x + y) + \sin(x - y)]$$

This is also obvious from the plot of Figure 6.2, where we observe that the net shaded area above and below the time axis is zero.

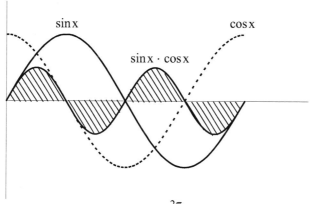

Figure 6.2. Graphical proof of $\displaystyle\int_0^{2\pi}(\sin mt)(\cos nt)dt = 0$

Moreover, if m and n are different integers, then,

$$\int_0^{2\pi}(\sin mt)(\sin nt)dt = 0 \qquad (6.6)$$

since

$$(\sin x)(\sin y) = \frac{1}{2}[\cos(x - y) - \cos(x - y)]$$

The integral of (6.6) can also be confirmed graphically as shown in Figure 6.3, where $m = 2$ and $n = 3$. We observe that the net shaded area above and below the time axis is zero.

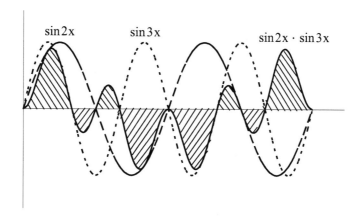

Figure 6.3. Graphical proof of $\int_0^{2\pi} (\sin mt)(\sin nt)dt = 0$ *for* m = 2 *and* n = 3

Also, if m and n are different integers, then,

$$\int_0^{2\pi} (\cos mt)(\cos nt)dt = 0 \qquad (6.7)$$

since

$$(\cos x)(\cos y) = \frac{1}{2}[\cos(x+y) + \cos(x-y)]$$

The integral of (6.7) can also be confirmed graphically as shown in Figure 6.4, where m = 2 and n = 3. We observe that the net shaded area above and below the time axis is zero.

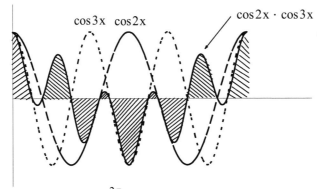

Figure 6.4. Graphical proof of $\int_0^{2\pi} (\cos mt)(\cos nt)dt = 0$ *for* m = 2 *and* n = 3

However, if in (6.6) and (6.7), m = n, then,

$$\int_0^{2\pi} (\sin mt)^2 dt = \pi \qquad (6.8)$$

and

$$\int_0^{2\pi} (\cos mt)^2 dt = \pi \qquad (6.9)$$

The integrals of (6.8) and (6.9) can also be seen to be true graphically with the plots of Figures 6.5 and 6.6.

It was stated earlier that the sine and cosine functions are orthogonal[*] to each other. The simpli-

---

* *We will discuss orthogonal functions in Chapter 14*

fication obtained by application of the orthogonality properties of the sine and cosine functions, becomes apparent in the discussion that follows.

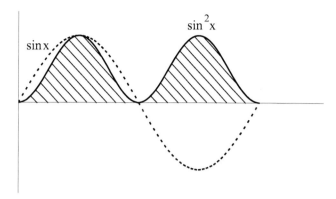

Figure 6.5. Graphical proof of $\int_{0}^{2\pi}(\sin mt)^2 dt = \pi$

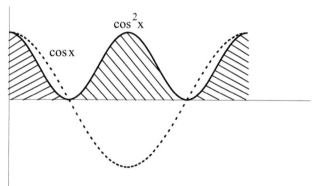

Figure 6.6. Graphical proof of $\int_{0}^{2\pi}(\cos mt)^2 dt = \pi$

In (6.1), for simplicity, we let $\omega = 1$. Then,

$$f(t) = \frac{1}{2}a_0 + a_1\cos t + a_2\cos 2t + a_3\cos 3t + a_4\cos 4t + \dots$$
$$+ b_1\sin t + b_2\sin 2t + b_3\sin 3t + b_4\sin 4t + \dots \tag{6.10}$$

To evaluate any coefficient, say b_2, we multiply both sides of (6.10) by $\sin 2t$. Then,

$$f(t)\sin 2t = \frac{1}{2}a_0\sin 2t + a_1\cos t\sin 2t + a_2\cos 2t\sin 2t + a_3\cos 3t\sin 2t + a_4\cos 4t\sin 2t + \dots$$

$$b_1\sin t\sin 2t + b_2(\sin 2t)^2 + b_3\sin 3t\sin 2t + b_4\sin 4t\sin 2t + \dots$$

Next, we multiply both sides of the above expression by dt, and we integrate over the period 0 to 2π. Then,

$$\int_0^{2\pi} f(t)\sin 2t dt = \frac{1}{2}a_0 \int_0^{2\pi} \sin 2t dt + a_1 \int_0^{2\pi} \cos t \sin 2t dt + a_2 \int_0^{2\pi} \cos 2t \sin 2t dt$$

$$+ a_3 \int_0^{2\pi} \cos 3t \sin 2t dt + a_4 \int_0^{2\pi} \cos 4t \sin 2t dt + \ldots$$

$$+ b_1 \int_0^{2\pi} \sin t \sin 2t dt + b_2 \int_0^{2\pi} (\sin 2t)^2 dt + b_3 \int_0^{2\pi} \sin 3t \sin 2t dt$$

$$+ b_4 \int_0^{2\pi} \sin 4t \sin 2t dt + \ldots \qquad (6.11)$$

We observe that every term on the right side of (6.11) except the term

$$b_2 \int_0^{2\pi} (\sin 2t)^2 dt$$

is zero as we found in (6.6) and (6.7). Therefore, (6.11) reduces to

$$\int_0^{2\pi} f(t)\sin 2t dt = b_2 \int_0^{2\pi} (\sin 2t)^2 dt = b_2 \pi$$

or

$$b_2 = \frac{1}{\pi} \int_0^{2\pi} f(t)\sin 2t dt$$

and thus we can evaluate this integral for any given function $f(t)$. The remaining coefficients can be evaluated similarly.

The coefficients a_0, a_n, and b_n are found from the following relations.

$$\boxed{\frac{1}{2}a_0 = \frac{1}{2\pi} \int_0^{2\pi} f(t)dt} \qquad (6.12)$$

$$\boxed{a_n = \frac{1}{\pi} \int_0^{2\pi} f(t)\cos nt dt} \qquad (6.13)$$

$$\boxed{b_n = \frac{1}{\pi} \int_0^{2\pi} f(t)\sin nt dt} \qquad (6.14)$$

The integral of (6.12) yields the average (DC) value of f(t).

6.3 Symmetry

With a few exceptions such as the waveform of Example 6.6, the most common waveforms used in science and engineering, do not have the average, cosine, and sine terms all present. Some waveforms have cosine terms only, while others have sine terms only. Still other waveforms have or have not DC components. Fortunately, it is possible to predict which terms will be present in the trigonometric Fourier series, by observing whether or not the given waveform possesses some kind of symmetry.

We will discuss three types of symmetry that can be used to facilitate the computation of the trigonometric Fourier series form. These are:

1. *Odd symmetry* – If a waveform has odd symmetry, that is, if it is an odd function, the series will consist of sine terms only. In other words, if f(t) is an odd function, all the a_i coefficients including a_0, will be zero.

2. *Even symmetry* – If a waveform has even symmetry, that is, if it is an even function, the series will consist of cosine terms only, and a_0 may or may not be zero. In other words, if f(t) is an even function, all the b_i coefficients will be zero.

3. *Half–wave symmetry* – If a waveform has half–wave symmetry (to be defined shortly), only odd (odd cosine and odd sine) harmonics will be present. In other words, all even (even cosine and even sine) harmonics will be zero.

We will now define even and odd functions and we should remember that even functions have nothing to do with even harmonics, and odd functions have nothing to do with odd harmonics.

A function f(t) is an *even function* of time if the following relation holds.

$$\boxed{f(-t) = f(t)} \tag{6.15}$$

that is, if in an even function we replace t with –t, the function f(t) does not change. Thus, polynomials with even exponents only, and with or without constants, are even functions. For instance, the cosine function is an even function because it can be written as the power series[*]

$$\cos t = 1 - \frac{t^2}{2!} + \frac{t^4}{4!} - \frac{t^6}{6!} + \ldots$$

Other examples of even functions are shown in Figure 6.7.

---

[*] *We will discuss power series later in this chapter.*

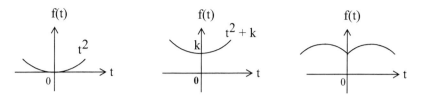

Figure 6.7. Examples of even functions

A function $f(t)$ is an *odd function* of time if the following relation holds.

$$\boxed{-f(-t) = f(t)}$$ (6.16)

that is, if in an odd function we replace t with $-t$, we obtain the negative of the function $f(t)$. Thus, polynomials with odd exponents only, and no constants are odd functions. For instance, the sine function is an odd function because it can be written as the power series

$$\sin t = t - \frac{t^3}{3!} + \frac{t^5}{5!} - \frac{t^7}{7!} + \dots$$

Other examples of odd functions are shown in Figure 6.8.

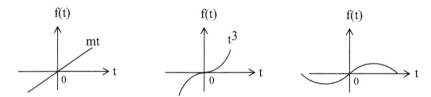

Figure 6.8. Examples of odd functions

We observe that for odd functions, $f(0) = 0$. However, the reverse is not always true; that is, if $f(0) = 0$, we should not conclude that $f(t)$ is an odd function. An example of this is the function $f(t) = t^2$ in Figure 6.7.

The product of *two even* or *two odd* functions is an even function, and the product of an even function times an odd function, is an odd function.

Henceforth, we will denote an even function with the subscript e, and an odd function with the subscript o. Thus, $f_e(t)$ and $f_o(t)$ will be used to represent even and odd functions of time respectively.

Also,

$$\int_{-T}^{T} f_e(t)dt = 2\int_{0}^{T} f_e(t)dt$$ (6.17)

and

$$\int_{-T}^{T} f_o(t)dt = 0 \tag{6.18}$$

A function $f(t)$ that is neither even nor odd can be expressed as

$$f_e(t) = \frac{1}{2}[f(t) + f(-t)] \tag{6.19}$$

or as

$$f_o(t) = \frac{1}{2}[f(t) - f(-t)] \tag{6.20}$$

By addition of (6.16) with (6.17), we get

$$f(t) = f_e(t) + f_o(t) \tag{6.21}$$

that is, any function of time can be expressed as the sum of an even and an odd function.

To understand half–wave symmetry, we recall that any periodic function with period T, is expressed as

$$f(t) = f(t + T) \tag{6.22}$$

that is, the function with value $f(t)$ at any time t, will have the same value again at a later time $t + T$.

A periodic waveform with period T, has half–wave symmetry if

$$-f(t + T/2) = f(t) \tag{6.23}$$

that is, the shape of the negative half–cycle of the waveform is the same as that of the positive half–cycle, but inverted.

We will test the waveforms of Figures 6.9 through 6.13 for any of the three types of symmetry.

1. Square waveform

For the waveform of Figure 6.9, the average value over one period T is zero, and therefore, $a_0 = 0$. It is also an odd function and has half–wave symmetry since $-f(-t) = f(t)$ and $-f(t + T/2) = f(t)$.

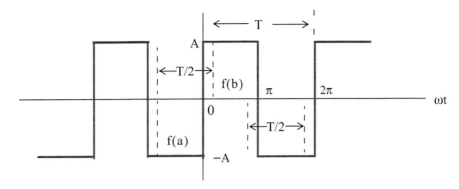

Figure 6.9. Square waveform test for symmetry

An easy method to test for half–wave symmetry is to choose any half–period $T/2$ length on the time axis as shown in Figure 6.9, and observe the values of $f(t)$ at the left and right points on the time axis, such as $f(a)$ and $f(b)$. If there is half–wave symmetry, these will always be equal but will have opposite signs as we slide the half–period $T/2$ length to the left or to the right on the time axis at non–zero values of $f(t)$.

2. Square waveform with ordinate axis shifted

If we shift the ordinate axis $\pi/2$ radians to the right, as shown in Figure 6.10, we see that the square waveform now becomes an even function and has half–wave symmetry since $f(-t) = f(t)$ and $-f(t + T/2) = f(t)$. Also, $a_0 = 0$.

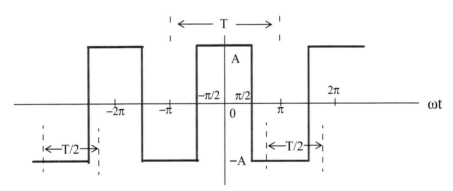

Figure 6.10. Square waveform with ordinate shifted by $\pi/2$

Obviously, if the ordinate axis is shifted by any other value other than an odd multiple of $\pi/2$, the waveform will have neither odd nor even symmetry.

3. Sawtooth waveform

For the sawtooth waveform of Figure 6.11, the average value over one period T is zero and therefore, $a_0 = 0$. It is also an odd function because $-f(-t) = f(t)$, but has no half–wave symmetry

since $-f(t + T/2) \neq f(t)$

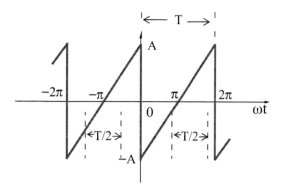

Figure 6.11. Sawtooth waveform test for symmetry

4. Triangular waveform

For this triangular waveform of Figure 6.12, the average value over one period T is zero and therefore, $a_0 = 0$. It is also an odd function since $-f(-t) = f(t)$. Moreover, it has half–wave symmetry because $-f(t + T/2) = f(t)$

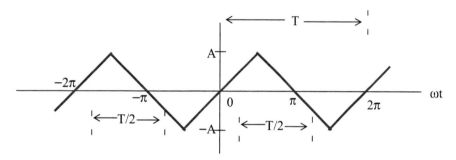

Figure 6.12. Triangular waveform test for symmetry

5. Fundamental, Second and Third Harmonics of a Sinusoid

Figure 6.13 shows a fundamental, second, and third harmonic of a typical sinewave where the half period $T/2$, is chosen as the half period of the period of the fundamental frequency. This is necessary in order to test the fundamental, second, and third harmonics for half–wave symmetry. The fundamental has half–wave symmetry since the a and –a values, when separated by $T/2$, are equal and opposite. The second harmonic has no half–wave symmetry because the ordinates b on the left and b on the right, although are equal, there are not opposite in sign. The third harmonic has half–wave symmetry since the c and –c values, when separated by $T/2$ are equal and opposite. These waveforms can be either odd or even depending on the position of the ordinate. Also, all three waveforms have zero average value unless the abscissa axis is shifted up or down.

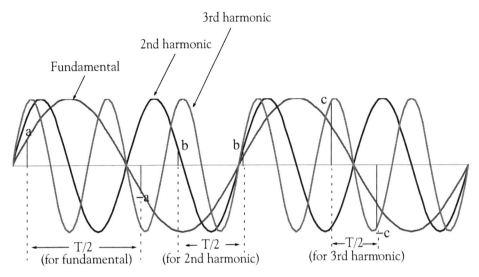

Figure 6.13. Fundamental, second, and third harmonic test for symmetry

In the expressions of the integrals in (6.12) through (6.14), Page 6–6, the limits of integration for the coefficients a_n and b_n are given as 0 to 2π, that is, one period T. Of course, we can choose the limits of integration as $-\pi$ to $+\pi$. Also, if the given waveform is an odd function, or an even function, or has half–wave symmetry, we can compute the non–zero coefficients a_n and b_n by integrating from 0 to π only, and multiply the integral by 2. Moreover, if the waveform has half–wave symmetry and is also an odd or an even function, we can choose the limits of integration from 0 to $\pi/2$ and multiply the integral by 4. The proof is based on the fact that, the product of two even functions is another even function, and also that the product of two odd functions results also in an even function. However, it is important to remember that when using these shortcuts, we must evaluate the coefficients a_n and b_n for the integer values of n that will result in non–zero coefficients. This point will be illustrated in Example 6.2.

6.4 Waveforms in Trigonometric Form of Fourier Series

We will now derive the trigonometric Fourier series of the most common periodic waveforms.

Example 6.1

Compute the trigonometric Fourier series of the square waveform of Figure 6.14.

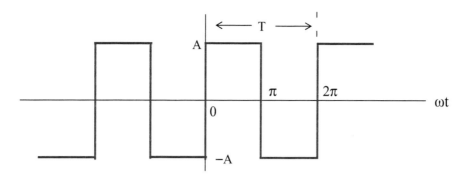

Figure 6.14. Square waveform for Example 6.1

Solution:

The trigonometric series will consist of sine terms only because, as we already know, this waveform is an odd function. Moreover, only odd harmonics will be present since this waveform has half–wave symmetry. However, we will compute all coefficients to verify this. Also, for brevity, we will assume that $\omega = 1$.

The a_i coefficients are found from

$$a_n = \frac{1}{\pi} \int_0^{2\pi} f(t) \cos nt\, dt = \frac{1}{\pi}\left[\int_0^{\pi} A \cos nt\, dt + \int_{\pi}^{2\pi} (-A) \cos nt\, dt \right] = \frac{A}{n\pi}\left(\sin nt \big|_0^{\pi} - \sin nt \big|_{\pi}^{2\pi} \right)$$

$$= \frac{A}{n\pi}\left(\sin n\pi - 0 - \sin n2\pi + \sin n\pi \right) = \frac{A}{n\pi}\left(2\sin n\pi - \sin n2\pi \right) \tag{6.24}$$

and since n is an integer (positive or negative) or zero, the terms inside the parentheses on the second line of (6.24) are zero and therefore, all a_i coefficients are zero, as expected, since the square waveform has odd symmetry. Also, by inspection, the average (DC) value is zero, but if we attempt to verify this using (6.24), we will get the indeterminate form $0/0$. To work around this problem, we will evaluate a_0 directly from (6.12). Then,

$$a_0 = \frac{1}{\pi}\left[\int_0^{\pi} A\, dt + \int_{\pi}^{2\pi} (-A)\, dt \right] = \frac{A}{\pi}(\pi - 0 - 2\pi + \pi) = 0 \tag{6.25}$$

The b_i coefficients are found from (6.14), that is,

$$b_n = \frac{1}{\pi} \int_0^{2\pi} f(t) \sin nt\, dt = \frac{1}{\pi}\left[\int_0^{\pi} A \sin nt\, dt + \int_{\pi}^{2\pi} (-A) \sin nt\, dt \right] = \frac{A}{n\pi}\left(-\cos nt \big|_0^{\pi} + \cos nt \big|_{\pi}^{2\pi} \right)$$

$$= \frac{A}{n\pi}\left(-\cos n\pi + 1 + \cos 2n\pi - \cos n\pi \right) = \frac{A}{n\pi}\left(1 - 2\cos n\pi + \cos 2n\pi \right) \tag{6.26}$$

For $n = even$, (6.26) yields

$$b_n = \frac{A}{n\pi}(1-2+1) = 0$$

as expected, since the square waveform has half–wave symmetry.

For $n = odd$, (6.21) reduces to

$$b_n = \frac{A}{n\pi}(1+2+1) = \frac{4A}{n\pi}$$

and thus

$$b_1 = \frac{4A}{\pi}$$

$$b_3 = \frac{4A}{3\pi}$$

$$b_5 = \frac{4A}{5\pi}$$

and so on.

Therefore, the trigonometric Fourier series for the *square waveform with odd symmetry* is

$$f(t) = \frac{4A}{\pi}\left(\sin\omega t + \frac{1}{3}\sin 3\omega t + \frac{1}{5}\sin 5\omega t + \dots\right) = \frac{4A}{\pi}\sum_{n=odd}\frac{1}{n}\sin n\omega t \qquad (6.27)$$

It was stated above that, if the given waveform has half–wave symmetry, and it is also an odd or an even function, we can integrate from 0 to $\pi/2$, and multiply the integral by 4. We will apply this property to the following example.

Example 6.2

Compute the trigonometric Fourier series of the square waveform of Example 1 by integrating from 0 to $\pi/2$, and multiplying the result by 4.

Solution:

Since the waveform is an odd function and has half–wave symmetry, we are only concerned with the odd b_n coefficients. Then,

$$b_n = 4\frac{1}{\pi}\int_0^{\pi/2} f(t)\sin nt\,dt = \frac{4A}{n\pi}(-\cos nt\Big|_0^{\pi/2}) = \frac{4A}{n\pi}\left(-\cos n\frac{\pi}{2}+1\right) \qquad (6.28)$$

For $n = odd$, (6.28) becomes

$$b_n = \frac{4A}{n\pi}(-0+1) = \frac{4A}{n\pi} \qquad (6.29)$$

as before, and thus the series is the same as in Example 1.

Example 6.3

Compute the trigonometric Fourier series of the square waveform of Figure 6.15.

Solution:

This is the same waveform as in Example 6.1, except that the ordinate has been shifted to the right by $\pi/2$ radians, and has become an even function. However, it still has half–wave symmetry. Therefore, the trigonometric Fourier series will consist of odd cosine terms only.

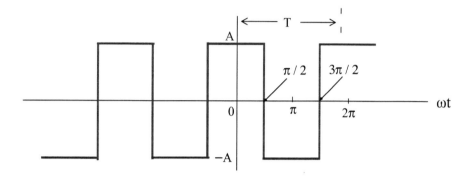

Figure 6.15. Waveform for Example 6.3

Since the waveform has half–wave symmetry and is an even function, it will suffice to integrate from 0 to $\pi/2$, and multiply the integral by 4. The a_n coefficients are found from

$$a_n = 4\frac{1}{\pi}\int_0^{\pi/2} f(t)\cos nt\, dt = \frac{4}{\pi}\left[\int_0^{\pi/2} A\cos nt\, dt\right] = \frac{4A}{n\pi}(\sin nt\big|_0^{\pi/2}) = \frac{4A}{n\pi}\left(\sin n\frac{\pi}{2}\right) \qquad (6.30)$$

We observe that for $n = even$, all a_n coefficients are zero, and thus all even harmonics are zero as expected. Also, by inspection, the average (DC) value is zero.

For $n = odd$, we observe from (6.30) that $\sin n\frac{\pi}{2}$, will alternate between +1 and –1 depending on the odd integer assigned to n. Thus,

$$a_n = \pm\frac{4A}{n\pi} \qquad (6.31)$$

For $n = 1, 5, 9, 13$, and so on, (6.30) becomes

$$a_n = \frac{4A}{n\pi}$$

and for $n = 3, 7, 11, 15$, and so on, it becomes

$$a_n = \frac{-4A}{n\pi}$$

Then, the trigonometric Fourier series for the *square waveform with even symmetry* is

$$f(t) = \frac{4A}{\pi}\left(\cos\omega t - \frac{1}{3}\cos 3\omega t + \frac{1}{5}\cos 5\omega t - ...\right) = \frac{4A}{\pi}\sum_{n = \text{odd}}(-1)^{\frac{(n-1)}{2}}\frac{1}{n}\cos n\omega t \qquad (6.32)$$

Alternate Solution:

Since the waveform of Example 6.3 is the same as of Example 6.1, but shifted to the right by $\pi/2$ radians, we can use the result of Example 6.1, i.e.,

$$f(t) = \frac{4A}{\pi}\left(\sin\omega t + \frac{1}{3}\sin 3\omega t + \frac{1}{5}\sin 5\omega t + ...\right) \qquad (6.33)$$

and substitute ωt with $\omega t + \pi/2$, that is, we let $\omega t = \omega\tau + \pi/2$. With this substitution, relation (6.33) becomes

$$f(\tau) = \frac{4A}{\pi}\left[\sin\left(\omega\tau + \frac{\pi}{2}\right) + \frac{1}{3}\sin 3\left(\omega\tau + \frac{\pi}{2}\right) + \frac{1}{5}\sin 5\left(\omega\tau + \frac{\pi}{2}\right) + ...\right]$$

$$= \frac{4A}{\pi}\left[\sin\left(\omega\tau + \frac{\pi}{2}\right) + \frac{1}{3}\sin\left(3\omega\tau + \frac{3\pi}{2}\right) + \frac{1}{5}\sin\left(5\omega\tau + \frac{5\pi}{2}\right) + ...\right] \qquad (6.34)$$

and using the identities $\sin(x + \pi/2) = \cos x$, $\sin(x + 3\pi/2) = -\cos x$, and so on, we rewrite (6.34) as

$$f(\tau) = \frac{4A}{\pi}\left[\cos\omega\tau - \frac{1}{3}\cos 3\omega\tau + \frac{1}{5}\cos 5\omega\tau - ...\right] \qquad (6.35)$$

and this is the same as (6.27).

Therefore, if we compute the trigonometric Fourier series with reference to one ordinate, and afterwards we want to recompute the series with reference to a different ordinate, we can use the above procedure to save time.

Example 6.4

Compute the trigonometric Fourier series of the sawtooth waveform of Figure 6.16.

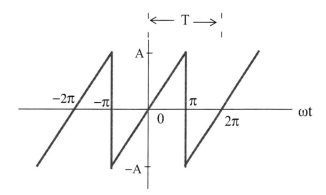

Figure 6.16. Sawtooth waveform

Solution:

This waveform is an odd function but has no half–wave symmetry; therefore, it contains sine terms only with both odd and even harmonics. Accordingly, we only need to evaluate the b_n coefficients. By inspection, the DC component is zero. As before, we will assume that $\omega = 1$.

If we choose the limits of integration from 0 to 2π we will need to perform two integrations since

$$f(t) = \begin{cases} \dfrac{A}{\pi}t & 0 < t < \pi \\[2ex] \dfrac{A}{\pi}t - 2A & \pi < t < 2\pi \end{cases}$$

However, we can choose the limits from $-\pi$ to $+\pi$, and thus we will only need one integration since

$$f(t) = \frac{A}{\pi}t \qquad -\pi < t < \pi$$

Better yet, since the waveform is an odd function, we can integrate from 0 to π, and multiply the integral by 2; this is what we will do.

From tables of integrals,

$$\int x \sin ax \, dx = \frac{1}{a^2} \sin ax - \frac{x}{a} \cos ax \qquad (6.36)$$

Then,

$$b_n = \frac{2}{\pi}\int_0^\pi \frac{A}{\pi}t\sin nt\,dt = \frac{2A}{\pi^2}\int_0^\pi t\sin nt\,dt = \frac{2A}{\pi^2}\left(\frac{1}{n^2}\sin nt - \frac{t}{n}\cos nt\right)\Big|_0^\pi$$

$$= \frac{2A}{n^2\pi^2}(\sin nt - nt\cos nt)\Big|_0^\pi = \frac{2A}{n^2\pi^2}(\sin n\pi - n\pi\cos n\pi)$$

(6.37)

We observe that:

1. If $n = even$, $\sin n\pi = 0$ and $\cos n\pi = 1$. Then, (6.37) reduces to

$$b_n = \frac{2A}{n^2\pi^2}(-n\pi) = -\frac{2A}{n\pi}$$

that is, the even harmonics have negative coefficients.

2. If $n = odd$, $\sin n\pi = 0$, $\cos n\pi = -1$. Then,

$$b_n = \frac{2A}{n^2\pi^2}(n\pi) = \frac{2A}{n\pi}$$

that is, the odd harmonics have positive coefficients.

Thus, the trigonometric Fourier series for the *sawtooth waveform with odd symmetry* is

$$f(t) = \frac{2A}{\pi}\left(\sin\omega t - \frac{1}{2}\sin 2\omega t + \frac{1}{3}\sin 3\omega t - \frac{1}{4}\sin 4\omega t + ...\right) = \frac{2A}{\pi}\sum(-1)^{n-1}\frac{1}{n}\sin n\omega t$$

(6.38)

Example 6.5

Find the trigonometric Fourier series of the triangular waveform of Figure 6.17. Assume $\omega = 1$.

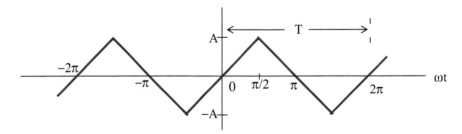

Figure 6.17. Triangular waveform for Example 6.5

Solution:

This waveform is an odd function and has half–wave symmetry; then, the trigonometric Fourier series will contain sine terms only with odd harmonics. Accordingly, we only need to evaluate the b_n coefficients. We will choose the limits of integration from 0 to $\pi/2$, and will multiply the integral by 4.

By inspection, the DC component is zero. From tables of integrals,

$$\int x \sin ax \, dx = \frac{1}{a^2} \sin ax - \frac{x}{a} \cos ax \tag{6.39}$$

Then,

$$b_n = \frac{4}{\pi} \int_0^{\pi/2} \frac{2A}{\pi} t \sin nt \, dt = \frac{8A}{\pi^2} \int_0^{\pi/2} t \sin nt \, dt = \frac{8A}{\pi^2} \left(\frac{1}{n^2} \sin nt - \frac{t}{n} \cos nt \right) \Bigg|_0^{\pi/2}$$

$$= \frac{8A}{n^2 \pi^2} \left(\sin nt - nt \cos nt \right) \Big|_0^{\pi/2} = \frac{8A}{n^2 \pi^2} \left(\sin n\frac{\pi}{2} - n\frac{\pi}{2} \cos n\frac{\pi}{2} \right) \tag{6.40}$$

We are only interested in the odd integers of n, and we observe that:

$$\cos n\frac{\pi}{2} = 0$$

For odd integers of n, the sine term yields

$$\sin n\frac{\pi}{2} = \begin{cases} 1 \text{ for } n = 1, 5, 9, \ldots \text{ then, } b_n = \dfrac{8A}{n^2 \pi^2} \\[2ex] -1 \text{ for } n = 3, 7, 11, \ldots \text{ then, } b_n = -\dfrac{8A}{n^2 \pi^2} \end{cases}$$

Thus, the trigonometric Fourier series for the *triangular waveform* with odd symmetry is

$$f(t) = \frac{8A}{\pi^2} \left(\sin \omega t - \frac{1}{9} \sin 3\omega t + \frac{1}{25} \sin 5\omega t - \frac{1}{49} \sin 7\omega t + \ldots \right) = \frac{8A}{\pi^2} \sum_{n = \text{odd}} (-1)^{\frac{(n-1)}{2}} \frac{1}{n^2} \sin n\omega t \tag{6.41}$$

Example 6.6

A *half–wave rectification waveform* is defined as

$$f(t) = \begin{cases} \sin \omega t & 0 < \omega t < \pi \\ 0 & \pi < \omega t < 2\pi \end{cases} \tag{6.42}$$

Express $f(t)$ as a trigonometric Fourier series. Assume $\omega = 1$.

Solution:

The waveform for this example is shown in Figure 6.18.

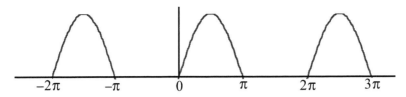

Figure 6.18. f(t) *for Example 6.6*

By inspection, the average is a non–zero value, and the waveform has neither odd nor even symmetry. Therefore, we expect all terms to be present. The a_n coefficients are found from

$$a_n = \frac{1}{\pi} \int_0^{2\pi} f(t)\cos nt\, dt$$

For this example,

$$a_n = \frac{A}{\pi} \int_0^{\pi} \sin t \cos nt\, dt + \frac{A}{\pi} \int_{\pi}^{2\pi} 0 \cos nt\, dt$$

and from tables of integrals

$$\int (\sin mx)(\cos nx)dx = -\frac{\cos(m-n)x}{2(m-n)} - \frac{\cos(m+n)x}{2(m+n)} \quad (m^2 \neq n^2)$$

Then,

$$a_n = \frac{A}{\pi}\left\{ -\frac{1}{2}\left[\frac{\cos(1-n)t}{1-n} + \frac{\cos(1+n)t}{1+n} \right]\Bigg|_0^{\pi} \right\}$$

$$= -\frac{A}{2\pi}\left\{ \left[\frac{\cos(\pi - n\pi)}{1-n} + \frac{\cos(\pi + n\pi)}{1+n} \right] - \left[\frac{1}{1-n} + \frac{1}{n+1} \right] \right\}$$

(6.43)

Using the trigonometric identities

$$\cos(x-y) = \cos x \cos y + \sin x \sin y$$

and

$$\cos(x+y) = \cos x \cos y - \sin x \sin y$$

we obtain

$$\cos(\pi - n\pi) = \cos\pi \cos n\pi + \sin\pi \sin n\pi = -\cos n\pi$$

and

$$\cos(\pi + n\pi) = \cos\pi \cos n\pi - \sin\pi \sin n\pi = -\cos n\pi$$

Then, by substitution into (6.43),

$$a_n = -\frac{A}{2\pi}\left\{ \left[\frac{-\cos n\pi}{1-n} + \frac{-\cos n\pi}{1+n} \right] - \frac{2}{1-n^2} \right\} = \frac{A}{2\pi}\left\{ \left[\frac{\cos n\pi}{1-n} + \frac{\cos n\pi}{1+n} \right] + \frac{2}{1-n^2} \right\}$$

(6.44)

$$= \frac{A}{2\pi}\left(\frac{\cos n\pi + n\cos n\pi + \cos n\pi - n\cos n\pi}{1-n^2} + \frac{2}{1-n^2} \right) = \frac{A}{\pi}\left(\frac{\cos n\pi + 1}{(1-n^2)} \right) \quad n \neq 1$$

Next, we can evaluate all the a_n coefficients, except a_1, from (6.44).

First, we will evaluate a_0 to obtain the DC value. By substitution of $n = 0$, we get $a_0 = 2A/\pi$. Therefore, the DC value is

$$\frac{1}{2}a_0 = \frac{A}{\pi} \tag{6.45}$$

We cannot use (6.44) to obtain the value of a_1; therefore, we will evaluate the integral

$$a_1 = \frac{A}{\pi} \int_0^\pi \sin t \cos t \, dt$$

From tables of integrals,

$$\int (\sin ax)(\cos ax)dx = \frac{1}{2a}(\sin ax)^2$$

and thus,

$$a_1 = \frac{A}{2\pi}(\sin t)^2 \Big|_0^\pi = 0 \tag{6.46}$$

From (6.44) with $n = 2, 3, 4, 5, \ldots$, we get

$$a_2 = \frac{A}{\pi}\left(\frac{\cos 2\pi + 1}{(1 - 2^2)}\right) = -\frac{2A}{3\pi} \tag{6.47}$$

$$a_3 = \frac{A(\cos 3\pi + 1)}{\pi(1 - 3^2)} = 0 \tag{6.48}$$

We see that for odd integers of n, $a_n = 0$. However, for $n = \text{even}$, we get

$$a_4 = \frac{A(\cos 4\pi + 1)}{\pi(1 - 4^2)} = -\frac{2A}{15\pi} \tag{6.49}$$

$$a_6 = \frac{A(\cos 6\pi + 1)}{\pi(1 - 6^2)} = -\frac{2A}{35\pi} \tag{6.50}$$

$$a_8 = \frac{A(\cos 8\pi + 1)}{\pi(1 - 8^2)} = -\frac{2A}{63\pi} \tag{6.51}$$

and so on.

Now, we need to evaluate the b_n coefficients. For this example,

$$b_n = A\frac{1}{\pi} \int_0^{2\pi} f(t)\sin nt \, dt = \frac{A}{\pi} \int_0^\pi \sin t \sin nt \, dt + \frac{A}{\pi} \int_\pi^{2\pi} 0 \sin nt \, dt$$

and from tables of integrals,

$$\int (\sin mx)(\sin nx)dx = \frac{\sin(m-n)x}{2(m-n)} - \frac{\sin(m+n)x}{2(m+n)} \quad (m^2 \neq n^2)$$

Therefore,

$$b_n = \frac{A}{\pi} \cdot \frac{1}{2} \left\{ \left[\frac{\sin(1-n)t}{1-n} - \frac{\sin(1+n)t}{1+n} \right] \Big|_0^{\pi} \right\}$$

$$= \frac{A}{2\pi} \left[\frac{\sin(1-n)\pi}{1-n} - \frac{\sin(1+n)\pi}{1+n} - 0 + 0 \right] = 0 \quad (n \neq 1)$$

that is, all the b_n coefficients, except b_1, are zero.

We will find b_1 by direct substitution into (6.14) for $n = 1$. Thus,

$$b_1 = \frac{A}{\pi} \int_0^{\pi} (\sin t)^2 dt = \frac{A}{\pi} \left[\frac{t}{2} - \frac{\sin 2t}{4} \right]\Big|_0^{\pi} = \frac{A}{\pi} \left[\frac{\pi}{2} - \frac{\sin 2\pi}{4} \right] = \frac{A}{2} \tag{6.52}$$

Combining (6.45) and (6.47) through (6.52), we find that the trigonometric Fourier series for the *half–wave rectification waveform with no symmetry* is

$$f(t) = \frac{A}{\pi} + \frac{A}{2}\sin t - \frac{A}{\pi}\left[\frac{\cos 2t}{3} + \frac{\cos 4t}{15} + \frac{\cos 6t}{35} + \frac{\cos 8t}{63} + \dots \right] \tag{6.53}$$

Example 6.7

A *full–wave rectification waveform* is defined as

$$f(t) = |A\sin \omega t| \tag{6.54}$$

Express $f(t)$ as a trigonometric Fourier series. Assume $\omega = 1$.

Solution:

The waveform is shown in Figure 6.19 where the ordinate was arbitrarily chosen as shown.

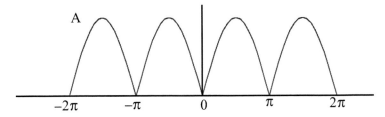

Figure 6.19. Full–wave rectified waveform with even symmetry

By inspection, the average is a non–zero value. We choose the period of the input sinusoid so that the output will be expressed in terms of the fundamental frequency. We also choose the limits of integration as $-\pi$ and $+\pi$, we observe that the waveform has even symmetry.

Therefore, we expect only cosine terms to be present. The a_n coefficients are found from

$$a_n = \frac{1}{\pi} \int_0^{2\pi} f(t) \cos nt \, dt$$

where for this example,

$$a_n = \frac{1}{\pi} \int_{-\pi}^{\pi} A \sin t \cos nt \, dt = \frac{2A}{\pi} \int_0^{\pi} \sin t \cos nt \, dt \tag{6.55}$$

and from tables of integrals,

$$\int (\sin mx)(\cos nx) dx = \frac{\cos(m-n)x}{2(n-m)} - \frac{\cos(m+n)x}{2(m+n)} \quad (m^2 \neq n^2)$$

Since

$$\cos(x-y) = \cos(y-x) = \cos x \cos y + \sin x \sin y$$

we express (6.55) as

$$a_n = \frac{2A}{\pi} \cdot \frac{1}{2} \left\{ \left[\frac{\cos(n-1)t}{n-1} - \frac{\cos(n+1)t}{n+1} \right] \Big|_0^\pi \right\}$$

$$= \frac{A}{\pi} \left\{ \left[\frac{\cos(n-1)\pi}{n-1} - \frac{\cos(n+1)\pi}{n+1} \right] - \left[\frac{1}{n-1} - \frac{1}{n+1} \right] \right\} \tag{6.56}$$

$$= \frac{A}{\pi} \left[\frac{1 - \cos(n\pi + \pi)}{n+1} + \frac{\cos(n\pi - \pi) - 1}{n-1} \right]$$

To simplify the last expression in (6.56), we make use of the trigonometric identities

$$\cos(n\pi + \pi) = \cos n\pi \cos \pi - \sin n\pi \sin \pi = -\cos n\pi$$

and

$$\cos(n\pi - \pi) = \cos n\pi \cos \pi + \sin n\pi \sin \pi = -\cos n\pi$$

Then, (6.56) simplifies to

$$a_n = \frac{A}{\pi} \left[\frac{1 + \cos n\pi}{n+1} - \frac{1 + \cos n\pi}{n-1} \right] = \frac{A}{\pi} \left[\frac{-2 + (n-1)\cos n\pi - (n+1)\cos n\pi}{n^2 - 1} \right]$$

$$= \frac{-2A(\cos n\pi + 1)}{\pi(n^2 - 1)} \quad n \neq 1 \tag{6.57}$$

Now, we can evaluate all the a_n coefficients, except a_1, from (6.57). First, we will evaluate a_0 to obtain the DC value. By substitution of $n = 0$, we get

$$a_0 = \frac{4A}{\pi}$$

Therefore, the DC value is

$$\frac{1}{2}a_0 = \frac{2A}{\pi} \tag{6.58}$$

From (6.57) we observe that for all $n = odd$, other than $n = 1$, $a_n = 0$.
To obtain the value of a_1, we must evaluate the integral

$$a_1 = \frac{1}{\pi} \int_0^\pi \sin t \cos t \, dt$$

From tables of integrals,

$$\int (\sin ax)(\cos ax) dx = \frac{1}{2a}(\sin ax)^2$$

and thus,

$$a_1 = \frac{1}{2\pi}(\sin t)^2 \Big|_0^\pi = 0 \tag{6.59}$$

For $n = even$, from (6.57) we get

$$a_2 = \frac{-2A(\cos 2\pi + 1)}{\pi(2^2 - 1)} = -\frac{4A}{3\pi} \tag{6.60}$$

$$a_4 = \frac{-2A(\cos 4\pi + 1)}{\pi(4^2 - 1)} = -\frac{4A}{15\pi} \tag{6.61}$$

$$a_6 = \frac{-2A(\cos 6\pi + 1)}{\pi(6^2 - 1)} = -\frac{4A}{35\pi} \tag{6.62}$$

$$a_8 = \frac{-2A(\cos 8\pi + 1)}{\pi(8^2 - 1)} = -\frac{4A}{63\pi} \tag{6.63}$$

and so on. Then, combining the terms of (6.58) and (6.60) through (6.63) we get

$$f(t) = \frac{2A}{\pi} - \frac{4A}{\pi} \left\{ \frac{\cos 2\omega t}{3} + \frac{\cos 4\omega t}{15} + \frac{\cos 6\omega t}{35} + \frac{\cos 8\omega t}{63} + \ldots \right\} \tag{6.64}$$

Therefore, the trigonometric form of the Fourier series for the *full–wave rectification waveform with even symmetry* is

$$f(t) = \frac{2A}{\pi} - \frac{4A}{\pi} \sum_{n=2,4,6,\ldots}^{\infty} \frac{1}{(n^2-1)} \cos n\omega t \qquad (6.65)$$

This series of (6.65) shows that there is no component of the fundamental frequency. This is because we chose the period to be from $-\pi$ and $+\pi$. Generally, the period is defined as the shortest period of repetition. In any waveform where the period is chosen appropriately, it is very unlikely that a Fourier series will consist of even harmonic terms only.

6.5 Alternate Forms of the Trigonometric Fourier Series

We recall that the trigonometric Fourier series is expressed as

$$f(t) = \frac{1}{2}a_0 + a_1 \cos\omega t + a_2 \cos 2\omega t + a_3 \cos 3\omega t + a_4 \cos 4\omega t + \ldots$$
$$+ b_1 \sin\omega t + b_2 \sin 2\omega t + b_3 \sin 3\omega t + b_4 \sin 4\omega t + \ldots \qquad (6.66)$$

If a given waveform does not have any kind of symmetry, it may be advantageous of using the *alternate form of the trigonometric Fourier series* where the cosine and sine terms of the same frequency are grouped together, and the sum is combined to a single term, either cosine or sine. However, we still need to compute the a_n and b_n coefficients separately.

We use the triangle shown in Figure 6.20 for the derivation of the alternate forms.

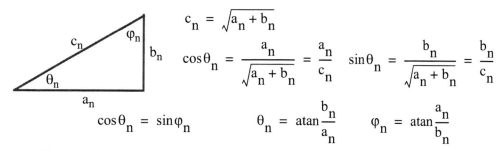

$$c_n = \sqrt{a_n + b_n}$$
$$\cos\theta_n = \frac{a_n}{\sqrt{a_n + b_n}} = \frac{a_n}{c_n} \qquad \sin\theta_n = \frac{b_n}{\sqrt{a_n + b_n}} = \frac{b_n}{c_n}$$
$$\cos\theta_n = \sin\varphi_n \qquad \theta_n = \operatorname{atan}\frac{b_n}{a_n} \qquad \varphi_n = \operatorname{atan}\frac{a_n}{b_n}$$

Figure 6.20. *Derivation of the alternate form of the trigonometric Fourier series*

We assume $\omega = 1$, and for $n = 1, 2, 3, \ldots$, we rewrite (6.66) as

$$f(t) = \frac{1}{2}a_0 + c_1\left(\frac{a_1}{c_1}\cos t + \frac{b_1}{c_1}\sin t\right) + c_2\left(\frac{a_2}{c_2}\cos 2t + \frac{b_2}{c_2}\sin 2t\right) + \dots$$

$$+ c_n\left(\frac{a_n}{c_n}\cos nt + \frac{b_n}{c_n}\sin nt\right)$$

$$= \frac{1}{2}a_0 + c_1\left(\frac{\cos\theta_1\cos t + \sin\theta_1\sin t}{\cos(t-\theta_1)}\right) + c_2\left(\frac{\cos\theta_2\cos 2t + \sin\theta_2\sin 2t}{\cos(2t-\theta_2)}\right) + \dots$$

$$+ c_n\left(\frac{\cos\theta_n\cos nt + \sin\theta_n\sin nt}{\cos(nt-\theta_n)}\right)$$

and, in general, for $\omega \neq 1$, we get

$$f(t) = \frac{1}{2}a_0 + \sum_{n=1}^{\infty} c_n\cos(n\omega t - \theta_n) = \frac{1}{2}a_0 + \sum_{n=1}^{\infty} c_n\cos\left(n\omega t - \operatorname{atan}\frac{b_n}{a_n}\right) \qquad (6.67)$$

Similarly,

$$f(t) = \frac{1}{2}a_0 + c_1\left(\frac{\sin\varphi_1\cos t + \cos\varphi_1\sin t}{\sin(t+\varphi_1)}\right)$$

$$c_2\left(\frac{\sin\varphi_2\cos 2t + \cos\varphi_2\sin 2t}{\sin(2t+\varphi_2)}\right) + \dots + c_n\left(\frac{\sin\varphi_n\cos nt + \cos\varphi_n\sin nt}{\sin(nt+\varphi_n)}\right)$$

and, in general, where $\omega \neq 1$, we get

$$f(t) = \frac{1}{2}a_0 + \sum_{n=1}^{\infty} c_n\sin(n\omega t + \varphi_n) = \frac{1}{2}a_0 + \sum_{n=1}^{\infty} c_n\sin\left(n\omega t + \operatorname{atan}\frac{a_n}{b_n}\right) \qquad (6.68)$$

The series of (6.67) and (6.68) can be expressed as phasors. Since it is customary to use the cosine function in the time domain to phasor transformation, we choose to use the transformation of (6.63) below.

$$\frac{1}{2}a_0 + \sum_{n=1}^{\infty} c_n\cos\left(n\omega t - \operatorname{atan}\frac{b_n}{a_n}\right) \Leftrightarrow \frac{1}{2}a_0 + \sum_{n=1}^{\infty} c_n\angle-\operatorname{atan}\frac{b_n}{a_n} \qquad (6.69)$$

Example 6.8

Find the first 5 terms of the alternate form of the trigonometric Fourier series for the waveform of Figure 6.21.

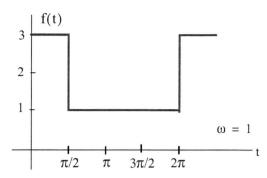

Figure 6.21. Waveform for Example 6.8

Solution:

The given waveform has no symmetry; thus, we expect both cosine and sine functions with odd and even terms present. Also, by inspection the DC value is not zero.

We will compute the a_n and b_n coefficients, the DC value, and we will combine them to get an expression in the form of (6.63). Then,

$$a_n = \frac{1}{\pi} \int_0^{\pi/2} (3)\cos nt\, dt + \frac{1}{\pi} \int_{\pi/2}^{2\pi} (1)\cos nt\, dt = \frac{3}{n\pi}\sin nt\Big|_0^{\pi/2} + \frac{1}{n\pi}\sin nt\Big|_{\pi/2}^{2\pi}$$

$$= \frac{3}{n\pi}\sin n\frac{\pi}{2} + \frac{1}{n\pi}\sin n2\pi - \frac{1}{n\pi}\sin n\frac{\pi}{2} = \frac{2}{n\pi}\sin n\frac{\pi}{2}$$

(6.70)

We observe that for $n = $ even, $a_n = 0$.

For $n = $ odd,

$$a_1 = \frac{2}{\pi}$$

(6.71)

and

$$a_3 = -\frac{2}{3\pi}$$

(6.72)

The DC value is

$$\frac{1}{2}a_0 = \frac{1}{2\pi}\int_0^{\pi/2}(3)dt + \frac{1}{2\pi}\int_{\pi/2}^{2\pi}(1)dt = \frac{1}{2\pi}\left(3t\Big|_0^{\pi/2} + t\Big|_{\pi/2}^{2\pi}\right)$$

$$= \frac{1}{2\pi}\left(\frac{3\pi}{2} + 2\pi - \frac{\pi}{2}\right) = \frac{1}{2\pi}(\pi + 2\pi) = \frac{3}{2}$$

(6.73)

The b_n coefficients are

$$b_n = \frac{1}{\pi} \int_0^{\pi/2} (3)\sin nt\, dt + \frac{1}{\pi} \int_{\pi/2}^{2\pi} (1)\sin nt\, dt = \frac{-3}{n\pi}\cos nt\Big|_0^{\pi/2} + \frac{-1}{n\pi}\cos nt\Big|_{\pi/2}^{2\pi}$$ (6.74)

$$= \frac{-3}{n\pi}\cos n\frac{\pi}{2} + \frac{3}{n\pi} + \frac{-1}{n\pi}\cos n2\pi + \frac{1}{n\pi}\cos n\frac{\pi}{2} = \frac{1}{n\pi}(3 - \cos n2\pi) = \frac{2}{n\pi}$$

Then,

$$b_1 = 2/\pi$$ (6.75)

$$b_2 = 1/\pi$$ (6.76)

$$b_3 = 2/3\pi$$ (6.77)

$$b_4 = 1/2\pi$$ (6.78)

From (6.69),

$$\frac{1}{2}a_0 + \sum_{n=1}^{\infty} c_n\cos\left(n\omega t - \operatorname{atan}\frac{b_n}{a_n}\right) \Leftrightarrow \frac{1}{2}a_0 + \sum_{n=1}^{\infty} c_n\angle -\operatorname{atan}\frac{b_n}{a_n}$$

where

$$c_n\angle -\operatorname{atan}\frac{b_n}{a_n} = \sqrt{a_n^2 + b_n^2}\angle -\operatorname{atan}\frac{b_n}{a_n} = \sqrt{a_n^2 + b_n^2}\angle -\theta_n = a_n - jb_n$$ (6.79)

Thus, for $n = 1, 2, 3,$ and 4, we get:

$$a_1 - jb_1 = \frac{2}{\pi} - j\frac{2}{\pi} = \sqrt{\left(\frac{2}{\pi}\right)^2 + \left(\frac{2}{\pi}\right)^2}\angle -45°$$

$$= \sqrt{\frac{8}{\pi^2}}\angle -45° = \frac{2\sqrt{2}}{\pi}\angle -45° \Leftrightarrow \frac{2\sqrt{2}}{\pi}\cos(\omega t - 45°)$$ (6.80)

Similarly,

$$a_2 - jb_2 = 0 - j\frac{1}{\pi} = \frac{1}{\pi}\angle -90° \Leftrightarrow \frac{1}{\pi}\cos(2\omega t - 90°)$$ (6.81)

$$a_3 - jb_3 = -\frac{2}{3\pi} - j\frac{2}{3\pi} = \frac{2\sqrt{2}}{3\pi}\angle -135° \Leftrightarrow \frac{2\sqrt{2}}{3\pi}\cos(3\omega t - 135°)$$ (6.82)

and

$$a_4 - jb_4 = 0 - j\frac{1}{2\pi} = \frac{1}{2\pi}\angle -90° \Leftrightarrow \frac{1}{2\pi}\cos(4\omega t - 90°)$$ (6.83)

Combining the terms of (6.73) and (6.80) through (6.83), we find that the alternate form of the trigonometric Fourier series representing the waveform of this example is

$$f(t) = \frac{3}{2} + \frac{1}{\pi} \left[2\sqrt{2} \cos(\omega t - 45°) + \cos(2\omega t - 90°) \right.$$

$$\left. + \frac{2\sqrt{2}}{3} \cos(3\omega t - 135°) + \frac{1}{2}\cos(4\omega t - 90°) + \dots \right] \qquad (6.84)$$

6.6 The Exponential Form of the Fourier Series

The Fourier series are often expressed in exponential form. The advantage of the exponential form is that we only need to perform one integration rather than two, one for the a_n, and another for the b_n coefficients in the trigonometric form of the series. Moreover, in most cases the integration is simpler.

The exponential form is derived from the trigonometric form by substitution of

$$\cos\omega t = \frac{e^{j\omega t} + e^{-j\omega t}}{2} \qquad (6.85)$$

and

$$\sin\omega t = \frac{e^{j\omega t} - e^{-j\omega t}}{j2} \qquad (6.86)$$

into $f(t)$. Thus,

$$f(t) = \frac{1}{2}a_0 + a_1\left(\frac{e^{j\omega t} + e^{-j\omega t}}{2}\right) + a_2\left(\frac{e^{j2\omega t} + e^{-j2\omega t}}{2}\right) + \qquad (6.87)$$

$$\dots + b_1\left(\frac{e^{j\omega t} - e^{-j\omega t}}{j2}\right) + b_2\left(\frac{e^{j2\omega t} - e^{-j2\omega t}}{j2}\right) + \dots$$

and grouping terms with same exponents, we get

$$f(t) = \dots + \left(\frac{a_2}{2} - \frac{b_2}{j2}\right)e^{-j2\omega t} + \left(\frac{a_1}{2} - \frac{b_1}{j2}\right)e^{-j\omega t} + \frac{1}{2}a_0 + \left(\frac{a_1}{2} + \frac{b_1}{j2}\right)e^{j\omega t} + \left(\frac{a_2}{2} + \frac{b_2}{j2}\right)e^{j2\omega t} \qquad (6.88)$$

The terms of (6.88) in parentheses are usually denoted as

$$C_{-n} = \frac{1}{2}\left(a_n - \frac{b_n}{j}\right) = \frac{1}{2}(a_n + jb_n) \qquad (6.89)$$

$$C_n = \frac{1}{2}\left(a_n + \frac{b_n}{j}\right) = \frac{1}{2}(a_n - jb_n) \qquad (6.90)$$

$$C_0 = \frac{1}{2}a_0 \qquad (6.91)$$

Then, (6.88) is written as

$$f(t) = \ldots + C_{-2}e^{-j2\omega t} + C_{-1}e^{-j\omega t} + C_0 + C_1 e^{j\omega t} + C_2 e^{j2\omega t} + \ldots \qquad (6.92)$$

We must remember that the C_i coefficients, except C_0, are complex and occur in complex conjugate pairs, that is,

$$C_{-n} = C_n^* \qquad (6.93)$$

We can derive a general expression for the complex coefficients C_n, by multiplying both sides of (6.92) by $e^{-jn\omega t}$ and integrating over one period, as we did in the derivation of the a_n and b_n coefficients of the trigonometric form. Then, with $\omega = 1$,

$$\int_0^{2\pi} f(t)e^{-jnt}dt = \ldots + \int_0^{2\pi} C_{-2}e^{-j2t}e^{-jnt}dt + \int_0^{2\pi} C_{-1}e^{-jt}e^{-jnt}dt \qquad (6.94)$$

$$+ \int_0^{2\pi} C_0 e^{-jnt}dt + \int_0^{2\pi} C_1 e^{jt}e^{-jnt}dt$$

$$+ \int_0^{2\pi} C_2 e^{j2t}e^{-jnt}dt + \ldots + \int_0^{2\pi} C_n e^{jnt}e^{-jnt}dt$$

We observe that all the integrals on the right side of (6.97) are zero except the last. Therefore,

$$\int_0^{2\pi} f(t)e^{-jnt}dt = \int_0^{2\pi} C_n e^{jnt}e^{-jnt}dt = \int_0^{2\pi} C_n dt = 2\pi C_n$$

or

$$C_n = \frac{1}{2\pi} \int_0^{2\pi} f(t)e^{-jnt}dt$$

and, in general, for $\omega \neq 1$,

$$C_n = \frac{1}{2\pi} \int_0^{2\pi} f(t)e^{-jn\omega t}d(\omega t) \qquad (6.95)$$

or

$$C_n = \frac{1}{T} \int_0^{T} f(t)e^{-jn\omega t}d(\omega t) \qquad (6.96)$$

We can derive the trigonometric Fourier series from the exponential series by addition and subtraction of the exponential form coefficients C_n and C_{-n}. Thus, from (6.89) and (6.90),

$$C_n + C_{-n} = \frac{1}{2}(a_n - jb_n + a_n + jb_n)$$

or

$$\boxed{a_n = C_n + C_{-n}}$$ (6.97)

Similarly,

$$C_n - C_{-n} = \frac{1}{2}(a_n - jb_n - a_n - jb_n)$$ (6.98)

or

$$\boxed{b_n = j(C_n - C_{-n})}$$ (6.99)

Symmetry in Exponential Series

1. *For even functions, all coefficients C_i are real*

We recall from (6.89) and (6.90) that

$$C_{-n} = \frac{1}{2}\left(a_n - \frac{b_n}{j}\right) = \frac{1}{2}(a_n + jb_n)$$ (6.100)

and

$$C_n = \frac{1}{2}\left(a_n + \frac{b_n}{j}\right) = \frac{1}{2}(a_n - jb_n)$$ (6.101)

Since even functions have no sine terms, the b_n coefficients in (6.100) and (6.101) are zero. Therefore, both C_{-n} and C_n are real.

2. *For odd functions, all coefficients C_i are imaginary*

Since odd functions have no cosine terms, the a_n coefficients in (6.100) and (6.101) are zero. Therefore, both C_{-n} and C_n are imaginary.

3. *If there is half–wave symmetry, $C_n = 0$ for $n = $ even*

We recall from the trigonometric Fourier series that if there is half–wave symmetry, all even harmonics are zero. Therefore, in (6.100) and (6.101) the coefficients a_n and b_n are both zero for $n = $ even, and thus, both C_{-n} and C_n are also zero for $n = $ even.

4. If there is no symmetry, $f(t)$ is complex.

5. $C_{-n} = C_n{}^*$ *always*

This can be seen in (6.100) and (6.101)

Example 6.9

Compute the exponential Fourier series for the square waveform of Figure 6.22 below. Assume that $\omega = 1$.

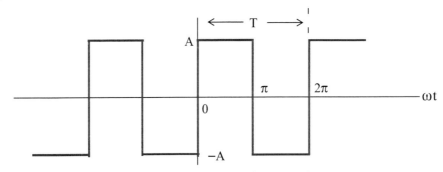

Figure 6.22. Waveform for Example 6.9

Solution:

This is the same waveform as in Example 6.1, and as we know, it is an odd function, has half–wave symmetry, and its DC component is zero. Therefore, the C_n coefficients will be imaginary, $C_n = 0$ for $n = $ even, and $C_0 = 0$. Using (6.95) with $\omega = 1$, we get

$$C_n = \frac{1}{2\pi} \int_0^{2\pi} f(t)e^{-jnt}dt = \frac{1}{2\pi}\int_0^\pi Ae^{-jnt}dt + \frac{1}{2\pi}\int_\pi^{2\pi} -Ae^{-jnt}dt$$

and for $n = 0$,

$$C_0 = \frac{1}{2\pi}\left[\int_0^\pi Ae^{-0}dt + \int_\pi^{2\pi}(-A)e^{-0}dt\right] = \frac{A}{2\pi}(\pi - 2\pi + \pi) = 0$$

as expected.

For $n \neq 0$,

$$C_n = \frac{1}{2\pi}\left[\int_0^\pi Ae^{-jnt}dt + \int_\pi^{2\pi} -Ae^{-jnt}dt\right] = \frac{1}{2\pi}\left[\frac{A}{-jn}e^{-jnt}\Big|_0^\pi + \frac{-A}{-jn}e^{-jnt}\Big|_\pi^{2\pi}\right]$$

$$= \frac{1}{2\pi}\left[\frac{A}{-jn}(e^{-jn\pi} - 1) + \frac{A}{jn}(e^{-jn2\pi} - e^{-jn\pi})\right] = \frac{A}{2j\pi n}(1 - e^{-jn\pi} + e^{-jn2\pi} - e^{-jn\pi}) \qquad (6.102)$$

$$= \frac{A}{2j\pi n}(1 + e^{-jn2\pi} - 2e^{-jn\pi}) = \frac{A}{2j\pi n}(e^{-jn\pi} - 1)^2$$

For $n = $ even, $e^{-jn\pi} = 1$; then,

$$C_n\Big|_{n\,=\,even} = \frac{A}{2j\pi n}(e^{-jn\pi} - 1)^2 = \frac{A}{2j\pi n}(1 - 1)^2 = 0 \qquad (6.103)$$

as expected.

For n = odd, $e^{-jn\pi}$ = -1. Therefore,

$$\underset{n\,=\,\text{odd}}{C_n} = \frac{A}{2j\pi n}(e^{-jn\pi} - 1)^2 = \frac{A}{2j\pi n}(-1 - 1)^2 = \frac{A}{2j\pi n}(-2)^2 = \frac{2A}{j\pi n} \qquad (6.104)$$

Using (6.92), that is,

$$f(t) = \ldots + C_{-2}e^{-j2\omega t} + C_{-1}e^{-j\omega t} + C_0 + C_1 e^{j\omega t} + C_2 e^{j2\omega t} + \ldots$$

we obtain the exponential Fourier series for the *square waveform with odd symmetry* as

$$f(t) = \frac{2A}{j\pi}\left(\ldots - \frac{1}{3}e^{-j3\omega t} - e^{-j\omega t} + e^{j\omega t} + \frac{1}{3}e^{j3\omega t}\right) = \frac{2A}{j\pi}\sum_{n\,=\,\text{odd}} \frac{1}{n}e^{jn\omega t} \qquad (6.105)$$

The minus ($-$) sign of the first two terms within the parentheses results from the fact that $C_{-n} = C_n{}^*$. For instance, since $C_3 = 2A/j3\pi$, it follows that $C_{-3} = C_3{}^* = -2A/j3\pi$. We observe that $f(t)$ is purely imaginary, as expected, since the waveform is an odd function.

To prove that (6.105) and (6.22) are the same, we group the two terms inside the parentheses of (6.105) for which n = 1; this will produce the fundamental frequency $\sin\omega t$. Then, we group the two terms for which n = 3, and this will produce the third harmonic $\sin 3\omega t$, and so on.

6.7 Line Spectra

When the Fourier series are known, it is useful to plot the amplitudes of the harmonics on a frequency scale that shows the first (fundamental frequency) harmonic, and the higher harmonics times the amplitude of the fundamental. Such a plot is known as *line spectrum* and shows the spectral lines that would be displayed by a *spectrum analyzer*[*].

Figure 6.23 shows the line spectrum of the square waveform of Example 6.1.

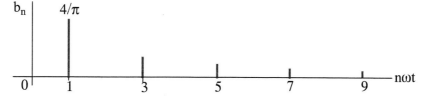

Figure 6.23. Line spectrum for square waveform of Example 6.1

---

[*] *An instrument that displays the spectral lines of a waveform.*

Figure 6.24 shows the line spectrum for the half–wave rectification waveform of Example 6.6.

Figure 6.24. Line spectrum for half–wave rectifier of Example 6.6

The line spectra of other waveforms can be easily constructed from the Fourier series.

Example 6.10

Compute the exponential Fourier series for the waveform of Figure 6.25, and plot its line spectra. Assume $\omega = 1$.

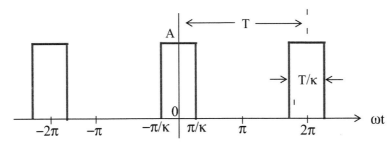

Figure 6.25. Waveform for Example 6.11

Solution:

This recurrent rectangular pulse is used extensively in digital communications systems. To determine how faithfully such pulses will be transmitted, it is necessary to know the frequency components.

As shown in Figure 6.25, the pulse duration is T/k. Thus, the recurrence interval (period) T, is k times the pulse duration. In other words, k is the ratio of the pulse repetition time to the duration of each pulse.

For this example, the components of the exponential Fourier series are found from

$$C_n = \frac{1}{2\pi} \int_{-\pi}^{\pi} A e^{-jnt} dt = \frac{A}{2\pi} \int_{-\pi/k}^{\pi/k} e^{-jnt} dt \tag{6.106}$$

The value of the average (DC component) is found by letting $n = 0$. Then, from (6.106) we get

$$C_0 = \frac{A}{2\pi}t\Big|_{-\pi/k}^{\pi/k} = \frac{A}{2\pi}\left(\frac{\pi}{k}+\frac{\pi}{k}\right) = \frac{A}{k} \tag{6.107}$$

For the values for $n \neq 0$, integration of (6.106) yields

$$C_n = \frac{A}{-jn2\pi}e^{-jnt}\Big|_{-\pi/k}^{\pi/k} = \frac{A}{n\pi}\cdot\frac{e^{jn\pi/k}-e^{-jn\pi/k}}{j2} = \frac{A}{n\pi}\cdot\sin\left(\frac{n\pi}{k}\right)$$

$$= A\frac{\sin(n\pi/k)}{n\pi} = \frac{A}{k}\cdot\frac{\sin(n\pi/k)}{n\pi/k} \tag{6.108}$$

and thus,

$$f(t) = \sum_{n=-\infty}^{\infty}\frac{A}{k}\cdot\frac{\sin(n\pi/k)}{n\pi/k} \tag{6.109}$$

The relation of (6.109) has the $\sin x/x$ form, and the line spectrum is shown in Figures 6.26 through 6.28, for $k = 2$, $k = 5$ and $k = 10$ respectively by using the MATLAB scripts below.

```
fplot('sin(2.*x)./(2.*x)',[-4  4  -0.4  1.2])
```

```
fplot('sin(5.*x)./(5.*x)',[-4  4  -0.4  1.2])
```

```
fplot('sin(10.*x)./(10.*x)',[-4  4 -0.4  1.2])
```

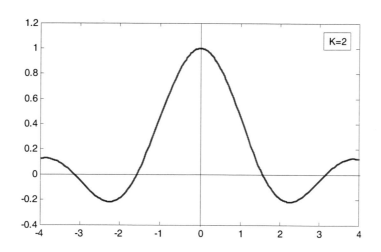

Figure 6.26. Line spectrum of (6.109) for $k = 2$

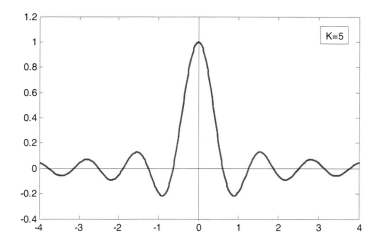

Figure 6.27. Line spectrum of (6.109) for k = 5

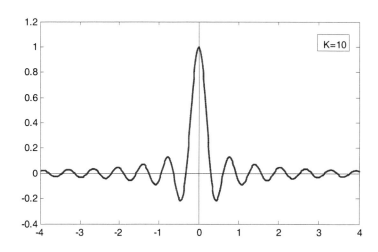

Figure 6.28. Line spectrum of (6.112) for k = 10

The spectral lines are separated by the distance $1/k$ and thus, as k gets larger, the lines get closer together while the lines are further apart as k gets smaller.

6.8 Numerical Evaluation of Fourier Coefficients

Quite often, it is necessary to construct the Fourier expansion of a function based on observed values instead of an analytic expression. Examples are meteorological or economic quantities whose period may be a day, a week, a month or even a year. In these situations, we need to evaluate the integral(s) using numerical integration.

The procedure presented here, will work for both the waveforms that have an analytical solution and those that do not. Even though we may already know the Fourier series from analytical methods, we can use this procedure to check our results.

Consider the waveform of f(x) shown in Figure 6.29, were we have divided it into small pulses of width Δx. Obviously, the more pulses we use, the better the approximation.

If the time axis is in degrees, we can choose Δx to be 2.5° and it is convenient to start at the zero point of the waveform. Then, using a spreadsheet, such as Microsoft Excel, we can divide the period 0° to 360° in 2.5° intervals, and enter these values in Column A of the spreadsheet.

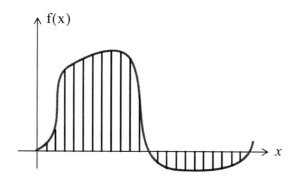

Figure 6.29. Waveform whose analytical expression is unknown

Since the arguments of the sine and the cosine are in radians, we multiply degrees by π (3.1459...) and divide by 180 to perform the conversion. We enter these in Column B and we denote them as x. In Column C we enter the corresponding values of $y = f(x)$ as measured from the waveform. In Columns D and E we enter the values of $\cos x$ and the product $y \cos x$ respectively. Similarly, we enter the values of $\sin x$ and $y \sin x$ in Columns F and G respectively.

Next, we form the sums of $y \cos x$ and $y \sin x$, we multiply these by Δx, and we divide by π to obtain the coefficients a_1 and b_1. To compute the coefficients of the higher order harmonics, we form the products $y \cos 2x$, $y \sin 2x$, $y \cos 3x$, $y \sin 3x$, and so on, and we enter these in subsequent columns of the spreadsheet.

Figure 6.30 is a partial table showing the computation of the coefficients of the square waveform, and Figure 6.31 is a partial table showing the computation of the coefficients of a clipped sine waveform. The complete tables extend to the seventh harmonic to the right and to 360° down.

Analytical:

$f(t)=4(\sin wt/p+\sin 3wt/3p+\sin 5wt/5p+ \ldots)$

Numerical:

DC=	0.000	
a1=	0.000	b1= 1.273
a2=	0.000	b2= 0.000
a3=	0.000	b3= 0.424
a4=	0.000	b4= 0.000
a5=	0.000	b5= 0.254
a6=	0.000	b6= 0.000
a7=	0.000	b7= 0.180

Square waveform

x(deg)	x(rad)	y=f(x)	0.5*a0	cosx	ycosx	sinx	ysinx	cos2x	ycox2x	sin2x	ysin2x	cos3x	ycos3x	sin3x	ysin3x
0.0	0.000	0.000	0.000	1.000	0.000	0.000	0.000	1.000	0.000	0.000	0.000	1.000	0.000	0.000	0.000
2.5	0.044	1.000	0.044	0.999	0.999	0.044	0.044	0.996	0.996	0.087	0.087	0.991	0.991	0.131	0.131
5.0	0.087	1.000	0.044	0.996	0.996	0.087	0.087	0.985	0.985	0.174	0.174	0.966	0.966	0.259	0.259
7.5	0.131	1.000	0.044	0.991	0.991	0.131	0.131	0.966	0.966	0.259	0.259	0.924	0.924	0.383	0.383
10.0	0.175	1.000	0.044	0.985	0.985	0.174	0.174	0.940	0.940	0.342	0.342	0.866	0.866	0.500	0.500
12.5	0.218	1.000	0.044	0.976	0.976	0.216	0.216	0.906	0.906	0.423	0.423	0.793	0.793	0.609	0.609
15.0	0.262	1.000	0.044	0.966	0.966	0.259	0.259	0.866	0.866	0.500	0.500	0.707	0.707	0.707	0.707
17.5	0.305	1.000	0.044	0.954	0.954	0.301	0.301	0.819	0.819	0.574	0.574	0.609	0.609	0.793	0.793
20.0	0.349	1.000	0.044	0.940	0.940	0.342	0.342	0.766	0.766	0.643	0.643	0.500	0.500	0.866	0.866
22.5	0.393	1.000	0.044	0.924	0.924	0.383	0.383	0.707	0.707	0.707	0.707	0.383	0.383	0.924	0.924
25.0	0.436	1.000	0.044	0.906	0.906	0.423	0.423	0.643	0.643	0.766	0.766	0.259	0.259	0.966	0.966
27.5	0.480	1.000	0.044	0.887	0.887	0.462	0.462	0.574	0.574	0.819	0.819	0.131	0.131	0.991	0.991
30.0	0.524	1.000	0.044	0.866	0.866	0.500	0.500	0.500	0.500	0.866	0.866	0.000	0.000	1.000	1.000
32.5	0.567	1.000	0.044	0.843	0.843	0.537	0.537	0.423	0.423	0.906	0.906	-0.131	-0.131	0.991	0.991
35.0	0.611	1.000	0.044	0.819	0.819	0.574	0.574	0.342	0.342	0.940	0.940	-0.259	-0.259	0.966	0.966
37.5	0.654	1.000	0.044	0.793	0.793	0.609	0.609	0.259	0.259	0.966	0.966	-0.383	-0.383	0.924	0.924
40.0	0.698	1.000	0.044	0.766	0.766	0.643	0.643	0.174	0.174	0.985	0.985	-0.500	-0.500	0.866	0.866
42.5	0.742	1.000	0.044	0.737	0.737	0.676	0.676	0.087	0.087	0.996	0.996	-0.609	-0.609	0.793	0.793
45.0	0.785	1.000	0.044	0.707	0.707	0.707	0.707	0.000	0.000	1.000	1.000	-0.707	-0.707	0.707	0.707
47.5	0.829	1.000	0.044	0.676	0.676	0.737	0.737	-0.087	-0.087	0.996	0.996	-0.793	-0.793	0.609	0.609
50.0	0.873	1.000	0.044	0.643	0.643	0.766	0.766	-0.174	-0.174	0.985	0.985	-0.866	-0.866	0.500	0.500

Figure 6.30. Numerical computation of the coefficients of the square waveform (partial listing)

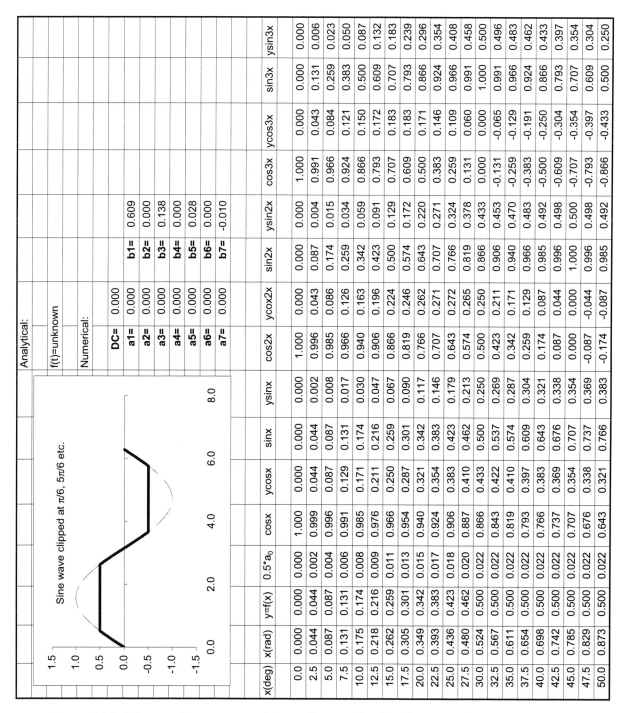

Analytical: f(t)=unknown

Numerical:

	b1= 0.609
DC= 0.000	
a1= 0.000	b1= 0.609
a2= 0.000	b2= 0.000
a3= 0.000	b3= 0.138
a4= 0.000	b4= 0.000
a5= 0.000	b5= 0.028
a6= 0.000	b6= 0.000
a7= 0.000	b7= -0.010

x(deg)	x(rad)	y=f(x)	0.5*a0	cosx	ycosx	sinx	ysinx	cos2x	ycox2x	sin2x	ysin2x	cos3x	ycos3x	sin3x	ysin3x
0.0	0.000	0.000	0.000	1.000	0.000	0.000	0.000	1.000	0.000	0.000	0.000	1.000	0.000	0.000	0.000
2.5	0.044	0.044	0.002	0.999	0.044	0.044	0.002	0.996	0.043	0.087	0.004	0.991	0.043	0.131	0.006
5.0	0.087	0.087	0.004	0.996	0.087	0.087	0.008	0.985	0.086	0.174	0.015	0.966	0.084	0.259	0.023
7.5	0.131	0.131	0.006	0.991	0.129	0.131	0.017	0.966	0.126	0.259	0.034	0.924	0.121	0.383	0.050
10.0	0.175	0.174	0.008	0.985	0.171	0.174	0.030	0.940	0.163	0.342	0.059	0.866	0.150	0.500	0.087
12.5	0.218	0.216	0.009	0.976	0.211	0.216	0.047	0.906	0.196	0.423	0.091	0.793	0.172	0.609	0.132
15.0	0.262	0.259	0.011	0.966	0.250	0.259	0.067	0.866	0.224	0.500	0.129	0.707	0.183	0.707	0.183
17.5	0.305	0.301	0.013	0.954	0.287	0.301	0.090	0.819	0.246	0.574	0.172	0.609	0.183	0.793	0.239
20.0	0.349	0.342	0.015	0.940	0.321	0.342	0.117	0.766	0.262	0.643	0.220	0.500	0.171	0.866	0.296
22.5	0.393	0.383	0.017	0.924	0.354	0.383	0.146	0.707	0.271	0.707	0.271	0.383	0.146	0.924	0.354
25.0	0.436	0.423	0.018	0.906	0.383	0.423	0.179	0.643	0.272	0.766	0.324	0.259	0.109	0.966	0.408
27.5	0.480	0.462	0.020	0.887	0.410	0.462	0.213	0.574	0.265	0.819	0.378	0.131	0.060	0.991	0.458
30.0	0.524	0.500	0.022	0.866	0.433	0.500	0.250	0.500	0.250	0.866	0.433	0.000	0.000	1.000	0.500
32.5	0.567	0.500	0.022	0.843	0.422	0.537	0.269	0.423	0.211	0.906	0.453	-0.131	-0.065	0.991	0.496
35.0	0.611	0.500	0.022	0.819	0.410	0.574	0.287	0.342	0.171	0.940	0.470	-0.259	-0.129	0.966	0.483
37.5	0.654	0.500	0.022	0.793	0.397	0.609	0.304	0.259	0.129	0.966	0.483	-0.383	-0.191	0.924	0.462
40.0	0.698	0.500	0.022	0.766	0.383	0.643	0.321	0.174	0.087	0.985	0.492	-0.500	-0.250	0.866	0.433
42.5	0.742	0.500	0.022	0.737	0.369	0.676	0.338	0.087	0.044	0.996	0.498	-0.609	-0.304	0.793	0.397
45.0	0.785	0.500	0.022	0.707	0.354	0.707	0.354	0.000	0.000	1.000	0.500	-0.707	-0.354	0.707	0.354
47.5	0.829	0.500	0.022	0.676	0.338	0.737	0.369	-0.087	-0.044	0.996	0.498	-0.793	-0.397	0.609	0.304
50.0	0.873	0.500	0.022	0.643	0.321	0.766	0.383	-0.174	-0.087	0.985	0.492	-0.866	-0.433	0.500	0.250

Figure 6.31. Numerical computation of the coefficients of a clipped sine waveform (partial listing)

6.9 Power Series Expansion of Functions

A *power series* has the form

$$\sum_{k=0}^{\infty} a_k x^k = a_0 + a_1 x + a_2 x^2 + \ldots \tag{6.110}$$

Some familiar power series expansions for real values of x are

$$e^x = 1 + x + \frac{x^2}{2!} + \frac{x^3}{3!} + \frac{x^4}{4!} + \ldots \tag{6.111}$$

$$\sin x = x - \frac{x^3}{3!} + \frac{x^5}{5!} - \frac{x^7}{7!} + \ldots \tag{6.112}$$

$$\cos x = 1 - \frac{x^2}{2!} + \frac{x^4}{4!} - \frac{x^6}{6!} + \ldots \tag{6.113}$$

The following example illustrates the fact that a power series expansion can lead us to a *Fourier Series*.

Example 6.11

If the applied voltage v is small (no greater than 5 volts), the current i in a semiconductor diode can be approximated by the relation

$$i = a(e^{kv} - 1) \tag{6.114}$$

where a and k are arbitrary constants, and the input voltage is a sinusoid, that is,

$$v = V_{max} \cos \omega t \tag{6.115}$$

Express the current i in (6.114) as a power series.

Solution:

The term e^{kv} inside the parentheses of (6.114) suggests the power series expansion of (6.111). Accordingly, we rewrite (6.114) as

$$i = a\left(1 + kv + \frac{(kv)^2}{2!} + \frac{(kv)^3}{3!} + \frac{(kv)^4}{4!} + \ldots - 1\right) = a\left(kv + \frac{(kv)^2}{2!} + \frac{(kv)^3}{3!} + \frac{(kv)^4}{4!} + \ldots\right) \tag{6.116}$$

Substitution of (6.115) into (6.116) yields,

$$i = a\left(kV_{max}\cos\omega t + \frac{(kV_p\cos\omega t)^2}{2!} + \frac{(kV_p\cos\omega t)^3}{3!} + \frac{(kV_p\cos\omega t)^4}{4!} + ...\right) \qquad (6.117)$$

This expression can be simplified with the use of the following trigonometric identities:

$$\cos^2 x = \frac{1}{2} + \frac{1}{2}\cos 2x$$

$$\cos^3 x = \frac{3}{4}\cos x + \frac{1}{4}\cos 3x \qquad (6.118)$$

$$\cos^4 x = \frac{3}{8} + \frac{1}{2}\cos 2x + \frac{1}{8}\cos 4x$$

Then, substitution of (6.118) into (6.117) and after simplification, we obtain a series of the following form:

$$i = A_0 + A_1\cos\omega t + A_2\cos 2\omega t + A_3\cos 3\omega t + A_4\cos 4\omega t + ... \qquad (6.119)$$

We recall that the series of (6.119) is the trigonometric series form of the Fourier series. We observe that it consists of a constant term, a term of the fundamental frequency, and terms of all harmonic frequencies, that is, higher frequencies which are multiples of the fundamental frequency.

6.10 Taylor and Maclaurin Series

A function $f(x)$ which possesses all derivatives up to order n at a point $x = x_0$ can be expanded in a *Taylor series* as

$$f(x) = f(x_0) + f'(x_0)(x - x_0) + \frac{f''(x_0)}{2!}(x - x_0)^2 + ... + \frac{f^{(n)}(x_0)}{n!}(x - x_0)^n \qquad (6.120)$$

If $x_0 = 0$, (6.120) reduces to

$$f(x) = f(0) + f'(0)x + \frac{f''(0)}{2!}x^2 + ... + \frac{f^{(n)}(0)}{n!}x^n \qquad (6.121)$$

Relation (6.121) is known as *Maclaurin series*, and has the form of power series of (6.110) with $a_n = f^{(n)}(0)/n!$.

To appreciate the usefulness and application of the Taylor series, we will consider the plot of Figure 6.32, where $i(v)$ represents some experimental data for the *current–voltage* ($i - v$) characteristics of a semiconductor diode operating at the $0 \le v \le 5$ volts region.

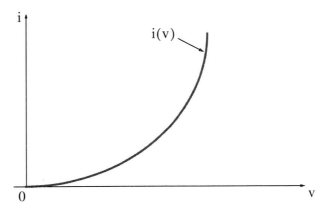

Figure 6.32. Current-voltage (i-v) characteristics for a typical semiconductor diode

Now, suppose that we want to approximate the function $i(v)$ by a power series, in the neighborhood of some arbitrary point $P(v_0, i_0)$ shown in Figure 6.33. We assume that the first n derivatives of the function $i(v)$ exist at this point.

We begin by referring to the power series of (6.110), where we observe that the first term on the right side is a constant. Therefore, we are seeking a constant that it will be the best approximation to the given curve in the vicinity of point P. Obviously, the horizontal line i_0 passes through point P, and we denote this first approximation as a_0 shown in Figure 6.34.

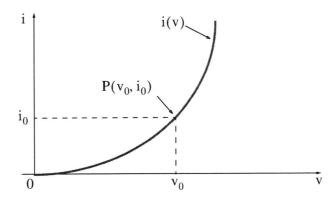

Figure 6.33. Approximation of the function $i(v)$ by a power series

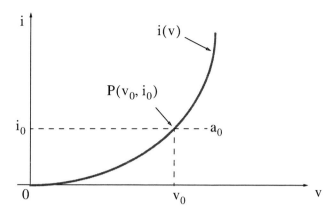

Figure 6.34. First approximation of $i(v)$

The next term in the power series is the linear term $a_1 x$. Thus, we seek a linear term of the form $a_0 + a_1 x$. But since we want the power series to be a good approximation to the given function for some distance on either side of point P, we are interested in the difference $v - v_0$. Accordingly, we express the desired power series as

$$f(v) = a_0 + a_1(v - v_0) + a_2(v - v_0)^2 + a_3(v - v_0)^3 + a_4(v - v_0)^4 + \ldots \tag{6.122}$$

Now, we want the linear term $a_0 + a_1(v - v_0)$ to be the best approximation to the function $i(v)$ in the vicinity of point P. This will be accomplished if the linear term has the same slope as the given function as shown in Figure 6.35.

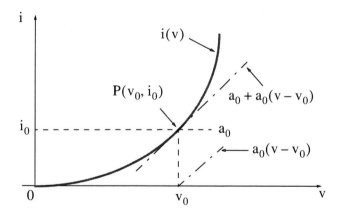

Figure 6.35. Second approximation of $i(v)$

It is evident that the slope of $i(v)$ at v_0 is $i'(v_0) = a_1$ and therefore, the linear term $a_0 + a_1(v - v_0)$ can be expressed as $i(v_0) + i'(v_0)(v - v_0)$.

The third term in (6.122), that is, $a_2(v - v_0)^2$ is a quadratic and therefore, we choose a_2 such that it matches the second derivative of the function $i(v)$ in the vicinity of point P as shown in Figure 6.36.

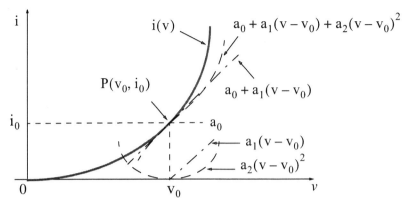

Figure 6.36. Third approximation of $i(v)$

Then, $2a_2 = i''(v_0)$ or $a_2 = i''(v_0)/2$. The remaining coefficients a_3, a_4, a_5, and so on of (6.122) are found by matching the third, fourth, fifth, and higher order derivatives of the given function with these coefficients. When this is done, we obtain the following Taylor series.

$$i(v) = i(v_0) + i'(v_0)(v - v_0) + \frac{i''(v_0)}{2!}(v - v_0)^2 + \frac{i'''(v_0)}{3!}(v - v_0)^3 + \dots \qquad (6.123)$$

We can also describe any function that has an analytical expression, by a Taylor series as illustrated by the following example.

Example 6.12

Compute the first three terms of the Taylor series expansion for the function

$$y = f(x) = \tan x \qquad (6.124)$$

at $a = \pi/4$.

Solution:

The Taylor series expansion about point a is given by

$$f_n(x) = f(a) + f'(a)(x - a) + \frac{f'(a)}{2!}(x - a)^2 + \frac{f''(a)}{3!}(x - a)^3 + \dots \qquad (6.125)$$

and since we are asked to compute the first three terms, we must find the first and second derivatives of $f(x) = \tan x$.

From math tables, $\frac{d}{dx}\tan x = \sec^2 x$, so $f'(x) = \sec^2 x$. To find $f''(x)$ we need to find the first

derivative of $\sec^2 x$, so we let $z = \sec^2 x$. Then, using $\frac{d}{dx}\sec x = \sec x \cdot \tan x$, we get

$$\frac{dz}{dx} = 2\sec x \frac{d}{dx}\sec x = 2\sec x(\sec x \cdot \tan x) = 2\sec^2 x \cdot \tan x \qquad (6.126)$$

Next, using the trigonometric identity

$$\sec^2 x = \tan^2 x + 1 \qquad (6.127)$$

and by substitution of (6.127) into (6.126), we get,

$$\frac{dz}{dx} = f''(x) = 2(\tan^2 x + 1)\tan x \qquad (6.128)$$

Now, at point $a = \pi/4$ we have:

$$f(a) = f\left(\frac{\pi}{4}\right) = \tan\left(\frac{\pi}{4}\right) = 1 \quad f'(a) = f'\left(\frac{\pi}{4}\right) = 1 + 1 = 2 \quad f''(a) = f''\left(\frac{\pi}{4}\right) = 2(1^2 + 1)1 = 4 \qquad (6.129)$$

and by substitution into (6.125),

$$f_n(x) = 1 + 2\left(x - \frac{\pi}{4}\right) + 2\left(x - \frac{\pi}{4}\right)^2 + \dots \qquad (6.130)$$

We can also obtain a Taylor series expansion with the MATLAB **taylor(f,n,a)** function where **f** is a symbolic expression, **n** produces the first n terms in the series, and **a** defines the Taylor approximation about point a. A detailed description can be displayed with the **help taylor** command. For example, the following MATLAB script computes the first 8 terms of the Taylor series expansion of $y = f(x) = \tan x$ about $a = \pi/4$.

```
x=sym('x'); y=tan(x); z=taylor(y,8,pi/4); pretty(z)
                       2                    3                    4
1 + 2x - 1/2 pi + 2(x - 1/4 pi) + 8/3(x - 1/4 pi) + 10/3(x - 1/4 pi)

     64                 5    244                 6    2176                 7
   + -- (x - 1/4 pi)  + --- (x - 1/4 pi)  + ---- (x - 1/4 pi)
     15                      45                       315
```

Example 6.13

Express the function

$$y = f(t) = e^t \tag{6.131}$$

in a Maclaurin's series.

Solution:

A Maclaurin's series has the form of (6.132), that is,

$$f(x) = f(0) + f'(0)x + \frac{f''(0)}{2!}x^2 + \ldots + \frac{f^{(n)}(0)}{n!}x^n \tag{6.132}$$

For this function, we have $f(t) = e^t$ and thus $f(0) = 1$. Since all derivatives are e^t, then, $f'(0) = f''(0) = f'''(0) = \ldots = 1$ and therefore,

$$f_n(t) = 1 + t + \frac{t^2}{2!} + \frac{t^3}{3!} + \ldots \tag{6.133}$$

MATLAB displays the same result.

t=sym('t'); fn=taylor(exp(t)); pretty(fn)

```
          2        3         4          5
1 + t + 1/2 t  + 1/6 t  + 1/24 t  + 1/120 t
```

Example 6.14

In a semiconductor diode D, the instantaneous current i_D and voltage v_D are related as

$$i_D(v_D) = I_D e^{v_D/nV_T} \tag{6.134}$$

where I_D is the DC (average) component of the current, the constant n has a value between 1 and 2 depending on the material and physical structure of the diode, and V_T is the *thermal voltage* which depends on the temperature, and its value at room temperature is approximately 25 mV.

Expand this relation into a power series that can be used to compute the current when the voltage is small and varies about $v_D = 0$.

Solution:

Since the voltage is small and varies about $v_D = 0$, we can use the following Maclaurin's series.

$$i_D(v_D) = i_D(0) + i'_D(0)v_D + \frac{i''_D(0)}{2!}v_D^2 + \frac{i'''_D(0)}{3!}v_D^3 + \ldots \tag{6.135}$$

The first term $i_D(0)$ on the right side of (6.135) is found by letting $v_D = 0$ in (6.134). Then,

$$i_D(0) = I_D \tag{6.136}$$

To compute the second and third terms of (6.135), we must find the first and second derivatives of (6.134). These are:

$$i'_D(v_D) = \frac{d}{dv_D}i_D = \frac{1}{nV_T} \cdot I_D e^{v_D/nV_T} \quad \text{and} \quad i'_D(0) = \frac{1}{nV_T} \cdot I_D \tag{6.137}$$

$$i''_D(v_D) = \frac{d^2}{d^2v_D}i_D = \frac{1}{n^2V_T^2} \cdot I_D e^{v_D/nV_T} \quad \text{and} \quad i''_D(0) = \frac{1}{n^2V_T^2} \cdot I_D \tag{6.138}$$

Then, by substitution of (6.136), (6.137), and (6.138) into (6.135) we get

$$i_D(v_D) = I_D\left(1 + \frac{1}{nV_T}v_D + \frac{1}{n^2V_T^2}v_D^2 + \ldots\right) \tag{6.139}$$

6.11 Summary

- Any periodic waveform f(t) can be expressed as

$$f(t) = \frac{1}{2}a_0 + \sum_{n=1}^{\infty} (a_n \cos n\omega t + b_n \sin n\omega t)$$

where the first term $a_0/2$ is a constant, and represents the DC (average) component of $f(t)$. The terms with the coefficients a_1 and b_1 together, represent the fundamental frequency component ω. Likewise, the terms with the coefficients a_2 and b_2 together, represent the second harmonic component 2ω, and so on. The coefficients a_0, a_n, and b_n are found from the following relations:

$$\frac{1}{2}a_0 = \frac{1}{2\pi}\int_0^{2\pi} f(t)dt$$

$$a_n = \frac{1}{\pi}\int_0^{2\pi} f(t)\cos nt\, dt$$

$$b_n = \frac{1}{\pi}\int_0^{2\pi} f(t)\sin nt\, dt$$

- If a waveform has odd symmetry, that is, if it is an odd function, the series will consist of sine terms only. Odd functions are those for which $-f(-t) = f(t)$.

- If a waveform has even symmetry, that is, if it is an even function, the series will consist of cosine terms only, and a_0 may or may not be zero. Even functions are those for which $f(-t) = f(t)$

- A periodic waveform with period T, has half–wave symmetry if

$$-f(t + T/2) = f(t)$$

that is, the shape of the negative half–cycle of the waveform is the same as that of the positive half–cycle, but inverted. If a waveform has half–wave symmetry only odd (odd cosine and odd sine) harmonics will be present. In other words, all even (even cosine and even sine) harmonics will be zero.

- The trigonometric Fourier series for the square waveform with odd symmetry is

$$f(t) = \frac{4A}{\pi}\left(\sin \omega t + \frac{1}{3}\sin 3\omega t + \frac{1}{5}\sin 5\omega t + \ldots\right) = \frac{4A}{\pi}\sum_{n=odd}\frac{1}{n}\sin n\omega t$$

- The trigonometric Fourier series for the square waveform with even symmetry is

$$f(t) = \frac{4A}{\pi}\left(\cos\omega t - \frac{1}{3}\cos 3\omega t + \frac{1}{5}\cos 5\omega t - \ldots\right) = \frac{4A}{\pi}\sum_{n=odd}(-1)^{\frac{(n-1)}{2}}\frac{1}{n}\cos n\omega t$$

- The trigonometric Fourier series for the sawtooth waveform with odd symmetry is

$$f(t) = \frac{2A}{\pi}\left(\sin\omega t - \frac{1}{2}\sin 2\omega t + \frac{1}{3}\sin 3\omega t - \frac{1}{4}\sin 4\omega t + \ldots\right) = \frac{2A}{\pi}\sum(-1)^{n-1}\frac{1}{n}\sin n\omega t$$

- The trigonometric Fourier series for the triangular waveform with odd symmetry is

$$f(t) = \frac{8A}{\pi^2}\left(\sin\omega t - \frac{1}{9}\sin 3\omega t + \frac{1}{25}\sin 5\omega t - \frac{1}{49}\sin 7\omega t + \ldots\right) = \frac{8A}{\pi^2}\sum_{n=odd}(-1)^{\frac{(n-1)}{2}}\frac{1}{n^2}\sin n\omega t$$

- The trigonometric Fourier series for the half–wave rectification waveform with no symmetry is

$$f(t) = \frac{A}{\pi} + \frac{A}{2}\sin t - \frac{A}{\pi}\left[\frac{\cos 2t}{3} + \frac{\cos 4t}{15} + \frac{\cos 6t}{35} + \frac{\cos 8t}{63} + \ldots\right]$$

- The trigonometric Fourier series for the full–wave rectification waveform with even symmetry is

$$f(t) = \frac{2A}{\pi} - \frac{4A}{\pi}\sum_{n=2,4,6,\ldots}^{\infty}\frac{1}{(n^2-1)}\cos n\omega t$$

- The Fourier series are often expressed in exponential form as

$$f(t) = \ldots + C_{-2}e^{-j2\omega t} + C_{-1}e^{-j\omega t} + C_0 + C_1 e^{j\omega t} + C_2 e^{j2\omega t} + \ldots$$

where the C_i coefficients are related to the trigonometric form coefficients as

$$C_{-n} = \frac{1}{2}\left(a_n - \frac{b_n}{j}\right) = \frac{1}{2}(a_n + jb_n)$$

$$C_n = \frac{1}{2}\left(a_n + \frac{b_n}{j}\right) = \frac{1}{2}(a_n - jb_n)$$

$$C_0 = \frac{1}{2}a_0$$

- The C_i coefficients, except C_0, are complex, and appear as complex conjugate pairs, that is,

$$C_{-n} = C_n{}^*$$

- In general, for $\omega \neq 1$,

$$C_n = \frac{1}{T} \int_0^T f(t)e^{-jn\omega t}d(\omega t) = \frac{1}{2\pi} \int_0^{2\pi} f(t)e^{-jn\omega t}d(\omega t)$$

- We can derive the trigonometric Fourier series from the exponential series from the relations

$$a_n = C_n + C_{-n}$$

and

$$b_n = j(C_n - C_{-n})$$

- For even functions, all coefficients C_i are real

- For odd functions, all coefficients C_i are imaginary

- If there is half–wave symmetry, $C_n = 0$ for $n = $ even

- $C_{-n} = C_n^*$ always

- A line spectrum is a plot that shows the amplitudes of the harmonics on a frequency scale.

- The frequency components of a recurrent rectangular pulse follow a $\sin x / x$ form.

- We can evaluate the Fourier coefficients of a function based on observed values instead of an analytic expression using numerical evaluations with the aid of a spreadsheet.

- A power series has the form

$$\sum_{k=0}^{\infty} a_k x^k = a_0 + a_1 x + a_2 x^2 + \dots$$

- A function $f(x)$ that possesses all derivatives up to order n at a point $x = x_0$ can be expanded in a Taylor series as

$$f(x) = f(x_0) + f'(x_0)(x - x_0) + \frac{f''(x_0)}{2!}(x - x_0)^2 + \dots + \frac{f^{(n)}(x_0)}{n!}(x - x_0)^n$$

If $x_0 = 0$, the series above reduces to

$$f(x) = f(0) + f'(0)x + \frac{f''(0)}{2!}x^2 + \dots + \frac{f^{(n)}(0)}{n!}x^n$$

and this relation is known as *Maclaurin series*

- We can also obtain a Taylor series expansion with the MATLAB **taylor(f,n,a)** function where **f** is a symbolic expression, **n** produces the first n terms in the series, and **a** defines the Taylor approximation about point a.

6.12 Exercises

1. Compute the first 5 components of the trigonometric Fourier series for the waveform below. Assume $\omega = 1$.

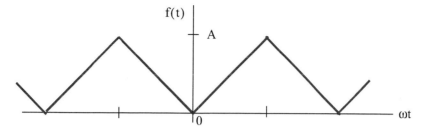

2. Compute the first 5 components of the trigonometric Fourier series for the waveform below. Assume $\omega = 1$.

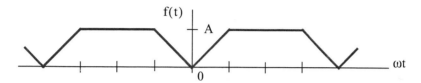

3. Compute the first 5 components of the exponential Fourier series for the waveform below. Assume $\omega = 1$.

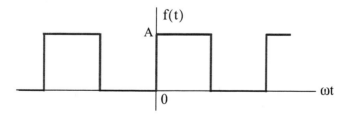

4. Compute the first 5 components of the exponential Fourier series for the waveform below. Assume $\omega = 1$.

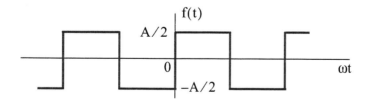

5. Compute the first 5 components of the exponential Fourier series for the waveform below. Assume $\omega = 1$.

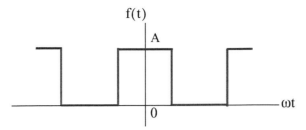

6. Compute the first 5 components of the exponential Fourier series for the waveform below. Assume $\omega = 1$.

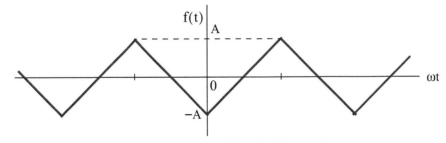

Figure 6.37. Waveform for Exercise 6

7. Compute the first 4 terms of the Maclaurin series for each of the following functions.

    a. $f(x) = e^{-x}$ b. $f(x) = \sin x$ c. $f(x) = \sinh x$

Confirm your answers with MATLAB.

8. Compute the first 4 terms of the Taylor series for each of the following functions.

    a. $f(x) = \dfrac{1}{x}$ about $a = -1$ b. $f(x) = \sin x$ about $a = -\dfrac{\pi}{4}$

Confirm your answers with MATLAB.

9. In a non–linear device, the voltage and current are related as

$$i(v) = k\left(1 + \frac{v}{V}\right)^{1.5}$$

where k is a constant and V is the DC component of the instantaneous voltage v. Expand this function into a power series that can be used to compute the current i, when the voltage v is small, and varies about $v = 0$.

6.13 Solutions to End–of–Chapter Exercises

1.

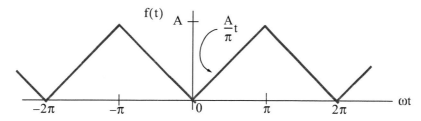

This is an even function; therefore, the series consists of cosine terms only. There is no half–wave symmetry and the average (DC component) is not zero. We will integrate from 0 to π and multiply by 2. Then,

$$a_n = \frac{2}{\pi}\int_0^\pi \frac{A}{\pi}t\cos nt\, dt = \frac{2A}{\pi^2}\int_0^\pi t\cos nt\, dt \quad (1)$$

From tables of integrals,

$$\int x\cos ax\, dx = \frac{1}{a^2}\cos ax + \frac{x}{a}\sin ax$$

and thus (1) becomes

$$a_n = \frac{2A}{\pi^2}\left(\frac{1}{n^2}\cos nt + \frac{t}{n}\sin nt\right)\Bigg|_0^\pi = \frac{2A}{\pi^2}\left(\frac{1}{n^2}\cos n\pi + \frac{t}{n}\sin nt\pi - \frac{1}{n^2} - 0\right)$$

and since $\sin nt\pi = 0$ for all integer n,

$$a_n = \frac{2A}{\pi^2}\left(\frac{1}{n^2}\cos n\pi - \frac{1}{n^2}\right) = \frac{2A}{n^2\pi^2}(\cos n\pi - 1) \quad (2)$$

We cannot evaluate the average$(1/2)/a_0$ from (2); we must use (1). Then, for $n = 0$,

$$\frac{1}{2}a_0 = \frac{2A}{2\pi^2}\int_0^\pi t\, dt = \frac{A}{\pi^2}\cdot\frac{t^2}{2}\Bigg|_0^\pi = \frac{A}{\pi^2}\cdot\frac{\pi^2}{2}$$

or

$$(1/2)/a_0 = A/2$$

We observe from (2) that for $n = \text{even}$, $a_{n=\text{even}} = 0$. Then,

$$\text{for } n = 1,\ a_1 = -\frac{4A}{\pi^2},\ \text{for } n = 3,\ a_3 = \frac{-4A}{3^2\pi^2},\ \text{for } n = 5,\ a_5 = -\frac{4A}{5^2\pi^2},\ \text{for } n = 7,\ a_3 = \frac{-4A}{7^2\pi^2}$$

and so on.

Therefore,

$$f(t) = \frac{1}{2}a_0 - \frac{4A}{\pi^2}\left(\cos t + \frac{1}{9}\cos 3t + \frac{1}{25}\cos 5t + \frac{1}{49}\cos 7t + \ldots\right) = \frac{A}{2} - \frac{4A}{\pi}\sum_{n=odd}^{\infty}\frac{1}{n^2}\cos nt$$

2.

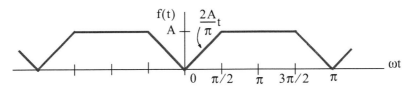

This is an even function; therefore, the series consists of cosine terms only. There is no half-wave symmetry and the average (DC component) is not zero.

$$\text{Average} = \frac{1}{2}a_0 = \frac{\text{Area}}{\text{Period}} = \frac{2\times[(A/2)\cdot(\pi/2)] + A\pi}{2\pi} = \frac{3A\cdot(\pi/2)}{2\pi} = \frac{3A}{4}$$

$$a_n = \frac{2}{\pi}\int_0^{\pi/2}\frac{2A}{\pi}t\cos nt\,dt + \frac{2}{\pi}\int_{\pi/2}^{\pi}A\cos nt\,dt \quad (1)$$

and with

$$\int x\cos ax\,dx = \frac{1}{a^2}\cos ax + \frac{x}{a}\sin ax = \frac{1}{a^2}(\cos ax + ax\sin ax)$$

(1) simplifies to

$$a_n = \frac{4A}{\pi^2}\left[\frac{1}{n^2}(\cos nt + nt\sin nt)\right]\Bigg|_0^{\pi/2} + \frac{2A}{n\pi}\sin nt\Bigg|_{\pi/2}^{\pi}$$

$$= \frac{4A}{n^2\pi^2}\left(\cos\frac{n\pi}{2} + \frac{n\pi}{2}\sin\frac{n\pi}{2} - 1 - 0\right) + \frac{2A}{n\pi}\left(\sin n\pi - \sin\frac{n\pi}{2}\right)$$

and since $\sin n\pi = 0$ for all integer n,

$$a_n = \frac{4A}{n^2\pi^2}\cos\frac{n\pi}{2} + \frac{2A}{n\pi}\sin\frac{n\pi}{2} - \frac{4A}{n^2\pi^2} - \frac{2A}{n\pi}\sin\frac{n\pi}{2} = \frac{4A}{n^2\pi^2}\left(\cos\frac{n\pi}{2} - 1\right)$$

for $n = 1$, $a_1 = \frac{4A}{\pi^2}(0-1) = -\frac{4A}{\pi^2}$, for $n = 2$, $a_2 = \frac{4A}{4\pi^2}(-1-1) = -\frac{2A}{\pi^2}$

for $n = 3$, $a_3 = \frac{4A}{9\pi^2}(0-1) = -\frac{4A}{9\pi^2}$, for $n = 4$, $a_4 = \frac{-4A}{7^2\pi^2}(1-1) = 0$

We observe that the fourth harmonic and all its multiples are zero. Therefore,

$$f(t) = \frac{3A}{4} - \frac{4A}{\pi^2}\left(\cos t + \frac{1}{2}\cos 2t + \frac{1}{9}\cos 3t + \ldots\right)$$

3.

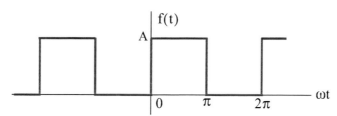

This is neither an even nor an odd function and has no half–wave symmetry; therefore, the series consists of both cosine and sine terms. The average (DC component) is not zero. Then,

$$C_n = \frac{1}{2\pi} \int_0^{2\pi} f(t)e^{-jn\omega t}d(\omega t)$$

and with $\omega = 1$

$$C_n = \frac{1}{2\pi} \int_0^{2\pi} f(t)e^{-jnt}dt = \frac{1}{2\pi}\left[\int_0^{\pi} Ae^{-jnt}dt + \int_{\pi}^{2\pi} 0e^{-jnt}dt \right] = \frac{A}{2\pi} \int_0^{\pi} e^{-jnt}dt$$

The DC value is

$$C_0 = \frac{A}{2\pi} \int_0^{\pi} e^0 dt = \frac{A}{2\pi}t\Big|_0^{\pi} = \frac{A}{2}$$

For $n \neq 0$

$$C_n = \frac{A}{2\pi} \int_0^{\pi} e^{-jnt}dt = \frac{A}{-j2n\pi}e^{-jnt}\Big|_0^{\pi} = \frac{A}{j2n\pi}(1 - e^{-jn\pi})$$

Recalling that

$$e^{-jn\pi} = \cos n\pi - j\sin n\pi$$

for $n = \text{even}$, $e^{-jn\pi} = 1$ and for $n = \text{odd}$, $e^{-jn\pi} = -1$. Then,

$$C_{n = \text{even}} = \frac{A}{j2n\pi}(1 - 1) = 0$$

and

$$C_{n = \text{odd}} = \frac{A}{j2n\pi}[1 - (-1)] = \frac{A}{jn\pi}$$

By substitution into the expression

$$f(t) = \ldots + C_{-2}e^{-j2\omega t} + C_{-1}e^{-j\omega t} + C_0 + C_1 e^{j\omega t} + C_2 e^{j2\omega t} + \ldots$$

we find that

$$f(t) = \frac{A}{2} + \frac{A}{j\pi}\left(\ldots - \frac{1}{3}e^{-j3\omega t} - e^{-j\omega t} + e^{j\omega t} + \frac{1}{3}e^{j3\omega t} + \ldots\right)$$

The minus (–) sign of the first two terms within the parentheses results from the fact that $C_{-n} = C_n{}^*$. For instance, since $C_1 = 2A/j\pi$, it follows that $C_{-1} = C_1{}^* = -2A/j\pi$. We observe that $f(t)$ is complex, as expected, since there is no symmetry.

4.

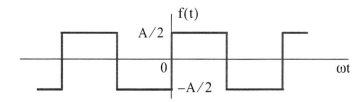

This is the same waveform as in Exercise 3 where the DC component has been removed. Then,

$$f(t) = \frac{A}{j\pi}\left(\ldots - \frac{1}{3}e^{-j3\omega t} - e^{-j\omega t} + e^{j\omega t} + \frac{1}{3}e^{j3\omega t} + \ldots\right)$$

It is also the same waveform as in Example 6.9, Page 6–32, except that the amplitude is halved. This waveform is an odd function and thus the expression for $f(t)$ is imaginary.

5.

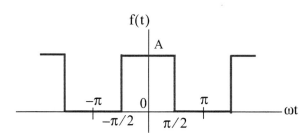

This is the same waveform as in Exercise 3 where the vertical axis has been shifted to make the waveform an even function. Therefore, for this waveform C_n is real. Then,

$$C_n = \frac{1}{2\pi} \int_{-\pi}^{\pi} f(t)e^{-jnt}dt = \frac{A}{2\pi} \int_{-\pi/2}^{\pi/2} e^{-jnt}dt$$

The DC value is

$$C_0 = \frac{A}{2\pi}t\Big|_{-\pi/2}^{\pi/2} = \frac{A}{2\pi}\left(\frac{\pi}{2} + \frac{\pi}{2}\right) = \frac{A}{2}$$

For $n \neq 0$

Numerical Analysis Using MATLAB® and Excel®, Third Edition
Copyright © Orchard Publications

$$C_n = \frac{A}{2\pi} \int_{-\pi/2}^{\pi/2} e^{-jnt}dt = \frac{A}{-j2n\pi} e^{-jnt}\Big|_{-\pi/2}^{\pi/2} = \frac{A}{-j2n\pi}(e^{-jn\pi/2} - e^{jn\pi/2})$$

$$= \frac{A}{j2n\pi}(e^{jn\pi/2} - e^{-jn\pi/2}) = \frac{A}{n\pi}\Big(\frac{e^{jn\pi/2} - e^{-jn\pi/2}}{j2}\Big) = \frac{A}{n\pi}\sin\frac{n\pi}{2}$$

and we observe that for $n = even$, $C_n = 0$

For $n = odd$, C_n alternates in plus (+) and minus (–) signs, that is,

$$C_n = \frac{A}{n\pi} \text{ if } n = 1, 5, 9, \ldots$$

$$C_n = -\frac{A}{n\pi} \text{ if } n = 3, 7, 11, \ldots$$

Thus,

$$f(t) = \frac{A}{2} + \sum_{n = odd} \Big(\pm\frac{A}{n\pi}e^{jn\omega t}\Big)$$

where the plus (+) sign is used with $n = 1, 5, 9, \ldots$ and the minus (–) sign is used with $n = 3, 7, 11, \ldots$. We can express $f(t)$ in a more compact form as

$$f(t) = \frac{A}{2} + \sum_{n = odd} (-1)^{(n-1)/2}\frac{A}{n\pi}e^{jn\omega t}$$

6.

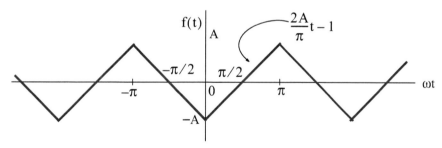

We will find the exponential form coefficients C_n from

$$C_n = \frac{1}{2\pi} \int_{-\pi}^{\pi} f(t)e^{-jnt}dt$$

From tables of integrals

$$\int xe^{ax}dx = \frac{e^{ax}}{a^2}(ax - 1)$$

Then,

$$C_n = \frac{1}{2\pi}\left[\int_{-\pi}^{0}\left(-\frac{2A}{\pi}t - 1\right)e^{-jnt}dt + \int_{0}^{\pi}\left(\frac{2A}{\pi}t - 1\right)e^{-jnt}dt\right]$$

Integrating and rearranging terms we get

$$C_n = \frac{1}{2\pi}\left[-\frac{4A}{n^2\pi} + \frac{4A}{n^2\pi}\left(n\pi \cdot \frac{e^{jn\pi} - e^{-jn\pi}}{j2} + \frac{e^{jn\pi} + e^{-jn\pi}}{2}\right) - \frac{2A}{n} \cdot \frac{e^{jn\pi} - e^{-jn\pi^-}}{j2}\right]$$

$$= \frac{4A}{2n^2\pi^2}\left(-1 + n\pi\sin n\pi + \cos n\pi - \frac{n\pi}{2}\sin n\pi\right)$$

and since $\sin n\pi = 0$ for all integer n,

$$C_n = \frac{2A}{n^2\pi^2}(\cos n\pi - 1)$$

For $n = even$, $C_n = 0$ and for $n = odd$, $\cos n\pi = -1$, and $C_n = \frac{-4A}{n^2\pi^2}$

Also, by inspection, the DC component $C_0 = 0$. Then,

$$f(t) = -\frac{4A}{\pi^2}\left(\ldots + \frac{1}{9}e^{-j3\omega t} + e^{-j\omega t} + e^{j\omega t} + \frac{1}{9}e^{j3\omega t} + \ldots\right)$$

The coefficients of the terms $e^{-j3\omega t}$ and $e^{-j\omega t}$ are positive because all coefficients of C_n are real. This is to be expected since $f(t)$ is an even function. It also has half–wave symmetry and thus $C_n = 0$ for $n = even$ as we've found.

7.

$$f(x) = f(0) + f'(0)x + \frac{f''(0)}{2!}x^2 + \ldots + \frac{f^{(n)}(0)}{n!}x^n$$

a. $f(x) = e^{-x}$, $f(0) = 1$, $f'(x) = -e^{-x}$, $f'(0) = -1$, $f''(x) = e^{-x}$, $f''(0) = 1$, $f'''(x) = -e^{-x}$, $f'''(0) = -1$, and so on. Therefore,

$$f_n(x) = 1 - x + \frac{x^2}{2!} - \frac{x^3}{3!} + \ldots$$

MATLAB displays the same result.

```
x=sym('x'); fn=taylor(exp(-x)); pretty(fn)
```

```
            2          3          4            5
  1 - x + 1/2 x   - 1/6 x   + 1/24 x   - 1/120 x
```

b. $f(x) = \sin x$, $f(0) = 0$, $f'(x) = \cos x$, $f'(0) = 1$, $f''(x) = -\sin x$, $f''(0) = 0$, $f'''(x) = -\cos x$, $f'''(0) = -1$, and so on. Therefore,

$$f_n(x) = x - \frac{x^3}{3!} + \frac{x^5}{5!} - \frac{x^7}{7!} + \ldots$$

MATLAB displays the same result.

x=sym('x'); fn=taylor(sin(x)); pretty(fn)

```
             3           5
    x  -  1/6 x   + 1/120 x
```

c. $f(x) = \sinh x$, $f(0) = 0$, $f'(x) = \cosh x$, $f'(0) = 1$, $f''(x) = \sinh x$, $f''(0) = 0$, $f'''(x) = \cosh x$, $f'''(0) = 1$, and so on. Therefore,

$$f_n(x) = x + \frac{x^3}{3!} + \frac{x^5}{5!} + \frac{x^7}{7!} + \ldots$$

MATLAB displays the same result.

x=sym('x'); fn=taylor(sinh(x)); pretty(fn)

```
             3           5
    x  + 1/6 x   + 1/120 x
```

8.

$$f_n(x) = f(a) + f'(a)(x-a) + \frac{f''(a)}{2!}(x-a)^2 + \frac{f'''(a)}{3!}(x-a)^3 + \ldots$$

a. $f(x) = 1/x$, $f(a) = f(-1) = -1$, $f'(x) = -1/x^2$, $f'(a) = f'(-1) = -1$, $f''(x) = 2/x^3$, $f''(a) = f''(-1) = -2$, $f'''(x) = -6/x^4$, $f'''(a) = f'''(-1) = -6$, and so on. Therefore,

$$f_n(x) = -1 - (x+1) - (x+1)^2 - (x+1)^3 + \ldots$$

or

$$f_n(x) = -2 - x - (x+1)^2 - (x+1)^3 + \ldots$$

MATLAB displays the same result.

x=sym('x'); y=1/x; z=taylor(y,4,-1); pretty(z)

```
                2           3
   -2 - x - (x + 1)   - (x + 1)
```

b. $f(x) = \sin x$, $f(a) = f(-\pi/4) = -\sqrt{2}/2$, $f'(x) = \cos x$, $f'(a) = f'(-\pi/4) = \sqrt{2}/2$,

$$f'(x) = -\sin x, \qquad f'(a) = f'(-\pi/4) = \sqrt{2}/2, \qquad f''(x) = -\cos x,$$

$$f'''(a) = f'''(-\pi/4) = -\sqrt{2}/2 \text{, and so on. Therefore,}$$

$$f_n(x) = -\sqrt{2}/2 + (\sqrt{2}/2)(x + \pi/4) + (\sqrt{2}/4)(x + \pi/4)^2 - (\sqrt{2}/12)(x + \pi/4)^3 + \dots$$

MATLAB displays the same result.

x=sym('x'); y=sin(x); z=taylor(y,4,–pi/4); pretty(z)

```
           1/2          1/2                         1/2                  2
    - 1/2 2      + 1/2 2      (x + 1/4 pi) + 1/4 2       (x + 1/4 pi)


           1/2                  3
    - 1/12 2      (x + 1/4 pi)
```

9.

$$i(v) = k\left(1 + \frac{v}{V}\right)^{1.5}$$

The Taylor series for this relation is

$$i(v) = i(v_0) + i'(v_0)(v - v_0) + \frac{i''(v_0)}{2!}(v - v_0)^2 + \frac{i'''(v_0)}{3!}(v - v_0)^3 + \dots$$

Since the voltage v is small, and varies about $v = 0$, we expand this relation about $v = 0$ and the series reduces to the Maclaurin series below.

$$i(v) = i(0) + i'(0)v + \frac{i''(0)}{2!}v^2 + \dots \quad (1)$$

By substitution of $v = 0$ into the given relation we get

$$i(0) = k$$

The first and second derivatives of i are

$$i'(v) = \frac{3k}{2V}\left(1 + \frac{v}{V}\right)^{1/2} \qquad i'(0) = \frac{3k}{2V}$$

$$i''(v) = \frac{3k}{4V^2}\left(1 + \frac{v}{V}\right)^{-1/2} \qquad i''(0) = \frac{3k}{4V^2}$$

and by substitution into (1)

$$i(v) = k + \frac{3k}{2V}v + \frac{3k}{8V^2}v^2 + \dots = k\left(1 + \frac{3}{2V}v + \frac{3}{8V^2}v^2 + \dots\right)$$

MATLAB displays the same result.

```
x=sym('x'); i=sym('i'); v=sym('v'); k=sym('k'); V=sym('V');...
i=k*(1+v/V)^1.5; z=taylor(i,4,0); pretty(z)
```

$$k + \frac{3}{2}\,\frac{k\,v}{V} + \frac{3}{8}\,\frac{k\,v^2}{V^2} - \frac{1}{16}\,\frac{k\,v^3}{V^3}$$

NOTES:

Chapter 7

Finite Differences and Interpolation

T his chapter begins with finite differences and interpolation which is one of its most important applications. Finite Differences form the basis of numerical analysis as applied to other numerical methods such as curve fitting, data smoothing, numerical differentiation, and numerical integration. These applications are discussed in this and the next three chapters.

7.1 Divided Differences

Consider the continuous function $y = f(x)$ and let x_0, x_1, x_2, ..., x_{n-1}, x_n be some values of x in the interval $x_0 \leq x \leq x_n$. It is customary to show the independent variable x, and its corresponding values of $y = f(x)$ in tabular form as in Table 7.1.

TABLE 7.1 *The variable x and* $y = f(x)$ *in tabular form*

x	$f(x)$
x_0	$f(x_0)$
x_1	$f(x_1)$
x_2	$f(x_2)$
...	...
x_{n-1}	$f(x_{n-1})$
x_n	$f(x_n)$

Let x_i and x_j be any two, not necessarily consecutive values of x, within this interval. Then, the *first divided difference* is defined as:

$$f(x_i, x_j) = \frac{f(x_i) - f(x_j)}{x_i - x_j} \tag{7.1}$$

Likewise, the *second divided difference* is defined as:

$$f(x_i, x_j, x_k) = \frac{f(x_i, x_j) - f(x_j, x_k)}{x_i - x_k} \tag{7.2}$$

The third, fourth, and so on divided differences, are defined similarly.

The divided differences are indicated in a difference table where each difference is placed between the values of the column immediately to the left of it as shown in Table 7.2.

TABLE 7.2 Conventional presentation of divided differences

x	$f(x)$			
x_0	$f(x_0)$			
		$f(x_0, x_1)$		
x_1	$f(x_1)$		$f(x_0, x_1, x_2)$	
		$f(x_1, x_2)$		$f(x_0, x_1, x_2, x_3)$
x_2	$f(x_2)$		$f(x_1, x_2, x_3)$	
		$f(x_2, x_3)$		
x_3	$f(x_3)$			

Example 7.1

Form a difference table showing the values of x given as 0, 1, 2, 3, 4, 7, and 9, the values of $f(x)$ corresponding to $y = f(x) = x^3$, and the first through the fourth divided differences.

Solution:

We construct Table 7.3 with six columns. The first column contains the given values of x, the second the values of $f(x)$, and the third through the sixth contain the values of the first through the fourth divided differences. These differences are computed from (7.1), (7.2), and other relations for higher order divided differences. For instance, the second value on the first divided difference is found from (7.1) as

$$\frac{1 - 27}{1 - 3} = 13$$

and third value on the second divided difference is found from (7.2) as

$$\frac{37 - 93}{3 - 7} = 14$$

Likewise, for the third divided difference we have

TABLE 7.3 *Divided differences for Example 7.1*

Function		Divided Differences			
x	$f(x) = x^3$	First	Second	Third	Fourth
0	0				
		1			
1	1		4		
		13		1	
3	27		8		0
		37		1	
4	64		14		0
		93		1	
7	343		20		
		193			
9	729				

$$\frac{4-8}{0-4} = 1$$

and for the fourth

$$\frac{1-1}{0-4} = 0$$

We observe that, if the values of the nth divided difference are the same, as in the fifth column (third divided differences for this example), all subsequent differences will be equal to zero.

In most cases, the values of x in a table are equally spaced. In this case, the differences are sets of consecutive values. Moreover, the denominators are all the same; therefore, they can be omitted. These values are referred to as just the *differences* of the function.

If the constant difference between successive values of x is h, the typical value of x_k is

$$x_k = x_0 + kh \quad \text{for} \quad k = \dots, -2, -1, 0, 1, 2, \dots \tag{7.3}$$

We can now express the first differences in terms of the difference operator Δ as

$$\Delta f_k = f_{k+1} - f_k \tag{7.4}$$

Likewise, the second differences are

$$\Delta^2 f_k = \Delta(\Delta f_k) = \Delta f_{k+1} - \Delta f_k \tag{7.5}$$

and, in general, for positive integer values of n

$$\Delta^n f_k = \Delta(\Delta^{n-1} f_k) = \Delta^{n-1} f_{k+1} - \Delta^{n-1} f_k \tag{7.6}$$

The difference operator Δ obeys the law of exponents, that is,

$$\Delta^m(\Delta^n f_k) = \Delta^{m+n} f_k \tag{7.7}$$

We construct the difference table in terms of the difference operator Δ as shown in Table 7.4.

TABLE 7.4 *Divided differences table in terms of the difference operator* Δ

Function		Differences				
x	f	First	Second	Third	Fourth	...
x_0	f_0					
		Δf_0				
x_1	f_1		$\Delta^2 f_0$			
		Δf_1		$\Delta^3 f_0$		
x_2	f_2		$\Delta^2 f_1$		$\Delta^4 f_0$	
		Δf_2		$\Delta^3 f_1$		
x_3	f_3		$\Delta^2 f_2$			
		Δf_3				
x_4	f_4					
...						
x_n	f_n					

Example 7.2

Construct a difference table showing the values of x given as $1, 2, 3, 4, 5, 6, 7$ and 8, the values of $f(x)$ corresponding to $y = f(x) = x^3$, and the first through the fourth differences.

Solution:

Following the same procedure as in the previous example, we construct Table 7.5.

TABLE 7.5 Difference table for Example 7.2

	Function		Differences				
k	x_k	f_k	Δf_k	$\Delta^2 f_k$	$\Delta^3 f_k$	$\Delta^4 f_k$	...
1	1	1					
			7				
2	2	8		12			
			19		6		
3	3	27		18		0	
			37		6		
4	4	64		24		0	
			61		6		
5	5	125		30		0	
			91		6		
6	6	216		36		0	
			127		6		
7	7	343		42			
			169				
8	8	512					

We observe that the fourth differences $\Delta^4 f_k$ are zero, as expected.

Using the binomial expansion

$$\binom{n}{j} = \frac{n!}{j!(n-j)!} \tag{7.8}$$

we can show that

$$\Delta^n f_k = f_{k+n} - nf_{k+n-1} + \frac{n(n-1)}{2!}f_{k+n-2} + \dots + (-1)^{n-1}nf_{k+1} + (-1)^n f_k \tag{7.9}$$

For $k = 0$, $n = 1, 2, 3$ and 4, relation (7.9) reduces to

$$\Delta f_0 = f_2 - f_1$$

$$\Delta^2 f_0 = f_2 - 2f_1 + f_0$$

$$\Delta^3 f_0 = f_3 - 3f_2 + 3f_1 - f_0 \tag{7.10}$$

$$\Delta^4 f_0 = f_4 - 4f_3 + 6f_2 - 4f_1 + f_0$$

It is interesting to observe that the first difference in (7.10), is the difference quotient whose limit defines the derivative of a continuous function that is defined as

$$\lim_{\Delta x \to 0} \frac{\Delta y}{\Delta x} = \lim_{\Delta x \to 0} \frac{f(x_1 + \Delta x) - f(x_1)}{\Delta x} \tag{7.11}$$

As with derivatives, *the* nth *differences of a polynomial of degree* n *are constant.*

7.2 Factorial Polynomials

The *factorial polynomials* are defined as

$$(x)^{(n)} = x(x-1)(x-2)...(x-n+1) \tag{7.12}$$

and

$$(x)^{-(n)} = \frac{1}{(x-1)(x-2)...(x+n)} \tag{7.13}$$

These expressions resemble the power functions x^n and x^{-n} in elementary algebra.

Using the difference operator Δ with (7.12) and (7.13) we obtain

$$\Delta(x)^{(n)} = n(x)^{(n-1)} \tag{7.14}$$

and

$$\Delta(x)^{-(n)} = -n(x)^{-(n-1)} \tag{7.15}$$

We observe that (7.14) and (7.15) are very similar to differentiation of x^n and x^{-n}.

Occasionally, it is desirable to express a polynomial $p_n(x)$ as a factorial polynomial. Then, in analogy with Maclaurin power series, we can express that polynomial as

$$p_n(x) = a_0 + a_1(x)^{(1)} + a_2(x)^{(2)} + ... + a_n(x)^{(n)} \tag{7.16}$$

and now our task is to compute the coefficients a_k.

For $x = 0$, relation (7.16) reduces to

$$a_0 = p_n(0) \tag{7.17}$$

To compute the coefficient a_1, we take the first difference of $p_n(x)$ in (7.16). Using (7.14) we obtain

$$\Delta p_n(x) = 1x^0 a_1 + 2a_2(x)^{(1)} + 3a_3(x)^{(2)} + ... + na_n(x)^{(n-1)} \tag{7.18}$$

and letting $x = 0$, we find that

$$a_1 = \Delta p_n(0) \tag{7.19}$$

Differencing again we obtain

$$\Delta^2 p_n(x) = 2 \cdot 1 a_2 + 3 \cdot 2 a_3(x)^{(1)} + \ldots + n(n-1)a_n(x)^{(n-2)} \tag{7.20}$$

and for $x = 0$,

$$a_2 = \frac{\Delta^2 p_n(0)}{2 \cdot 1} = \frac{\Delta^2 p_n(0)}{2!} \tag{7.21}$$

In general,

$$a_j = \frac{\Delta^j p_n(0)}{j!} \quad \text{for} \quad j = 0, 1, 2, \ldots, n \tag{7.22}$$

Factorial polynomials provide an easier method of constructing a difference table. With this method we perform the following steps:

1. We divide $p_n(x)$ in (7.16) by x to obtain a quotient $q_0(x)$ and a remainder r_0 which turns out to be the constant term a_0. Then, we express (7.16) as

$$p_n(x) = r_0 + xq_0(x) \tag{7.23}$$

2. We divide $q_0(x)$ in (7.23) by $(x-1)$ to obtain a quotient $q_1(x)$ and a remainder r_1 which turns out to be the constant term a_1. Then,

$$q_0(x) = r_1 + (x-1)q_1(x) \tag{7.24}$$

By substitution of (7.24) into (7.23), and using the form of relation (7.16), we obtain

$$p_n(x) = r_0 + x[r_1 + (x-1)q_1(x)] = r_0 + r_1(x)^{(1)} + x(x-1)q_1(x) \tag{7.25}$$

3. We divide $q_1(x)$ in (7.25) by $(x-2)$ to obtain a quotient $q_2(x)$ and a remainder r_2 which turns out to be the constant term a_2, and thus

$$q_1(x) = r_2 + (x-2)q_2(x) \tag{7.26}$$

By substitution of (7.26) into (7.25), we obtain

$$p_n(x) = r_0 + r_1(x)^{(1)} + x(x-1)[r_2 + (x-2)q_2(x)]$$
$$= r_0 + r_1(x)^{(1)} + r_2(x)^{(2)} + x(x-1)(x-2)q_2(x) \tag{7.27}$$

Continuing with the above procedure, we obtain a new quotient whose degree is one less than preceding quotient and therefore, the process of finding new quotients and remainders terminates after $(n+1)$ steps.

The general form of a factorial polynomial is

$$p_n(x) = r_0 + r_1(x)^{(1)} + r_2(x)^{(2)} + \ldots + r_{n-1}(x)^{(n-1)} + r_n(x)^{(n)} \qquad (7.28)$$

and from (7.16) and (7.22),

$$r_j = a_j = \frac{\Delta^j p_n(0)}{j!} \qquad (7.29)$$

or

$$\Delta^j p_n(0) = j! r_j \qquad (7.30)$$

Example 7.3

Express the algebraic polynomial

$$p(x) = x^4 - 5x^3 + 3x + 4 \qquad (7.31)$$

as a factorial polynomial. Then, construct the difference table with $h = 1$.

Solution:

Since the highest power of the given polynomial $p(x)$ is 4, we must evaluate the remainders r_0, r_1, r_2, r_3 and r_4; then, we will use (7.28) to determine $p_n(x)$. We can compute the remainders by long division, but for convenience, we will use the MATLAB **deconv(p,q)** function which divides the polynomial **p** by **q**.

The MATLAB script is as follows:

```
px=[1 –5 0 3 4];          % Coefficients of given polynomial
d0=[1 0];                 % Coefficients of first divisor, i.e, x
[q0,r0]=deconv(px,d0)     % Computation of first quotient and remainder
d1=[1 –1];                % Coefficients of second divisor, i.e, x–1
[q1,r1]=deconv(q0,d1)     % Computation of second quotient and remainder
d2=[1 –2];                % Coefficients of third divisor, i.e, x–2
[q2,r2]=deconv(q1,d2)     % Computation of third quotient and remainder
d3=[1 –3];                % Coefficients of fourth divisor, i.e, x–3
[q3,r3]=deconv(q2,d3)     % Computation of fourth quotient and remainder
d4=[1 –4];                % Coefficients of fifth (last) divisor, i.e, x–4
[q4,r4]=deconv(q3,d4)     % Computation of fifth (last) quotient and remainder

q0 =
```

```
          1      -5       0       3
  r0  =
          0       0       0       0       4
  q1  =
          1      -4      -4
  r1  =
          0       0       0      -1
  q2  =
          1      -2
  r2  =
          0       0      -8
  q3  =
          1
  r3  =
          0       1
  q4  =
          0
  r4  =
          1
```

Therefore, with reference to (7.28), the factorial polynomial is

$$p_n(x) = 4 - (x)^{(1)} - 8(x)^{(2)} + (x)^{(3)} + (x)^{(4)} \tag{7.32}$$

We can verify that (7.32) is the same polynomial as (7.31), by expansion of the factorials using (7.12). This can be easily done with the MATLAB **collect('s_expr')** function, where **'s_expr'** is a symbolic expression. For this example, the MATLAB script is

syms x; px=collect((x*(x–1)*(x–2)*(x–3))+(x*(x–1)*(x–2))–(8*x*(x–1))–x+4)

px =
x^4–5*x^3+3*x+4

We observe that this is the same algebraic polynomial as in (7.31).

We will now compute the leading entries for the difference table using (7.30) and (7.32). Then,

$$\Delta^0 p(0) = 0! \cdot 4 = 4$$
$$\Delta^1 p(0) = 1! \cdot (-1) = -1$$
$$\Delta^2 p(0) = 2! \cdot (-8) = -16$$
$$\Delta^3 p(0) = 3! \cdot 1 = 6$$
$$\Delta^4 p(0) = 4! \cdot 1 = 24$$
$$\Delta^5 p(0) = 5! \cdot 0 = 0$$

(7.33)

1. We enter the values of (7.33) in the appropriate spaces as shown in Table 7.6.

2. We obtain the next set of values by crisscross addition as shown in Table 7.7.

3. The second crisscross addition extends the difference table as shown in Table 7.8.

TABLE 7.6 Leading entries of (7.33) in table form

TABLE 7.7 Crisscross addition to find second set of values

x	p(x)	Δ	Δ^2	Δ^3	Δ^4	Δ^5
	4					
		−1				
	3		−16			
		−17		6		
			−10		24	
				30		0
					24	

TABLE 7.8 Second crisscross addition to find third set of values

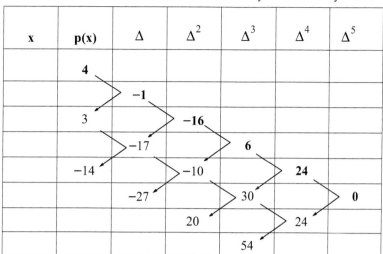

x	p(x)	Δ	$Δ^2$	$Δ^3$	$Δ^4$	$Δ^5$
	4					
		−1				
	3		−16			
		−17		6		
	−14		−10		24	
		−27		30		0
			20		24	
				54		

4. Continuation of this procedure produces the complete difference table. This is shown in Table 7.9.

TABLE 7.9 Complete difference table for Example 7.3

x	p(x)	Δ	$Δ^2$	$Δ^3$	$Δ^4$	$Δ^5$
	4					
		−1				
	3		−16			
		−17		6		
	−14		−10		24	
		−27		30		0
	−41		20		24	
		−7		54		
	−48		74			
		67				
	19					

7.3 Antidifferences

We recall from elementary calculus that when we know the first derivative of a function, we can *integrate* or *antidifferentiate* to find the function. By a similar method, we can find the *antidifference* of a factorial polynomial. We denote the antidifference as $\Delta^{-1} p_n(x)$. It is computed from

$$\Delta^{-1}(x)^{(n)} = \frac{(x)^{(n+1)}}{(n+1)} \tag{7.34}$$

Example 7.4

Compute the antidifference of the algebraic polynomial

$$p(x) = x^4 - 5x^3 + 3x + 4 \tag{7.35}$$

Solution:

This is the same algebraic polynomial as that of the previous example, where we found that the corresponding factorial polynomial is

$$p_n(x) = 4 - (x)^{(1)} - 8(x)^{(2)} + (x)^{(3)} + (x)^{(4)} \tag{7.36}$$

Then, by (7.34), its antidifference is

$$\Delta^{-1} p_n(x) = \frac{(x)^{(5)}}{5} + \frac{(x)^{(4)}}{4} - 8\frac{(x)^{(3)}}{3} - \frac{(x)^{(2)}}{2} + 4(x)^{(1)} + C \tag{7.37}$$

where C is an arbitrary constant.

Antidifferences are very useful in finding sums of series. Before we present an example, we need to review the definite sum and the fundamental theorem of sum calculus. These are discussed below.

In analogy with definite integrals for continuous functions, in finite differences we have the *definite sum* of $p_n(x)$ which for the interval $a \le x \le a + (n-1)h$ is denoted as

$$\sum_{x=\alpha}^{\alpha+(n-1)h} p_n(x) = p_n(\alpha) + p_n(\alpha+h) + p_n(\alpha+2h) + \ldots + p_n[\alpha+(n-1)h] \tag{7.38}$$

Also, in analogy with the *fundamental theorem of integral calculus* which states that

$$\int_a^b f(x)dx = f(b) - f(a) \tag{7.39}$$

we have the *fundamental theorem of sum calculus* which states that

$$\sum_{x=\alpha}^{\alpha+(n-1)h} p_n(x) = \Delta^{-1}p_n(x)\Big|_\alpha^{\alpha+nh} \tag{7.40}$$

Example 7.5

Derive a simple expression, in closed form, that computes the sum of the cubes of the first n odd integers.

Solution:

An odd number can be expressed as $2m-1$, and thus its cube is $(2m-1)^3$. To use (7.40), we must express this term as a factorial polynomial. Recalling from (7.12) that

$$(x)^{(n)} = x(x-1)(x-2)...(x-n+1) \tag{7.41}$$

and using the MATLAB **expand(f)** function where **f** is a symbolic expression, we execute

```
syms m; f = (2*m-1)^3; expand(f)
```

and we obtain

```
ans =
8*m^3-12*m^2+6*m-1
```

Thus

$$p(m) = (2m-1)^3 = 8m^3 - 12m^2 + 6m - 1 \tag{7.42}$$

Following the procedure of Example 7.3, we find $p_n(m)$ with MATLAB as

```
pm=[8 -12 6 -1];
d0=[1 0];
[q0,r0]=deconv(pm,d0)
d1=[1 -1];
[q1,r1]=deconv(q0,d1)
d2=[1 -2];
[q2,r2]=deconv(q1,d2)
d3=[1 -3];
[q3,r3]=deconv(q2,d3)

q0 =
     8    -12     6
```

```
r0 =
        0       0       0      -1
q1 =
        8      -4
r1 =
        0       0       2
q2 =
        8
r2 =
        0      12
q3 =
        0
r3 =
        8
```

Therefore,

$$p_n(m) = 8(m)^{(3)} + 12(m)^{(2)} + 2(m)^{(1)} - 1 \tag{7.43}$$

Taking the antidifference of (7.43) we obtain

$$\Delta^{-1} p_n(m) = \frac{8(m)^{(4)}}{4} + \frac{12(m)^{(3)}}{3} + \frac{2(m)^{(2)}}{2} - (m)^{(1)}$$
$$= 2(m)^{(4)} + 4(m)^{(3)} + (m)^{(2)} - (m)^{(1)} \tag{7.44}$$

and with (7.40)

$$\sum \text{cubes} = 2(m)^{(4)} + 4(m)^{(3)} + (m)^{(2)} - (m)^{(1)} \Big|_{m=1}^{n+1}$$
$$= 2(n+1)n(n-1)(n-2) + 4(n+1)n(n-1) + (n+1)n - (n+1)$$
$$-2(1)^{(4)} - 4(1)^{(3)} - (1)^{(2)} + (1)^{(1)} \tag{7.45}$$

Since

$$(1)^{(4)} = 1(1-1)(1-2)(1-3) = 0$$
$$(1)^{(3)} = 1(1-1)(1-2) = 0$$
$$(1)^{(2)} = 1(1-1) = 0 \tag{7.46}$$
$$(1)^{(1)} = 1$$

relation (7.45) reduces to

$$\sum \text{cubes} = 2(n+1)n(n-1)(n-2) + 4(n+1)n(n-1) + (n+1)n - (n+1) + 1 \tag{7.47}$$

and this can be simplified with the MATLAB **collect(f)** function as follows.

```
syms n; sum=collect(2*(n+1)*n*(n–1)*(n–2)+4*(n+1)*n*(n–1)+(n+1)*n–(n+1)+1)
sum =
2*n^4-n^2
```

that is,

$$\sum \text{cubes} = 2n^4 - n^2 = n^2(2n^2 - 1) \tag{7.48}$$

We can verify that this is the correct expression by considering the first 4 odd integers $1, 3, 5,$ and 7. The sum of their cubes is

$$1 + 27 + 125 + 343 = 496$$

This is verified with (7.48) since

$$n^2(2n^2 - 1) = 4^2(2 \cdot 4^2 - 1) = 16 \cdot 31 = 496$$

One important application of finite differences is *interpolation*. Newton's divided–difference interpolation method, Lagrange's interpolation method, Gregory–Newton forward, and Gregory-Newton backward interpolation methods are discussed in Sections 7.4 through 7.7 below. We will use spreadsheets to facilitate the computations. Interpolation using MATLAB is discussed in Section 7.8 below.

7.4 Newton's Divided Difference Interpolation Method

This method, has the advantage that the values $x_0, x_1, x_2, ..., x_n$ need not be equally spaced, or taken in consecutive order. It uses the formula

$$\begin{aligned} f(x) = {}& f(x_0) + (x - x_0)f(x_0, x_1) + (x - x_0)(x - x_1)f(x_0, x_1, x_2) \\ & + (x - x_0)(x - x_1)(x - x_2)f(x_0, x_1, x_2, x_3) \end{aligned} \tag{7.49}$$

where $f(x_0, x_1)$, $f(x_0, x_1, x_2)$, and $f(x_0, x_1, x_2, x_3)$ are the first, second, and third divided differences respectively.

Example 7.6

Use Newton's divided–difference method to compute $f(2)$ from the experimental data shown in Table 7.10.

TABLE 7.10 Data for Example 7.6

x	−1.0	0.0	0.5	1.0	2.5	3.0
y = f(x)	3.0	−2.0	−0.375	3.0	16.125	19.0

Solution:

We must compute the first, second, and third divided differences as required by (7.49).

The first divided differences are:

$$\frac{-2.000 - 3.000}{0 - (-1.0)} = -5.000$$

$$\frac{-0.375 - (-2.000)}{0.5 - 0.0} = 3.250$$

$$\frac{3.000 - (-0.375)}{1.0 - 0.5} = 6.750 \qquad (7.50)$$

$$\frac{16.125 - 3.000}{2.5 - 1.0} = 8.750$$

$$\frac{19.000 - 16.125}{3.0 - 2.5} = 5.750$$

The second divided differences are:

$$\frac{3.250 - (-5.000)}{0.5 - (-1.0)} = 5.500$$

$$\frac{6.750 - 3.250}{1.0 - 0.0} = 3.500$$

$$\frac{8.750 - 6.750}{2.5 - 0.5} = 1.000 \qquad (7.51)$$

$$\frac{5.750 - 8.750}{3.0 - 1.0} = -1.500$$

and the third divided differences are:

$$\frac{3.500 - 5.500}{1.0 - (-1.0)} = -1.000$$

$$\frac{1.000 - 3.500}{2.5 - 0.0} = -1.000 \qquad (7.52)$$

$$\frac{-1.500 - 1.000}{3.0 - 0.5} = -1.000$$

With these values, we construct the difference Table 7.11.

TABLE 7.11 Difference table for Example 7.6

x	f(x)	1st Divided Difference $f(x_0, x_1)$	2nd Divided Difference $f(x_0, x_1, x_2)$	3rd Divided Difference $f(x_0, x_1, x_2, x_3)$
−1.0	3.000			
		−5.000		
0.0	−2.000		5.500	
		3.250		−1.000
0.5	−0.375		3.500	
		6.750		−1.000
1.0	3.000		1.000	
		8.750		−1.000
2.5	16.125		−1.500	
		5.750		
3.0	19.000			

Now, we have all the data that we need to find $f(2)$. We start with $x_0 = 0.00$,[*] and for x in (7.49), we use $x = 2$. Then,

$$f(2) = -2.0 + (2-0)(3.250) + (2-0)(2-0.5)(3.500) + (2-0)(2-0.5)(2-1)(-1.000)$$
$$= -2.0 + 6.5 + 10.5 - 3$$
$$= 12$$

This, and other interpolation problems, can also be solved with a spreadsheet. The Excel spreadsheet for this example is shown in Figure 7.1.

7.5 Lagrange's Interpolation Method

Lagrange's interpolation method uses the formula

$$f(x) = \frac{(x-x_1)(x-x_2)...(x-x_n)}{(x_0-x_1)(x_0-x_2)...(x_0-x_n)}f(x_0) + \frac{(x-x_0)(x-x_2)...(x-x_n)}{(x_1-x_0)(x_1-x_2)...(x_1-x_n)}f(x_1)$$
$$+ \frac{(x-x_0)(x-x_1)...(x-x_{n-1})}{(x_n-x_0)(x_n-x_2)...(x_n-x_{n-1})}f(x_n)$$

(7.53)

and, like Newton's divided difference method, has the advantage that the values $x_0, x_1, x_2, ..., x_n$ need not be equally spaced or taken in consecutive order.

* We chose this as our starting value so that $f(2)$ will be between $f(1)$ and $f(2.5)$

		1st divided	*2nd divided*	*3rd divided*		
		difference	*difference*	*difference*		

Interpolation with Newton's Divided Difference Formula

$f(x) = f(x_0)+(x-x_0)f(x_0,x_1)+(x-x_0)(x-x_1)f(x_0,x_1,x_2)+(x-x_0)(x-x_1)(x-x_2)f(x_0,x_1,x_2,x_3)$

In this example, we want to evaluate $f(x)$ at x= 2

x	f(x)	$f(x_0, x_1)$	$f(x_0, x_1, x_2)$	$f(x_0,x_1,x_2,x_3)$		
-1.00	3.000					
		-5.000				
0.00	-2.000		5.500			
		3.250		-1.000		
0.50	-0.375		3.500			
		6.750		-1.000		
1.00	3.000		1.000			
		8.750		-1.000		
2.50	16.125		-1.500			
		5.750				
3.00	19.000					

We use the above formula with starting value > 0.00

f(2)=B12+(E3-E18)*C13+(E3-E18)*(E3-A14)*D14+(E3-E18)*(E3-A14)*(E3-A16)*E15

or f(2)= 12.00

The plot below verifies that our answer is correct

-1.000	3.000
0.000	-2.000
0.500	-0.375
1.000	3.000
2.500	16.125
3.000	19.000

Figure 7.1. Spreadsheet for Example 7.6

Example 7.7

Repeat Example 7.6 using Lagrange's interpolation formula.

Solution:

All computations appear in the spreadsheet of Figure 7.2 where we have used relation (7.53).

	A	B	C	D	E	F	G	H	I	J	K	L
1	Lagrange's Interpolation Method											
2										Numer.	Denom.	Division
3	Interpol. at x=			2						Partial	Partial	of Partial
4										Prods	Prods	Prods
5		x	f(x)	$x-x_1$	$x-x_2$	$x-x_3$	$x-x_4$	$x-x_5$	$f(x_0)$			
6	x_0	-1.00	3.000	2.000	1.500	1.000	-0.500	-1.000	3.000	4.500		
7	x_1	0.00	-2.000	x_0-x_1	x_0-x_2	x_0-x_3	x_0-x_4	x_0-x_5				-0.107
8	x_2	0.50	-0.375	-1.000	-1.500	-2.000	-3.500	-4.000			-42.000	
9	x_3	1.00	3.000	$x-x_0$	$x-x_2$	$x-x_3$	$x-x_4$	$x-x_5$	$f(x_1)$			
10	x_4	2.50	16.125	3.000	1.500	1.000	-0.500	-1.000	-2.000	-4.500		
11	x_5	3.00	19.000	x_1-x_0	x_1-x_2	x_1-x_3	x_1-x_4	x_1-x_5				-1.200
12				1.000	-0.500	-1.000	-2.500	-3.000			3.750	
13				$x-x_0$	$x-x_1$	$x-x_3$	$x-x_4$	$x-x_5$	$f(x_2)$			
14				3.000	2.000	1.000	-0.500	-1.000	-0.375	-1.125		
15				x_2-x_0	x_2-x_1	x_2-x_3	x_2-x_4	x_2-x_5				0.600
16				1.500	0.500	-0.500	-2.000	-2.500			-1.875	
17				$x-x_0$	$x-x_1$	$x-x_2$	$x-x_4$	$x-x_5$	$f(x_3)$			
18				3.000	2.000	1.500	-0.500	-1.000	3.000	13.500		
19				x_3-x_0	x_3-x_1	x_3-x_2	x_3-x_4	x_3-x_5				4.500
20				2.000	1.000	0.500	-1.500	-2.000			3.000	
21				$x-x_0$	$x-x_1$	$x-x_2$	$x-x_3$	$x-x_5$	$f(x_4)$			
22				3.000	2.000	1.500	1.000	-1.000	16.125	-145.125		
23				x_4-x_0	x_4-x_1	x_4-x_2	x_4-x_3	x_4-x_5				11.057
24				3.500	2.500	2.000	1.500	-0.500			-13.125	
25				$x-x_0$	$x-x_1$	$x-x_2$	$x-x_3$	$x-x_4$	$f(x_5)$			
26				3.000	2.000	1.500	1.000	-0.500	19.000	-85.500		
27				x_5-x_0	x_5-x_1	x_5-x_2	x_5-x_3	x_5-x_4				-2.850
28				4.000	3.000	2.500	2.000	0.500			30.000	
29												
30										f(2)=	Sum=	12

Figure 7.2. Spreadsheet for Example 7.7

7.6 Gregory–Newton Forward Interpolation Method

This method uses the formula

$$f(x) = f_0 + r\Delta f_0 + \frac{r(r-1)}{2!}\Delta^2 f_0 + \frac{r(r-1)(r-2)}{3!}\Delta^3 f_0 + \dots \qquad (7.54)$$

where f_0 is the first value of the data set, Δf_0, $\Delta^2 f_0$, and $\Delta^3 f_0$ are the first, second, and third forward* differences respectively.

The variable r is the difference between an unknown point x and a known point x_1 divided by the interval h, that is,

$$r = \frac{(x - x_1)}{h} \qquad (7.55)$$

---

* This is an expression to indicate that we use the differences in a forward sequence, that is, the first entries on the columns where the differences appear.

The formula of (7.54) is valid only when the values $x_0, x_1, x_2, ..., x_n$ are equally spaced with interval h. It is used to interpolate values near the smaller values of x, that is, the values near the beginning of the given data set. The formula that we will study on the next section, is used to interpolate values near the larger values of x, that is, the values near the end of the given data set.

Example 7.8

Use the Gregory–Newton forward interpolation formula to compute $f(1.03)$ from the following data.

TABLE 7.12 Table for Example 7.8

x	1.00	1.05	1.10	1.15	1.20	1.25
y = f(x)	1.000000	1.257625	1.531000	1.820875	2.128000	2.453125

Solution:

We enter the given x and $f(x)$ values in a difference table; then, we compute the first, second, and third differences. *These are not divided differences* and therefore, we simply subtract the second value of $f(x)$ from the first, the third from the second, and so on, as shown in Table 7.13.

For this example,

$$f_0 = f(1.00) = 1.000000$$

$$h = x_1 - x_0 = 1.05 - 1.00 = 0.05$$

$$r = \frac{x - x_1}{h} = \frac{1.03 - 1.00}{0.05} = 0.60$$

(7.56)

and with the values shown in Table 7.13 and using (7.54), we obtain

$$f(1.03) = 1.000000 + (0.60) \cdot (0.257625) + \frac{(0.60) \cdot (0.60 - 1)}{2!}$$

$$+ \frac{(0.60) \cdot (0.60 - 1)(0.60 - 2)}{3!} \cdot (0.000750) = 1.152727$$

(7.57)

The spreadsheet of Figure 7.3 shows the layout and computations for this example.

TABLE 7.13 Difference table for Example 7.8

x	f(x)	1st Difference $f(x_0, x_1)$	2nd Difference $f(x_0, x_1, x_2)$	3rd Difference $f(x_0, x_1, x_2, x_3)$
1.00	1.000000			
		0.257625		
1.05	1.257625		0.015750	
		0.273375		0.000750
1.10	1.531000		0.016500	
		0.289875		0.000750
1.15	1.820875		0.017250	
		0.307125		0.000750
1.20	2.128000		0.018000	
		0.325125		
1.25	2.453125			

7.7 Gregory–Newton Backward Interpolation Method

This method uses the formula

$$f(x) = f_0 + r\Delta f_{-1} + \frac{r(r+1)}{2!}\Delta^2 f_{-2} + \frac{r(r+1)(r+2)}{3!}\Delta^3 f_{-3} + \dots \tag{7.58}$$

where f_0 is the first value of the data set, Δf_{-1}, $\Delta^2 f_{-2}$, and $\Delta^3 f_{-3}$ are the first, second and third backward differences, and

$$r = \frac{(x - x_1)}{h}$$

Expression (7.58) is valid only when the values $x_0, x_1, x_2, \dots, x_n$ are equally spaced with interval h. It is used to interpolate values near the end of the data set, that is, the larger values of x. Backward interpolation is an expression to indicate that we use the differences in a backward sequence, that is, the last entries on the columns where the differences appear.

Example 7.9

Use the Gregory–Newton backward interpolation formula to compute f(1.18) from the data set of Table 7.14.

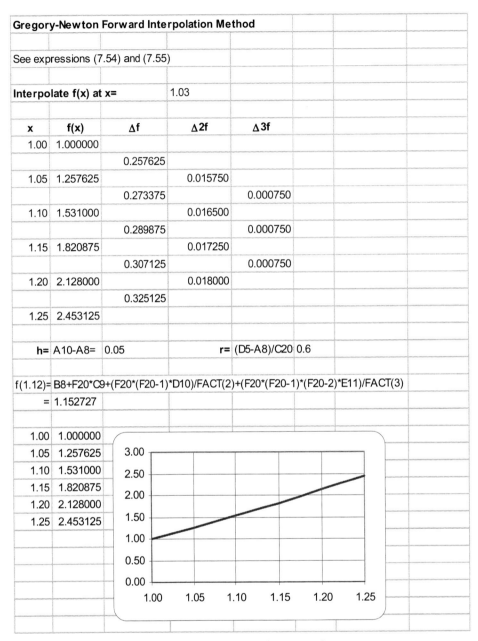

Gregory-Newton Forward Interpolation Method

See expressions (7.54) and (7.55)

Interpolate f(x) at x=   1.03

x	f(x)	Δf	Δ2f	Δ3f
1.00	1.000000			
		0.257625		
1.05	1.257625		0.015750	
		0.273375		0.000750
1.10	1.531000		0.016500	
		0.289875		0.000750
1.15	1.820875		0.017250	
		0.307125		0.000750
1.20	2.128000		0.018000	
		0.325125		
1.25	2.453125			

h= A10-A8= 0.05   r= (D5-A8)/C20 0.6

f(1.12)= B8+F20*C9+(F20*(F20-1)*D10)/FACT(2)+(F20*(F20-1)*(F20-2)*E11)/FACT(3)

= 1.152727

1.00	1.000000
1.05	1.257625
1.10	1.531000
1.15	1.820875
1.20	2.128000
1.25	2.453125

Figure 7.3. Spreadsheet for Example 7.8

TABLE 7.14 Data for Example 7.9

x	1.00	1.05	1.10	1.15	1.20	1.25
y = f(x)	1.000000	1.257625	1.531000	1.820875	2.128000	2.453125

Solution:

We arbitrarily choose $f_0 = 2.128000$ as our starting point since $f(1.18)$ lies between $f(1.15)$ and $f(1.20)$. Then,

$$h = 1.20 - 1.15 = 0.05$$

and

$$r = (x - x_1)/h = (1.18 - 1.20)/0.05 = -0.4$$

Now, by (7.58) we have:

$$f(1.18) = 2.128 + (-0.4)(0.307125) + \frac{(-0.4)(-0.4+1)}{2!}(0.01725)$$
$$+ \frac{(-0.4)(-0.4+1)(-0.4+2)}{3!}(0.00075) = 2.003032$$

The computations were made with the spreadsheet of Figure 7.4.

If the increments in x values are small, we can use the Excel **VLOOKUP** function to perform interpolation. The syntax of this function is as follows.

VLOOKUP(lookup_value, table_array, col_index_num, range lookup)

where:

lookup_value is the value being searched in the first column of the lookup table

table_array are the columns forming a rectangular range or array

col_index_num is the column where the answer will be found

range lookup is a logical value (**TRUE** or **FALSE**) that specifies whether we require **VLOOKUP** to find an exact or an approximate match. If **TRUE** is omitted, an approximate match is returned. In other words, if an exact match is not found, the next largest value that is less than the *lookup_value* is returned. If **FALSE** is specified, **VLOOKUP** will attempt to find an exact match, and if one is not found, the error value **#N/A** will be returned.

A sample spreadsheet is shown in Figure 7.5 where the values of x extend from –5 to +5 volts. Only a partial table is shown.

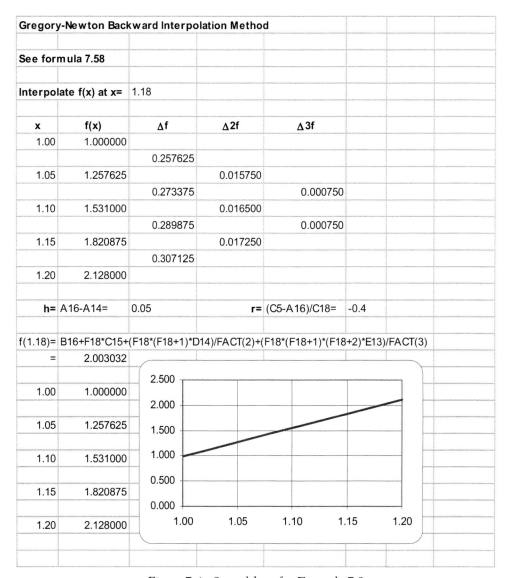

Gregory-Newton Backward Interpolation Method				
See formula 7.58				
Interpolate f(x) at x=	1.18			
x	**f(x)**	**Δf**	**Δ2f**	**Δ3f**
1.00	1.000000			
		0.257625		
1.05	1.257625		0.015750	
		0.273375		0.000750
1.10	1.531000		0.016500	
		0.289875		0.000750
1.15	1.820875		0.017250	
		0.307125		
1.20	2.128000			
h= A16-A14=	0.05		**r=** (C5-A16)/C18=	-0.4
f(1.18)= B16+F18*C15+(F18*(F18+1)*D14)/FACT(2)+(F18*(F18+1)*(F18+2)*E13)/FACT(3)				
=	2.003032			
1.00	1.000000			
1.05	1.257625			
1.10	1.531000			
1.15	1.820875			
1.20	2.128000			

Figure 7.4. Spreadsheet for Example 7.9

7.8 Interpolation with MATLAB

MATLAB has several functions that perform interpolation of data. We will study the following:

1. **interp1(x,y,x$_i$)** performs one dimensional interpolation where x and y are related as $y = f(x)$ and x_i is some value for which we want to find $y(x_i)$ by linear interpolation, i.e., "table lookup". This command will search the x vector to find two consecutive entries between which the desired value falls. It then performs linear interpolation to find the corresponding value of y. To obtain a correct result, the components of the x vector must be monotonic, that is, either in ascending or descending order.

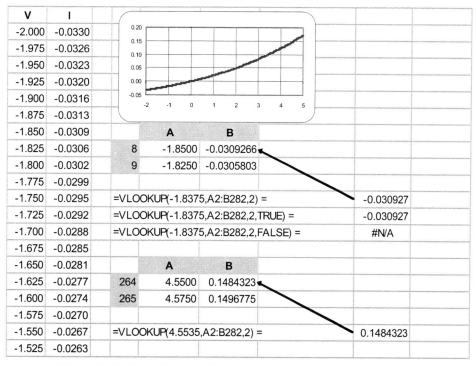

Figure 7.5. Using the Excel VLOOKUP function for interpolation

2. **interp1(x,y,x$_i$,'method')** performs the same operation as **interp1(x,y,x$_i$)** where the string **method** allows us to specify one of the methods listed below.

   nearest – nearest neighbor interpolation

   linear – linear interpolation; this is the default interpolation

   spline – cubic spline interpolation; this does also extrapolation

   cubic – cubic interpolation; this requires equidistant values of x

3. **interp2(x,y,z,x$_i$,y$_i$)** is similar to **interp1(x,y,x$_i$)** but performs two dimensional interpolation;

4. **interp2(x,y,z,x$_i$,y$_i$,'method')** is similar to **interp1(x,y,x$_i$,'method')** but performs two dimensional interpolation. The default is **linear**. The **spline** method does not apply to two dimensional interpolation.

We will illustrate the applications of these functions with the examples below.

Example 7.10

The $i - v$ (current–voltage) relation of a non–linear electrical device is given by

$$i(t) = 0.1(e^{0.2v(t)} - 1) \qquad (7.59)$$

where v is in volts and i in milliamperes. Compute i for 30 data points of v within the interval $-(2 \leq v \leq 5)$, plot i versus v in this range, and using linear interpolation compute i when $v = 1.265$ volts.

Solution:

We are required to use 30 data points within the given range; accordingly, we will use the MATLAB **linspace(first_value, last_value, number_of_values)** command. The script below produces 30 values in volts, the corresponding values in milliamperes, and plots the data for this range. Then, we use the **interp1(x,y,x$_i$)** command to interpolate at the desired value.

```
% This script is for Example_7_10.m
% It computes the values of current (in milliamps) vs. voltage (volts)
% for a diode whose v–i characteristics are i=0.1(exp(0.2v)–1).
% We can use the MATLAB function 'interp1' to linearly interpolate
% the value of milliamps for any value of v within the specified interval.
%
v=linspace(–2, 5, 30);       % Specify 30 intervals in the –2<=v<=5 interval
a=0.1.* (exp(0.2 .* v)–1);   % We use "a" for current instead of "i" to avoid conflict
                             % with imaginary numbers
v_a=[v;a]';                  % Define "v_a" as a two–column matrix to display volts
                             % and amperes side–by–side.
plot(v,a); grid;
title('volt–ampere characteristics for a junction diode');
xlabel('voltage (volts)');
ylabel('current (milliamps)');
fprintf('   volts   milliamps \n');   % Heading of the two–column matrix
fprintf(' \n');
disp(v_a);                   % Display values of volts and amps below the heading
ma=interp1(v,a,1.265);       % Linear (default) interpolation
fprintf('current (in milliamps) @ v=1.265 is %2.4f \n', ma)
```

The data and the value obtained by interpolation are shown below.

```
    volts    milliamps

   -2.0000    -0.0330
   -1.7586    -0.0297
   -1.5172    -0.0262
   -1.2759    -0.0225
```

-1.0345	-0.0187
-0.7931	-0.0147
-0.5517	-0.0104
-0.3103	-0.0060
-0.0690	-0.0014
0.1724	0.0035
0.4138	0.0086
0.6552	0.0140
0.8966	0.0196
1.1379	0.0256
1.3793	0.0318
1.6207	0.0383
1.8621	0.0451
2.1034	0.0523
2.3448	0.0598
2.5862	0.0677
2.8276	0.0760
3.0690	0.0847
3.3103	0.0939
3.5517	0.1035
3.7931	0.1135
4.0345	0.1241
4.2759	0.1352
4.5172	0.1468
4.7586	0.1590
5.0000	0.1718

```
current (in milliamps) @ v=1.265 is 0.0288
```

The plot for this example is shown in Figure 7.6.

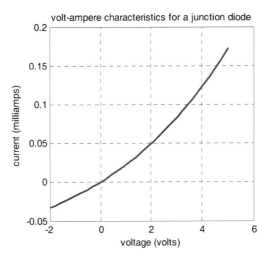

Figure 7.6. Plot for Example 7.10

Example 7.11

Plot the function

$$y = f(x) = \cos^5 x \qquad (7.60)$$

in the interval $0 \le x \le 2\pi$ with 120 intermediate values. Then, use the MATLAB **interp1(x,y,x$_i$,'method')** function to interpolate at $\pi/8$, $\pi/4$, $3\pi/5$, and $3\pi/7$. Compare the values obtained with the linear, cubic, and spline methods, with the analytical values.

Solution:

The script below plots (7.60) and produces the values of analytical values, for comparison with the linear, cubic, and spline interpolation methods.

```
% This is the script for Example_7_11
%
x=linspace(0,2*pi,120);        % We need these two
y=(cos(x)) .^ 5;               % statements for the plot
%
analytic=(cos([pi/8 pi/4 3*pi/5 3*pi/7]').^ 5);
%
plot(x,y); grid; title('y=cos^5(x)'); xlabel('x'); ylabel('y');
%
linear_int=interp1(x,y,[pi/8 pi/4 3*pi/5 3*pi/7]', 'linear');
% The label 'linear' on the right side of the above statement
% could be have been omitted since the default is linear
%
cubic_int=interp1(x,y,[pi/8 pi/4 3*pi/5 3*pi/7]', 'cubic');
```

```
%
spline_int=interp1(x,y,[pi/8 pi/4 3*pi/5 3*pi/7]','spline');
%
y=zeros(4,4);% Construct a 4 x 4 matrix of zeros
y(:,1)=analytic;                % 1st column of matrix
y(:,2)=linear_int;              % 2nd column of matrix
y(:,3)=cubic_int;               % 3rd column of matrix
y(:,4)=spline_int;              % 4th column of matrix
fprintf(' \n');                 % Insert line
fprintf('Analytic \t Linear Int \t Cubic Int \t Spline Int \n')
fprintf(' \n');
fprintf('%8.5f\t %8.5f\t %8.5f\t %8.5f\n',y')
fprintf(' \n');
%
% The statements below compute the percent error for the three
% interpolation methods as compared with the exact (analytic) values
%
error1=(linear_int-analytic).*100 ./ analytic;
error2=(cubic_int-analytic).*100 ./ analytic;
error3=(spline_int-analytic).*100 ./ analytic;
%
z=zeros(4,3);                   % Construct a 4 x 3 matrix of zeros
z(:,1)=error1;                  % 1st column of matrix
z(:,2)=error2;                  % 2nd column of matrix
z(:,3)=error3;                  % 3rd column of matrix
% fprintf(' \n');               % Insert line
disp('The percent errors for each interpolation method are:')
fprintf(' \n');
fprintf('Linear Int \t Cubic Int \t Spline Int \n')
fprintf(' \n');
fprintf('%8.5f\t %8.5f\t %8.5f\n',z')
fprintf(' \n');
```

The plot for the function of this example is shown in Figure 7.7.

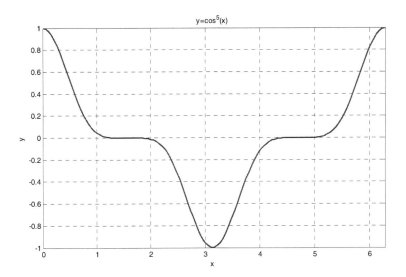

Figure 7.7. Plot the function of Example 7.11

The analytical and interpolated values are shown below for comparison.

```
Analytic   Linear Int   Cubic Int   Spline Int

 0.67310     0.67274     0.67311     0.67310
 0.17678     0.17718     0.17678     0.17678
-0.00282    -0.00296    -0.00281    -0.00282
 0.00055     0.00062     0.00054     0.00055
```

The percent errors for each interpolation method are:

```
Linear Int   Cubic Int   Spline Int

-0.05211      0.00184      0.00002
 0.22707     -0.00012      0.00011
 5.09681     -0.40465     -0.01027
13.27678     -0.64706     -0.07445
```

Example 7.12

For the impedance example of Section 1.7 in Chapter 1 whose script and plot are shown below, use the spline method of interpolation to find the magnitude of the impedance at $\omega = 792$ rad/s.

Solution:

% The file is Example_7_12.m
% It calculates and plots the impedance Z(w) versus radian frequency w.

```
%
% Use the following five statements to obtain |Z| versus radian frequency w
w=300:100:3000;
z=zeros(28,2);
z(:,1)=w';
z(:,2)=(10+(10.^4–j.*10.^6./w)./(10+j.*(0.1.*w–10.^5./w)))';
fprintf('%2.0f\t %10.3f\n',abs(z)')
%
w=[300 400 500 600 700 800 900 1000 1100 1200 1300 1400 1500....
            1600 1700 1800 1900 2000 2100 2200 2300....
            2400 2500 2600 2700 2800 2900 3000];
z=[39.339 52.789 71.104 97.665 140.437 222.182 436.056 1014.938...
            469.830 266.032 187.052 145.751 120.353...
            103.111 90.603 81.088 73.588 67.513 62.481...
            58.240 54.611 51.468 48.717 46.286 44.122...
            42.182 40.432 38.845];
semilogx(w,z); grid;
title('Magnitude of Impedance vs. Radian Frequency');
xlabel('w in rads/sec'); ylabel('|Z| in Ohms');
%
zi=interp1(w,z,792,'spline');
fprintf(' \n')
fprintf('Magnitude of Z at w=792 rad/s is %6.3f Ohms \n', zi)
fprintf(' \n')
```

The plot for the function of this example is shown in Figure 7.8.

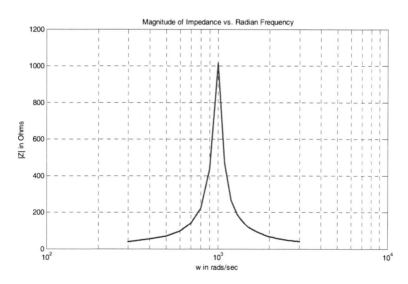

Figure 7.8. Plot for the function of Example 7.12

MATLAB interpolates the impedance at $\omega = 792$ rad/s and displays the following message:

```
Magnitude of Z at w=792 rad/s is 217.034 Ohms
```

Two–dimensional plots were briefly discussed in Chapter 1. For convenience, we will review the following commands which can be used for two–dimensional interpolation.

1. **mesh(Z)** – Plots the values in the matrix **Z** as height values above a rectangular grid, and connects adjacent points to form a mesh surface.

2. **[X,Y]=meshgrid(x,y)** – Generates interpolation arrays which contain all combinations of the **x** and **y** points which we specify. **X** and **Y** comprise a pair of matrices representing a rectangular grid of points in the $x - y$ plane. Using these points, we can form a function $z = f(x, y)$ where z is a matrix.

Example 7.13

Generate the plot of the function

$$Z = \frac{\sin R}{R} \qquad (7.61)$$

in three dimensions x, y, and z. This function is the equivalent of the function $y = \sin x / x$ in two dimensions. Here, R is a matrix that contains the distances from the origin to each point in the pair of $[X, Y]$ matrices that form a rectangular grid of points in the $x - y$ plane.

Solution:

The matrix R that contains the distances from the origin to each point in the pair of $[X, Y]$ matrices, is

$$R = \sqrt{X^2 + Y^2} \qquad (7.62)$$

We let the origin be at $(x_0, y_0) = (0, 0)$, and the plot in the intervals $-2\pi \leq x \leq 2\pi$ and $-2\pi \leq y \leq 2\pi$. Then, we write and execute the following MATLAB script.

```
% This is the script for Example_7_13
x=-2*pi: pi/24: 2*pi;          % Define interval in increments of pi/24
y=x;                           % y must have same number of points as x
[X,Y]=meshgrid(x,y);           % Create X and Y matrices
R=sqrt(X.^ 2 + Y.^ 2);         % Compute distances from origin (0,0) to x-y points
Z=sin(R)./ (R+eps);            % eps prevents division by zero
mesh(X,Y,Z);                   % Generate mesh plot for Z=sin(R)/R
xlabel('x'); ylabel('y'); zlabel('z');
```

title('Plot for the Three–dimensional sin(R) / R Function')

The plot for the function of this example is shown in Figure 7.9.

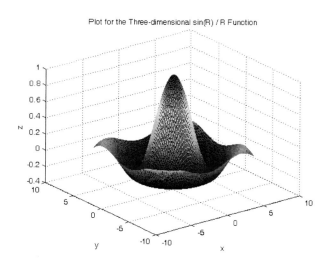

Plot for the Three-dimensional sin(R) / R Function

Figure 7.9. Plot for Example 7.13

Example 7.14

Generate the plot of the function

$$z = x^3 + y^3 - 3xy \tag{7.63}$$

in three dimensions x, y, and z. Use the cubic method to interpolate the value of z at $x = -1$ and $y = 2$.

Solution:

We let the origin be at $(x_0, y_0) = (0, 0)$, and the plot in the intervals $-10 \leq x \leq 10$ and $-10 \leq y \leq 10$. Then, we write and execute the following script.

```
% This is the script for Example_7_14
x=-10: 0.25: 10;              % Define interval in increments of 0.25
y=x;                          % y must have same number of points as x
[X,Y]=meshgrid(x,y);         % Create X and Y matrices
Z=X.^3+Y.^3-3.*X.*Y;
mesh(X,Y,Z);                 % Generate mesh plot
xlabel('x'); ylabel('y'); zlabel('z');
title('Plot for the Function of Example 7.14');
z_int=interp2(X,Y,Z, -1,2,'cubic');
fprintf(' \n')
```

fprintf('Interpolated Value of z at x = –1 and y = 2 is z = %4.2f \n',z_int)
fprintf(' \n')

The plot for the function of this example is shown in Figure 7.10.

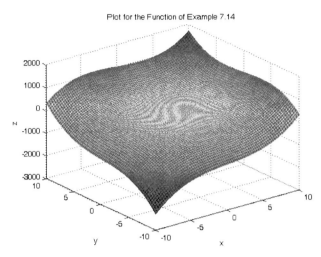

Figure 7.10. Plot for Example 7.14

```
Interpolated Value of z at x = -1 and y = 2 is z = 13.00
```

Example 7.15

A land surveyor measured and recorded the data below for a rectangular undeveloped land which lies approximately 500 meters above sea level.

500.08 500.15 500.05 500.08 500.14 500.13 500.09 500.15
500.12 500.01 500.11 500.18 500.15 500.12 500.05 500.15
500.13 500.12 500.09 500.11 500.11 500.05 500.15 500.02
500.09 500.17 500.17 500.14 500.16 500.09 500.02 500.11
500.08 500.09 500.13 500.18 500.14 500.14 500.14 500.15
500.15 500.10 500.11 500.11 500.12 500.13 500.14 500.12
500.17 500.12 500.13 500.18 500.13 500.15 500.17 500.11
500.13 500.14 500.13 500.09 500.14 500.16 500.17 500.14
500.15 500.09 500.14 500.18 500.17 500.08 500.13 500.09
500.12 500.15 500.14 500.01 500.16 500.12 500.11 500.10
500.02 500.19 500.01 500.08 500.12 500.02 500.16 500.12
500.19 500.21 500.17 500.03 500.17 500.09 500.14 500.17

This rectangular land parcel is 175 meters wide and 275 meters deep. The measurements shown above were made at points 25 meters apart.

a. Denoting the width as the $x-axis$, the depth as the $y-axis$ and the height as the $z-axis$, plot the given data to form a rectangular grid.

b. Interpolate the value of z at x = 108 m, and y = 177 m.

c. Compute the maximum height and its location on the $x-y$ plane.

Solution:

The MATLAB script and plot are shown below and explanations are provided with comment statements.

```
% This script is for Example_7_15
%
x=0: 25: 175;  % x-axis varies across the rows of z
y=0: 25: 275; % y-axis varies down the columns of z
z=[500.08 500.15 500.05 500.08 500.14 500.13 500.09 500.15;
   500.12 500.01 500.11 500.18 500.15 500.12 500.05 500.15;
   500.13 500.12 500.09 500.11 500.11 500.05 500.15 500.02;
   500.09 500.17 500.17 500.14 500.16 500.09 500.02 500.11;
   500.08 500.09 500.13 500.18 500.14 500.14 500.14 500.15;
   500.15 500.10 500.11 500.11 500.12 500.13 500.14 500.12;
   500.17 500.12 500.13 500.18 500.13 500.15 500.17 500.11;
   500.13 500.14 500.13 500.09 500.14 500.16 500.17 500.14;
   500.15 500.09 500.14 500.18 500.17 500.08 500.13 500.09;
   500.12 500.15 500.14 500.01 500.16 500.12 500.11 500.10;
   500.02 500.19 500.01 500.08 500.12 500.02 500.16 500.12;
   500.19 500.21 500.17 500.03 500.17 500.09 500.14 500.17];
%
mesh(x,y,z); axis([0 175  0 275  500 502]); grid off; box off
xlabel('x-axis, m'); ylabel('y-axis, m'); zlabel('Height, meters above sea level'); title('Parcel map')
% The pause command below stops execution of the program for 10 seconds
%  so that we can see the mesh plot
pause(10);
z_int=interp2(x,y,z,108,177,'cubic');
disp('Interpolated z is:'); z_int
[xx,yy]=meshgrid(x,y);
xi=0: 2.5: 175; % Make x-axis finer
% size(xi); % Returns a row vector containing the size of xi where the
% first element denotes the number of rows and the second is the number
% of columns. Here, size(xi) = 1  71
disp('size(xi)'); size(xi)
yi=0: 2.5: 275; % Make y-axis finer
disp('size(yi)'); size(yi)
[xxi,yyi]=meshgrid(xi,yi);               % Forms grid of all combinations of xi and yi
```

```
% size(xxi) = size(yyi) = size(zzi) = 111  71
disp('size(xxi)'); size(xxi); disp('size(yyi)'); size(yyi); disp('size(zzi)'); size(zzi)
size(xxi), size(yyi), size(zzi)
zzi=interp2(x,y,z,xxi,yyi,'cubic');   % Cubic interpolation – interpolates
% all combinations of xxi and yyi and constructs the matrix zzi
mesh(xxi,yyi,zzi);                     % Plot smoothed data
hold on;
[xx,yy]=meshgrid(x,y);                 % Grid with original data
plot3(xx,yy,z,'*k'); axis([0 175  0 275  500 503]); grid off; box off
xlabel('x–axis, m'); ylabel('y–axis, m'); zlabel('Height, meters above sea level');
title('Map of Rectangular Land Parcel')
hold off;
% max(x) returns the largest element of vector x
% max(A) returns a row vector which contains the maxima of the columns
% in matrix A. Likewise max(zzi) returns a row vector which contains the
% maxima of the columns in zzi. Observe that size(max(zzi)) = 1  71
% and size(max(max(zzi))) = 1  1
zmax=max(max(zzi))              % Estimates the peak of the terrain
% The 'find' function returns the subscripts where a relational expression
% is true. For Example,
% A=[a11 a12 a13; a21 a22 a23; a31 a32 a33] or
% A=[–1 0 3; 2 3 –4; –2 5 6];
% [i,j]=find(A>2)
% returns
% i =
%
%    2
%    3
%    1
%    3
%
%
% j =
%
%    2
%    2
%    3
%    3
% That is, the elements a22=3, a32=5, a13=3 and a33=6
%  satisfy the condition A>2
% The == operator compares two variables and returns ones when they
%  are equal, and zeros when they are not equal
%
[m,n]=find(zmax==zzi)
% m =
```

```
%
%   65
%
% n =
%
%   36
%
% that is, zmax is located at zzi = Z(65)(36)
%
% the x–cordinate is found from
xmax=xi(n)
% xmax =
%
%   1.7500 % Column 36; size(xi) = 1  71
% and the y–coordinate is found from
ymax=yi(m)
% ymax =
%
%   3.2000 % Row 65; size(yi) = 1  111
% Remember that i is the row index, j is the column index, and x–axis
% varies across the rows of z and y–axis varies down the columns of z

Interpolated z is:
z_int =
   500.1492
```

size(xi)

```
ans =
     1     71
```

size(yi)

```
ans =
     1    111
```

size(xxi)

```
ans =
   111     71
```

size(yyi)

```
ans =
   111     71
```

```
zzi=interp2(x,y,z,xxi,yyi,'cubic');   % Cubic interpolation – interpolates
% all combinations of xxi and yyi and constructs the matrix zzi
```

```
size(zzi)

ans =
    111      71
```

zmax=max(max(zzi)) % Estimates the peak of the terrain

```
zmax =
    500.2108

m =
    111
n=
      9
xmax =
     20

ymax =
    275
```

These values indicate that z_{max} = 500.21 where the x and y coordinates are x = 20 and y = 275. The interpolated value of z at x = 108 m and y = 177 m is z = 500.192. The plot is shown in Figure 7.11.

Map of Rectangular Land Parcel

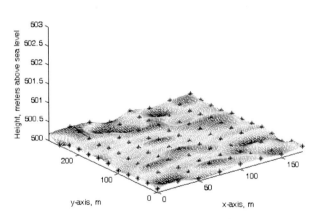

Figure 7.11. Plot for Example 7.15

7.9 Summary

- The first divided difference is defined as:

$$f(x_i, x_j) = \frac{f(x_i) - f(x_j)}{x_i - x_j}$$

where x_i and x_j are any two, not necessarily consecutive values of x, within an interval.

- Likewise, the *second divided difference* is defined as:

$$f(x_i, x_j, x_k) = \frac{f(x_i, x_j) - f(x_j, x_k)}{x_i - x_k}$$

and the third, fourth, and so on divided differences are defined similarly.

- If the values of x are equally spaced and the denominators are all the same, these values are referred to as the differences of the function.

- If the constant difference between successive values of x is h, the typical value of x_k is

$$x_k = x_0 + kh \quad \text{for} \quad k = \ldots, -2, -1, 0, 1, 2, \ldots$$

- We can now express the first differences are usually expressed in terms of the difference operator Δ as

$$\Delta f_k = f_{k+1} - f_k$$

- Likewise, the second differences are expressed as

$$\Delta^2 f_k = \Delta(\Delta f_k) = \Delta f_{k+1} - \Delta f_k$$

and, in general, for positive integer values of n

$$\Delta^n f_k = \Delta(\Delta^{n-1} f_k) = \Delta^{n-1} f_{k+1} - \Delta^{n-1} f_k$$

- The difference operator Δ obeys the law of exponents which states that

$$\Delta^m(\Delta^n f_k) = \Delta^{m+n} f_k$$

- The nth differences $\Delta^n f_k$ are found from the relation

$$\Delta^n f_k = f_{k+n} - n f_{k+n-1} + \frac{n(n-1)}{2!} f_{k+n-2} + \ldots + (-1)^{n-1} n f_{k+1} + (-1)^n f_k$$

For $k = 0$, $n = 1, 2, 3$ and 4, the above relation reduces to

$$\Delta f_0 = f_2 - f_1$$

$$\Delta^2 f_0 = f_2 - 2f_1 + f_0$$

$$\Delta^3 f_0 = f_3 - 3f_2 + 3f_1 - f_0$$

$$\Delta^4 f_0 = f_4 - 4f_3 + 6f_2 - 4f_1 + f_0$$

- As with derivatives, the nth differences of a polynomial of degree n are constant.

- The factorial polynomials are defined as

$$(x)^{(n)} = x(x-1)(x-2)\ldots(x-n+1)$$

and

$$(x)^{-(n)} = \frac{1}{(x-1)(x-2)\ldots(x+n)}$$

Using the difference operator Δ with the above relations we obtain

$$\Delta(x)^{(n)} = n(x)^{(n-1)}$$

and

$$\Delta(x)^{-(n)} = -n(x)^{-(n-1)}$$

These are very similar to differentiation of x^n and x^{-n}.

- We can express any algebraic polynomial $f_n(x)$ as a factorial polynomial $p_n(x)$. Then, in analogy with Maclaurin power series, we can express that polynomial as

$$p_n(x) = a_0 + a_1(x)^{(1)} + a_2(x)^{(2)} + \ldots + a_n(x)^{(n)}$$

where

$$a_j = \frac{\Delta^j p_n(0)}{j!} \quad \text{for} \quad j = 0, 1, 2, \ldots, n$$

- Factorial polynomials provide an easier method of constructing a difference table. The procedure is as follows:

1. We divide $p_n(x)$ by x to obtain a quotient $q_0(x)$ and a remainder r_0 which turns out to be the constant term a_0. Then, the factorial polynomial reduces to

$$p_n(x) = r_0 + x q_0(x)$$

2. We divide $q_0(x)$ by $(x - 1)$, to obtain a quotient $q_1(x)$ and a remainder r_1 which turns out to be the constant term a_1. Then,

$$q_0(x) = r_1 + (x - 1)q_1(x)$$

and by substitution we obtain

$$p_n(x) = r_0 + x[r_1 + (x - 1)q_1(x)] = r_0 + r_1(x)^{(1)} + x(x - 1)q_1(x)$$

3. We divide $q_1(x)$ by $(x - 2)$, to obtain a quotient $q_2(x)$ and a remainder r_2 which turns out to be the constant term a_2, and thus

$$q_1(x) = r_2 + (x - 2)q_2(x)$$

and by substitution we obtain

$$p_n(x) = r_0 + r_1(x)^{(1)} + x(x - 1)[r_2 + (x - 2)q_2(x)]$$
$$= r_0 + r_1(x)^{(1)} + r_2(x)^{(2)} + x(x - 1)(x - 2)q_2(x)$$

and in general,

$$p_n(x) = r_0 + r_1(x)^{(1)} + r_2(x)^{(2)} + \ldots + r_{n-1}(x)^{(n-1)} + r_n(x)^{(n)}$$

where

$$r_j = a_j = \frac{\Delta^j p_n(0)}{j!}$$

- The antidifference of a factorial polynomial is analogous to integration in elementary calculus. It is denoted as $\Delta^{-1}p_n(x)$, and it is computed from

$$\Delta^{-1}(x)^{(n)} = \frac{(x)^{(n+1)}}{(n+1)}$$

- Antidifferences are very useful in finding sums of series.

- The definite sum of $p_n(x)$ for the interval $a \le x \le a + (n - 1)h$ is

$$\sum_{x = \alpha}^{\alpha + (n-1)h} p_n(x) = p_n(\alpha) + p_n(\alpha + h) + p_n(\alpha + 2h) + \ldots + p_n[\alpha + (n - 1)h]$$

- In analogy with the fundamental theorem of integral calculus which states that

$$\int_a^b f(x)dx = f(b) - f(a)$$

we have the fundamental theorem of sum calculus which states that

$$\sum_{x=\alpha}^{\alpha+(n-1)h} p_n(x) = \Delta^{-1} p_n(x)\Big|_{\alpha}^{\alpha+nh}$$

- One important application of finite differences is interpolation.

- Newton's Divided Difference Interpolation Method uses the formula

$$f(x) = f(x_0) + (x - x_0)f(x_0, x_1) + (x - x_0)(x - x_1)f(x_0, x_1, x_2)$$
$$+ (x - x_0)(x - x_1)(x - x_2)f(x_0, x_1, x_2, x_3)$$

where $f(x_0, x_1)$, $f(x_0, x_1, x_2)$, and $f(x_0, x_1, x_2, x_3)$ are the first, second, and third divided differences respectively. This method has the advantage that the values $x_0, x_1, x_2, ..., x_n$ need not be equally spaced, or taken in consecutive order.

- Lagrange's Interpolation Method uses the formula

$$f(x) = \frac{(x - x_1)(x - x_2)...(x - x_n)}{(x_0 - x_1)(x_0 - x_2)...(x_0 - x_n)} f(x_0) + \frac{(x - x_0)(x - x_2)...(x - x_n)}{(x_1 - x_0)(x_1 - x_2)...(x_1 - x_n)} f(x_1)$$
$$+ \frac{(x - x_0)(x - x_1)...(x - x_{n-1})}{(x_n - x_0)(x_n - x_2)...(x_n - x_{n-1})} f(x_n)$$

and, like Newton's divided difference method, has the advantage that the values $x_0, x_1, x_2, ..., x_n$ need not be equally spaced or taken in consecutive order.

- The Gregory–Newton Forward Interpolation method uses the formula

$$f(x) = f_0 + r\Delta f_0 + \frac{r(r-1)}{2!}\Delta^2 f_0 + \frac{r(r-1)(r-2)}{3!}\Delta^3 f_0 + ...$$

where f_0 is the first value of the data set, Δf_0, $\Delta^2 f_0$, and $\Delta^3 f_0$ are the first, second, and third forward differences respectively. The variable r is the difference between an unknown point x and a known point x_1 divided by the interval h, that is,

$$r = \frac{(x - x_1)}{h}$$

This formula is valid only when the values $x_0, x_1, x_2, ..., x_n$ are equally spaced with interval h. It is used to interpolate values near the smaller values of x, that is, the values near the beginning of the given data set, hence the name forward interpolation.

- The Gregory–Newton Backward Interpolation method uses the formula

$$f(x) = f_0 + r\Delta f_{-1} + \frac{r(r+1)}{2!}\Delta^2 f_{-2} + \frac{r(r+1)(r+2)}{3!}\Delta^3 f_{-3} + \dots$$

where f_0 is the first value of the data set, Δf_{-1}, $\Delta^2 f_{-2}$, and $\Delta^3 f_{-3}$ are the first, second and third backward differences, and

$$r = \frac{(x - x_1)}{h}$$

This formula is valid only when the values $x_0, x_1, x_2, \dots, x_n$ are equally spaced with interval h. It is used to interpolate values near the end of the data set, that is, the larger values of x. Backward interpolation is an expression to indicate that we use the differences in a backward sequence, that is, the last entries on the columns where the differences appear.

- If the increments in x values are small, we can use the Excel **VLOOKUP** function to perform interpolation.

- We can perform interpolation to verify our results with the MATLAB functions **interp1(x,y,x$_i$), interp1(x,y,x$_i$,'method')** where **method** allows us to specify **nearest** (nearest neighbor interpolation), **linear** (linear interpolation, the default interpolation), **spline** (cubic spline interpolation which does also extrapolation), **cubic** (cubic interpolation which requires equidistant values of x), and **interp2(x,y,z,x$_i$,y$_i$)** which is similar to **interp1(x,y,x$_i$)** but performs two dimensional interpolation;

7.10 Exercises

1. Express the given polynomial $f(x)$ below as a factorial polynomial $p(x)$, calculate the leading differences, and then construct the difference table with $h = 1$.

$$f(x) = x^5 - 2x^4 + 4x^3 - x + 6$$

2. Use the data of the table below and the appropriate (forward or backward) Gregory–Newton formula, to compute:

a. $\sqrt{50.2}$

b. $\sqrt{55.9}$

x	50	51	52	53	54	55	56
$\sqrt{x}$	7.071	7.141	7.211	7.280	7.348	7.416	7.483

3. Use the data of the table below and Newton's divided difference formula to compute:

a. $f(1.3)$

b. $f(1.95)$

x	1.1	1.2	1.5	1.7	1.8	2.0
y=f(x)	1.112	1.219	1.636	2.054	2.323	3.011

7.11 Solutions to End–of–Chapter Exercises

1.

$$f(x) = x^5 - 2x^4 + 4x^3 - x + 6$$

The highest power of the given polynomial $f(x)$ is 5, we must evaluate the remainders r_0, r_1, r_2, r_3, r_4 and r_5; then, we will use (7.28), repeated below, to determine $p_n(x)$.

$$p_n(x) = r_0 + r_1(x)^{(1)} + r_2(x)^{(2)} + \dots + r_{n-1}(x)^{(n-1)} + r_n(x)^{(n)}$$

We can compute the remainders by long division but, for convenience, we will use the MAT-LAB **deconv(p,q)** function which divides the polynomial **p** by **q**.

The MATLAB script is as follows:

```
px=[1 -2  4  0 -1  6];        % Coefficients of given polynomial
d0=[1  0];                    % Coefficients of first divisor, i.e, x
[q0,r0]=deconv(px,d0)         % Computation of first quotient and remainder
d1=[1 -1];                    % Coefficients of second divisor, i.e, x-1
[q1,r1]=deconv(q0,d1)         % Computation of second quotient and remainder
d2=[1 -2];                    % Coefficients of third divisor, i.e, x-2
[q2,r2]=deconv(q1,d2)         % Computation of third quotient and remainder
d3=[1 -3];                    % Coefficients of fourth divisor, i.e, x-3
[q3,r3]=deconv(q2,d3)         % Computation of fourth quotient and remainder
d4=[1 -4];                    % Coefficients of fifth divisor, i.e, x-4
[q4,r4]=deconv(q3,d4)         % Computation of fifth quotient and remainder
d5=[1 -5];                    % Coefficients of sixth (last) divisor, i.e, x-5
[q5,r5]=deconv(q4,d5)         % Computation of sixth (last) quotient and remainder
```

```
q0 =
      1     -2      4      0     -1

r0 =
      0      0      0      0      0      6

q1 =
      1     -1      3      3
r1 =
      0      0      0      0      2

q2 =
      1      1      5

r2 =
      0      0      0     13
```

```
q3 =
      1       4

r3 =
      0       0      17

q4 =
      1

r4 =
      0       8

q5 =
      0

r5 =
      1
```

Therefore, with reference to (7.28), the factorial polynomial is

$$p_n(x) = 6 + 2(x)^{(1)} + 13(x)^{(2)} + 17(x)^{(3)} + 8(x)^{(4)} + (x)^{(5)}$$

We will verify that $p_n(x)$ above is the same polynomial as the given $f_n(x)$ by expansion of the factorials using (7.12), i.e.,

$$(x)^{(n)} = x(x-1)(x-2)...(x-n+1)$$

with the MATLAB **collect('s_expr')** function.

syms x; px=collect((x*(x–1)*(x–2)*(x–3)*(x-4)+(8*x*(x–1)*(x–2)*(x-3))+(17*x*(x–1)*(x–2))+...

(13*x*(x–1))+2*x+6))

```
px =
x^5-2*x^4+4*x^3-x+6
```

We observe that this is the same algebraic polynomial as $f(x)$.

We will now compute the leading entries for the difference table using (7.30), i.e, $\Delta^j p_n(0) = j! r_j$ and $p_n(x)$ above

$$\Delta^0 p(0) = 0! \cdot 6 = 6 \qquad \Delta^1 p(0) = 1! \cdot 2 = 2 \qquad \Delta^2 p(0) = 2! \cdot 13 = 26$$

$$\Delta^3 p(0) = 3! \cdot 17 = 102 \qquad \Delta^4 p(0) = 4! \cdot 8 = 192 \qquad \Delta^5 p(0) = 5! \cdot 1 = 120$$

$$\Delta^6 p(0) = 6! \cdot 0 = 0$$

We enter these values in the appropriate spaces as shown on the table below.

x	p(x)	Δ	Δ²	Δ³	Δ⁴	Δ⁵	Δ⁶
	6						
		2					
			26				
				102			
					192		
						120	
							0

We obtain the remaining set of values by crisscross addition as shown on the table below.

x	p(x)	Δ	Δ²	Δ³	Δ⁴	Δ⁵	Δ⁶
	6						
		2					
	8		**26**				
		28		**102**			
	36		128		**192**		
		156		294		**120**	
	192		422		312		**0**
		578		606		120	
	770		1028		432		
		1606		1038			
	2376		2066				
		3672					
	6048						

2.

x	50	51	52	53	54	55	56
$\sqrt{x}$	7.071	7.141	7.211	7.280	7.348	7.416	7.483

a. We will use the differences in a forward sequence, that is, the first entries on the columns where the differences appear. This is because the value of $\sqrt{50.2}$ should be in the interval $50 \leq x \leq 51$. We enter the given x and $f(x)$ values in a difference table; then, we compute the first, second, and third differences. These are not divided differences and therefore, we simply subtract the second value of $f(x)$ from the first, the third from the second, and so on, as shown below.

x	f(x)	1st Difference $f(x_0, x_1)$	2nd Difference $f(x_0, x_1, x_2)$	3rd Difference $f(x_0, x_1, x_2, x_3)$
50	7.071			
		0.070		
51	7.141		0.000	
		0.070		-0.001
52	7.211		-0.001	
		0.069		0.000
53	7.280		-0.001	
		0.068		0.001
54	7.348		0.000	
		0.068		-0.001
55	7.416		-0.001	
		0.067		
56	7.483			

$$f_0 = f(50) = 7.071$$

$$h = x_1 - x_0 = 51 - 50 = 1$$

$$r = \frac{x - x_1}{h} = \frac{50.2 - 50.0}{1} = 0.20$$

and with these values, using (7.54), we obtain

$$f(50.2) = 7.071 + (0.20) \cdot (7.071) + \frac{(0.20) \cdot (0.20 - 1)}{2!}$$

$$+ \frac{(0.20) \cdot (0.20 - 1)(0.20 - 2)}{3!} \cdot (0.001) = 7.085$$

The spreadsheet below shows the layout and computations for Part (a).

Check with MATLAB:

```
x =[50 51 52 53 54 55 56];
fx=[7.071 7.141 7.211 7.280 7.348 7.416 7.483];
spline_interp=interp1(x,fx,[50.2]','spline'); fprintf('\n');...
fprintf('spline interpolation yields f(50.2) = \n'); disp(spline_interp)

spline interpolation yields f(50.2) =
  7.0849
```

Gregory-Newton Forward Interpolation Method for Exercise 7.2(a)

See expressions (7.54) and (7.55)

Interpolate f(x) at x=   50.2

x	f(x)	Δf	Δ2f	Δ3f	Δ4f	Δ5f	Δ6f
50.0	7.071						
		0.070					
51.0	7.141		0.000				
		0.070		-0.001			
52.0	7.211		-0.001		0.001		
		0.069		0.000		0.000	
53.0	7.280		-0.001		0.001		-0.003
		0.068		0.001		-0.003	
54.0	7.348		0.000		-0.002		
		0.068		-0.001			
55.0	7.416		-0.001				
		0.067					
56.0	7.483						

h= A10-A8=	1.00		**r=** (D5-A8)/C22=	0.2

f(50.2)= round(B8+F22*C9+(F20*(F20-1)*D10)/FACT(2)+(F20*(F20-1)*(F20-2)*E11)/FACT(3),3)

= 7.085

50	7.071
51	7.141
52	7.211
53	7.280
54	7.348
55	7.416
56	7.483

b. Since the value of $\sqrt{55.9}$ is very close to the last value in the given range, we will use the backward interpolation formula

$$f(x) = f_0 + r\Delta f_{-1} + \frac{r(r+1)}{2!}\Delta^2 f_{-2} + \frac{r(r+1)(r+2)}{3!}\Delta^3 f_{-3} + \ldots$$

where f_0 is the first value of the data set, Δf_{-1}, $\Delta^2 f_{-2}$, and $\Delta^3 f_{-3}$ are the first, second and third backward differences, and

$$r = \frac{(x - x_1)}{h}$$

We arbitrarily choose $f_0 = 7.483$ as our starting point since $f(55.9)$ lies between $f(55)$ and $f(56)$. Then,

$$h = 56 - 55 = 1$$

and

$$r = (x - x_1)/h = (55.9 - 56.0)/1 = -0.1$$

Now, by (7.58) we have:

$$f(55.9) = 7.483 + (-0.1)(0.070) + \frac{(-0.1)(-0.1 + 1)}{2!}(0.000)$$
$$+ \frac{(-0.1)(-0.1 + 1)(-0.1 + 2)}{3!}(-0.001) = 7.476$$

Check with MATLAB:

```
x =[50 51 52 53 54 55 56];
fx=[7.071 7.141 7.211 7.280 7.348 7.416 7.483];
spline_interp=interp1(x,fx,[55.9]','spline'); fprintf('\n');...
fprintf('spline interpolation yields f(55.9) = \n'); disp(spline_interp)
```

```
spline interpolation yields f(55.9) =
    7.4764
```

3.

x	1.1	1.2	1.5	1.7	1.8	2.0
y=f(x)	1.112	1.219	1.636	2.054	2.323	3.011

a. The first divided differences are:

$$\frac{1.219 - 1.112}{1.2 - 1.1} = 1.070 \qquad \frac{1.636 - 1.219}{1.5 - 1.2} = 1.390 \qquad \frac{2.054 - 1.636}{1.7 - 1.5} = 2.090$$

$$\frac{2.323 - 2.054}{1.8 - 1.7} = 2.690 \qquad \frac{3.011 - 2.323}{2.0 - 1.8} = 3.440$$

The second divided differences are:

$$\frac{1.390 - 1.070}{1.5 - 1.1} = 0.800 \qquad \frac{2.090 - 1.390}{1.7 - 1.2} = 1.400$$

$$\frac{2.690 - 2.090}{1.8 - 1.5} = 2.000 \qquad \frac{3.440 - 2.690}{2.0 - 1.7} = 2.500$$

and the third divided differences are:

$$\frac{1.400 - 0.800}{1.7 - 1.1} = 1.000 \qquad \frac{2.000 - 1.400}{1.8 - 1.2} = 1.000 \qquad \frac{2.500 - 2.000}{2.0 - 1.5} = 1.000$$

With these values, we construct the difference table below.

x	f(x)	1st Divided Difference $f(x_0, x_1)$	2nd Divided Difference $f(x_0, x_1, x_2)$	3rd Divided Difference $f(x_0, x_1, x_2, x_3)$
1.1	1.112			
		1.070		
1.2	1.219		0.800	
		1.390		1.000
1.5	1.636		1.400	
		2.090		1.000
1.7	2.054		2.000	
		2.690		1.000
1.8	2.323		2.500	
		3.440		
2.0	3.011			

To find $f(1.3)$. We start with $x_0 = 1.1$ and for x in (7.49), we use $x = 1.3$. Then,

$$f(x) = f(x_0) + (x - x_0)f(x_0, x_1) + (x - x_0)(x - x_1)f(x_0, x_1, x_2)$$
$$+ (x - x_0)(x - x_1)(x - x_2)f(x_0, x_1, x_2, x_3)$$

$$f(1.3) = 1.112 + (1.3 - 1.1)(1.07) + (1.3 - 1.1)(1.3 - 1.2)(1.4) + (1.3 - 1.1)(1.3 - 1.2)(1.3 - 1.5)(1)$$
$$= 1.112 + 0.214 + 0.028 - 0.004$$
$$= 1.350$$

b. To find $f(1.95)$ we start with $x_0 = 2.0$ and for x in (7.49), we use $x = 1.95$. Then,

$$f(x) = f(x_0) + (x - x_0)f(x_0, x_1) + (x - x_0)(x - x_1)f(x_0, x_1, x_2)$$
$$+ (x - x_0)(x - x_1)(x - x_2)f(x_0, x_1, x_2, x_3)$$

$$f(1.95) = 3.011 + (1.95 - 2)(3.44) + (1.95 - 2)(1.95 - 1.8)(2.5) + (1.95 - 2)(1.95 - 1.8)(1.95 - 1.7)(1)$$
$$= 3.011 - 0.172 - 0.019 - 0.002$$
$$= 2.818$$

The spreadsheet below verifies our calculated values.

x	f(x)	1st divided difference f(x0, x1)	2nd divided difference f(x0, x1, x2)	3rd divided difference f(x0,x1,x2,x3)		
1.1	1.112					
		1.070				
1.2	1.219		0.800			
		1.390		1.000		
1.5	1.636		1.400			
		2.090		1.000		
1.7	2.054		2.000			
		2.690		1.000		
1.8	2.323		2.500			
		3.440				
2.0	3.011					

x	f(x)
1.1	1.112
1.2	1.219
1.5	1.636
1.7	2.054
1.8	2.323
2	3.011

Check with MATLAB:

```
x=[ 1.1    1.2    1.5    1.7    1.8    2.0];
fx=[ 1.112 1.219 1.636 2.054 2.323 3.011];
spline_interp=interp1(x,fx,[1.3]','spline'); fprintf('\n');...
fprintf('spline interpolation value of f(1.3): \n\n'); disp(spline_interp)

spline interpolation value of f(1.3):
    1.3380

spline_interp=interp1(x,fx,[1.95]','spline'); fprintf('\n');...
fprintf('spline interpolation value of f(1.95): \n\n'); disp(spline_interp)

spline interpolation value of f(1.95):
    2.8184
```

Chapter 8

T his chapter is an introduction to regression and procedures for finding the best curve to fit a set of data. We will discuss linear and parabolic regression, and regression with power series approximations. We will illustrate their application with several examples.

8.1 Curve Fitting

Curve fitting is the process of finding equations to approximate straight lines and curves that best fit given sets of data. For example, for the data of Figure 8.1, we can use the equation of a straight line, that is,

$$y = mx + b \tag{8.1}$$

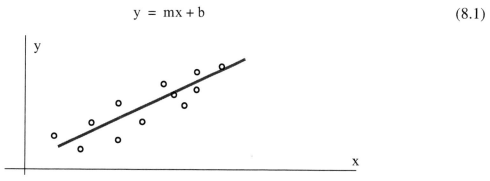

Figure 8.1. Straight line approximation.

For Figure 8.2, we can use the equation for the quadratic or parabolic curve of the form

$$y = ax^2 + bx + c \tag{8.2}$$

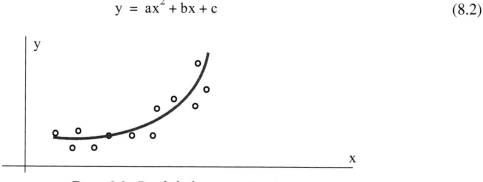

Figure 8.2. Parabolic line approximation

In finding the best line, we normally assume that the data, shown by the small circles in Figures 8.1 and 8.2, represent the independent variable x, and our task is to find the dependent variable y. This process is called *regression*.

Regression can be linear (straight line) or curved (quadratic, cubic, etc.) and it is not restricted to engineering applications. Investment corporations use regression analysis to compare a portfolio's past performance versus index figures. Financial analysts in large corporations use regression to forecast future costs, and the Census Bureau use it for population forecasting.

Obviously, we can find more than one straight line or curve to fit a set of given data, but we interested in finding the most suitable.

Let the distance of data point x_1 from the line be denoted as d_1, the distance of data point x_2 from the same line as d_2, and so on. The best fitting straight line or curve has the property that

$$d_1^2 + d_2^2 + \ldots + d_3^2 = \text{minimum} \tag{8.3}$$

and it is referred to as the *least-squares curve*. Thus, a straight line that satisfies (8.3) is called a *least squares line*. If it is a parabola, we call it a *least-squares parabola*.

8.2 Linear Regression

We perform linear regression with the *method of least squares*. With this method, we compute the coefficients m (slope) and b (y-intercept) of the straight line equation

$$y = mx + b \tag{8.4}$$

such that the sum of the squares of the errors will be minimum. We derive the values of m and b, that will make the equation of the straight line to best fit the observed data, as follows:

Let x and y be two related variables, and assume that corresponding to the values $x_1, x_2, x_3, \ldots, x_n$, we have observed the values $y_1, y_2, y_3, \ldots, y_n$. Now, let us suppose that we have plotted the values of y versus the corresponding values of x, and we have observed that the points $(x_1, y_1), (x_2, y_2), (x_3, y_3), \ldots, (x_n, y_n)$ approximate a straight line. We denote the straight line equations passing through these points as

$$\begin{aligned} y_1 &= mx_1 + b \\ y_2 &= mx_2 + b \\ y_3 &= mx_3 + b \\ &\ldots \\ y_n &= mx_n + b \end{aligned} \tag{8.5}$$

In (8.5), the slope m and y-intercept b are the same in all equations since we have assumed that all points lie close to one straight line. However, we need to determine the values of the unknowns m and b from all n equations; we will not obtain valid values for all points if we solve just two equations with two unknowns. [*]

The *error* (difference) between the observed value y_1, and the value that lies on the straight line, is $y_1 - (mx_1 + b)$. This difference could be positive or negative, depending on the position of the observed value, and the value at the point on the straight line. Likewise, the error between the observed value y_2 and the value that lies on the straight line is $y_2 - (mx_2 + b)$ and so on. The straight line that we choose must be a straight line such that the distances between the observed values, and the corresponding values on the straight line, will be minimum. This will be achieved if we use the magnitudes (absolute values) of the distances; if we were to combine positive and negative values, some may cancel each other and give us an erroneous sum of the distances. Accordingly, we find the sum of the squared distances between observed points and the points on the straight line. For this reason, this method is referred to as the *method of least squares*.

Let the sum of the squares of the errors be

$$\sum \text{squares} = [y_1 - (mx_1 + b)]^2 + [y_2 - (mx_2 + b)]^2 + \ldots \qquad (8.6)$$
$$+ [y_n - (mx_n + b)]^2$$

Since $\sum \text{squares}$ is a function of two variables m and b, to minimize (8.6) we must equate to zero its two partial derivatives with respect to m and b. Then,

$$\frac{\partial}{\partial m} \sum \text{squares} = -2x_1[y_1 - (mx_1 + b)] - 2x_2[y_2 - (mx_2 + b)] - \ldots \qquad (8.7)$$
$$-2x_n[y_n - (mx_n + b)] = 0$$

and

$$\frac{\partial}{\partial b} \sum \text{squares} = -2[y_1 - (mx_1 + b)] - 2[y_2 - (mx_2 + b)] - \ldots \qquad (8.8)$$
$$-2[y_n - (mx_n + b)] = 0$$

The second derivatives of (8.7) and (8.8) are positive and thus $\sum \text{squares}$ will have its minimum value.

Collecting like terms, and simplifying (8.7) and (8.8) we obtain

---

[*] *A linear system of independent equations that has more equations than unknowns is said to be* **overdetermined** *and no exact solution exists. On the contrary, a system that has more unknowns than equations is said to be* **underdetermined** *and these systems have infinite solutions.*

$$(\Sigma x^2)m + (\Sigma x)b = \Sigma xy$$
$$(\Sigma x)m + nb = \Sigma y$$

(8.9)

where

Σx = sum of the numbers x

Σy = sum of the numbers y

Σxy = sum of the numbers of the product xy

Σx^2 = sum of the numbers x squared

n = number of data x

We can solve the equations of (8.9) simultaneously by Cramer's rule, or with Excel, or with MATLAB using matrices.

With Cramer's rule, m and b are computed from

$$m = \frac{D_1}{\Delta} \qquad b = \frac{D_2}{\Delta}$$

(8.10)

where

$$\Delta = \begin{vmatrix} \Sigma x^2 & \Sigma x \\ \Sigma x & n \end{vmatrix} \qquad D_1 = \begin{vmatrix} \Sigma xy & \Sigma x \\ \Sigma y & n \end{vmatrix} \qquad D_2 = \begin{vmatrix} \Sigma x^2 & \Sigma xy \\ \Sigma x & \Sigma y \end{vmatrix}$$

(8.11)

Example 8.1

In a typical resistor, the resistance R in Ω (denoted as y in the equations above) increases with an increase in temperature T in °C (denoted as x). The temperature increments and the observed resistance values are shown in Table 8-1. Compute the straight line equation that best fits the observed data.

TABLE 8.1 Data for Example 8.1 - Resistance versus Temperature

T (°C)	x	0	10	20	30	40	50	60	70	80	90	100
R (Ω)	y	27.6	31.0	34.0	37	40	42.6	45.5	48.3	51.1	54	56.7

Solution:

There are 11 sets of data and thus n = 11. For convenience, we use the spreadsheet of Figure 8.3 where we enter the given values and we perform the computations using spreadsheet formulas.

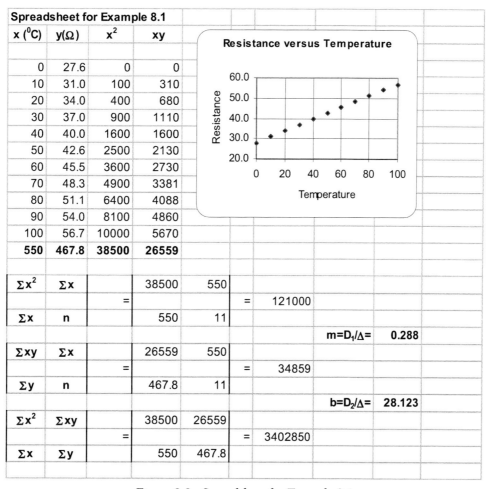

Spreadsheet for Example 8.1

x (^{0}C)	y(Ω)	x^2	xy
0	27.6	0	0
10	31.0	100	310
20	34.0	400	680
30	37.0	900	1110
40	40.0	1600	1600
50	42.6	2500	2130
60	45.5	3600	2730
70	48.3	4900	3381
80	51.1	6400	4088
90	54.0	8100	4860
100	56.7	10000	5670
550	**467.8**	**38500**	**26559**

Σx^2	Σx		38500	550			
		=			=	121000	
Σx	n		550	11			
						m=D_1/Δ=	0.288
Σxy	Σx		26559	550			
		=			=	34859	
Σy	n		467.8	11			
						b=D_2/Δ=	28.123
Σx^2	Σxy		38500	26559			
		=			=	3402850	
Σx	Σy		550	467.8			

Figure 8.3. Spreadsheet for Example 8.1

Accordingly, we enter the x (temperature) values in Column A, and y (the measured resistance corresponding to each temperature value) in Column B. Columns C and D show the x^2 and xy products. Then, we compute the sums so they can be used with (8.10) and (8.11). All work is shown on the spreadsheet of Figure 8.3. The values of m and b are shown in cells I20 and I24 respectively. Thus, the straight line equation that best fits the given data is

$$y = mx + b = 0.288x + 28.123 \tag{8.12}$$

We can use Excel's *Add Trendline* feature to produce quick answers to regression problems. We will illustrate the procedure with the following example.

Example 8.2

Repeat Example 8.1 using Excel's *Add Trendline* feature.

Solution:

We first enter the given data in columns A and B as shown on the spreadsheet of Figure 8.4.

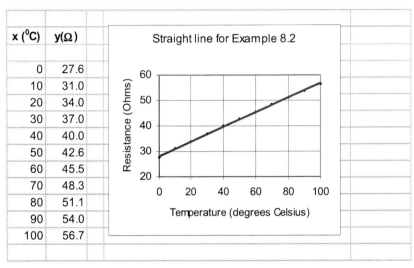

Figure 8.4. Plot of the straight line for Example 8.2

To produce the plot of Figure 8.4, we perform the following steps:

1. We click on the *Chart Wizard* icon. The displayed chart types appear on the *Standard Types* tab. We click on *XY (Scatter) Type*. On the *Chart sub-types* options, we click on the top (scatter) sub-type. Then, we click on *Next>Next> Next>Finish*, and we observe that the plot appears next to the data. We click on the *Series 1 block* inside the Chart box, and we press the *Delete* key to delete it.

2. To change the plot area from gray to white, we choose *Plot Area* from the taskbar below the main taskbar, we click on the small (with the hand) box, on the *Patterns* tab we click on the white box (below the selected gray box), and we click on *OK*. We observe now that the plot area is white. Next, we click anywhere on the perimeter of the Chart area, and observe six square handles (small black squares) around it. We click on *Chart* on the main taskbar, and on the *Gridlines* tab. Under the *Value (Y) axis*, we click on the *Major gridlines* box to deselect it.

3. We click on the *Titles* tab, and on the *Chart title* box, we type *Straight line for Example 8.2*, on the *Value X-axis*, we type *Temperature (degrees Celsius)*, and on the *Value Y-axis*, we type *Resistance (Ohms)*. We click anywhere on the x-axis to select it, and we click on the small (with the hand) box. We click on the *Scale* tab, we change the maximum value from 150 to 100, and we click *OK*. We click anywhere on the y-axis to select it, and we click on the small (with the hand) box. We click on the *Scale* tab, we change the minimum value from 0 to 20, we change the *Major Unit* to 10, and we click on *OK*.

4. To make the plot more presentable, we click anywhere on the perimeter of the Chart area, and we observe the six handles around it. We place the cursor near the center handle of the upper side of the graph, and when the two-directional arrow appears, we move it upwards by moving the mouse in that direction. We can also stretch (or shrink) the height of the Chart area by placing the cursor near the center handle of the lower side of the graph, and move it downwards with the mouse. Similarly, we can stretch or shrink the width of the plot to the left or to the right, by placing the cursor near the center handle of the left or right side of the Chart area.

5. We click anywhere on the perimeter of the *Chart* area to select it, and we click on *Chart* above the main taskbar. On the pull-down menu, we click on *Add Trendline*. On the *Type* tab, we click on the first (*Linear*), and we click on *OK*. We now observe that the points on the plot have been connected by a straight line.

We can also use Excel to compute and display the equation of the straight line. This feature will be illustrated in Example 8.4. The *Data Analysis Toolpack* in Excel includes the *Regression Analysis* tool which performs linear regression using the least squares method. It provides a wealth of information for statisticians, and contains several terms used in probability and statistics.

8.3 Parabolic Regression

We find the *least-squares parabola* that fits a set of sample points with

$$y = ax^2 + b + c \tag{8.13}$$

where the coefficients a, b, and c are found from

$$(\Sigma x^2)a + (\Sigma x)b + nc = \Sigma y$$
$$(\Sigma x^3)a + (\Sigma x^2)b + (\Sigma x)c = \Sigma xy \tag{8.14}$$
$$(\Sigma x^4)a + (\Sigma x^3)b + (\Sigma x^2)c = \Sigma x^2 y$$

where n = number of data points.

Example 8.3

Find the least–squares parabola for the data shown in Table 8.2.

TABLE 8.2 Data for Example 8.3

x	1.2	1.5	1.8	2.6	3.1	4.3	4.9	5.3
y	4.5	5.1	5.8	6.7	7.0	7.3	7.6	7.4
x	5.7	6.4	7.1	7.6	8.6	9.2	9.8	
y	7.2	6.9	6.6	5.1	4.5	3.4	2.7	

Solution:

We construct the spreadsheet of Figure 8.5, and from the data of Columns A and B, we compute the values shown in Columns C through G. The sum values are shown in Row 18, and from these we form the coefficients of the unknown a, b, and c.

	A	B	C	D	E	F	G
1	x	y	x^2	x^3	x^4	xy	x^2y
2	1.2	4.5	1.44	1.73	2.07	5.40	6.48
3	1.5	5.1	2.25	3.38	5.06	7.65	11.48
4	1.8	5.8	3.24	5.83	10.50	10.44	18.79
5	2.6	6.7	6.76	17.58	45.70	17.42	45.29
6	3.1	7.0	9.61	29.79	92.35	21.70	67.27
7	4.3	7.3	18.49	79.51	341.88	31.39	134.98
8	4.9	7.6	24.01	117.65	576.48	37.24	182.48
9	5.3	7.4	28.09	148.88	789.05	39.22	207.87
10	5.7	7.2	32.49	185.19	1055.60	41.04	233.93
11	6.4	6.9	40.96	262.14	1677.72	44.16	282.62
12	7.1	6.6	50.41	357.91	2541.17	46.86	332.71
13	7.6	5.1	57.76	438.98	3336.22	38.76	294.58
14	8.6	4.5	73.96	636.06	5470.08	38.70	332.82
15	9.2	3.4	84.64	778.69	7163.93	31.28	287.78
16	9.8	2.7	96.04	941.19	9223.68	26.46	259.31
17	$\Sigma x=$	$\Sigma y=$	$\Sigma x^2=$	$\Sigma x^3=$	$\Sigma x^4=$	$\Sigma xy=$	$\Sigma x^2y=$
18	79.1	87.8	530.15	4004.50	32331.49	437.72	2698.37

Figure 8.5. Spreadsheet for Example 8.3

By substitution into (8.14),

$$530.15a + 79.1b + 15c = 87.8$$
$$4004.50a + 530.15b + 79.1c = 437.72 \tag{8.15}$$
$$32331.49a + 4004.50b + 530.15c = 2698.37$$

We solve the equations of (8.15) with matrix inversion and multiplication, as shown in Figure 8.6. The procedure was presented in Chapter 4.

	A	B	C	D	E	F	G
1	Matrix Inversion and Matrix Multiplication for Example 8.3						
2							
3		530.15	79.10	15.00		$\Sigma y=$	87.80
4	A=	4004.50	530.15	79.10		$\Sigma xy=$	437.72
5		32331.49	4004.50	530.15		$\Sigma x^2 y=$	2698.37
6							
7		0.032	-0.016	0.002		a=	-0.20
8	$A^{-1}=$	-0.385	0.181	-0.016		b=	1.94
9		0.979	-0.385	0.032		c=	2.78

Figure 8.6. Spreadsheet for the solution of the equations of (8.15)

Therefore, the least–squares parabola is

$$y = -0.20x^2 + 1.94x + 2.78$$

The plot for this parabola is shown in Figure 8.7.

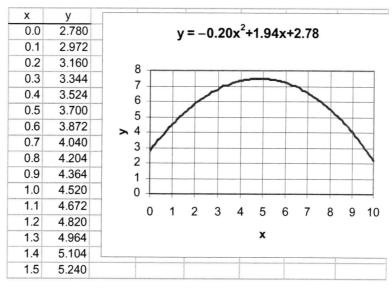

x	y
0.0	2.780
0.1	2.972
0.2	3.160
0.3	3.344
0.4	3.524
0.5	3.700
0.6	3.872
0.7	4.040
0.8	4.204
0.9	4.364
1.0	4.520
1.1	4.672
1.2	4.820
1.3	4.964
1.4	5.104
1.5	5.240

$$y = -0.20x^2 + 1.94x + 2.78$$

Figure 8.7. Parabola for Example 8.3

Example 8.4

The voltages (volts) shown on Table 8.3 were applied across the terminal of a non–linear device and the current ma (milliamps) values were observed and recorded. Use Excel's *Add Trendline* feature to derive a polynomial that best approximates the given data.

Solution:

We enter the given data on the spreadsheet of Figure 8.8 where, for brevity, only a partial list of the given data is shown. However, to obtain the plot, we need to enter all data in Columns A and B.

TABLE 8.3 Data for Example 8.4

Experimental Data										
Volts	0.00	0.25	0.50	0.75	1.00	1.25	1.50	1.75	2.00	2.25	2.50
ma	0.00	0.01	0.03	0.05	0.08	0.11	0.14	0.18	0.23	0.28	0.34
Volts	2.75	3.00	3.25	3.50	3.75	4.00	4.25	4.50	4.75	5.00	
ma	0.42	0.50	0.60	0.72	0.85	1.00	1.18	1.39	1.63	1.91	

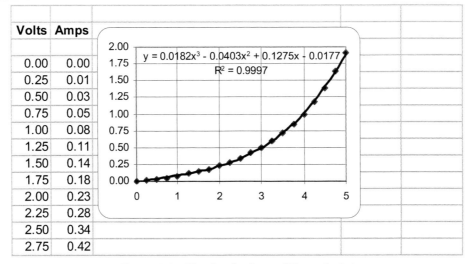

Figure 8.8. Plot for the data of Example 8.4

Following the steps of Example 8.2, we create the plot shown next to the data. Here, the smooth curve was chosen from the *Add trendline* feature, but we clicked on the *polynomial order 3* on the *Add trendline Type* tab. On the *Options tab*, we clicked on *Display equation on chart*, we clicked on *Display R squared value on chart*, and on OK. The quantity R^2 is a measure of the goodness of fit for a straight line or, as in this example, for parabolic regression. This is the *Pearson correlation coefficient* R; it is discussed in probability and statistics textbooks.[*]

---

[*] *It is also discussed in Mathematics for Business, Science, and Technology, ISBN 0-9709511-0-8.*

The correlation coefficient can vary from 0 to 1. When $R^2 \approx 0$, there is no relationship between the dependent y and independent x variables. When $R^2 \approx 1$, there is a nearly perfect relationship between these variables. Thus, the result of Example 8.4 indicates that there is a strong relationship between the variables x and y, that is, there is a nearly perfect fit between the cubic polynomial and the experimental data.

With MATLAB, regression is performed with the **polyfit(x,y,n)** command, where **x** and **y** are the coordinates of the data points, and **n** is the degree of the polynomial. Thus, if n = 1, MATLAB computes the best straight line approximation, that is, linear regression, and returns the coefficients m and b. If n = 2, it computes the best quadratic polynomial approximation and returns the coefficients of this polynomial. Likewise, if n = 3, it computes the best cubic polynomial approximation, and so on.

Let p denote the polynomial (linear, quadratic, cubic, or higher order) approximation that is computed with the MATLAB **polyfit(x,y,n)** function. Suppose we want to evaluate the polynomial p at one or more points. We can use the **polyval(p,x)** function to evaluate the polynomial. If **x** is a scalar, MATLAB returns the value of the polynomial at point **x**. If **x** is a row vector, the polynomial is evaluated for all values of the vector **x**.

***

Example 8.5

Repeat Example 8.1 using the MATLAB's **polyfit(x,y,n)** function. Use n = 1 to compute the best straight line approximation. Plot resistance R versus temperature T in the range $-10 \le T \le 110$ °C. Use also the **polyval(p,x)** command to evaluate the best line approximation **p** in the $0 \le T \le 100$ range in ten degrees increments, and compute the percent error (difference between the given values and the polynomial values).

Solution:

The following MATLAB script will do the computations and plot the data.

```
% This is the script for Example 8.5
%
T=[ 0   10   20  30   40   50    60    70    80    90  100];     % x-axis data
R=[27.6  31   34  37   40  42.6  45.5  48.3  51.1   54  56.7];    % y-axis data
axis([-10 110 20 60]);              % Establishes desired x and y axes limits
plot(T,R,'*b');                     % Display experimental (given) points with asterisk
                                    % and smoothed data with blue line
grid; title('R (Ohms) vs T (deg Celsius, n=1'); xlabel('T'); ylabel('R');
hold                                % Hold current plot so we can add other data
p=polyfit(T,R,1);       % Fits a first degree polynomial (straight line since n =1) and returns
                        % the coefficients m and b of the straight line equation y = mx + b
```

```
a=0: 10: 100;              % Define range to plot the polynomial
q=polyval(p,a);            % Compute p for each value of a
plot(a,q)                  % Plot the polynomial
                           % Display the coefficients m and b
fprintf('\n')              % Insert line
disp('Coefficients m and b are:'); fprintf('\n'); disp(p);
format bank                            % Two decimal place display will be sufficient
disp('Smoothed R values evaluated from straight line are:');
R_smoothed=polyval(p,T)    % Compute and display the values of the fitted
                           % polynomial at same points as given
                           % (experimental) values of R
R_exper = R                % Display the experimental values of R for comparison
                           % The statement below computes the percent error between
                           % the fitted polynomial and the experimental data
disp('% Error at points of given values is:')
                           % The percent error is computed with the following statement
error=(R_smoothed–R_exper).*100./R_exper
format short               % Return to default format
```

The plot for the data of this example is shown in Figure 8.9.

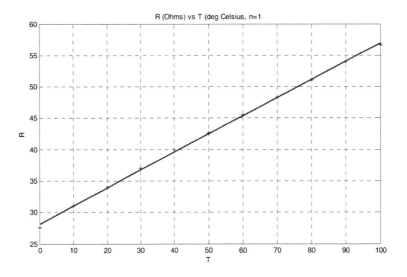

Figure 8.9. Plot for Example 8.5

MATLAB also displays the following data:

```
Coefficients m and b are:
    0.2881    28.1227

Smoothed R values evaluated from straight line are:

R_smoothed =
```

```
Columns 1 through 5
  28.12   31.00   33.88   36.77   39.65

Columns 6 through 10
  42.53   45.41   48.29   51.17   54.05

Column 11
  56.93
```

R_exper =
```
Columns 1 through 5
  27.60   31.00   34.00   37.00   40.00

Columns 6 through 10
  42.60   45.50   48.30   51.10   54.00

Column 11
       56.70
```

% Error at points of given values is:

error =
```
Columns 1 through 5
   1.89    0.01   -0.34   -0.63   -0.88

Columns 6 through 10
  -0.17   -0.20   -0.02    0.14    0.09

Column 11
   0.41
```

We can make the displayed data more presentable by displaying the values in four columns. The following MATLAB script will do that and will display the error in absolute values.

```
T= [0 10 20 30 40 50 60 70 80 90 100];                  % x–axis data
R=[27.6 31.0 34.0 37.0 40.0 42.6 45.5 48.3 51.1 54.0 56.7];   % y–axis data
p=polyfit(T,R,1); R_smoothed=polyval(p,T); R_exper = R;
error=(R_smoothed–R_exper).*100./R_exper;
y=zeros(11,4);                          % Construct an 11 x 4 matrix of zeros
y(:,1)=T';                              % 1st column of matrix
y(:,2)=R_exper';                        % 2nd column of matrix
y(:,3)=R_smoothed';                     % 3rd column of matrix
y(:,4)=abs(error)';                     % 4th column of matrix
fprintf(' \n');                         % Insert line
fprintf('Temp \t Exper R\t Smoothed R \t |Error| \n')
fprintf(' \n');                         % Insert line
fprintf('%3.0f\t                        %5.4f\t %5.4f\t %5.4f\n',y')
```

fprintf(' \n');   % Insert line

When this script is executed, MATLAB displays the following where the error is in percent.

```
Temp   Exper R  Smoothed R  |Error|

   0   27.6000   28.1227   1.8939
  10   31.0000   31.0036   0.0117
  20   34.0000   33.8845   0.3396
  30   37.0000   36.7655   0.6339
  40   40.0000   39.6464   0.8841
  50   42.6000   42.5273   0.1707
  60   45.5000   45.4082   0.2018
  70   48.3000   48.2891   0.0226
  80   51.1000   51.1700   0.1370
  90   54.0000   54.0509   0.0943
 100   56.7000   56.9318   0.4089
```

8.4 Regression with Power Series Approximations

In cases where the observed data deviate significantly from the points of a straight line, we can draw a smooth curve and compute the coefficients of a power series by approximating the derivatives di/dv with finite differences $\Delta i/\Delta v$. The following example illustrates the procedure.

Example 8.6

The voltages (volts) shown in Table 8.4, were applied across the terminal of a non–linear device, and the current ma (milliamps) values were observed and recorded. Use the power series method to derive a polynomial that best approximates the given data.

TABLE 8.4 Data for Example 8.6

Experimental Data										
Volts	0.00	0.25	0.50	0.75	1.00	1.25	1.50	1.75	2.00	2.25	2.50
ma	0.00	0.01	0.03	0.05	0.08	0.11	0.14	0.18	0.23	0.28	0.34
Volts	2.75	3.00	3.25	3.50	3.75	4.00	4.25	4.50	4.75	5.00	
ma	0.42	0.50	0.60	0.72	0.85	1.00	1.18	1.39	1.63	1.91	

Solution:

We begin by plotting the given data and we draw a smooth curve as shown in spreadsheet of Figure 8.10.

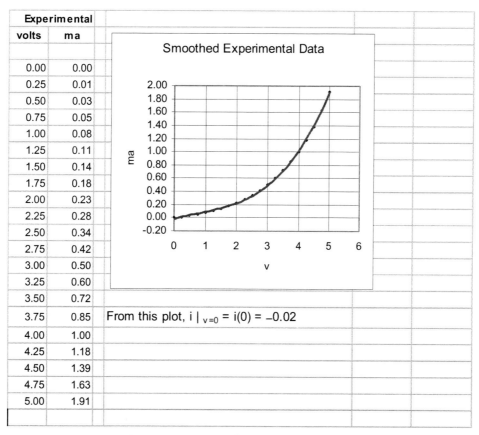

Experimental	
volts	ma
0.00	0.00
0.25	0.01
0.50	0.03
0.75	0.05
1.00	0.08
1.25	0.11
1.50	0.14
1.75	0.18
2.00	0.23
2.25	0.28
2.50	0.34
2.75	0.42
3.00	0.50
3.25	0.60
3.50	0.72
3.75	0.85
4.00	1.00
4.25	1.18
4.50	1.39
4.75	1.63
5.00	1.91

From this plot, $i \mid_{v=0} = i(0) = -0.02$

Figure 8.10. Spreadsheet for Example 8.6

Using the plot of Figure 8.10 we read the voltmeter reading and the corresponding smoothed ma readings and enter the values in Table 8.5.

TABLE 8.5 Data for the first derivative

Smoothed Data for Computation of $\Delta i / \Delta v$											
Volts	0.00	0.25	0.50	0.75	1.00	1.25	1.50	1.75	2.00	2.25	2.50
ma	−0.02	0.01	0.04	0.06	0.09	0.11	0.14	0.18	0.22	0.27	0.33
Volts	2.75	3.00	3.25	3.50	3.75	4.00	4.25	4.50	4.75	5.00	
ma	0.41	0.49	0.60	0.72	0.85	1.01	1.20	1.40	1.63	1.89	

Next, we compute Δ_i / Δ_v for $i = 1, 2, \ldots 20$

To facilitate the computations, we enter these values in the spreadsheet of Figure 8.11. In cell E4 we enter the formula =(B5-B4)/(A5-A4) and we copy it down to E5:E23.

	A	B	C	D	E
1		Smoothed			Computed
2	Volts	ma			Δi / Δv
3					
4	0.00	-0.02	Δi₁ / Δv₁=	(0.01-(-0.02))/(0.25-0.00)=	0.12
5	0.25	0.01	Δi₂ / Δv₂=	(0.04-(0.01))/(0.25-0.00)=	0.12
6	0.50	0.04	...	...	0.08
7	0.75	0.06	...	...	0.12
8	1.00	0.09	...	...	0.08
9	1.25	0.11	...	...	0.12
10	1.50	0.14	...	...	0.16
11	1.75	0.18	...	...	0.16
12	2.00	0.22	...	...	0.20
13	2.25	0.27	...	...	0.24
14	2.50	0.33	...	...	0.32
15	2.75	0.41	...	...	0.32
16	3.00	0.49	...	...	0.44
17	3.25	0.60	...	...	0.48
18	3.50	0.72	...	...	0.52
19	3.75	0.85	...	...	0.64
20	4.00	1.01	...	...	0.76
21	4.25	1.20	...	...	0.80
22	4.50	1.40	...	...	0.92
23	4.75	1.63	Δi₂₀ / Δv₂₀=	(1.89-(1.63))/(0.25-0.00)=	1.04
24	5.00	1.89			

Figure 8.11. Spreadsheet for computation of $\Delta i / \Delta v$ in Example 8.6

Next, we plot the computed values of $\Delta i / \Delta v$ versus v and again we smooth the data as shown in the spreadsheet of Figure 8.12. The smoothed values of the plot of Figure 8.12 are shown in Figure 8.13, and from these we compute $\Delta^2 i / \Delta^2 v$. Finally, we plot $\Delta^2 i / \Delta^2 v$ versus *volts* and again we smooth the data as shown in Figure 8.14.

Following the same procedure we can find higher order derivatives. However, for this example we will consider only the first three terms of the polynomial whose coefficients i, $\Delta i / \Delta v$ and $\Delta i^2 / \Delta v^2$, all three evaluated at v = 0 and are read from the plots. Therefore, the polynomial that best fits the given data is

$$i(v) = i(0) + i'(0) + \frac{1}{2!}i''(0) + ... = -0.02 + 0.12v + 0.5(-0.08)v^2$$

$$= -0.04v^2 + 0.12v - 0.02$$

(8.16)

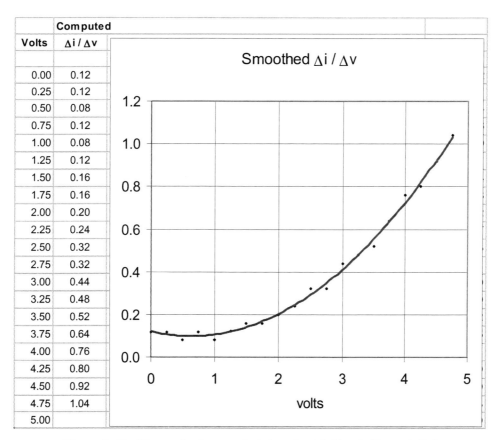

Volts	Computed $\Delta i / \Delta v$
0.00	0.12
0.25	0.12
0.50	0.08
0.75	0.12
1.00	0.08
1.25	0.12
1.50	0.16
1.75	0.16
2.00	0.20
2.25	0.24
2.50	0.32
2.75	0.32
3.00	0.44
3.25	0.48
3.50	0.52
3.75	0.64
4.00	0.76
4.25	0.80
4.50	0.92
4.75	1.04
5.00	

Figure 8.12. Plot to obtain smoothed data for $\Delta i / \Delta v$ in Example 8.6

Example 8.7

Repeat Example 8.4 using the MATLAB **polyfit(x,y,n)** function. Use n = 3 to compute the best cubic polynomial approximation.

Solution:

With MATLAB, higher degree polynomial regression is also performed with the **polyfit(x,y,n)** function, where $n \geq 2$. In this example we will use n = 3 as we did with Excel. The MATLAB script below computes the smoothed line and produces the plot shown on Figure 8.15.

	A	B	C	D	E
1		**Smoothed**			**Computed**
2	**Volts**	$\Delta i / \Delta v$			$\Delta i^2 / \Delta v^2$
3					
4	0.00	0.12	$\Delta i^2{}_1 / \Delta v^2{}_1=$	(0.11-0.12)/(0.25-0.00)=	-0.04
5	0.25	0.11	$\Delta i^2{}_2 / \Delta v^2{}_2=$	(0.10-011)/(0.25-0.00)=	-0.04
6	0.50	0.10	...	...	0.00
7	0.75	0.10	...	...	0.04
8	1.00	0.11	...	...	0.04
9	1.25	0.12	...	...	0.08
10	1.50	0.14	...	...	0.12
11	1.75	0.17	...	...	0.12
12	2.00	0.20	...	...	0.16
13	2.25	0.24	...	...	0.20
14	2.50	0.29	...	...	0.24
15	2.75	0.35	...	...	0.24
16	3.00	0.41	...	...	0.28
17	3.25	0.48	...	...	0.32
18	3.50	0.56	...	...	0.32
19	3.75	0.64	...	...	0.36
20	4.00	0.73	...	...	0.36
21	4.25	0.82	...	...	0.40
22	4.50	0.92	...	...	0.44
23	4.75	1.03	$\Delta i^2{}_{20} / \Delta v^2{}_{20}=$	(1.03-0.92)/(0.25-0.00)=	0.44
24	5.00				

Figure 8.13. Spreadsheet for computation of $\Delta i^2 / \Delta v^2$ in Example 8.6

```
v=[0 0.25 0.5 0.75 1 1.25 1.5 1.75 2 2.25 2.5 2.75 3....
    3.25 3.5 3.75 4 4.25 4.5 4.75 5];              % x–axis data
ma=[0 0.01 0.03 0.05 0.08 0.11 0.14 0.18 0.23 0.28....
    0.34 0.42 0.50 0.60 0.72 0.85 1.00 1.18  1.39 1.63 1.91];    % y–axis data
axis([–1 6 –1  2]);              % Establishes desired x and y axes limits
plot(v,ma,'+r'); grid            % Indicate data points with + and straight line in red
%
hold                             % hold current plot so we can add other data
disp('Polynomial coefficients in descending order are: ')
%
p=polyfit(v,ma,3)                % Fits a third degree polynomial to
                                 % the data and returns the coefficients
                                 % of the polynomial (cubic equation for
                                 % this example since n=3)
a=0:0.25:5;                      % Define range to plot the polynomial
q=polyval(p,a);% Calculate p at each value of a
% continued on the next page
```

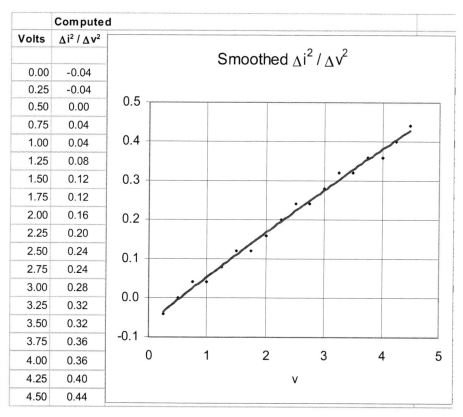

	Computed	
Volts	$\Delta i^2 / \Delta v^2$	
0.00	-0.04
0.25	-0.04
0.50	0.00
0.75	0.04
1.00	0.04
1.25	0.08
1.50	0.12
1.75	0.12
2.00	0.16
2.25	0.20
2.50	0.24
2.75	0.24
3.00	0.28
3.25	0.32
3.50	0.32
3.75	0.36
4.00	0.36
4.25	0.40
4.50	0.44

Figure 8.14. Plot to obtain smoothed data of $\Delta i^2 / \Delta v^2$ in Example 8.6

```
%
plot(a,q); title('milliamps vs volts, n=3');...
xlabel('v'); ylabel('ma')              % Plot the polynomial
% Display actual, smoothed and % error values
ma_smooth=polyval(p,v);                % Calculate the values of the fitted polynomial
ma_exper = ma;
% The following statement computes the percent error between the
%  smoothed polynomial and the experimental (given) data
error=(ma_smooth–ma_exper).*100./(ma_exper+eps);
%
y=zeros(21,4);                         % Construct a 21 x 4 matrix of zeros
y(:,1)=v';                             % 1st column of matrix
y(:,2)=ma_exper';                      % 2nd column of matrix
y(:,3)=ma_smooth';                     % 3rd column of matrix
y(:,4)=abs(error)';                    % 4th column of matrix
fprintf(' \n');                        % Insert line
% continued on the next page
```

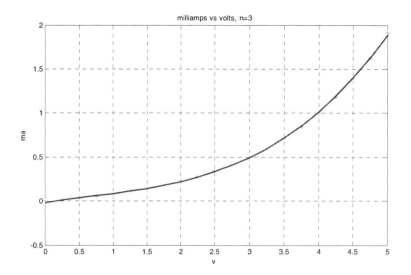

Figure 8.15. Plot for Example 8.7

```
fprintf('volts \t Exper ma\t Smoothed ma \t |%%Error| \n');
fprintf(' \n');
fprintf('%3.2f\t %7.5f\t %7.5f\t %7.5f\n',y')
fprintf(' \n');
```

MATLAB computes and displays the following data.

```
Polynomial coefficients in descending order are:
p =
    0.0182    -0.0403    0.1275    -0.0177

volts  Exper ma Smoothed ma  |%Error|

0.00   0.00000  -0.01766    7955257388080461.00000
0.25   0.01000   0.01197    19.74402
0.50   0.03000   0.03828    27.61614
0.75   0.05000   0.06298    25.95052
1.00   0.08000   0.08775     9.69226
1.25   0.11000   0.11433     3.93513
1.50   0.14000   0.14441     3.14852
1.75   0.18000   0.17970     0.16677
2.00   0.23000   0.22191     3.51632
2.25   0.28000   0.27275     2.58785
2.50   0.34000   0.33393     1.78451
2.75   0.42000   0.40716     3.05797
3.00   0.50000   0.49413     1.17324
3.25   0.60000   0.59657     0.57123
```

3.50	0.72000	0.71618	0.53040
3.75	0.85000	0.85467	0.54911
4.00	1.00000	1.01374	1.37399
4.25	1.18000	1.19511	1.28020
4.50	1.39000	1.40048	0.75362
4.75	1.63000	1.63155	0.09538
5.00	1.91000	1.89005	1.04436

We will conclude this chapter with one more example to illustrate the uses of the MATLAB **polyfit(x,y,n)** and **polyval(p,x)** functions.

Example 8.8

Use MATLAB to

a. plot the function

$$y = \sin x / x \tag{8.17}$$

in the interval $0 \le x \le 16$ radians.

b. compute $y(0)$, $y(2)$, $y(4)$, $y(6)$, $y(8)$, $y(10)$, $y(12)$, $y(14)$, $y(16)$

c. plot y versus x for these values and use the MATLAB **polyfit(x,y,n)** and **polyval(p,x)** functions to find a suitable polynomial that best fits the x and y data.

Solution:

a. The **fplot** function below plots $y = \sin x / x$. We added **eps** to avoid division by zero.

```
fplot('sin(x)./(x+eps)',[0  16  −0.5  1]); grid;...
title('(sinx)/x curve for x > 0')
```

The plot for the function of (8.17) is shown in Figure 8.16.

b. We use the script below to evaluate y at the specified points.

```
x=0:2:16; y=sin(x)./(x+eps)
p7=polyfit(x,y,7);              %  of x and y with fifth, seventh,
p9=polyfit(x,y,9); %  and ninth degree polynomials
% continued on the next page
```

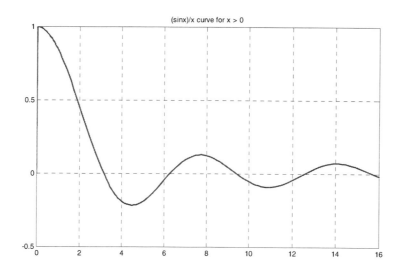

Figure 8.16. Plot for Example 8.8

c. The script for finding a suitable polynomial is listed below.

```
x=[0 2 4 6 8 10 12 14 16];
y=[1 0.4546 –0.1892 –0.0466 0.1237 –0.0544 –0.0447 0.0708 –0.0180];
p5=polyfit(x,y,5);              % Fits the polynomial to the data
x_span=0: 0.1: 16;             % Specifies values for x–axis
p5_pol=polyval(p5, x_span);  % Compute the polynomials for this range of x values.
p7_pol=polyval(p7, x_span); p9_pol=polyval(p9, x_span);
plot(x_span,p5_pol,'––', x_span,p7_pol,'–.', x_span,p9_pol,'–',x,y,'*');
% The following two statements establish coordinates for three legends
% in x and y directions to indicate degree of polynomials
x_ref=[2 5.3]; y_ref=[1.3,1.3];
hold on;
% The following are line legends for each curve
plot(x_ref,y_ref,'––',x_ref,y_ref–0.2,'–.',x_ref,y_ref–0.4,'–');
% The following are text legends for each curve
text(5.5,1.3, '5th degree polynomial');
text(5.5,1.1, '7th degree polynomial');
text(5.5,0.9, '9th degree polynomial'); grid;
hold off
format short e               % Exponential short format
disp('The coefficients of 5th order polynomial in descending order are:')
p5_coef=polyfit(x,y,5)
disp('The coefficients of 7th order polynomial in descending order are:')
p7_coef=polyfit(x,y,7)
disp('The coefficients of 9th order polynomial in descending order are:')
p9_coef=polyfit(x,y,9)
format short                 % We could just type format only since it is the default
```

The 5th, 7th, and 9th order polynomials are shown in Figure 8.17.

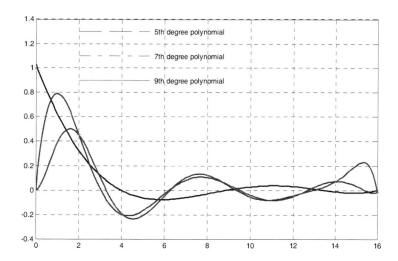

Figure 8.17. Polynomials for Example 8.8

The coefficients of the 5th, 7th, and 9th order polynomials are shown below.

The coefficients of 5th order polynomial in descending order are:

```
p5_coef =
6.5865e-006       -1.4318e-004       -1.5825e-003
   6.0067e-002       -4.6529e-001       1.0293e+000
```

The coefficients of 7th order polynomial in descending order are:

```
p7_coef =
  Columns 1 through 6
2.6483e-006    -1.6672e-004    4.1644e-003
 -5.2092e-002     3.3560e-001   -9.9165e-001
  Columns 7 through 8
  7.2508e-001     9.9965e-001
```

The coefficients of 9th order polynomial in descending order are:

```
p9_coef =
 Columns 1 through 6
  -1.0444e-008     1.1923e-006    -4.8340e-005
   9.5032e-004    -9.7650e-003     4.9437e-002

Columns 7 through 10
  -8.4572e-002    -1.0057e-001        0       1.0000e+000
```

8.5 Summary

- *Curve fitting* is the process of finding equations to approximate straight lines and curves that best fit given sets of data.

- *Regression* is the process of finding the dependent variable y from some data of the independent variable x. Regression can be linear (straight line) or curved (quadratic, cubic, etc.)

- The best fitting straight line or curve has the property that $d_1^2 + d_2^2 + \ldots + d_3^2 = \text{minimum}$ and it is referred to as the *least–squares curve*. A straight line that satisfies this property is called a *least squares line*. If it is a parabola, we call it a *least–squares parabola*.

- We perform linear regression with the *method of least squares*. With this method, we compute the coefficients m (slope) and b (y-intercept) of the straight line equation $y = mx + b$ such that the sum of the squares of the errors will be minimum. The values of m and b can be found from the relations

$$(\Sigma x^2)m + (\Sigma x)b = \Sigma xy$$
$$(\Sigma x)m + nb = \Sigma y$$

where

Σx = sum of the numbers x, Σy = sum of the numbers y

Σxy = sum of the numbers of the product xy, Σx^2 = sum of the numbers x squared

n = number of data x

The values of m and b are computed from

$$m = \frac{D_1}{\Delta} \qquad b = \frac{D_2}{\Delta}$$

where

$$\Delta = \begin{vmatrix} \Sigma x^2 & \Sigma x \\ \Sigma x & n \end{vmatrix} \qquad D_1 = \begin{vmatrix} \Sigma xy & \Sigma x \\ \Sigma y & n \end{vmatrix} \qquad D_2 = \begin{vmatrix} \Sigma x^2 & \Sigma xy \\ \Sigma x & \Sigma y \end{vmatrix}$$

- We find the *least–squares parabola* that fits a set of sample points with $y = ax^2 + b + c$ where the coefficients a, b, and c are found from

$$(\Sigma x^2)a + (\Sigma x)b + nc = \Sigma y$$
$$(\Sigma x^3)a + (\Sigma x^2)b + (\Sigma x)c = \Sigma xy$$
$$(\Sigma x^4)a + (\Sigma x^3)b + (\Sigma x^2)c = \Sigma x^2 y$$

where n = number of data points.

- With MATLAB, regression is performed with the **polyfit(x,y,n)** command, where **x** and **y** are the coordinates of the data points, and **n** is the degree of the polynomial. Thus, if n = 1, MATLAB computes the best straight line approximation, that is, linear regression, and returns the coefficients m and b. If n = 2, it computes the best quadratic polynomial approximation and returns the coefficients of this polynomial. Likewise, if n = 3, it computes the best cubic polynomial approximation, and so on.

- In cases where the observed data deviate significantly from the points of a straight line, we can draw a smooth curve and compute the coefficients of a power series by approximating the derivatives dy/dx with finite differences $\Delta y/\Delta x$.

8.6 Exercises

1. Consider the system of equations below derived from some experimental data.

$$2x + y = -1$$
$$x - 3y = -4$$
$$x + 4y = 3$$
$$3x - 2y = -6$$
$$-x + 2y = 3$$
$$x + 3y = 2$$

Using the relations (8.10) and (8.11), find the values of x and y that best fit this system of equations.

2. In a non–linear device, measurements yielded the following sets of values:

millivolts	100	120	140	160	180	200
milliamps	0.45	0.55	0.60	0.70	0.80	0.85

Use the procedure of Example 8.1 to compute the straight line equation that best fits the given data.

3. Repeat Exercise 2 above using Excel's *Trendline* feature.

4. Repeat Exercise 2 above using the MATLAB's **polyfit(x,y,n)** and **polyval(p,x)** functions.

5. A sales manager wishes to forecast sales for the next three years for a company that has been in business for the past 15 years. The sales during these years are shown on the next page.

Year	Sales
1	$9,149,548
2	13,048,745
3	19,147,687
4	28,873,127
5	39,163,784
6	54,545,369
7	72,456,782
8	89,547,216
9	112,642,574
10	130,456,321
11	148,678,983
12	176,453,837
13	207,547,632
14	206,147,352
15	204,456,987

Using Excel's *Trendline* feature, choose an appropriate polynomial to smooth the given data and using the polynomial found, compute the sales for the next three years. You may round the sales to the nearest thousand.

6. Repeat Exercise 5 above using the MATLAB **polyfit(x,y,n)** and **polyval(p,x)** functions.

8.7 Solutions to End–of–Chapter Exercises

1. We construct the spreadsheet below by entering the given values and computing the values from the formulas given.

	A	B	C	D	E	F	G	H	I	J
1	**Spreadsheet for Exercise 8.1**									
2										
3		a	b	c	a^2	ab	b^2	ac	bc	
4										
5		2	1	-1	4	2	1	-2	-1	
6		1	-3	-4	1	-3	9	-4	12	
7		1	4	3	1	4	16	3	12	
8		3	-2	-6	9	-6	4	-18	12	
9		-1	2	3	1	-2	4	-3	6	
10		1	3	2	1	3	9	2	6	
11										
12	Σ	7	5	-3	17	-2	43	-22	47	
13										
14		Σa^2	Σab		17	-2				
15	Δ			=			=	727		
16		Σab	Σb^2		-2	43				
17									x=D_1/Δ=	-1.172
18		Σac	Σab		-22	-2				
19	D_1			=			=	-852		
20		Σbc	Σb^2		47	43				
21									y=D_2/Δ=	1.039
22		Σa^2	Σac		17	-22				
23	D_2			=			=	755		
24		Σab	Σbc		-2	47				

Thus, $x = -1.172$ and $y = 1.039$

2. We construct the spreadsheet below by entering the given values and computing the values from the given formulas.

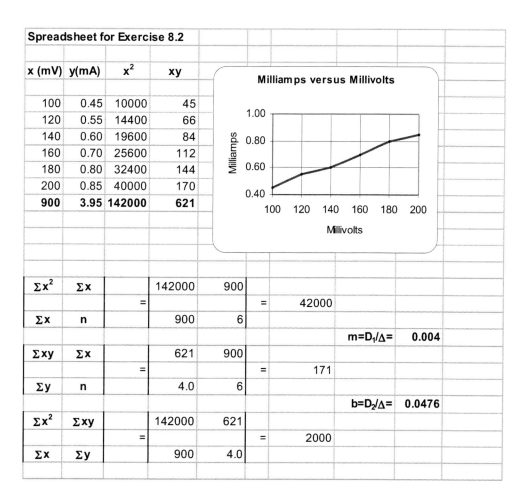

Spreadsheet for Exercise 8.2

x (mV)	y(mA)	x^2	xy
100	0.45	10000	45
120	0.55	14400	66
140	0.60	19600	84
160	0.70	25600	112
180	0.80	32400	144
200	0.85	40000	170
900	3.95	142000	621

Σx^2	Σx		142000	900				
		=			=	42000		
Σx	n		900	6				
							$m=D_1/\Delta=$	0.004
Σxy	Σx		621	900				
		=			=	171		
Σy	n		4.0	6				
							$b=D_2/\Delta=$	0.0476
Σx^2	Σxy		142000	621				
		=			=	2000		
Σx	Σy		900	4.0				

Thus, $y = mx + b = 0.004x + 0.0476$

3. Following the procedure of Example 8.2, we obtain the trendline shown below.

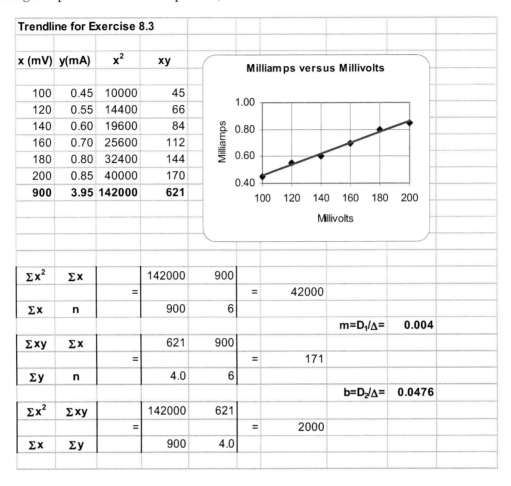

Trendline for Exercise 8.3

x (mV)	y(mA)	x^2	xy
100	0.45	10000	45
120	0.55	14400	66
140	0.60	19600	84
160	0.70	25600	112
180	0.80	32400	144
200	0.85	40000	170
900	**3.95**	**142000**	**621**

Σx^2	Σx		142000	900		
		=			=	42000
Σx	n		900	6		

m=D_1/Δ= 0.004

Σxy	Σx		621	900		
		=			=	171
Σy	n		4.0	6		

b=D_2/Δ= 0.0476

Σx^2	Σxy		142000	621		
		=			=	2000
Σx	Σy		900	4.0		

4.

```
mv= [100   120   140   160   180   200];   % x-axis data
ma=[0.45   0.55   0.60   0.70   0.80   0.85];   % y-axis data
axis([100  200  0  1]);                    % Establishes desired x and y axes limits
plot(mv,ma,'*b');                          % Display experimental (given) points with
                                           % asterisk and smoothed data with blue line
grid; title('ma (milliamps) vs mv (millivolts, n=1'); xlabel('mv'); ylabel('ma');
hold                                       % Hold current plot so we can add other data
p=polyfit(mv,ma,1);                        % Fits a first degree polynomial (straight line since n =1) and returns
                                           % the coefficients m and b of the straight line equation y = mx + b
a=0: 10: 200;                              % Define range to plot the polynomial
q=polyval(p,a);                            % Compute p for each value of a
plot(a,q)                                  % Plot the polynomial
                                           % Display the coefficients m and b
fprintf('\n')                              % Insert line
```

```
disp('Coefficients m and b are:'); fprintf('\n'); disp(p);
format bank              % Two decimal place display will be sufficient
ma_smoothed=polyval(p,mv);       % Compute the values of the fitted polynomial at
                                 %  same points as given (experimental) values of ma
ma_exper = ma;          % Display the experimental values of ma for comparison
                        % The statement below computes the percent error between
                        % the fitted polynomial and the experimental data
                        % disp('% Error at points of given values is:');
                        % The percent error is computed with the following statement
error=(ma_smoothed-ma_exper).*100./ma_exper;
format short            % Return to default format
y=zeros(6,4);           % Construct an 6 x 4 matrix of zeros
y(:,1)=mv';             % 1st column of matrix
y(:,2)=ma_exper';       % 2nd column of matrix
y(:,3)=ma_smoothed';    % 3rd column of matrix
y(:,4)=abs(error)';     % 4th column of matrix
fprintf(' \n');         % Insert line
fprintf('mv \t Exper ma\t Smoothed ma \t |Error| percent \n')
fprintf(' \n');         % Insert line
fprintf('%3.0f\t %5.4f\t   %5.4f\t    %5.4f\n',y')
fprintf(' \n');         % Insert line
```

```
Coefficients m and b are:

    0.0041      0.0476

mv   Exper ma  Smoothed ma    |Error|  percent

100   0.4500     0.4548        1.0582
120   0.5500     0.5362        2.5108
140   0.6000     0.6176        2.9365
160   0.7000     0.6990        0.1361
180   0.8000     0.7805        2.4405
200   0.8500     0.8619        1.4006
```

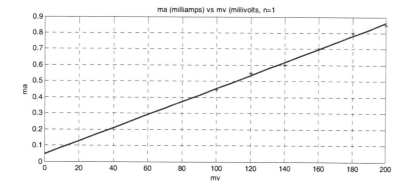

5. Following the procedure of Example 8.4, we choose Polynomial 4 and we obtain the trendline shown below.

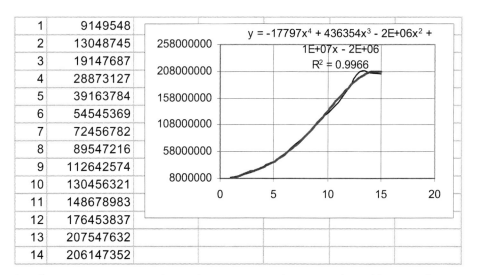

1	9149548
2	13048745
3	19147687
4	28873127
5	39163784
6	54545369
7	72456782
8	89547216
9	112642574
10	130456321
11	148678983
12	176453837
13	207547632
14	206147352

The sales for the next 3 years are from the equation above produced by Excel.

$$y_{16} = -17797x^4 + 436354x^3 - 2 \times 10^6 x^2 + 10^7 x - 2 \times 10^6 \big|_{x = 16} = 266961792$$

$$y_{17} = -17797x^4 + 436354x^3 - 2 \times 10^6 x^2 + 10^7 x - 2 \times 10^6 \big|_{x = 17} = 247383965$$

$$y_{18} = -17797x^4 + 436354x^3 - 2 \times 10^6 x^2 + 10^7 x - 2 \times 10^6 \big|_{x = 18} = 206558656$$

These results indicate that non-linear interpolation is, in most cases, unreliable. We will compare these values with the results of Exercise 6.

6.
```
year= [1  2  3  4  5  6  7  8  9  10  11  12  13  14  15];          % x-axis data
sales=[9149548  13048745  19147687  28873127  39163784 ...
    54545369  72456782  89547216  112642574  130456321 ...
    148678983  176453837  207547632  206147352 204456987];      % y-axis data
plot(year,sales,'*b');              % Display experimental (given) points with
                                    % asterisk and smoothed data with blue line
hold                                % Hold current plot so we can add other data
grid; title('Yearly Sales vs Years, n=4'); xlabel('Years'); ylabel('Yearly Sales');
p=polyfit(year,sales,4);            % Fits a first degree polynomial (n=4) and returns
                                    % the coefficients of the polynomial
a=linspace(0, 15, 15);              % Define range to plot the polynomial
q=polyval(p,a);                     % Compute p for each value of a
plot(a,q)                           % Plot the polynomial
                                    % Display coefficients ofpolynomial
fprintf('\n')                       % Insert line
disp('Coefficients are:'); fprintf('\n'); disp(p);
```

Numerical Analysis Using MATLAB® and Excel®, Third Edition
Copyright © Orchard Publications

```
sales_smoothed=polyval(p,year);    % Compute the values of the fitted polynomial at
                                   %  same points as given (experimental) values of ma
sales_exper = sales;               % Display the experimental values of ma for comparison
% The statement below computes the percent error between
% the fitted polynomial and the experimental data
% The percent error is computed with the following statement
error=(sales_smoothed-sales_exper).*100./sales_exper;
y=zeros(15,4);                     % Construct an 15 x 4 matrix of zeros
y(:,1)=year';                      % 1st column of matrix
y(:,2)=sales_exper';               % 2nd column of matrix
y(:,3)=sales_smoothed';            % 3rd column of matrix
y(:,4)=abs(error)';                % 4th column of matrix
fprintf(' \n');
fprintf('year\t Exper sales\t Smoothed sales \t |Error| percent \n')
fprintf(' \n');
fprintf('%2.0f\t   %9.0f\t   %9.0f\t    %5.2f\n',y')
fprintf(' \n');
```

Coefficients are:

   1.0e+007 *

     -0.0018   0.0436   -0.2386   1.1641   -0.2415

Yearly Sales vs Years, n=4

year	Exper sales	Smoothed sales	|Error| percent
1	9149548	7258461	20.67
2	13048745	14529217	11.35
3	19147687	21374599	11.63
4	28873127	29344934	1.63

5	39163784	39563426	1.02
6	54545369	52726163	3.34
7	72456782	69102111	4.63
8	89547216	88533118	1.13
9	112642574	110433913	1.96
10	130456321	133792104	2.56
11	148678983	157168183	5.71
12	176453837	178695519	1.27
13	207547632	196080363	5.53
14	206147352	206601848	0.22
15	204456987	207111986	1.30

From the coefficients produced by MATLAB, shown on the previous page, we form the polynomial

$$y = -1.8 \times 10^4 x^4 + 4.36 \times 10^5 x^3 - 2.386 \times 10^6 x^2 + 1.1641 \times 10^7 x - 2.415 \times 10^6$$

and from it we find the values of y (the yearly sales) as follows:

```
x=16; y16=-1.8*10^4*x^4+4.36*10^5*x^3-2.386*10^6*x^2+1.1641*10^7*x-2.415*10^6;
x=17; y17=-1.8*10^4*x^4+4.36*10^5*x^3-2.386*10^6*x^2+1.1641*10^7*x-2.415*10^6;
x=18; y18=-1.8*10^4*x^4+4.36*10^5*x^3-2.386*10^6*x^2+1.1641*10^7*x-2.415*10^6;

y16, y17, y18

y16 =

   1.7923e+008

y17 =

   1.4462e+008

y18 =

   8.7243e+007
```

These values vary significantly from those of Exercise 5. As stated above, non-linear interpolation especially for polynomials of fourth degree and higher give inaccurate results. We should remember that the equations produced by both Excel and MATLAB represent the equations that best fit the experimental values. For extrapolation, linear regression gives the best approximations.

Chapter 9

Solution of Differential Equations by Numerical Methods

This chapter is an introduction to several methods that can be used to obtain approximate solutions of differential equations. Such approximations are necessary when no exact solution can be found. The Taylor, Runge–Kutta, Adams', and Milne's methods are discussed.

9.1 Taylor Series Method

We recall from Chapter 6 that the Taylor series expansion about point a is

$$y_n = f(x) = f(a) + f'(a)(x-a) + \frac{f''(a)}{2!}(x-a)^2 + \dots + \frac{f^{(n)}(a)}{n!}(x-a)^n \tag{9.1}$$

Now, if $x_1 > a$ is a value close to a, we can find the approximate value y_1 of $f(x_1)$ by using the first $k+1$ terms in the Taylor expansion of $f(x_1)$ about $x = a$. Letting $h_1 = x - a$ in (9.1), we obtain:

$$y_1 = y_0 + y'_0 h_1 + \frac{1}{2!}y''_0 h_1^2 + \frac{1}{3!}y'''_0 h_1^3 + \frac{1}{4!}y_0^{(4)}h_1^4 + \dots \tag{9.2}$$

Obviously, to minimize the error $f(x_1) - y_1$ we need to keep h_1 sufficiently small.

For another value $x_2 > x_1$, close to x_1, we repeat the procedure with $h_2 = x_2 - x_1$; then,

$$y_2 = y_1 + y'_1 h_2 + \frac{1}{2!}y''_1 h_2^2 + \frac{1}{3!}y'''_1 h_2^3 + \frac{1}{4!}y_1^{(4)}h_2^4 + \dots \tag{9.3}$$

In general,

$$\boxed{y_{i+1} = y_i + y'_i h_{i+1} + \frac{1}{2!}y''_i h_{i+1}^2 + \frac{1}{3!}y'''_i h_{i+1}^3 + \frac{1}{4!}y_i^{(4)}h_{i+1}^4 + \dots} \tag{9.4}$$

Example 9.1

Use the Taylor series method to obtain a solution of

$$y' = -xy \tag{9.5}$$

correct to four decimal places for values $x_0 = 0.0$, $x_1 = 0.1$, $x_2 = 0.2$, $x_3 = 0.3$, $x_4 = 0.4$, and $x_5 = 0.5$ with the initial condition $y(0) = 1$.

Solution:

For this example,

$$h = x_1 - x_0 = 0.1 - 0.0 = 0.1$$

and by substitution into (9.4),

$$y_{i+1} = y_i + 0.1y'_i + 0.005y''_i + 0.000167y'''_i + 0.000004y_i^{(4)} \qquad (9.6)$$

for $i = 0, 1, 2, 3,$ and 4.

The first through the fourth derivatives of (9.5) are:

$$
\begin{aligned}
y' &= -xy \\
y'' &= -xy' - y = -x(-xy) - y = (x^2 - 1)y \\
y''' &= (x^2 - 1)y' + 2xy = (x^2 - 1)(-xy)2xy = (-x^3 + 3x)y \\
y^{(4)} &= (-x^3 + 3x)(-xy) + (-3x^2 + 3)y = (x^4 - 6x^2 + 3)y
\end{aligned}
\qquad (9.7)
$$

We use the subscript i to express them as

$$
\begin{aligned}
y'_i &= -x_i y_i \\
y''_i &= (x_i^2 - 1)y_i \\
y'''_i &= (-x_i^3 + 3x_i)y_i \\
y_i^{(4)} &= (x_i^4 - 6x_i^2 + 3)y_i
\end{aligned}
\qquad (9.8)
$$

where x_i represents $x_0 = 0.0$, $x_1 = 0.1$, $x_2 = 0.2$, $x_3 = 0.3$, and $x_4 = 0.4$.

Using the values of the coefficients of y_i in (9.8), we construct the spreadsheet of Figure 9.1.

	A	B	C	D	E	F	G	H
1	**Differential Equation is y' = –xy**							
2	Numerical solution by Taylor method follows							
3								
4	x_i	x_i^2	x_i^3	x_i^4	$-x_i$	$x_i^2 -1$	$-x_i^3+3x_i$	$x_i^4-6x_i^2+3$
5								
6	0.0	0.00	0.0000	0.0000	0.0	-1.00	0.000	3.0000
7	0.1	0.01	0.0010	0.0001	-0.1	-0.99	0.299	2.9401
8	0.2	0.04	0.0080	0.0016	-0.2	-0.96	0.592	2.7616
9	0.3	0.09	0.0270	0.0081	-0.3	-0.91	0.873	2.4681
10	0.4	0.16	0.0640	0.0256	-0.4	-0.84	1.136	2.0656

Figure 9.1. Spreadsheet for Example 9.1

The values in E6:E10, F6:F10, G6:G10, and H6:H10 of the spreadsheet of Figure 9.1, are now substituted into (9.8), and we obtain the following relations:

$$
\begin{aligned}
y'_0 &= -x_0 y_0 = -0 y_0 = 0 \\
y'_1 &= -x_1 y_1 = -0.1 y_1 \\
y'_2 &= -x_2 y_2 = -0.2 y_1 \\
y'_3 &= -x_3 y_3 = -0.3 y_1 \\
y'_4 &= -x_4 y_4 = -0.4 y_1
\end{aligned}
\tag{9.9}
$$

$$
\begin{aligned}
y''_0 &= (x_0^2 - 1) y_0 = -y_0 \\
y''_1 &= (x_1^2 - 1) y_1 = -0.99 y_1 \\
y''_2 &= (x_2^2 - 1) y_2 = -0.96 y_2 \\
y''_3 &= (x_3^2 - 1) y_3 = -0.91 y_3 \\
y''_4 &= (x_4^2 - 1) y_4 = -0.84 y_1
\end{aligned}
\tag{9.10}
$$

$$
\begin{aligned}
y'''_0 &= (-x_0^3 + 3x_0) y_0 = 0 \\
y'''_1 &= (-x_1^3 + 3x_1) y_1 = 0.299 y_1 \\
y'''_2 &= (-x_2^3 + 3x_2) y_2 = 0.592 y_2 \\
y'''_3 &= (-x_3^3 + 3x_3) y_3 = 0.873 y_3 \\
y'''_4 &= (-x_4^3 + 3x_4) y_4 = 1.136 y_4
\end{aligned}
\tag{9.11}
$$

$$
\begin{aligned}
y_0^{(4)} &= (x_0^4 - 6x_0^2 + 3) y_0 = 3 y_0 \\
y_1^{(4)} &= (x_1^4 - 6x_1^2 + 3) y_1 = 2.9401 y_1 \\
y_2^{(4)} &= (x_2^4 - 6x_2^2 + 3) y_2 = 2.7616 y_2 \\
y_3^{(4)} &= (x_3^4 - 6x_3^2 + 3) y_3 = 2.4681 y_3 \\
y_4^{(4)} &= (x_4^4 - 6x_4^2 + 3) y_4 = 2.0656 y_4
\end{aligned}
\tag{9.12}
$$

By substitution of (9.9) through (9.12) into (9.6), and using the given initial condition $y_0 = 1$, we obtain:

$$y_1 = y_0 + 0.1y_0' + 0.005y_0'' + 0.000167y_0''' + 0.000004y_0^{(4)}$$
$$= 1 + 0.1(0) + 0.005(-1) + 0.00167(0) + 0.000004(3)$$
$$= 1 - 0.005 + 0.000012$$
$$= 0.99501$$

(9.13)

Similarly,

$$y_2 = y_1 + 0.1y_1' + 0.005y_1'' + 0.000167y_1''' + 0.000004y_1^{(4)}$$
$$= (1 - 0.01 - 0.00495 + 0.00005 + 0.00001)y_1$$
$$= 0.98511(0.99501)$$
$$= 0.980194$$

(9.14)

$$y_3 = y_2 + 0.1y_2' + 0.005y_2'' + 0.000167y_2''' + 0.000004y_2^{(4)}$$
$$= (1 - 0.02 - 0.0048 + 0.0001 + 0.00001)y_2$$
$$= 0.97531(0.980194)$$
$$= 0.955993$$

(9.15)

$$y_4 = y_3 + 0.1y_3' + 0.005y_3'' + 0.000167y_3''' + 0.000004y_3^{(4)}$$
$$= (1 - 0.03 - 0.00455 + 0.00015 + 0.00001)y_3$$
$$= 0.9656(0.955993)$$
$$= 0.923107$$

(9.16)

$$y_5 = y_4 + 0.1y_4' + 0.005y_4'' + 0.000167y_4''' + 0.000004y_4^{(4)}$$
$$= (1 - 0.04 - 0.0042 + 0.00019 + 0.00001)y_4$$
$$= 0.95600(0.923107)$$
$$= 0.88249$$

(9.17)

The differential equation $\frac{dy}{dx} = -xy$ of this example can be solved analytically as follows:

$$\frac{dy}{y} = -xdx \qquad \int\frac{dy}{y} = -\int xdx \qquad \ln y = -\frac{1}{2}x^2 + C$$

and with the initial condition $y = 1$ when $x = 0$,

$$\ln 1 = -\frac{1}{2}(0) + C \qquad C = 0 \qquad \ln y = -\frac{1}{2}x^2$$

or

$$y = e^{-x/2}$$

(9.18)

For $x_5 = 0.5$ (9.18) yields

$$y = e^{-0.125} = 0.8825$$

and we observe that this value is in close agreement with the value of (9.17).

We can verify the analytical solution of Example 9.1 with MATLAB's dsolve(s) function using the following script:

```
syms x y z
z=dsolve('Dy=–x*y','y(0)=1','x')

z =
exp(-1/2*x^2)
```

The procedure used in this example, can be extended to apply to a second order differential equation

$$y'' = f(x, y, y') \tag{9.19}$$

In this case, we need to apply the additional formula

$$y'_{i+1} = y'_i + y''_i h + \frac{1}{2!} y'''_i h^2 + \frac{1}{3!} y^{(4)}_i h^3 + \dots \tag{9.20}$$

9.2 Runge–Kutta Method

The *Runge–Kutta method* is the most widely used method of solving differential equations with numerical methods. It differs from the Taylor series method in that we use values of the first derivative of $f(x, y)$ at several points instead of the values of successive derivatives at a single point.

For a Runge–Kutta method of order 2, the following formulas are applicable.

$$\begin{array}{c} k_1 = hf(x_n, y_n) \\ k_2 = hf(x_n + h, y_n + h) \\ y_{n+1} = y_n + \frac{1}{2}(k_1 + k_2) \\ \hline \text{For Runge-Kutta Method of Order 2} \end{array} \tag{9.21}$$

When higher accuracy is desired, we can use order 3 or order 4. The applicable formulas are as follows:

$$l_1 = hf(x_n, y_n) = k_1$$

$$l_2 = hf\left(x_n + \frac{h}{2}, y_n + \frac{l_1}{2}\right)$$

$$l_3 = hf(x_n + h, y_n + 2l_2 - l_1)$$

$$y_{n+1} = y_n + \frac{1}{6}(l_1 + 4l_2 + l_3)$$

For Runge-Kutta Method of Order 3

(9.22)

$$m_1 = hf(x_n, y_n) = l_1 = k_1$$

$$m_2 = hf\left(x_n + \frac{h}{2}, y_n + \frac{m_1}{2}\right) = l_2$$

$$m_3 = hf\left(x_n + \frac{h}{2}, y_n + \frac{m_2}{2}\right)$$

$$m_4 = hf(x_n + h, y_n + m_3)$$

$$y_{n+1} = y_n + \frac{1}{6}(m_1 + 2m_2 + 2m_3 + m_4)$$

For Runge-Kutta Method of Order 4

(9.23)

Example 9.2

Compute the approximate value of y at $x = 0.2$ from the solution $y(x)$ of the differential equation

$$y' = x + y^2 \qquad (9.24)$$

given the initial condition $y(0) = 1$. Use order 2, 3, and 4 Runge–Kutta methods with $h = 0.2$.

Solution:

a. For order 2, we use (9.21). Since we are given that $y(0) = 1$, we begin with $x = 0$, and $y = 1$. Then,

$$k_1 = hf(x_n, y_n) = 0.2(0 + 1^2) = 0.2$$

$$k_2 = hf(x_n + h, y_n + h) = 0.2[0 + 0.2 + (1 + 0.2^2)] = 0.328$$

and

$$y_1 = y_0 + \frac{1}{2}(k_1 + k_2) = 1 + \frac{1}{2}(0.2 + 0.328) = 1.264 \qquad (9.25)$$

b. For order 3, we use (9.22). Then,

$$l_1 = hf(x_n, y_n) = k_1 = 0.2$$

$$l_2 = hf\left(x_n + \frac{h}{2}, y_n + \frac{l_1}{2}\right) = 0.2\left[\left(0 + \frac{1}{2} \cdot 0.2\right) + \left(1 + \frac{1}{2} \cdot 0.2\right)^2\right] = 0.262 \tag{9.26}$$

$$l_3 = hf(x_n + h, y_n + 2l_2 - l_1) = 0.2[(0 + 0.2) + (1 + 2 \times 0.262 - 0.2)^2] = 0.391$$

and

$$y_1 = y_0 + \frac{1}{6}(l_1 + 4l_2 + l_3) = 1 + \frac{1}{6}(0.2 + 4 \times 0.262 + 0.391) = 1.273 \tag{9.27}$$

c. For order 4, we use (9.23). Then,

$$m_1 = hf(x_n, y_n) = l_1 = k_1 = 0.2$$

$$m_2 = hf\left(x_n + \frac{h}{2}, y_n + \frac{m_1}{2}\right) = l_2 = 0.262$$

$$m_3 = hf\left(x_n + \frac{h}{2}, y_n + \frac{m_2}{2}\right) = 0.2\left[0 + \frac{0.2}{2} + \left(1 + \frac{0.262}{2}\right)^2\right] = 0.276 \tag{9.28}$$

$$m_4 = hf(x_n + h, y_n + m_3) = 0.2[0 + 0.2 + (1 + 0.276)^2] = 0.366$$

and

$$y_1 = y_0 + \frac{1}{6}(m_1 + 2m_2 + 2m_3 + m_4)$$

$$= 1 + \frac{1}{6}(0.2 + 2 \times 0.262 + 2 \times 0.276 + 0.366) = 1.274 \tag{9.29}$$

The Runge–Kutta method can also be used for second order differential equations of the form

$$y'' = f(x, y, y') \tag{9.30}$$

For second order differential equations, the pair of 3rd–order formulas[*] are:

---

[*] *Third and fourth order formulas can also be used, but these will not be discussed in this text. They can be found in differential equations and advanced mathematics texts.*

$$l_1 = hy'_n$$

$$l'_1 = hf(x_n, y_n, y'_n)$$

$$l_2 = h\left(y'_n + \frac{l'_1}{2}\right)$$

$$l'_2 = hf\left(x_n + \frac{h}{2},\ y_n + \frac{l_1}{2},\ y'_n + \frac{l'_1}{2}\right)$$

$$l_3 = h\,(y'_n + 2l'_2 - l'_1)$$

$$l'_3 = hf(x_n + h,\ y_n + 2l_2 - l_1,\ y'_n + 2l'_2 - l'_1)$$

$$y_{n+1} = y_n + \frac{1}{6}(l_1 + 4l_2 + l_3)$$

$$y'_{n+1} = y'_n + \frac{1}{6}(l'_1 + 4l'_2 + l'_3)$$

For Runge-Kutta Method of Order 3

2nd Order Differential Equation

(9.31)

Example 9.3

Given the 2nd order non–linear differential equation

$$y'' - 2y^3 = 0 \qquad (9.32)$$

with the initial conditions $y(0) = 1$, $y'(0) = -1$, compute the approximate values of y and y' at $x = 0.2$. Use $h = 0.2$.

Solution:

We are given the values of $x_0 = 0$, $y_0 = 0$, $y'_0 = -1$ and we are seeking the values of y_1 and y'_1 at $x_1 = 0.2$. We will use $h = x_1 - x_0 = 0.2$.

We rewrite the given equation as

$$y'' = 2y^3 = 0 \cdot x + 2y^3 \qquad (9.33)$$

and using (9.31) we obtain:

$$l_1 = hy'_0 = 0.2(-1) = -0.2$$

$$l'_1 = hf(x_0, y_0, y'_0) = 0.2(0 + 2 \times 1^3 + 0) = 0.4$$

$$l_2 = h\left(y'_0 + \frac{l'_1}{2}\right) = 0.2\left(-1 + \frac{0.4}{2}\right) = -0.16$$

$$l'_2 = hf\left(x_0 + \frac{h}{2}, y_0 + \frac{l_1}{2}, y'_0 + \frac{l'_1}{2}\right) = 0.2\left[0 + 2\left(1 + \frac{-0.2}{2}\right)^3 + 0\right] \quad (9.34)$$

$$= 0.2[2(1 - 0.1)^3] = 0.2(1.458) = 0.2916$$

$$l_3 = h(y'_0 + 2l'_2 - l'_1) = 0.2(-1 + 2 \times 0.2916 - 0.4) = -0.1634$$

$$l'_3 = hf(x_0 + h, y_0 + 2l_2 - l_1, y'_0 + 2l'_2 - l'_1)$$

$$= 0.2\{0 + 2[1 + 2(-0.16) - (-0.2)]^3 + 0\}$$

$$= 0.2[2(1 - 0.32 + 0.2)^3] = 0.2[2(0.88)^3] = 0.2726$$

By substitution into the last two formulas of (9.31), we obtain:

$$y_1 = y_0 + \frac{1}{6}(l_1 + 4l_2 + l_3) = 1 + \frac{1}{6}(-0.2 + 4(0.16) - 0.1634) = 0.8328$$

$$\quad (9.35)$$

$$y'_1 = y'_0 + \frac{1}{6}(l'_1 + 4l'_2 + l'_3) = -1 + \frac{1}{6}(0.4 + 4(0.2916) + 0.2726) = -0.6935$$

MATLAB has two functions for computing numerical solutions of Ordinary Differential Equations (ODE). The first, **ode23**, uses second and third–order Runge–Kutta methods. The second, **ode45**, uses fourth and fifth–order Runge–Kutta methods. Both have the same syntax; therefore, we will use the **ode23** function in our subsequent discussion.

The syntax for **ode23** is **ode23('f',tspan,y0)**. The first argument, **f**, in single quotation marks, is the name of the user defined MATLAB function. The second, **tspan**, defines the desired time span of the interval over which we want to evaluate the function $y = f(x)$. The third argument, **y0**, represents the initial condition or boundary point that is needed to determine a unique solution. This function produces two outputs, a set of x values and the corresponding set of y values that represent points of the function $y = f(x)$.

Example 9.4

Use the MATLAB **ode23** function to find the analytical solution of the second order nonlinear equation

$$y'' - 2y^3 = 0 \quad (9.36)$$

with the initial conditions $y(0) = 1$ and $y'(0) = -1$. Then, plot the numerical solution using the function **ode23** for the **tspan** interval $0 \le x \le 1$. Compare values with those of Example 9.3, at points $y(0.2)$ and $y'(0.2)$.

Solution:

If we attempt to find the analytical solution with the following MATLAB script

```
syms x y
y=dsolve('D2y=2*y^3,y(0)=1,Dy(0)=-1','x')
```

MATLAB displays the following message:

```
Warning: Explicit solution could not be found.
```

This warning indicates that MATLAB could not find a closed–form solution for this non–linear differential equation. This is because, in general, non–linear differential equations cannot be solved analytically, although few methods are available for special cases. These can be found in differential equations textbooks.

The numerical solution for this non–linear differential equation is obtained and plotted with the following script, by first writing a user defined m–file which we denote as **fex9_4**. The script is shown below.

```
function d2y=fex9_4(x,y);
d2y=[y(2);2*y(1)^3]; % Output must be a column
```

This file is saved as **fex9_4**. Next, we write and execute the script below to obtain the plots for y and y'.

```
tspan=[0 1]; % Interval over which we want to evaluate y=f(x)
y0=[1;-1]; % Given initial conditions
[x,y]=ode23('fex9_4', tspan, y0); % Use 2nd and 3rd Order Runge–Kutta
% Plot numeric values with the statements below
plot(x, y(:,1), '+r-', x, y(:,2), 'Ob--')
title('Numerical Solution for Differential Equation of Example 9.4'),...
xlabel('x'), ylabel('y (upper curve), yprime (lower curve)'), grid
```

The plots for y and y' are shown in Figure 9.2. We observe that the values at points $y(0.2)$ and $y'(0.2)$, compare favorably with those that we found in Example 9.3.

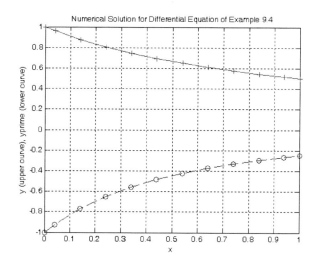

Figure 9.2. Plot for Example 9.4

Example 9.5

Use MATLAB to find the analytical solution of

$$x^2 y'' - xy' - 3y = x^2 \ln x \tag{9.37}$$

with the initial conditions $y(1) = -1$ and $y'(1) = 0$. Then, compute and plot the numerical solution using the command **ode23** along with points of the analytical solution, to verify the accuracy of the numerical solution.

Solution:

The analytical solution of (9.37) with the given initial conditions is found with MATLAB as follows:

```
syms x y
y=dsolve('x^2*D2y-x*Dy-3*y=x^2*log(x), Dy(1)=0, y(1)=-1', 'x')

y =
1/9*(-3*x^3*log(x)-2*x^3-7)/x
y=simple(y)

y =
(-1/3*log(x)-2/9)*x^2-7/9/x

pretty(y)
                          2
       (- 1/3 log(x) - 2/9)  x  - 7/9 1/x
```

and therefore, the analytical solution of (9.37) is

$$y = \left(-\frac{1}{3}\ln x - \frac{2}{9}\right)x^2 - \frac{7}{9x} \qquad (9.38)$$

Next, we create and save a user defined m–file, **fex9_5**.

```
function d2y=fex9_5(x,y);              % Produces the derivatives of Example 9.5
% x^2*y''-x*y'-3*y=x^2*log(x) where y''=2nd der, y'=1st der, logx=lnx
%
% we let y(1) = y and y(2)=y', then y(1)'=y(2)
%
% and y(2)'=y(2)/x^2+3*y(1)/x^2+log(x)
%
d2y=[y(2); y(2)/x+3*y(1)/x^2+log(x)];  % output must be a column
```

The following MATLAB script computes and plots the numerical solution values for the interval $1 \le x \le 4$ and compares these with the actual values obtained from the analytical solutions.

```
tspan=[1 4]; % Interval over which we want to evaluate y=f(x)
y0=[-1;0]; % Given initial conditions
[x,y]=ode23('fex9_5', tspan, y0); % Use 2nd and 3rd Order Runge-Kutta
anal_y=((-1./3).*log(x)-2./9).*x.^2-7./(9.*x); % This is the...
% analytic solution of the 2nd order differential equation of (9.38)
anal_yprime=((-2./3).*log(x)-7./9).*x+7./(9.*x.^2); % This is the first derivative of (9.38)
% Plot numeric and analytic values with the statements below
plot(x, y(:,1), '+', x, anal_y, '-', x, y(:,2), 'O', x, anal_yprime, '-'),...
title('Numeric and Analytic Solutions of Differential Equation of Example 9.5'),...
xlabel('x'), ylabel('y (line with +), yprime (line with O)'), grid
```

The numeric and analytical solutions are shown in Figure 9.3.

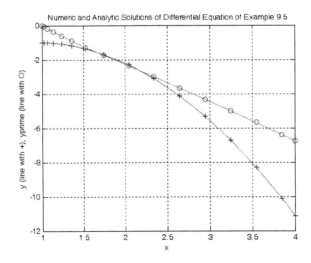

Figure 9.3. Plot for Example 9.5

9.3 Adams' Method

In this method, the step from y_n to y_{n+1} is performed by a formula expressed in terms of differences of $f(x, y)$.

Adams' method uses the formula

$$y_{n+1} = y_n + h\left[f_n + \frac{1}{2}\Delta f_n + \frac{5}{12}\Delta^2 f_n + \frac{3}{8}\Delta^3 f_n + \ldots\right] \qquad (9.39)$$

where

$$h = x_{n+1} - x_n$$

$$f_n = (x_n, y_n)$$

$$\Delta f_n = f_n - f_{n-1}$$

$$\Delta^2 f_n = \Delta f_n - \Delta f_{n-1}$$

and so on.

Obviously, to form a table of differences, it is necessary to have several (4 or more) approximate values of $y(x)$ in addition to the given initial condition $y(0)$. These values can be found by other methods such as the Taylor series or Runge–Kutta methods.

Example 9.6

Given the differential equation

$$y' = 2y + x \qquad (9.40)$$

with the initial condition $y(0) = 1$,

compute the approximate values of y for $x = 0.1, 0.2, 0.3, 0.4$ and 0.5 by the third–order Runge–Kutta method. Then, find the value of y corresponding to $x = 0.6$ correct to three decimal places using Adams' method.

Solution:

The spreadsheet of Figure 9.4 shows the results of the computations of $y_1, y_2, y_3, y_4,$ and y_5 using the third–order Runge–Kutta method as in Example 9.2.

	A	B	C	D	E	F
1	Differential Equation **y' = x + 2y**					
2	Numerical solution by Runga-Kutte method follows					
4	h=	x(1)-x(0)=		0.1000	x_0=	0.0
5	x(0)=	(given)		0.0000	x_1=	0.1
6	y(0)=	Initial condition (given)		1.0000	x_2=	0.2
7					x_3=	0.3
8					x_4=	0.4
9	L(1)=	h*f(x_n,y_n)	h*(0+2*1)=	0.2000	x_5=	0.5
10	L(2)=	h*f(x_n+0.5*h,y_n+0.5*L(1))	h*[(0+0.5*h)+2*(1+0.5*L(1))]=	0.2250		
11	L(3)=	h*f(x_n+h,y_n+2*L(2)-L(1))	h*[(0+h)+2*(1+2*L(2)-L(1))]=	0.2600		
13	y(1)=	y(0) +(L(1) + 4*L(2) + L(3))/6		**1.2267**		
15	h=	x(2)-x(1)=		0.1000		
16	x(0.1)=	Next value x(0) + h		0.1000		
17	y(1)=	From previous computation		1.2267		
19	L(1)=	h*f(x_n,y_n)	h*(0.1+2*1.2267)=	0.2553		
20	L(2)=	h*f(x_n+0.5*h,y_n+0.5*L(1))	h*[(0+0.5*h)+2*(1+0.5*L(1))]=	0.2859		
21	L(3)=	h*f(x_n+h,y_n+2*L(2)-L(1))	h*[(0+h)+2*(1+2*L(2)-L(1))]=	0.3286		
23	y(2)=	y(1) +(L(1) + 4*L(2) + L(3))/6		**1.5146**		
25	h=	x(2)-x(1)=		0.1000		
26	x(0.2)=	Next value x(0) + 2*h		0.2000		
27	y(2)=	From previous computation		1.5146		
29	L(1)=	h*f(x_n,y_n)	h*(0.2+2*1.5146)=	0.3229		
30	L(2)=	h*f(x_n+0.5*h,y_n+0.5*L(1))	h*[(0+0.5*h)+2*(1+0.5*L(1))]=	0.3602		
31	L(3)=	h*f(x_n+h,y_n+2*L(2)-L(1))	h*[(0+h)+2*(1+2*L(2)-L(1))]=	0.4124		
33	y(3)=	y(2) +(L(1) + 4*L(2) + L(3))/6		**1.8773**		
35	h=	x(3)-x(2)=		0.1000		
36	x(0.3)=	Next value x(0) + 3*h		0.3000		
37	y(3)=	From previous computation		1.8773		
39	L(1)=	h*f(x_n,y_n)	h*(0.3+2*1.8773)=	0.4055		
40	L(2)=	h*f(x_n+0.5*h,y_n+0.5*L(1))	h*[(0+0.5*h)+2*(1+0.5*L(1))]=	0.4510		
41	L(3)=	h*f(x_n+h,y_n+2*L(2)-L(1))	h*[(0+h)+2*(1+2*L(2)-L(1))]=	0.5148		
43	y(4)=	y(2) +(L(1) + 4*L(2) + L(3))/6		**2.3313**		
45	h=	x(4)-x(3)=		0.1000		
46	x(0.4)=	Next value x(0) + 4*h		0.4000		
47	y(4)=	From previous computation		2.3313		
49	L(1)=	h*f(x_n,y_n)	h*(0.3+2*1.8773)=	0.5063		
50	L(2)=	h*f(x_n+0.5*h,y_n+0.5*L(1))	h*[(0+0.5*h)+2*(1+0.5*L(1))]=	0.5619		
51	L(3)=	h*f(x_n+h,y_n+2*L(2)-L(1))	h*[(0+h)+2*(1+2*L(2)-L(1))]=	0.6398		
53	y(5)=	y(2) +(L(1) + 4*L(2) + L(3))/6		**2.8969**		

Figure 9.4. Spreadsheet for Example 9.6

Next, we compute the following values to be used in Adams' formula of (9.39). These are shown below.

x_n	y_n	$f_n = x_n + 2y_n$	Δf_n	$\Delta^2 f_n$	$\Delta^3 f_n$
0.0	1.0000	2.0000			
			0.5534		
0.1	1.2267	2.5534		0.1224	
			0.6758		0.0272
0.2	1.5146	3.2292		0.1496	
			0.8254		0.0330
0.3	1.8773	4.0546		0.1826	
			1.0080		0.0406
0.4	2.3313	5.0626		0.2232	
			1.2312		
0.5	2.8969	6.2938			

and by substitution into (7.39)

$$y_6 = 2.8969 + 0.1\left[6.2638 + \frac{1}{2}(1.2312) + \frac{5}{12}(0.2232) + \frac{3}{8}(0.0406)\right] = 3.599 \qquad (9.41)$$

As with the other methods, Adams' method can also be applied to second order differential equations of the form $y'' = f(x, y, y')$ with initial conditions $y(x_0) = y_0$ and $y'(x_0) = y_0'$.

9.4 Milne's Method

Milne's method also requires prior knowledge of several values of y. It uses the predictor–corrector pair

$$\boxed{y_{n+1} = y_{n-3} + \frac{4}{3}h[2f_n - f_{n-1} + 2f_{n-2}]} \qquad (9.42)$$

and

$$\boxed{Y_{n+1} = y_{n-1} + \frac{1}{3}h[f_{n+1} + 4f_n + f_{n-1}]} \qquad (9.43)$$

The corrector formula of (9.43) serves as a check for the value

$$y_{n+1} = f(x_{n+1}, y_{n+1}) \qquad (9.44)$$

If y_{n+1} and Y_{n+1} in (9.42) and (9.43) respectively, do not differ considerably, we accept Y_{n+1} as the best approximation. If they differ significantly, we must reduce the interval h.

Example 9.7

Use Milne's method to find the value of y corresponding to $x = 0.6$ for the differential equation

$$y' = 2y + x \qquad (9.45)$$

with the initial condition $y(0) = 1$.

Solution:

This is the same differential equation as in Example 9.6 where we found the following values:

TABLE 9.1 Table for Example 9.7

n	x_n	y_n	$f_n = x_n + 2y_n$
2	0.2	1.5146	3.2292
3	0.3	1.8773	4.0546
4	0.4	2.3313	5.0626
5	0.5	2.8969	6.2938

and using the predictor formula we find

$$y_6 = y_2 + \frac{4}{3}(0.1)[2f_5 - f_4 + 2f_3]$$

$$= 1.5146 + \frac{4}{3} \times 0.1(2 \times 6.2938 - 5.0626 + 2 \times 4.0546) = 3.599 \qquad (9.46)$$

Before we use the corrector formula of (9.43), we must find the value of f_6; this is found from

$$f_6 = x_6 + 2y_6$$

where $x_6 = 0.6$, and from Example 9.5 $y_6 = 3.599$. Then,

$$f_6 = x_6 + 2y_6 = 0.6 + 2 \times 3.599 = 7.7984$$

and

$$Y_6 = y_4 + \frac{1}{3}0.1(f_6 + 4f_5 + f_4)$$

$$= 2.3313 + \frac{1}{3} \times 0.1(7.7984 + 4 \times 6.2938 + 5.0626) = 3.599 \qquad (9.47)$$

We see from (9.46) and (9.47) that the predictor–corrector pair is in very close agreement.

Milne's method can also be extended to second order differential equations of the form $y'' = f(x, y, y')$ with initial conditions $y(x_0) = y_0$ and $y'(x_0) = y'_0$.

9.5 Summary

- The Taylor series method uses values of successive derivatives at a single point. We can use this series method to obtain approximate solutions of differential equations with the relation

$$y_{i+1} = y_i + y'_i h_{i+1} + \frac{1}{2!} y''_i h_{i+1}^2 + \frac{1}{3!} y'''_i h_{i+1}^3 + \frac{1}{4!} y_i^{(4)} h_{i+1}^4 + \dots$$

provided that h is sufficiently small such as $h = 0.1$.

- The Taylor series method can also be extended to apply to a second order differential equation

$$y'' = f(x, y, y')$$

using the relation

$$y'_{i+1} = y'_i + y''_i h + \frac{1}{2!} y'''_i h^2 + \frac{1}{3!} y_i^{(4)} h^3 + \dots$$

- The Runge–Kutta method uses values of the first derivative of $f(x, y)$ at several points. it is the most widely used method of solving differential equations using numerical methods.

- For a Runge–Kutta method of order 2 we use the relations

$$k_1 = hf(x_n, y_n) \qquad k_2 = hf(x_n + h, y_n + h) \qquad y_{n+1} = y_n + \frac{1}{2}(k_1 + k_2)$$

provided that h is sufficiently small such as $h = 0.1$.

- For a Runge–Kutta method of order 3 we use the relations

$$l_1 = hf(x_n, y_n) = k_1 \qquad l_2 = hf\left(x_n + \frac{h}{2}, y_n + \frac{l_1}{2}\right) \qquad l_3 = hf(x_n + h, y_n + 2l_2 - l_1)$$

$$y_{n+1} = y_n + \frac{1}{6}(l_1 + 4l_2 + l_3)$$

- For a Runge–Kutta method of order 4 we use the relations

$$m_1 = hf(x_n, y_n) = l_1 = k_1 \qquad m_2 = hf\left(x_n + \frac{h}{2}, y_n + \frac{m_1}{2}\right) = l_2$$

$$m_3 = hf\left(x_n + \frac{h}{2}, y_n + \frac{m_2}{2}\right) \qquad m_4 = hf(x_n + h, y_n + m_3)$$

$$y_{n+1} = y_n + \frac{1}{6}(m_1 + 2m_2 + 2m_3 + m_4)$$

- The Runge–Kutta method can also be used for second order differential equations of the form

$$y'' = f(x, y, y')$$

- For second order differential equations, the pair of 3rd–order relations are:

$$l_1 = hy'_n \qquad l'_1 = hf(x_n, y_n, y'_n)$$

$$l_2 = h\left(y'_n + \frac{l'_1}{2}\right) \qquad l'_2 = hf\left(x_n + \frac{h}{2}, \ y_n + \frac{l_1}{2}, y'_n + \frac{l'_1}{2}\right)$$

$$l_3 = h(y'_n + 2l'_2 - l'_1) \qquad l'_3 = hf(x_n + h, \ y_n + 2l_2 - l_1, y'_n + 2l'_2 - l'_1)$$

$$y_{n+1} = y_n + \frac{1}{6}(l_1 + 4l_2 + l_3) \qquad y'_{n+1} = y'_n + \frac{1}{6}(l'_1 + 4l'_2 + l'_3)$$

Third and fourth order formulas can also be used but they were not be discussed in this text. They can be found in differential equations texts.

- MATLAB has two functions for computing numerical solutions of Ordinary Differential Equations (ODE). The first, **ode23**, uses second and third–order Runge–Kutta methods. The second, **ode45**, uses fourth and fifth–order Runge–Kutta methods. Both have the same syntax.

- The syntax for **ode23** is **ode23('f',tspan,y0)**. The first argument, **f**, in single quotation marks, is the name of the user defined MATLAB function. The second, **tspan**, defines the desired time span of the interval over which we want to evaluate the function $y = f(x)$. The third argument, **y0**, represents the initial condition or boundary point that is needed to determine a unique solution. This function produces two outputs, a set of x values and the corresponding set of y values that represent points of the function $y = f(x)$.

- Adams' method provides the transition from y_n to y_{n+1} and the step is performed by a formula expressed in terms of differences of $f(x, y)$. This method uses the formula

$$y_{n+1} = y_n + h\left[f_n + \frac{1}{2}\Delta f_n + \frac{5}{12}\Delta^2 f_n + \frac{3}{8}\Delta^3 f_n + \ldots\right]$$

where

$$h = x_{n+1} - x_n$$

$$f_n = (x_n, y_n)$$

$$\Delta f_n = f_n - f_{n-1}$$

$$\Delta^2 f_n = \Delta f_n - \Delta f_{n-1}$$

and so on. To use this method, it is necessary to have several (4 or more) approximate values of $y(x)$ in addition to the given initial condition $y(0)$. These values can be found by other methods such as the Taylor series or Runge–Kutta methods.

- Milne's method also requires prior knowledge of several values of y. It uses the predictor–corrector pair

$$y_{n+1} = y_{n-3} + \frac{4}{3}h[2f_n - f_{n-1} + 2f_{n-2}]$$

and

$$Y_{n+1} = y_{n-1} + \frac{1}{3}h[f_{n+1} + 4f_n + f_{n-1}]$$

where y_{n+1} is the predictor formula and Y_{n+1} is the corrector formula. The corrector formula serves as a check for the value

$$y_{n+1} = f(x_{n+1}, y_{n+1})$$

If y_{n+1} and Y_{n+1} do not differ considerably, we accept Y_{n+1} as the best approximation. If they differ significantly, we must reduce the interval h.

Milne's method can also be extended to second order differential equations of the form $y'' = f(x, y, y')$ with initial conditions $y(x_0) = y_0$ and $y'(x_0) = y'_0$. The procedure for this method was not discussed. It can be found in differential equations texts.

9.6 Exercises

1. Use the MATLAB **ode23** function to verify the analytical solution of Example 9.1.

2. Construct a spreadsheet for the numerical solutions of Example 9.2.

3. Use MATLAB to find the analytical solution of

$$y' = f(x) = 3x^2$$

with the initial condition $y(2) = 0.5$. Then, compute and plot the numerical solution using the MATLAB function **ode23** along with points of the analytical solution to verify the accuracy of the numerical solution for the interval $2 \le x \le 4$.

4. Use MATLAB to plot the numerical solution of the non–linear differential equation

$$y' = -y^3 + 0.2\sin x$$

with the initial condition $y(0) = 0.707$ using the command **ode23** for the interval $0 \le x \le 10$.

5. Given the differential equation

$$y' = x^2 - y$$

with the initial condition $y(0) = 1$ and $x_0 = 0.0$ find the values of y corresponding to the values of $x_0 + 0.1$ and $x_0 + 0.2$ correct to four decimal places using the third–order Runge-Kutta method. It is suggested that a spreadsheet is used to do all computations.

6. Given the differential equation

$$y'' + y' = xy$$

compute the approximate values of y and y' at $x_0 + 0.1$ and $x_0 + 0.2$ given that $y(0) = 1$, $y'(0) = -1$, and $x_0 = 0.0$ correct to four decimal places, using the third–order Runge–Kutta method. It is suggested that a spreadsheet is used to do all computations.

9.7 Solutions to End–of–Chapter Exercises

1.

We write and save the following function file:

```
function dy = func_exer9_1(x,y)
dy = -x*y;
```

Next, we write and execute the MATLAB script below.

```
tspan=[0 3]; % Interval over which we want to evaluate y=f(x)
y0=[1;-1]; % Given initial conditions
[x,y]=ode23('func_exer9_1', tspan, y0); % Use 2nd and 3rd Order Runge-Kutta
% Plot numeric values with the statements below
plot(x, y(:,1), '+r-', x, y(:,2), 'Ob--')
title('Numeric Solution of Differential Equation of Exercise 9.1'),...
xlabel('x'), ylabel('y (upper curve), yprime (lower curve)'), grid
```

The plot below shows the function $y = f(x)$ and its derivative dy/dx.

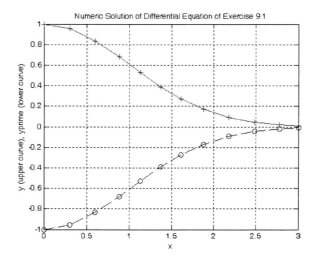

2.

	A	B	C	D
1	Differential Equation $y' = x + y^2$			
2	Numerical solution by Runga-Kutte method follows			
3				
4	h=	(given)	0.2000	
5	x(0)=	(given)	0.0000	
6	y(0)=	Initial condition (given)	1.0000	
7				
8	k(1)=	$h*f(x_n,y_n)$	$h*(0+1^2)=$	0.2000
9	k(2)=	$h*f(x_n+h,y_n+h)$	$h*(0+0.2+(1+0.2)^2)=$	0.3280
10	y(1)=	y(0)+0.5(k(1)+k(2))	y(0)+0.5*(D11+D12)	1.2640
11				
12	L(1)=	$h*f(x_n,y_n)$	$h*(0+1^2)=$	0.2000
13	L(2)=	$h*f(x_n+0.5*h,y_n+0.5*L(1))$	$h*[(0+0.5*h)+(1+0.5*L(1)^2)]=$	0.2620
14	L(3)=	$h*f(x_n+h,y_n+2*L(2)-L(1))$	$h*[(0+h)+(1+2*L(2)-L(1))^2]=$	0.3906
15	y(1)=	y(0) +(L(1) + 4*L(2) + L(3))/6		1.2731
16				
17	m(1)=	$h*f(x_n,y_n)$	$h*(0+1^2)=$	0.2000
18	m(2)=	$h*f(x_n+0.5*h,y_n+0.5*m(1))$	$h*[(0+0.5*h)+(1+0.5*m(1))^2]=$	0.2620
19	m(3)=	$h*f(x_n+0.5*h,y_n+0.5*m(2))$	$h*[(0+0.5*h)+(1+0.5*m(2))^2]=$	0.2758
20	m(4)=	$h*f(x_n+h,y_n+m(3))$	$h*[(0+h)+(1+m(3))^2]=$	0.3655
21	**y(1)=**	y(0) +(1/6)*(m(1) + 2*m(2)+2*m(3) + m(4))		**1.2735**

3.

The analytical solution is found with

```
syms x y
y=dsolve('Dy=3*x^2,y(2)=0.5','x')
```

and MATLAB displays

```
y  =
x^3-15/2
```

Next, we write and save the following statements as function file **fexer9_3**

```
function Dy=fexer9_3(x,y);
Dy=3*x^2;
```

The MATLAB script for the numerical solution is as follows:

```
tspan=[2 4];      % Interval over which we want to evaluate y=f(x)
y0=7.5;           % Initial condition: Since y=x^3-15/2 and y(2) = 0.5, it follows that y(0) = 7.5
[x,y]=ode23('fexer9_3', tspan, y0);      % Use 2nd and 3rd Order Runge-Kutta
                  % Plot numeric values with the statements below
plot(x, y, '+r-')
title('Numeric Solution of Differential Equation of Exercise 9.3'),...
xlabel('x'), ylabel('y'), grid
```

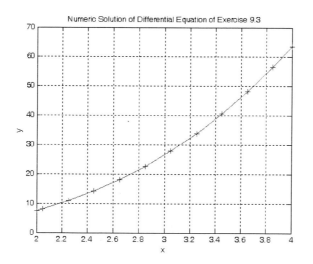

4.

We write and save the following statements as function file **fexer4**

```
function Dy=fexer4(x,y);
Dy=−y^3+0.2*sin(x);
```

The MATLAB script and the plot for the numerical solution are as follows:

```
tspan=[0 10]; x0=[0.707];
[x,num_x]=ode23('fexer4',tspan,x0); plot(x,num_x,'+', x,num_x, '−'),...
title('Numeric solution of non−linear differential equation dy/dx=−x^3+0.2sinx'),...
xlabel('x'), ylabel('y=f(x)'), grid
```

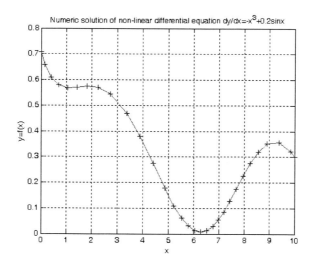

5. The spreadsheet is shown on the following two pages.

	A	B	C	D	E	F
1	Differential Equation is $y' = x^2 - y$					
2	Numerical solution by Runga-Kutte method follows					
3						
4	h=	x(1)-x(0)=	0.1000		x_0=	0.0
5	x(0)=	(given)	0.0000		x_1=	0.1
6	y(0)=	Initial condition (given)	1.0000		x_2=	0.2
7						
8						
9	L(1)=	h*f(x_n,y_n)	h*(0-1)=	-0.1000		
10	L(2)=	h*f(x_n+0.5*h,y_n+0.5*L(1))	h*[(0+0.5*h)^2-(1+0.5*L(1))]=	-0.0948		
11	L(3)=	h*f(x_n+h,y_n+2*L(2)-L(1))	h*[(0+h)^2-(1+2*L(2)-L(1))]=	-0.0901		
12						
13	y(1)=	y(0) +(L(1) + 4*L(2) + L(3))/6		**0.9052**		
14						
15	h=	x(2)-x(1)=	0.1000			
16	x(0.1)=	Next value x(0) + h	0.1000			
17	y(1)=	From previous computation	0.9052			
18						
19	L(1)=	h*f(x_n,y_n)	h*(0.1^2-0.9052)=	-0.0895		
20	L(2)=	h*f(x_n+0.5*h,y_n+0.5*L(1))	h*[(0+0.5*h)^2-(0.9052+0.5*L(1))]=	-0.0838		
21	L(3)=	h*f(x_n+h,y_n+2*L(2)-L(1))	h*[(0.1+h)^2-(0.9052+2*L(2)-L(1))]=	-0.0787		
22						
23	y(2)=	y(1) +(L(1) + 4*L(2) + L(3))/6		**0.8213**		
24						
25	h=	x(2)-x(1)=	0.1000			
26	x(0.2)=	Next value x(0) + 2*h	0.2000			
27	y(2)=	From previous computation	0.8213			
28						
29	L(1)=	h*f(x_n,y_n)	h*(0.2+2*1.5146)=	0.1843		
30	L(2)=	h*f(x_n+0.5*h,y_n+0.5*L(1))	h*[(0+0.5*h)+2*(1+0.5*L(1))]=	0.2077		
31	L(3)=	h*f(x_n+h,y_n+2*L(2)-L(1))	h*[(0+h)+2*(1+2*L(2)-L(1))]=	0.2405		
32						
33	y(3)=	y(2) +(L(1) + 4*L(2) + L(3))/6		**1.0305**		
34						
35	h=	x(3)-x(2)=	-0.2000			
36	x(0.3)=	Next value x(0) + 3*h	-0.6000			
37	y(3)=	From previous computation	1.0305			
38						
39	L(1)=	h*f(x_n,y_n)	h*(0.3+2*1.8773)=	-0.2922		
40	L(2)=	h*f(x_n+0.5*h,y_n+0.5*L(1))	h*[(0+0.5*h)+2*(1+0.5*L(1))]=	-0.2138		
41	L(3)=	h*f(x_n+h,y_n+2*L(2)-L(1))	h*[(0+h)+2*(1+2*L(2)-L(1))]=	-0.1981		
42						
43	y(4)=	y(2) +(L(1) + 4*L(2) + L(3))/6		**0.8063**		

continued on next page

	A	B	C	D	E	F
44						
45	h=	x(4)-x(3)=	0.0000			
46	x(0.4)=	Next value x(0) + 4*h	0.0000			
47	y(3)=	From previous computation	0.8063			
48						
49	L(1)=	h*f(x$_n$,y$_n$)	h*(0.3+2*1.8773)=	0.0000		
50	L(2)=	h*f(x$_n$+0.5*h,y$_n$+0.5*L(1))	h*[(0+0.5*h)+2*(1+0.5*L(1))]=	0.0000		
51	L(3)=	h*f(x$_n$+h,y$_n$+2*L(2)-L(1))	h*[(0+h)+2*(1+2*L(2)-L(1))]=	0.0000		
52						
53	y(4)=	y(2) +(L(1) + 4*L(2) + L(3))/6		**0.8063**		

6.

	A	B	C
1	Differential Equation is y''+y'=xy or **y''=xy-y'**		
2	Numerical solution by Runga-Kutte method follows		
3			
4	h=	x(1)-x(0)=	0.1000
5	x(0)=	Initial condition (given)	0.0000
6	y(0)=	Initial condition (given)	1.0000
7	y'(0)=	Initial condition (given)	-1.0000
8			
9	L(1)=	h*y'(0)=	-0.1000
10	L'(1)=	h*f(x(0), y(0), y'(0))=	0.1000
11	L(2)=	h*(y'(0) + 0.5*L'(1))=	-0.0950
12	L'(2)=	h*f(x(0) + 0.5*h, y(0) + 0.5*L(1), y'(0) + 0.5*L(1))=	0.0998
13	L(3)=	h*(y'(0) + 2*L(2) - L'(1))=	-0.0901
14	L'(3)=	h*f(x(0) + h, y(0) + 2*L(2) - L(1), y'(0) + 2*L'(2) - L'(1))=	0.0992
15			
16	y(1)=	y(0) +(L(1) + 4*L(2) + L(3))/6	**0.9050**
17	y'(1)=	y'(0) + (L'(1) + 4*L'(2) + L'(3))/6=	**-0.9003**
18			
19	h=	x(1)-x(0)=	0.1000
20	x(0.1)=	Next value x(0) + h	0.1000
21	y(1)=	From previous computation	0.9050
22	y'(1)=	From previous computation	-0.9003
23			

continued on next page

	A	B	C
24	L(1)=	h*y'(1)=	-0.0900
25	L'(1)=	h*f(x(1), y(1), y'(1))=	0.0991
26	L(2)=	h*(y'(1) + 0.5*L'(1))=	-0.0851
27	L'(2)=	h*f(x(1) + 0.5*h, y(1) + 0.5*L(1), y'(1) + 0.5*L'(1))=	0.0980
28	L(3)=	h*(y'(1) + 2*L'(2) - L'(1))=	-0.0803
29	L'(3)=	h*f(x(1) + h, y(1) + 2*L(2) - L(1), y'(1) + 2*L'(2) - L'(1))=	0.0968
30			
31	y(2)=	y(0) +(L(1) + 4*L(2) + L(3))/6	**0.8199**
32	y'(2)=	y'(0) + (L'(1) + 4*L'(2) + L'(3))/6=	**-0.8023**
33			
34	h=	x(2)-x(1)=	0.1000
35	x(0.2)=	Next value x(0) + 2*h	0.2000
36	y(2)=	From previous computation	0.8199
37	y'(2)=	From previous computation	-0.8023
38			
39	L(1)=	h*y'(0)=	-0.0802
40	L'(1)=	h*f(x(0), y(0), y'(0))=	0.0966
41	L(2)=	h*(y'(0) + 0.5*L'(1))=	-0.0754
42	L'(2)=	h*f(x(0) + 0.5*h, y(0) + 0.5*L(1), y'(0) + 0.5*L(1))=	0.0949
43	L(3)=	h*(y'(0) + 2*L'(2) - L'(1))=	-0.0709
44	L'(3)=	h*f(x(0) + h, y(0) + 2*L(2) - L(1), y'0 + 2*L'(2) - L'(1))=	0.0934
45			
46	y(3)=	y(0) +(L(1) + 4*L(2) + L(3))/6	**0.7444**
47	y'(3)=	y'(0) + (L'(1) + 4*L'(2) + L'(3))/6=	**-0.7074**

Chapter 10

T his chapter is an introduction to numerical methods for integrating functions which are very difficult or impossible to integrate using analytical means. We will discuss the trapezoidal rule that computes a function $f(x)$ with a set of linear functions, and Simpson's rule that computes a function $f(x)$ with a set of quadratic functions.

10.1 The Trapezoidal Rule

Consider the function $y = f(x)$ for the interval $a \leq x \leq b$, shown in Figure 10.1.

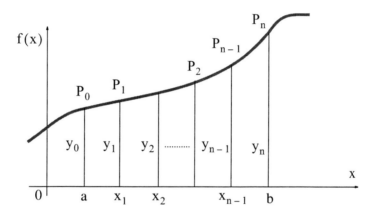

Figure 10.1. Integration by the trapezoidal rule

To evaluate the definite integral $\int_a^b f(x)dx$, we divide the interval $a \leq x \leq b$ into n subintervals each of length $\Delta x = \dfrac{b-a}{n}$. Then, the number of points between $x_0 = a$ and $x_n = b$ is $x_1 = a + \Delta x$, $x_2 = a + 2\Delta x$, ..., $x_{n-1} = a + (n-1)\Delta x$. Therefore, the integral from *a* to *b* is the sum of the integrals from a to x_1, from x_1 to x_2, and so on, and finally from x_{n-1} to b. The total area is

$$\int_a^b f(x)dx = \int_a^{x_1} f(x)dx + \int_{x_1}^{x_2} f(x)dx + \ldots + \int_{x_{n-1}}^b f(x)dx = \sum_{k=1}^n \int_{x_{k-1}}^{x_k} f(x)dx$$

The integral over the first subinterval, can now be approximated by the area of the trapezoid

$aP_0P_1x_1$ that is equal to $\frac{1}{2}(y_0 + y_1)\Delta x$ plus the area of the trapezoid $x_1P_1P_2x_2$ that is equal to

$\frac{1}{2}(y_1 + y_2)\Delta x$, and so on. Then, the trapezoidal approximation becomes

$$T = \frac{1}{2}(y_0 + y_1)\Delta x + \frac{1}{2}(y_1 + y_2)\Delta x + \ldots + \frac{1}{2}(y_{n-1} + y_n)\Delta x$$

or

$$\boxed{T = \left(\frac{1}{2}y_0 + y_1 + y_2 + \ldots + y_{n-1} + \frac{1}{2}y_n\right)\Delta x}$$

$$\text{Trapezoidal Rule}$$

(10.1)

Example 10.1

Using the trapezoidal rule with $n = 4$, estimate the value of the definite integral

$$\int_1^2 x^2 dx$$

(10.2)

Compare with the exact value, and compute the percent error.

Solution:

The exact value of this integral is

$$\int_1^2 x^2 dx = \left.\frac{x^3}{3}\right|_1^2 = \frac{8}{3} - \frac{1}{3} = \frac{7}{3} = 2.33333$$

(10.3)

For the trapezoidal rule approximation we have

$$x_0 = a = 1$$

$$x_n = b = 2$$

$$n = 4$$

$$\Delta x = \frac{b-a}{n} = \frac{2-1}{4} = \frac{1}{4}$$

$$y = f(x) = x^2$$

Then,

$$x_0 = a = 1 \qquad y_0 = f(x_0) = 1^2 = \frac{16}{16}$$

$$x_1 = a + \Delta x = \frac{5}{4} \qquad y_1 = f(x_1) = \left(\frac{5}{4}\right)^2 = \frac{25}{16}$$

$$x_2 = a + 2\Delta x = \frac{6}{4} \qquad y_2 = f(x_2) = \left(\frac{6}{4}\right)^2 = \frac{36}{16}$$

$$x_3 = a + 3\Delta x = \frac{7}{4} \qquad y_3 = f(x_3) = \left(\frac{7}{4}\right)^2 = \frac{49}{16}$$

$$x_4 = b = 2 \qquad y_4 = f(x_4) = \left(\frac{8}{4}\right)^2 = \frac{64}{16}$$

and by substitution into (10.1),

$$T = \left(\frac{1}{2} \times \frac{16}{16} + \frac{25}{16} + \frac{36}{16} + \frac{49}{16} + \frac{1}{2} \times \frac{64}{16}\right) \times \frac{1}{4} = \frac{150}{16} \times \frac{1}{4} = \frac{75}{32} = 2.34375 \qquad (10.4)$$

From (10.3) and (10.4), we find that the percent error is

$$\% \text{ Error} = \frac{2.34375 - 2.33333}{2.33333} \times 100 = 0.45\ \% \qquad (10.5)$$

The MATLAB function **trapz(x,y,n)** where **y** is the integral with respect to **x**, approximates the integral of a function $y = f(x)$ using the trapezoidal rule, and **n** (optional) performs integration along dimension n.

Example 10.2

Use the MATLAB function **trapz(x,y)** to approximate the value of the integral

$$\int_1^2 \frac{1}{x} dx \qquad (10.6)$$

and by comparison with the exact value, compute the percent error when $n = 5$ and $n = 10$

Solution:

The exact value is found from

$$\int_1^2 \frac{1}{x} dx = \ln x \Big|_1^2 = \ln 2 - \ln 1 = 0.6931 - 0.0000 = 0.6931$$

For the approximation using the trapezoidal rule, we let x_5 represent the row vector with $n = 5$, and x_{10} the vector with $n = 10$, that is, $\Delta x = 1/5$ and $\Delta x = 1/10$ respectively. The corresponding values of y are denoted as y_5 and y_{10}, and the areas under the curve as area5 and area10 respectively. We use the following MATLAB script.

```
x5=linspace(1,2,5); x10=linspace(1,2,10);
y5=1./x5; y10=1./x10;
area5=trapz(x5,y5), area10=trapz(x10,y10)

area5 =
    0.6970

area10 =
    0.6939
```

The percent error when $\Delta x = 1/5$ is used is

$$\% \text{ Error} = \frac{0.6970 - 0.6931}{0.6931} \times 100 = 0.56 \%$$

and the percent error when $\Delta x = 1/10$ is used is

$$\% \text{ Error} = \frac{0.6939 - 0.6931}{0.6931} \times 100 = 0.12 \%$$

Example 10.3

The integral

$$f(t) = \int_0^t e^{-\tau^2} d\tau \qquad (10.7)$$

where τ is a dummy variable of integration, is called *the error function*[*] and it is used extensively in communications theory. Use the MATLAB **trapz(x,y)** function to find the area under this integral with $n = 10$ when the upper limit of integration is $t = 2$.

Solution:

We use the same procedure as in the previous example. The MATLAB script for this example is

```
t=linspace(0,2,10); y=exp(-t.^2); area=trapz(t,y)
```

MATLAB displays the following result.

---

[*] *The formal definition of the error function is* $\text{erf}(u) = \dfrac{2}{\sqrt{\pi}} \displaystyle\int_0^u e^{-\tau^2} d\tau$

```
area =
    0.8818
```

Example 10.4

The $i - v$ (current–voltage) relation of a non–linear electrical device is given by

$$i(t) = 0.1(e^{0.2v(t)} - 1) \qquad (10.8)$$

where $v(t) = \sin 3t$.

By any means, find

a. The instantaneous power $p(t)$

b. The energy $W(t_0, t_1)$ dissipated in this device from $t_0 = 0$ to $t_1 = 10$ s.

Solution:

a. The instantaneous power is

$$p(t) = v(t)i(t) = 0.1\sin 3t(e^{0.2\sin 3t} - 1) \qquad (10.9)$$

b. The energy is the integral of the instantaneous power, that is,

$$W(t_0, t_1) = \int_{t_0}^{t_1} p(t)dt = 0.1\int_{0}^{10\,s} \sin 3t(e^{0.2\sin 3t} - 1)dt \qquad (10.10)$$

An analytical solution of the last integral is possible using integration by parts, but it is not easy. We can try the MATLAB **int(f,a,b)** function where **f** is a symbolic expression, and **a** and **b** are the lower and upper limits of integration respectively.

When MATLAB cannot find a solution, it returns a warning. For this example, MATLAB returns the following message when integration is attempted with the symbolic expression of (10.10).

```
t=sym('t');
s=int(0.1*sin(3*t)*(exp(0.2*sin(3*t))−1),0,10)
```

When this script is executed, MATLAB displays the following message.

```
Warning: Explicit integral could not be found.
```

Next, we will find and sketch the power and energy by the trapezoidal rule using the MATLAB **trapz(x,y)** function. For this example, we choose $n = 100$, so that $\Delta x = 1/100$. The MATLAB script below will compute and plot the power.

```
t=linspace(0,10,100);
v=sin(3.*t); i=0.1.*(exp(0.2.*v)−1); p=v.*i;
```

plot(t,p); grid; title('Power vs Time'); xlabel('seconds'); ylabel('watts')

The power varies in a uniform fashion as shown by the plot of Figure 10.2.

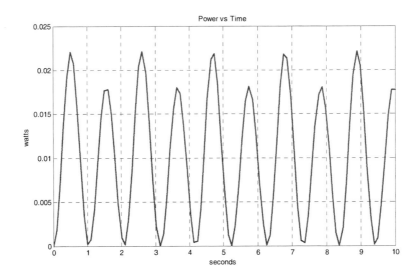

Figure 10.2. Plot for the power variation in Example 10.4

The plot of Figure 10.2 shows that the power is uniform for all time, and thus we expect the energy to be constant.

The MATLAB script below computes and plots the energy.

energy=trapz(t,p), plot(t,energy, '+'); grid; title('Energy vs Time');...
xlabel('seconds'); ylabel('joules')

```
energy =
    0.1013
```

Thus, the value of the energy is 0.1013 joule. The energy is shown in Figure 10.3.

10.2 Simpson's Rule

The trapezoidal and Simpson's rules are special cases of the *Newton–Cote rules* which use higher degree functions for numerical integration.

Let the curve of Figure 10.4 be represented by the parabola

$$y = \alpha x^2 + \beta x + \gamma \tag{10.11}$$

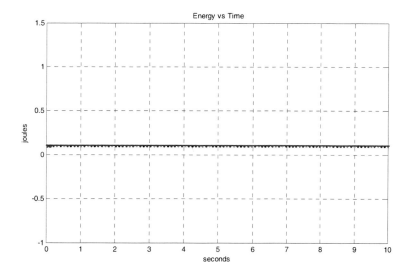

Figure 10.3. Plot for the energy of Example 10.4

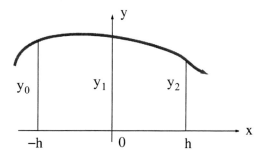

Figure 10.4. Simpson's rule of integration

The area under this curve for the interval $-h \le x \le h$ is

$$
\begin{aligned}
\text{Area}\Big|_{-h}^{h} = \int_{-h}^{h} (\alpha x^2 + \beta x + \gamma)dx &= \frac{\alpha x^3}{3} + \frac{\beta x^2}{2} + \gamma x \Big|_{-h}^{h} \\
&= \frac{\alpha h^3}{3} + \frac{\beta h^2}{2} + \gamma h - \left(-\frac{\alpha h^3}{3} + \frac{\beta h^2}{2} - \gamma h \right) = \frac{2\alpha h^3}{3} + 2\gamma h \\
&= \frac{1}{3}h(2\alpha h^3 + 6\gamma)
\end{aligned}
\tag{10.12}
$$

The curve passes through the three points $(-h, y_0)$, $(0, y_1)$, and (h, y_2). Then, by (10.11) we have:

$$y_0 = \alpha h^2 - \beta h + \gamma \qquad \text{(a)}$$

$$y_1 = \gamma \qquad \text{(b)} \qquad\qquad (10.13)$$

$$y_2 = \alpha h^2 + \beta h + \gamma \qquad \text{(c)}$$

We can now evaluate the coefficients α, β, γ and express (10.12) in terms of h, y_0, y_1 and y_2. This is done with the following procedure.

By substitution of (b) of (10.13) into (a) and (c) and rearranging we obtain

$$\alpha h^2 - \beta h = y_0 - y_1 \qquad\qquad (10.14)$$

$$\alpha h^2 + \beta h = y_2 - y_1 \qquad\qquad (10.15)$$

Addition of (10.14) with (10.15) yields

$$2\alpha h^2 = y_0 - 2y_1 + y_2 \qquad\qquad (10.16)$$

and by substitution into (10.12) we obtain

$$\text{Area}\Big|_{-h}^{h} = \frac{1}{3}h(2\alpha h^3 + 6\gamma) = \frac{1}{3}h[(y_0 - 2y_1 + y_2) + 6y_1] \qquad\qquad (10.17)$$

or

$$\text{Area}\Big|_{-h}^{h} = \frac{1}{3}h(y_0 + 4y_1 + y_2) \qquad\qquad (10.18)$$

Now, we can apply (10.18) to successive segments of any curve $y = f(x)$ in the interval $a \le x \le b$ as shown on the curve of Figure 10.5.

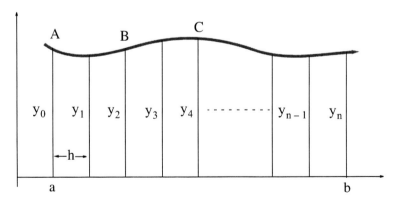

Figure 10.5. Simpson's rule of integration by successive segments

From Figure 10.5, we observe that each segment of width $2h$ of the curve can be approximated by a parabola through its ends and its midpoint. Thus, the area under segment AB is

$$\text{Area}\big|_{AB} = \frac{1}{3}h(y_0 + 4y_1 + y_2) \tag{10.19}$$

Likewise, the area under segment BC is

$$\text{Area}\big|_{BC} = \frac{1}{3}h(y_2 + 4y_3 + y_4) \tag{10.20}$$

and so on. When the areas under each segment are added, we obtain

$$\text{Area} = \frac{1}{3}h(y_0 + 4y_1 + 2y_2 + 4y_3 + 2y_4 + \dots + 2y_{n-2} + 4y_{n-1} + y_n) \tag{10.21}$$

Simpson's Rule of Numerical Integration

Since each segment has width $2h$, *to apply Simpson's rule of numerical integration, the number* n *of subdivisions must be even.* This restriction does not apply to the trapezoidal rule of numerical integration. The value of h for (10.21) is found from

$$h = \frac{b-a}{n} \quad n = \text{even} \tag{10.22}$$

Example 10.5

Using Simpson's rule with 4 subdivisions ($n = 4$), compute the approximate value of

$$\int_1^2 \frac{1}{x}dx \tag{10.23}$$

Solution:

This is the same integral as that of Example 10.2 where we found that the analytical value of this definite integral is $\ln = 0.6931$. We can also find the analytical value with MATLAB's **int(f,a,b)** function where **f** is a symbolic expression, and **a** and **b** are the lower and upper limits of integration respectively. For this example,

```
syms x
Area=int(1/x,1,2)

Area =
log(2)
```

We recall that **log(x)** in MATLAB is the natural logarithm.

To use Simpson's rule, for convenience, we construct the following table using the spreadsheet of Figure 10.6.

	A	B	C	D	E
1	**Example 10.5**				
2	$\int (1/x)dx$ evaluated from a = 1 to b = 2 with n = 4				
3	Numerical integration by Simpson's method follows				
4	Given	a=	1		
5		b=	2		
6		n=	4		
7	Then,	h = (b-a)/n =	0.2500		
8				Multiplier	Products
9		x_0=a=	1.00000		
10		y_0=1/x_0=	1.00000	1	1.00000
11		x_1=a+h=	1.25000		
12		y_1=1/x_1=	0.80000	4	3.20000
13		x_2=a+2h=	1.50000		
14		y_2=1/x_2=	0.66667	2	1.33333
15		x_3=a+3h=	1.75000		
16		y_3=1/x_3=	0.57143	4	2.28571
17		x_4=b=	2.00000		
18		y_4=1/x_4=	0.50000	1	0.50000
19			Sum of Products =		8.31905
20	Area = (h/3)*(Sum of Products) = (1/12)*8.31905 =				**0.69325**

Figure 10.6. Spreadsheet for numerical integration of (10.23)

By comparison of the numerical with the exact value, we observe that the error is very small when Simpson's method is applied.

MATLAB has two quadrature functions for performing numerical integration, the **quad** and **quad8**. The description of these can be seen by typing help quad or help quad8. Both of these functions use *adaptive quadrature methods*; this means that these methods can handle irregularities such as singularities. When such irregularities occur, MATLAB displays a warning message but still provides an answer.

The **quad** function uses an adaptive form of Simpson's rule, while the **quad8** function uses the so–called *Newton–Cotes 8–panel rule*. The **quad8** function detects and handles irregularities more efficiently.

Both functions have the same syntax, that is, **q=quad('f',a,b,tol)**, and integrate to a relative error **tol** which we must specify. If **tol** is omitted, it is understood to be the *standard tolerance* of 10^{-3}. The string **'f'** is the name of a user defined function, and **a** and **b** are the lower and upper limits of integration respectively.

Example 10.6

Given the definite integral

$$y = f(x) = \int_0^2 e^{-x^2} dx \tag{10.24}$$

a. Use MATLAB's symbolic **int** function to obtain the value of this integral

b. Obtain the value of this integral with the **q=quad('f',a,b)** function

c. Obtain the value of this integral with the **q=quad('f',a,b,tol)** function where tol = 10^{-10}

d. Obtain the value of this integral with the **q=quad8('f',a,b)** function

e. Obtain the value of this integral with the **q=quad8('f',a,b,tol)** function where tol = 10^{-10}

Solution:

a.

```
syms x; y=int(exp(-x^2),0,2)  % Define symbolic variable x and integrate

y =
1/2*erf(2)*pi^(1/2)
pretty(y)
                    1/2
          1/2 erf(2) pi
```

`erf` is an acronym for the error function and we can obtain its definition with help erf

b. First, we need to create and save a function m–file. We name it errorfcn1.m as shown below. We will use format long to display the values with 15 digits.

```
function y = errorfcn1(x)
y = exp(-x.^2);
```

With this file saved as errorfcn1.m, we write and execute the following MATLAB script.

```
format long
y_std=quad('errorfcn1',0,2)
```

We obtain the answer in standard tolerance form as

```
y_std =
      0.88211275610253
```

c. With the specified tolerance, the script and the answer are as follows:

```
y_tol=quad('errorfcn1',0,2,10^-10)

y_tol =
      0.88208139076242
```

d. With the standard tolerance,

```
y_std8=quad8('errorfcn1',0,2)

y_std8 =
     0.88208139076194
```

e. With the specified tolerance,

```
y_tol8=quad8('errorfcn1',0,2,10^-10)

y_tol8 =
     0.88208139076242
```

We observe that with the 10^{-10} tolerance, both **quad** and **quad8** produce the same result.

Example 10.7

Using the **quad** and **quad8** functions with standard tolerance, evaluate the integral

$$y = f(x) = \int_a^b \sqrt{x}\,dx \tag{10.25}$$

at $((a, b) = (0.2, 0.8), (1.4, 2.3))$, and $(3,8)$. Use the **fprintf** function to display first the analytical values, then, the numerical values produced by the **quad** and **quad8** functions for each set of data.

Solution:

Evaluating the given integral, we obtain

$$y = \int_a^b x^{1/2}dx = \left.\frac{x^{3/2}}{3/2}\right|_a^b = \frac{2}{3}(b^{3/2} - a^{3/2}) \tag{10.26}$$

where a and b are non–negative values. Substitution of the values of the given values of a and b will be included in the MATLAB script below.

The **sqrt** function in a built–in function and therefore, we need not write a user defined m–file. We will include the **input** function in the script. The script is then saved as *Example_10_7*.

```
% This script displays the approximations obtained with the quad and quad8 functions
% with the analytical results for the integration of the square root of x over the
% interval (a,b) where a and b are non–negative.
%
fprintf(' \n'); % Insert line
a=input('Enter first point  "a" (non–negative): ');
b=input('Enter second point "b" (non–negative): ');
```

```
k=2/3.*(b.^(1.5)-a.^(1.5));
kq=quad('sqrt',a,b);
kq8=quad8('sqrt',a,b);
fprintf(' \n');... % Insert line
fprintf(' Analytical: %f \n Numerical quad, quad8: %f  %f \n',k,kq,kq8);...
fprintf(' \n'); fprintf(' \n') % Insert two lines
```

Now, we execute this saved file by typing its name, that is,

Example_10_7

Enter first point "a" (non-negative): 0.2
Enter second point "b" (non-negative): 0.8

Analytical: 0.417399

Numerical quad, quad8: 0.417396 0.417399

Example_10_7

Enter first point "a" (non-negative): 1.4
Enter second point "b" (non-negative): 2.3

Analytical: 1.221080

Numerical quad, quad8: 1.221080 1.221080

Example_10_7

Enter first point "a" (non-negative): 3
Enter second point "b" (non-negative): 8

Analytical: 11.620843

Numerical quad, quad8: 11.620825 11.620843

10.3 Summary

- We can evaluate a definite integral $\int_a^b f(x)dx$ with the trapezoidal approximation

$$T = \left(\frac{1}{2}y_0 + y_1 + y_2 + \ldots + y_{n-1} + \frac{1}{2}y_n\right)\Delta x$$

by dividing interval $a \le x \le b$ into n subintervals each of length $\Delta x = \frac{b-a}{n}$. The number n of subdivisions can be even or odd.

- The MATLAB function **trapz(x,y,n)** where **y** is the integral with respect to **x**, approximates the integral of a function $y = f(x)$ using the trapezoidal rule, and **n** (optional) performs integration along dimension n.

- We can perform numerical integration with the MATLAB function **int(f,a,b)** function where **f** is a symbolic expression, and **a** and **b** are the lower and upper limits of integration respectively.

- We can evaluate a definite integral $\int_a^b f(x)dx$ with Simpson's rule of numerical integration using the expression

$$\text{Area} = \frac{1}{3}h(y_0 + 4y_1 + 2y_2 + 4y_3 + 2y_4 + \ldots + 2y_{n-2} + 4y_{n-1} + y_n)$$

where the number n of subdivisions must be even.

- The trapezoidal and Simpson's rules are special cases of the *Newton–Cote rules* which use higher degree functions for numerical integration.

- MATLAB has two quadrature functions for performing numerical integration, the **quad** and **quad8**. Both of these functions use adaptive quadrature methods. The **quad** function uses an adaptive form of Simpson's rule, while the **quad8** function uses the so–called *Newton–Cotes 8–panel rule*. The **quad8** function detects and handles irregularities more efficiently. Both functions have the same syntax, that is, **q=quad('f',a,b,tol)**, and integrate to a relative error **tol** which we must specify. If **tol** is omitted, it is understood to be the *standard tolerance* of 10^{-3}. The string **'f'** is the name of a user defined function, and **a** and **b** are the lower and upper limits of integration respectively.

10.4 Exercises

1. Use the trapezoidal approximation to compute the values the following definite integrals and compare your results with the analytical values. Verify your answers with the MATLAB **trapz(x,y,n)** function.

a. $\displaystyle\int_0^2 x\, dx \qquad n = 4$

b. $\displaystyle\int_0^2 x^3\, dx \qquad n = 4$

c. $\displaystyle\int_0^2 x^4\, dx \qquad n = 4$

d. $\displaystyle\int_1^2 \frac{1}{x^2}\, dx \qquad n = 4$

2. Use Simpson's rule to approximate the following definite integrals and compare your results with the analytical values. Verify your answers with the MATLAB **quad('f',a,b)** function.

a. $\displaystyle\int_0^2 x^2\, dx \qquad n = 4$

b. $\displaystyle\int_0^\pi \sin x\, dx \qquad n = 4$

c. $\displaystyle\int_0^1 \frac{1}{x^2 + 1}\, dx \qquad n = 4$

10.5 Solution to End–of–Chapter Exercises

1.

$$T = \left(\frac{1}{2}y_0 + y_1 + y_2 + \ldots + y_{n-1} + \frac{1}{2}y_n\right)\Delta x$$

a. The exact value is

$$\int_0^2 x\,dx = \left.\frac{x^2}{2}\right|_0^2 = 2$$

For the trapezoidal rule approximation we have

$$x_0 = a = 0$$
$$x_n = b = 2$$
$$n = 4$$
$$\Delta x = \frac{b-a}{n} = \frac{2-0}{4} = \frac{1}{2}$$
$$y = f(x) = x$$

$$x_0 = a = 0 \qquad\qquad y_0 = f(x_0) = 0$$
$$x_1 = a + \Delta x = \frac{1}{2} \qquad y_1 = f(x_1) = \frac{1}{2}$$
$$x_2 = a + 2\Delta x = 1 \qquad y_2 = f(x_2) = 1$$
$$x_3 = a + 3\Delta x = \frac{3}{2} \qquad y_3 = f(x_3) = \frac{3}{2}$$
$$x_4 = b = 2 \qquad\qquad y_4 = f(x_4) = 2$$

$$T = \left(\frac{1}{2}\times 0 + \frac{1}{2} + 1 + \frac{3}{2} + \frac{1}{2}\times 2\right)\times\frac{1}{2} = 4\times\frac{1}{2} = 2$$

x=linspace(0,2,4); y=x; area=trapz(x,y)

```
area =
      2
```

b. The exact value is

$$\int_0^2 x^3\,dx = \left.\frac{x^4}{4}\right|_0^2 = 4$$

For the trapezoidal rule approximation we have

$$x_0 = a = 0$$

$$x_n = b = 2$$

$$n = 4$$

$$\Delta x = \frac{b-a}{n} = \frac{2-0}{4} = \frac{1}{2}$$

$$y = f(x) = x^3$$

$x_0 = a = 0$	$y_0 = f(x_0) = 0$
$x_1 = a + \Delta x = \dfrac{1}{2}$	$y_1 = f(x_1) = \dfrac{1}{8}$
$x_2 = a + 2\Delta x = 1$	$y_2 = f(x_2) = 1$
$x_3 = a + 3\Delta x = \dfrac{3}{2}$	$y_3 = f(x_3) = \dfrac{27}{8}$
$x_4 = b = 2$	$y_4 = f(x_4) = 8$

$$T = \left(\frac{1}{2} \times 0 + \frac{1}{8} + 1 + \frac{27}{8} + \frac{1}{2} \times 8 \right) \times \frac{1}{2} = \left(5 + \frac{7}{2} \right) \times \frac{1}{2} = 4.25$$

```
x=linspace(0,2,4); y=x.^3; area=trapz(x,y)

area =
    4.4444
```

The deviations from the exact value are due to the small number of divisions n we chose.

c. The exact value is

$$\int_0^2 x^4 dx = \left. \frac{x^5}{5} \right|_0^2 = \frac{32}{5} = 6.4$$

For the trapezoidal rule approximation we have

$$x_0 = a = 0$$

$$x_n = b = 2$$

$$n = 4$$

$$\Delta x = \frac{b-a}{n} = \frac{2-0}{4} = \frac{1}{2}$$

$$y = f(x) = x^4$$

$$x_0 = a = 0 \qquad\qquad y_0 = f(x_0) = 0$$

$$x_1 = a + \Delta x = \frac{1}{2} \qquad\qquad y_1 = f(x_1) = \frac{1}{16}$$

$$x_2 = a + 2\Delta x = 1 \qquad\qquad y_2 = f(x_2) = 1$$

$$x_3 = a + 3\Delta x = \frac{3}{2} \qquad\qquad y_3 = f(x_3) = \frac{81}{8}$$

$$x_4 = b = 2 \qquad\qquad y_4 = f(x_4) = 16$$

$$T = \left(\frac{1}{2} \times 0 + \frac{1}{16} + 1 + \frac{81}{16} + \frac{1}{2} \times 16 \right) \times \frac{1}{2} = \left(9 + \frac{41}{8} \right) \times \frac{1}{2} = 7.0625$$

x=linspace(0,2,4); y=x.^4; area=trapz(x,y)

```
area =
    7.5720
```

d. The exact value is

$$\int_1^2 \frac{1}{x^2} dx = -\frac{1}{x} \Big|_1^2 = \frac{1}{2}$$

For the trapezoidal rule approximation we have

$$x_0 = a = 1$$

$$x_n = b = 2$$

$$n = 4$$

$$\Delta x = \frac{b-a}{n} = \frac{2-1}{4} = \frac{1}{4}$$

$$y = f(x) = 1/x^2$$

$$x_0 = a = 1 \qquad\qquad y_0 = f(x_0) = 1$$

$$x_1 = a + \Delta x = \frac{5}{4} \qquad\qquad y_1 = f(x_1) = \frac{16}{25}$$

$$x_2 = a + 2\Delta x = \frac{3}{2} \qquad\qquad y_2 = f(x_2) = \frac{4}{9}$$

$$x_3 = a + 3\Delta x = \frac{7}{4} \qquad\qquad y_3 = f(x_3) = \frac{16}{49}$$

$$x_4 = b = 2 \qquad\qquad y_4 = f(x_4) = \frac{1}{4}$$

$$T = \left(\frac{1}{2} \times 1 + \frac{16}{25} + \frac{4}{9} + \frac{16}{49} + \frac{1}{2} \times \frac{1}{4} \right) \times \frac{1}{4} = \left(\frac{3905}{1918} \right) \times \frac{1}{4} = 0.5090$$

```
x=linspace(1,2,4); y=1./x.^2; area=trapz(x,y)
```

```
area =
    0.5158
```

2.

$$\text{Area} = \frac{1}{3}h(y_0 + 4y_1 + 2y_2 + 4y_3 + 2y_4 + \ldots + 2y_{n-2} + 4y_{n-1} + y_n)$$

a. The exact value is

$$\int_0^2 x^2 dx = \left.\frac{x^3}{3}\right|_0^2 = \frac{8}{3} = 2.6667$$

To use Simpson's rule we construct the following table using a spreadsheet.

	A	B	C	D	E
1	Exercise 10.2.a				
2	$\int x^2 \, dx$ evaluated from a = 0 to b = 2 with n = 4				
3	Numerical integration by Simpson's method follows				
4	Given	a=	0		
5		b=	2		
6		n=	4		
7	Then,	h = (b-a)/n =	0.5000		
8				Multiplier	Products
9		x_0=a=	0.00000		
10		y_0=x_0^2=	0.00000	1	0.0000
11		x_1=a+h=	0.50000		
12		y_1=x_1^2=	0.25000	4	1.0000
13		x_2=a+2h=	1.00000		
14		y_2=x_2^2=	1.00000	2	2.0000
15		x_3=a+3h=	1.50000		
16		y_3=x_3^2=	2.25000	4	9.0000
17		x_4=b=	2.00000		
18		y_4=x_4^2=	4.00000	1	4.0000
19			Sum of Products =		16.0000
20	Area = (h/3)*(Sum of Products) = (1/12)*8.31905 =				2.6667

We create and save a function m–file. We name it **exer_10_2_a.m** as shown below.

```
function y = exer_10_2_a(x)
y = x.^2;
```

We write and execute the following MATLAB script:

```
y_std=quad('exer_10_2_a',0,2)
```

```
y_std =
    2.6667
```

b. The exact value is

$$\int_0^\pi \sin x \, dx = -\cos x \Big|_0^\pi = -(-1-1) = 2$$

To use Simpson's rule we construct the following table using a spreadsheet.

	A	B	C	D	E
1	**Exercise 10.2.b**				
2	$\int \sin x \, dx$ evaluated from a = 0 to b = π with n = 4				
3	Numerical integration by Simpson's method follows				
4	Given	a=	0		
5		b=	3.14159		
6		n=	4		
7	Then,	h = (b-a)/n =	0.7854		
8				Multiplier	Products
9		x_0=a=	0.00000		
10		y_0=sinx_0=	0.00000	1	0.0000
11		x_1=a+h=	0.78540		
12		y_1=sinx_1=	0.70711	4	2.8284
13		x_2=a+2h=	1.57080		
14		y_2=sinx_2=	1.00000	2	2.0000
15		x_3=a+3h=	2.35619		
16		y_3=sinx_3=	0.70711	4	2.8284
17		x_4=b=	3.14159		
18		y_4=sinx_4=	0.00000	1	0.0000
19			Sum of Products =		7.6569
20	Area = (h/3)*(Sum of Products) = (1/12)*8.31905 =				**2.0046**

We create and save a function m–file. We name it **exer_10_2_b.m** as shown below.

```
function y = exer_10_2_b(x)
y = sin(x);
```

We write and execute the following MATLAB script:

```
y_std=quad('exer_10_2_b',0,pi)

y_std =
    2.0000
```

c. The exact value is

$$\int_0^1 \frac{1}{x^2+1} dx = \tan^{-1}x \Big|_0^1 = \frac{\pi}{4} = 0.7854$$

To use Simpson's rule we construct the following table using a spreadsheet.

Numerical Analysis Using MATLAB® and Excel®, Third Edition
Copyright © Orchard Publications

	A	B	C	D	E
1	**Exercise 10.2.c**				
2	$\int(1/(x^2+1))dx$ evaluated from a = 0 to b = 1 with n = 4				
3	Numerical integration by Simpson's method follows				
4	Given	a=	0		
5		b=	1		
6		n=	4		
7	Then,	h = (b-a)/n =	0.2500		
8				Multiplier	Products
9		x_0=a=	0.00000		
10		y_0=1/(x_0^2+1)=	1.00000	1	1.0000
11		x_1=a+h=	0.25000		
12		y_1=1/(x_1^2+1)=	0.94118	4	3.7647
13		x_2=a+2h=	0.50000		
14		y_2=1/(x_2^2+1)=	0.80000	2	1.6000
15		x_3=a+3h=	0.75000		
16		y_3=1/(x_3^2+1)=	0.64000	4	2.5600
17		x_4=b=	1.00000		
18		y_4=1/(x_4^2+1)=	0.50000	1	0.5000
19			Sum of Products =		9.4247
20	Area = (h/3)*(Sum of Products) = (1/12)*8.31905 =				**0.7854**

We create and save a function m–file. We name it **exer_10_2_c.m** as shown below.

```
function y = exer_10_2_c(x)
y = 1./(x.^2+1);
```

We write and execute the following MATLAB script:

```
y_std=quad('exer_10_2_c',0,1)
```

```
y_std =
    0.7854
```

NOTES:

Numerical Analysis Using MATLAB® and Excel®, Third Edition
Copyright © Orchard Publications

Chapter 11

Difference Equations

T his chapter is an introduction to difference equations based on finite differences. The discussion is limited to linear difference equations with constant coefficients. The Fibonacci numbers are defined, and a practical example in electric circuit theory is given at the end of this chapter.

11.1 Introduction

In mathematics, a *recurrence relation* is an equation which defines a sequence recursively: each term of the sequence is defined as a function of the preceding terms. A *difference equation* is a specific type of recurrence relation, and this type is discussed in this chapter. Difference equations as used with discrete type systems, are discussed in Appendix A.

11.2 Definition, Solutions, and Applications

The difference equations discussed in this chapter, are used in numerous applications such as engineering, mathematics, physics, and other sciences.

The general form of a linear, constant coefficient difference equation has the form

$$(a_r E^r + a_{r-1} E^{r-1} + \ldots + a_1 E + a_0) y = \phi(x) \tag{11.1}$$

where a_k represents a constant coefficient and E is an operator similar to the D operator in ordinary differential equations. The E operator increases the argument of a function by one interval h, and r is a positive integer that denotes the order of the difference equation.

In terms of the interval h, the difference operator E is

$$Ef(x_k) = f(x_k + h) = f(x_{k+h}) \tag{11.2}$$

The interval h is usually unity, i.e., $h = 1$, and the subscript k is normally omitted. Thus, (11.2) is written as

$$Ef(x) = f(x+1) = f_{x+1} \tag{11.3}$$

If, in (11.3), we increase the argument of f by another unit, we obtain the second order operator E^2, that is,

$$E^2 f(x) = E[Ef(x)] = Ef(x+1) = f(x+2) = f_{x+2} \qquad (11.4)$$

and in general,

$$E^r f(x) = f(x+r) = f_{x+r} \qquad (11.5)$$

As with ordinary differential equations, the right side of (11.3) is a linear combination of terms such as kx, $coskx$, and x^n, where k is a non–zero constant and n is a non–negative integer. Moreover, if, in (11.1), $\varphi(x) = 0$, the equation is referred to as a *homogeneous difference equation*, and if $\varphi(x) \neq 0$, it is a *non–homogeneous difference equation*.

If, in (11.1), we let $r = 2$, we obtain the second order difference equation

$$(a_2 E^2 + a_1 E + a_0)y = \phi(x) \qquad (11.6)$$

and if the right side is zero, it reduces to

$$(a_2 E^2 + a_1 E + a_0)y = 0 \qquad (11.7)$$

If $y_1(x)$ and $y_2(x)$ are any two solutions of (11.7), the linear combination $k_1 y_1(x) + k_2 y_1(x)$ is also a solution. Also, if the *Casorati determinant*, analogous to the *Wronskian determinant* in ordinary differential equations, is non–zero, that is, if

$$C[y_1(x), y_2(x)] = \begin{bmatrix} y_1(x) & y_2(x) \\ Ey_1(x) & Ey_2(x) \end{bmatrix} \neq 0 \qquad (11.8)$$

then, any other solution of (11.7) can be expressed as

$$y_3(x) = k_1 y_1(x) + k_2 y_2(x) \qquad (11.9)$$

where k_1 and k_2 are constants.

For the non–homogeneous difference equation

$$(a_2 E^2 + a_1 E + a_0)y = \phi(x) \qquad (11.10)$$

where $\varphi(x) \neq 0$, if $Y(x)$ is any solution of (11.10), then the complete solution is

$$y = k_1 y_1(x) + k_2 y_2(x) + Y(x) \qquad (11.11)$$

As with ordinary differential equations, we first find the solution of the homogeneous difference equation; then, we add the particular solution $Y(x)$ to it to obtain the total solution. We find $Y(x)$ by the *Method of Undetermined Coefficients*.

We have assumed that the coefficients a_i in (11.10) are constants; then, in analogy with the solution of the differential equation of the form $y = ke^{ax}$, for the homogeneous difference equation we assume a solution of the form

$$y = M^x \tag{11.12}$$

By substitution into (11.7), and recalling that $Ef(x) = f(x + 1)$, we obtain

$$a_2 M^{x+2} + a_1 M^{x+1} + a_0 M^x = 0 \tag{11.13}$$

and this is the *characteristic equation of a second order difference equation.*

As with algebraic quadratic equations, the roots of (11.13) can be real and unequal, real and equal, or complex conjugates depending on whether the discriminant $a_1^2 - 4a_2 a_0$ is positive, zero, or negative. These cases are summarized in Table 11.1.

TABLE 11.1 *Roots of the characteristic equation in difference equations*

Characteristic equation $a_2 M^2 + a_1 M + a_0 = 0$ *of* $(a_2 E^2 + a_1 E + a_0)y = 0$		
Roots M_1 and M_2	Discriminant	General Solution
Real and Unequal $M_1 \neq M_2$	$a_1^2 - 4a_2 a_0 > 0$	$y = k_1 M_1^x + k_2 M_2^x$ k_1 and k_2 constants
Real and Equal $M_1 = M_2$	$a_1^2 - 4a_2 a_0 = 0$	$y = k_1 M_1^x + k_2 x M_2^x$ k_1 and k_2 constants
Complex Conjugates $M_1 = \alpha + j\beta$ $M_2 = \alpha - j\beta$	$a_1^2 - 4a_2 a_0 < 0$	$y = r^x (C_1 \cos\theta x + C_2 \sin\theta x)$ $r = \sqrt{\alpha^2 + \beta^2} \quad \theta = \tan^{-1}\frac{\beta}{\alpha}$

Example 11.1

Find the solution of the difference equation

$$(E^2 - 6E + 8)y = 0 \tag{11.14}$$

with initial conditions $y_0 = y(0) = 3$ and $y_1 = y(1) = 2$. Then, compute $y_5 = y(5)$.

Solution:

The characteristic equation of (11.14) is

$$M^2 - 6M + 8 = 0 \tag{11.15}$$

and its roots are $M_1 = 2$ and $M_2 = 4$. Therefore, with reference to Table 11.1, we obtain the solution

$$y_x = y(x) = k_1 2^x + k_2 4^x \tag{11.16}$$

To make use of the first initial condition, we let $x = 0$. Then, (11.16) becomes

$$y_0 = 3 = k_1 2^0 + k_2 4^0$$

or

$$k_1 + k_2 = 3 \tag{11.17}$$

For the second initial condition, we let $x = 1$. Then, (11.16) becomes

$$y_1 = 2 = k_1 2^1 + k_2 4^1$$

or

$$2k_1 + 4k_2 = 2 \tag{11.18}$$

Simultaneous solution of (11.17) and (11.18) yields $k_1 = 5$ and $k_2 = -2$. Thus, the solution is

$$y_x = 5 \cdot 2^x - 2 \cdot 4^x \tag{11.19}$$

For $x = 5$, we obtain

$$y_5 = 5 \cdot 2^5 - 2 \cdot 4^5 = 5 \times 32 - 2 \times 1024 = -1888$$

Example 11.2

Find the solution of the difference equation

$$(E^2 + 2E + 4)y = 0 \tag{11.20}$$

Solution:

The characteristic equation of (11.20) is

$$M^2 + 2M + 4 = 0 \tag{11.21}$$

and its roots are $M_1 = -1 + j\sqrt{3}$ and $M_2 = -1 - j\sqrt{3}$. From Table 11.1, $r = \sqrt{(-1)^2 + (\sqrt{3})^2} = 2$ and $\theta = \tan^{-1}(\sqrt{3}/(-1)) = 2\pi/3$. Therefore, the solution is

$$y = 2^x \left(C_1 \cos \frac{2\pi}{3} x + C_2 \sin \frac{2\pi}{3} x \right) \qquad (11.22)$$

The constants C_1 and C_2 can be evaluated from the initial conditions.

For non–homogeneous difference equations of the form of (11.10), we combine the particular solution with the solution of the homogeneous equation shown in (11.11). For the particular solution, we start with a linear combination of all the terms of the right side, that is, $\phi(x)$, and we apply the operator E. If any of the terms in the initial choice duplicates a term in the solution of the homogeneous equation, this choice must be multiplied by x until there is no duplication of terms.

Table 11.2 shows the form of the particular solution for different terms of $\phi(x)$.

TABLE 11.2 *Form of the particular solution for a non–homogeneous difference equation*

Non–homogeneous difference equation $(a_2 E^2 + a_1 E + a_0)y = \phi(x)$	
$\phi(x)$	**Form of Particular Solution $Y(x)$**
α (constant)	A (constant)
αx^k (k = positive integer)	$A_k x^k + A_{k-1} x^{k-1} + \ldots + A_1 x + A_0$
αk^x	$A k^x$
$\alpha \cos mx$ or $\alpha \sin mx$	$A_1 \cos mx + A_2 \sin mx$
$\alpha x^k l^x \cos mx$ or $\alpha x^k l^x \sin mx$	$(A_k x^k + A_{k-1} x^{k-1} + \ldots + A_1 x + A_0) l^x \cos mx$ $+ (B_k x^k + B_{k-1} x^{k-1} + \ldots + B_1 x + B_0) l^x \sin mx$

Example 11.3

Find the solution of the difference equation

$$(E^2 - 5E + 6)y = x + 2^x \qquad (11.23)$$

Solution:

The characteristic equation of (11.23) is

$$M^2 - 5M + 6 = 0 \qquad (11.24)$$

and its roots are $M_1 = 2$ and $M_2 = 3$. From Table 11.1, the solution Y_H of the homogeneous difference equation is

$$Y_H = k_1 2^x + k_2 3^x \tag{11.25}$$

For the particular solution we refer Table 11.2. For the first term x of the right side of (11.23), we use the term $A_1 x + A_0$, or $Ax + B$. For the second term 2^x, we obtain $A2^x$ or $C2^x$, and thus, the particular solution has the form

$$Y_P = Ax + B + C2^x \tag{11.26}$$

But the term $C2^x$ in (11.26), is also a term in (11.25). Therefore, to eliminate the duplication, we multiply the term $C2^x$ by x. Thus, the correct form of the particular solution is

$$Y_P = Ax + B + Cx2^x \tag{11.27}$$

To evaluate the constants A, B, and C, we substitute (11.27) into (11.23). Then,

$$[A(x+2) + B + C(x+2) \cdot 2^{x+2}] - 5[A(x+1) + B + C(x+1) \cdot 2^{x+1}] \tag{11.28}$$
$$+ 6[Ax + B + Cx2^x] = x + 2^x$$

Using the law of exponents $W^{m+n} = W^m \times W^n$, simplifying, and equating like terms, we obtain

$$2Ax + (-3A + 2B) - 2C2^x = x + 2^x \tag{11.29}$$

Relation (11.29) will be true if

$$2A = 1 \qquad -3A + 2B = 0 \qquad -2C = 1$$

or

$$A = 0.5 \qquad B = 0.75 \qquad C = -0.5$$

By substitution into (11.28), we obtain the particular solution

$$Y_P = 0.5x + 0.75 - 0.5x2^x \tag{11.30}$$

Therefore, the total solution is the sum of (11.25) and (11.30), that is,

$$y_{total} = Y_H + Y_P = k_1 2^x + k_2 3^x + 0.5x + 0.75 - 0.5x2^x \tag{11.31}$$

11.3 Fibonacci Numbers

The *Fibonacci numbers* are solutions of the difference equation

$$y_{x+2} = y_{x+1} + y_x \qquad (11.32)$$

that is, in a series of numbers, each number after the second, is the sum of the two preceding numbers.

Example 11.4

Given that $y_0 = 0$ and $y_1 = 1$, compute the first 12 Fibonacci numbers.

Solution:

For $x = 0, 1, 2, 3$ and so on, we obtain the Fibonacci numbers

$$1, 2, 3, 5, 8, 13, 21, 34, 55, 89, 144, 233, \ldots$$

We will conclude this chapter with an application to electric circuit analysis.

Example 11.5

For the electric network of Figure 11.1, derive an expression for the voltage V_x at each point P_x

where $x = 0, 1, 2, \ldots, n$, given that the voltage V_0 at point P_0 is known.

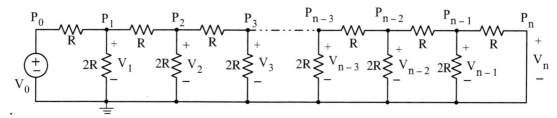

Figure 11.1. Electric network for Example 11.5

Solution:

We need to derive a difference equation that relates the unknown voltage V_x to the known voltage V_0. We start by drawing part of the circuit as shown in Figure 11.2, and we denote the voltages and currents as indicated.

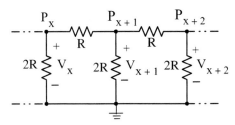

Figure 11.2. Part of the circuit of Figure 11.1

By application of Kirchoff's Current Law (KCL) at node P_{x+1} of Figure 11.2, we obtain

$$\frac{V_{x+1} - V_x}{R} + \frac{V_{x+1}}{2R} + \frac{V_{x+1} - V_{x+2}}{R} = 0 \tag{11.33}$$

and after simplification,

$$\frac{2}{R}(V_{x+2} - 2.5V_{x+1} + V_x) = 0$$

Of course, the term $2/R$ cannot be zero. Therefore, we must have

$$V_{x+2} - 2.5V_{x+1} + V_x = 0 \tag{11.34}$$

Relation (11.34) is valid for all points except P_1 and P_{n-1}* as shown in Figure 11.1; therefore, we must find the current relations at these two points.

Also, by application of Kirchoff's current law (KCL) at node P_1 of Figure 11.1, we obtain

$$\frac{V_1 - V_0}{R} + \frac{V_1}{2R} + \frac{V_1 - V_2}{R} = 0$$

and after simplification,

$$V_2 - 2.5V_1 + V_0 = 0 \tag{11.35}$$

Likewise, at node P_{n-1} of Figure 11.1, we obtain

$$\frac{V_{n-1} - V_{n-2}}{R} + \frac{V_{n-1}}{2R} + \frac{V_{n-1} - V_n}{R} = 0$$

Observing that $V_n = 0$, and simplifying, we obtain

$$2.5V_{n-1} - V_{n-2} = 0 \tag{11.36}$$

---

* *The voltages at nodes P_0 and P_n are V_0 and V_n respectively.*

Relation (11.35) is a difference equation of the form

$$(E^2 - 2.5E + 1)y = 0$$

where $y = V_x$. Its characteristic equation is

$$M^2 - 2.5M + 1 = 0 \tag{11.37}$$

The roots of the characteristic equation of (11.37) are $M_1 = 0.5$ and $M_2 = 2$. Thus, the solution is

$$y = V_x = k_1(0.5)^x + k_2(2)^x \tag{11.38}$$

The constant coefficients k_1 and k_2 in (11.38), are found by substitution of this relation into (11.35) and (11.36). Thus, from (11.37) and (11.38), we obtain

$$k_1(0.5)^2 + k_2(2)^2 - 2.5(k_1(0.5)^1 + k_2(2)^1) + V_0 = 0$$

or

$$0.25k_1 + 4k_2 - 1.25k_1 - 5k_2 + V_0 = 0$$

or

$$k_1 + k_2 = V_0 \tag{11.39}$$

Likewise, from (11.38) and (11.36) we obtain

$$2.5\left(k_1\left(\frac{1}{2}\right)^{n-1} + k_2(2)^{n-1}\right) - k_1\left(\frac{1}{2}\right)^{n-2} - k_2(2)^{n-2} = 0$$

or

$$\frac{2.5k_1}{2^{n-1}} + 2.5k_2(2)^{n-1} - \frac{k_1}{2^{n-2}} - k_2(2)^{n-2} = 0$$

or

$$\frac{2(2.5)k_1}{2^n} + \frac{2.5k_2(2)^n}{2} - \frac{4k_1}{2^n} - \frac{k_2(2)^n}{4} = 0$$

or

$$\frac{k_1}{2^n} + k_2(2)^n = 0 \tag{11.40}$$

Simultaneous solution of (11.39) and (11.40) yields

$$k_1 = \frac{2^{2n}}{2^{2n} - 1}V_0 \qquad k_2 = \frac{-1}{2^{2n} - 1}V_0 \tag{11.41}$$

Finally, substitution of (11.41) into (11.38) yields a solution of the difference equation in terms of V_0, that is,

$$y = V_x = \frac{2^{2n}}{2^{2n}-1}V_0\left(\frac{1}{2}\right)^x + \frac{-1}{2^{2n}-1}V_0(2)^x$$

or

$$y = V_x = \left(\frac{2^{2n}}{2^x} - 2^x\right)\frac{V_0}{2^{2n}-1} \qquad (11.42)$$

We observe that when $x = 0$,

$$y = V_x = \left(\frac{2^{2n}}{1} - 1\right)\frac{V_0}{2^{2n}-1} = V_0$$

and when $x = n$,

$$y = V_x = \left(\frac{2^{2n}}{2^n} - 2^n\right)\frac{V_0}{2^{2n}-1} = (2^n - 2^n)\frac{V_0}{2^{2n}-1} = 0$$

11.4 Summary

- The general form of a linear, constant coefficient difference equation has the form

$$(a_r E^r + a_{r-1} E^{r-1} + \dots + a_1 E + a_0)y = \phi(x)$$

where a_k represents a constant coefficient and E is an operator similar to the D operator in ordinary differential equations. As with ordinary differential equations, the right side is a linear combination of terms such as kx, $\cos kx$, and x^n, where k is a non–zero constant and n is a non–negative integer. If $\varphi(x) = 0$, the equation is referred to as a homogeneous difference equation, and if $\varphi(x) \neq 0$, it is a non–homogeneous difference equation.

- The difference operator E is

$$Ef(x_k) = f(x_k + h) = f(x_{k+h})$$

The interval h is usually unity, i.e., $h = 1$, and the subscript k is normally omitted. Thus, (11.3) is written as

$$Ef(x) = f(x+1) = f_{x+1}$$

and in general,

$$E^r f(x) = f(x+r) = f_{x+r}$$

- If $y_1(x)$ and $y_2(x)$ are any two solutions of a homogeneous difference equation, the linear combination $k_1 y_1(x) + k_2 y_1(x)$, where k_1 and k_2 are constants, is also a solution.

- If the Casorati determinant, analogous to the *Wronskian determinant* in ordinary differential equations, is non–zero, that is, if

$$C[y_1(x), y_2(x)] = \begin{bmatrix} y_1(x) & y_2(x) \\ Ey_1(x) & Ey_2(x) \end{bmatrix} \neq 0$$

then, any other solution of the homogeneous difference equation can be expressed as

$$y_3(x) = k_1 y_1(x) + k_2 y_2(x)$$

- For the non–homogeneous difference equation

$$(a_2 E^2 + a_1 E + a_0)y = \phi(x)$$

where $\varphi(x) \neq 0$, if $Y(x)$ is any solution of (11.11), then the complete solution is

$$y = k_1 y_1(x) + k_2 y_2(x) + Y(x)$$

As with ordinary differential equations, we first find the solution of the homogeneous difference equation; then, we add the particular solution $Y(x)$ to it to obtain the total solution. We find $Y(x)$ by the Method of Undetermined Coefficients.

- In analogy with the solution of the differential equation of the form $y = ke^{ax}$, for the homogeneous difference equation, we assume a solution of the form

$$y = M^x$$

- Since $Ef(x) = f(x + 1)$, the characteristic equation of a second order difference equation is

$$a_2 M^{x+2} + a_1 M^{x+1} + a_0 M^x = 0$$

and as with algebraic quadratic equations, the roots can be real and unequal, real and equal, or complex conjugates depending on whether the discriminant $a_1^2 - 4a_2 a_0$ is positive, zero, or negative. These cases are summarized in Table 11.1.

- For non–homogeneous difference equations we combine the particular solution with the solution of the homogeneous equation. For the particular solution, we start with a linear combination of all the terms of the right side, that is, $\phi(x)$, and we apply the operator E. If any of the terms in the initial choice duplicates a term in the solution of the homogeneous equation, this choice must be multiplied by x until there is no duplication of terms. The form of the particular solution for different terms of $\phi(x)$ is shown in Table 11.2.

- The Fibonacci numbers are solutions of the difference equation

$$y_{x+2} = y_{x+1} + y_x$$

that is, in a series of numbers, each number after the second, is the sum of the two preceding numbers.

11.5 Exercises

Find the total solution of the following difference equations.

1. $(E^2 + 7E + 12)y = 0$

2. $(E^2 + 2E + 2)y = 0$

3. $(E^2 - E - 6)y = x + 3^x$

4. $(E^2 + 1)y = \sin x$

11.6 Solutions to End–of–Chapter Exercises

1.

$$(E^2 + 7E + 12)y = 0$$

The characteristic equation is

$$M^2 + 7M + 12 = 0$$

and its roots are $M_1 = -3$ and $M_2 = -4$. Therefore, with reference to Table 11.1, we obtain the solution

$$y_x = y(x) = k_1(-3)^x + k_2(-4)^x \quad (1)$$

The constants k_1 and k_2 can be evaluated from the initial conditions. Since they were not given, let us assume that $y_0 = y(0) = 1$ and $y_1 = y(1) = 2$.

To make use of the first initial condition, we let $x = 0$. Then, (1) becomes

$$y_0 = 1 = k_1(-3)^0 + k_2(-4)^0$$

or

$$k_1 + k_2 = 1 \quad (2)$$

For the second initial condition, we let $x = 1$. Then, (1) becomes

$$y_1 = 2 = k_1(-3)^1 + k_2(-4)^1$$

or

$$-3k_1 - 4k_2 = 2 \quad (3)$$

Simultaneous solution of (2) and (3) yields $k_1 = 6$ and $k_2 = -5$. Thus, the solution is

$$y_x = y(x) = 6 \times (-3)^x - 5 \times (-4)^x$$

2.

$$(E^2 + 2E + 2)y = 0$$

The characteristic equation is

$$M^2 + 2M + 2 = 0$$

and its roots are $M_1 = -1 + j$ and $M_2 = -1 - j$. From Table 11.1, $r = \sqrt{(-1)^2 + (1)^2} = \sqrt{2}$ and

$\theta = \tan^{-1} 1/(-1) = -\pi/4$. Therefore, the solution is

$$y = \sqrt{2}^x(C_1\cos(-\pi/4)x + C_2\sin(-\pi/4)x)$$

The constants C_1 and C_2 can be evaluated from the initial conditions. For this exercise, they were not given.

3.

$$(E^2 - E - 6)y = x + 3^x$$

The characteristic equation is

$$M^2 - M - 6 = 0$$

and its roots are $M_1 = -2$ and $M_2 = 3$

From Table 11.1, the solution Y_H of the homogeneous difference equation is

$$Y_H = k_1(-2)^x + k_2 3^x \quad (1)$$

For the particular solution we refer Table 11.2. For the first term x of the right side of the given equation we use the term $A_1 x + A_0$, or $Ax + B$. For the second term 3^x, we obtain $A3^x$ or $C3^x$, and thus, the particular solution has the form

$$Y_P = Ax + B + C3^x$$

But the term $C3^x$ is also a term in the given equation. Therefore, to eliminate the duplication, we multiply the term $C3^x$ by x. Thus, the correct form of the particular solution is

$$Y_P = Ax + B + Cx3^x \quad (2)$$

To evaluate the constants A, B, and C, we substitute the last expression above into the given equation. Then,

$$[A(x + 2) + B + C(x + 2) \cdot 3^{x+2}] - [A(x + 1) + B + C(x + 1) \cdot 3^{x+1}]$$
$$-6[Ax + B + Cx3^x] = x + 3^x$$

Using the law of exponents $W^{m+n} = W^m \times W^n$, simplifying, and equating like terms, we obtain

$$-6Ax + (A - 6B) + 15C3^x = x + 3^x$$

This relation will be true if

$$-6A = 1 \qquad A - 6B = 0 \qquad 15C = 1$$

or

$$A = -1/6 \qquad B = 1/36 \qquad C = 1/15$$

By substitution into (2), we obtain the particular solution

$$Y_P = (-1/6)x + 1/36 + (1/15)x2^x \quad (3)$$

Therefore, the total solution is the sum of (1) and (3), that is,

$$y_{total} = Y_H + Y_P = k_1(-2)^x + k_2 3^x + (-1/6)x + 1/36 + (1/15)x2^x$$

4.

$$(E^2 + 1)y = \sin x \quad (1)$$

The characteristic equation is

$$M^2 + 1 = 0$$

and its roots are $M_1 = j$ and $M_2 = -j$

From Table 11.1, $r = \sqrt{(1)^2} = 1$ and $\theta = \tan^{-1} 1/0 = \pi/2$. Therefore, the homogeneous part of the solution is

$$Y_H = C_1 \cos(\pi/2)x + C_2 \sin(\pi/2)x \quad (2)$$

For the particular solution we refer Table 11.2 where we find that the solution has the form $A_1 \cos mx + A_2 \sin mx$, and for this exercise

$$Y_P = A \cos x + B \sin x$$

Since the cosine and sine terms appear in the complimentary solution, we multiply the terms of the particular solution by x and we obtain

$$Y_P = A x \cos x + B x \sin x \quad (3)$$

To evaluate the constants A, B, and C, we substitute the last expression above into (1) and we obtain

$$A(x + 2)\cos(x + 2) + B(x + 2)\sin(x + 2) + A x \cos x + B x \sin x = \sin x$$

Using the trig identities

$$\cos(a + b) = \cos a \cos b - \sin a \sin b$$

$$\sin(a + b) = \sin a \cos b - \sin b \cos a$$

expanding, rearranging, equating like terms, and combining the complimentary and particular solutions we obtain

$$y = C_1 \cos\frac{\pi}{2}x + C_2 \sin\frac{\pi}{2}x + \frac{\sin x + \sin(x - 2)}{2(1 + \cos 2)}$$

Chapter 12

T his chapter is an introduction to partial fraction expansion methods. In elementary algebra we learned how to combine fractions over a common denominator. Partial fraction expansion is the reverse process and splits a rational expression into a sum of fractions having simpler denominators.

12.1 Partial Fraction Expansion

The partial fraction expansion method is used extensively in integration and in finding the inverses of the Laplace, Fourier, and $\mathcal{Z}$ transforms. This method allows us to decompose a rational polynomial into smaller rational polynomials with simpler denominators, from which we can easily recognize their integrals or inverse transformations. In the subsequent discussion we will discuss the partial fraction expansion method and we will illustrate with several examples. We will also use the MATLAB **residue(r,p,k)** function which returns the residues (coefficients) **r** of a partial fraction expansion, the poles **p** and the direct terms **k**. There are no direct terms if the highest power of the numerator is less than that of the denominator.

Let

$$F(s) = \frac{N(s)}{D(s)} \qquad (12.1)$$

where $N(s)$ and $D(s)$ are polynomials and thus (12.1) can be expressed as

$$F(s) = \frac{N(s)}{D(s)} = \frac{b_m s^m + b_{m-1} s^{m-1} + b_{m-2} s^{m-2} + \ldots + b_1 s + b_0}{a_n s^n + a_{n-1} s^{n-1} + a_{n-2} s^{n-2} + \ldots + a_1 s + a_0} \qquad (12.2)$$

The coefficients a_k and b_k for $k = 0, 1, 2, \ldots, n$ are real numbers and, for the present discussion, we have assumed that the highest power of $N(s)$ is less than the highest power of $D(s)$, i.e., $m < n$. In this case, $F(s)$ is a *proper rational function*. If $m \geq n$, $F(s)$ is an *improper rational function*.

It is very convenient to make the coefficient a_n of s^n in (12.2) unity; to do this, we rewrite it as

$$F(s) = \frac{N(s)}{D(s)} = \frac{\dfrac{1}{a_n}(b_m s^m + b_{m-1} s^{m-1} + b_{m-2} s^{m-2} + \ldots + b_1 s + b_0)}{s^n + \dfrac{a_{n-1}}{a_n} s^{n-1} + \dfrac{a_{n-2}}{a_n} s^{n-2} + \ldots + \dfrac{a_1}{a_n} s + \dfrac{a_0}{a_n}} \qquad (12.3)$$

The roots of the numerator are called the *zeros* of $F(s)$, and are found by letting $N(s) = 0$ in (12.3). The roots of the denominator are called the *poles* of $F(s)$ and are found by letting $D(s) = 0$.

The zeros and poles of (12.3) can be real and distinct, or repeated, or complex conjugates, or combinations of real and complex conjugates. However, in most engineering applications we are interested in the nature of the poles. We will consider the nature of the poles for each case.

Case I: Distinct Poles

If all the poles p_1, p_2, p_3, ..., p_n of $F(s)$ are *distinct* (different from each another), we can factor the denominator of $F(s)$ in the form

$$F(s) = \frac{N(s)}{(s - p_1) \cdot (s - p_2) \cdot (s - p_3) \cdot \ldots \cdot (s - p_n)} \tag{12.4}$$

where p_k is distinct from all other poles. Then, the partial fraction expansion method allows us to express (12.4) as

$$F(s) = \frac{r_1}{(s - p_1)} + \frac{r_2}{(s - p_2)} + \frac{r_3}{(s - p_3)} + \ldots + \frac{r_n}{(s - p_n)} \tag{12.5}$$

where r_1, r_2, r_3, ..., r_n are the *residues* of $F(s)$.

To evaluate the residue r_k, we multiply both sides of (12.5) by $(s - p_k)$; then, we let $s \to p_k$, that is,

$$r_k = \lim_{s \to p_k} (s - p_k)F(s) = (s - p_k)F(s)\Big|_{s = p_k} \tag{12.6}$$

Example 12.1

Use partial fraction expansion to simplify $F_1(s)$ of (12.7) below.

$$F_1(s) = \frac{3s + 2}{s^2 + 3s + 2} \tag{12.7}$$

Solution:

$$F_1(s) = \frac{3s + 2}{s^2 + 3s + 2} = \frac{3s + 2}{(s + 1)(s + 2)} = \frac{r_1}{(s + 1)} + \frac{r_2}{(s + 2)} \tag{12.8}$$

$$r_1 = \lim_{s \to -1} (s + 1)F(s) = \frac{3s + 2}{(s + 2)}\bigg|_{s = -1} = -1$$

$$r_2 = \lim_{s \to -2} (s+2)F(s) = \left.\frac{3s+2}{(s+1)}\right|_{s=-2} = 4$$

Therefore, by substitution into (12.8), we obtain

$$F_1(s) = \frac{3s+2}{s^2+3s+2} = \frac{-1}{(s+1)} + \frac{4}{(s+2)} \qquad (12.9)$$

We can us the MATLAB **residue(r,p,k)** function to verify our answers with the following script:

Ns = [3, 2]; Ds = [1, 3, 2]; [r, p, k] = residue(Ns, Ds)

```
r =
      4
     -1
p =
     -2
     -1
k =
     [ ]
```

where we have denoted Ns and Ds as two vectors that contain the numerator and denominator coefficients of $F_1(s)$. MATLAB displays the r, p, and k vectors; these represent the residues, poles, and direct terms respectively. The first value of the vector r is associated with the first value of the vector p, the second value of r is associated with the second value of p, and so on. The vector k is referred to as the *direct term*, and it is always empty (has no value) whenever F(s) is a proper rational function. For this example, we observe that the highest power of the denominator is s^2 whereas the highest power of the numerator is s and therefore, the direct term k is empty.

Example 12.2

Use partial fraction expansion to simplify $F_2(s)$ of (12.10) below.

$$F_2(s) = \frac{3s^2+2s+5}{s^3+12s^2+44s+48} \qquad (12.10)$$

Solution:

First, we will use the MATLAB function **factor(s)** to express the denominator polynomial of $F_2(s)$ in factored form.[*] This function returns an expression that contains the prime factors of a polynomial. However, this function is used with *symbolic expressions*. These expressions are

---

* *Of course, we can use the* roots(p) *function. The* factor(s) *function is a good alternative.*

---

explained below.

The functions, like **roots(p)**, which we have used before, are display *numeric expressions*, that is, they produce numerical results. Symbolic expressions, on the other hand, can manipulate mathematical expressions without using actual numbers. Some examples of symbolic expressions are given below.

$$\sin^2 x \qquad e^{-\alpha t} \qquad y = \frac{d^2}{dt^2}(3t^3 - 4t^2 + 5t + 8) \qquad u = \int \frac{1}{x} dx$$

MATLAB contains the so-called *Symbolic Math Toolbox*. This is a collection of *tools* (functions) which are used in solving symbolic expressions; they are discussed in detail in MATLAB User's Manual. For the present, our interest is in using the **factor(s)** to express the denominator of (12.10) as a product of simple factors.

Before using symbolic expressions, we must create a symbolic variable x, y, s, t etc. This is done with the **sym** function. For example, $s = $ **sym** ('s') creates the symbolic variable s. Alternately, we can use the **syms** function to define one or more symbolic variables with a single statement. For example,

syms x y z a1 k2

defines the symbolic variables x, y, z, $a1$ and $k2$.

Returning to Example 12.2 and using MATLAB we have:

syms s; den=s^3+12*s^2+44*s+48; factor(den)

ans =
(s+4)*(s+2)*(s+6)

and thus,

$$F_2(s) = \frac{3s^2 + 2s + 5}{s^3 + 12s^2 + 44s + 48} = \frac{3s^2 + 2s + 5}{(s+2)(s+4)(s+6)} = \frac{r_1}{(s+2)} + \frac{r_2}{(s+4)} + \frac{r_3}{(s+6)}$$

Next, we find the residues r_1, r_2, and r_3. These are

$$r_1 = \left.\frac{3s^2 + 2s + 5}{(s+4)(s+6)}\right|_{s=-2} = \frac{9}{8} \qquad r_2 = \left.\frac{3s^2 + 2s + 5}{(s+2)(s+6)}\right|_{s=-4} = -\frac{37}{4} \qquad r_3 = \left.\frac{3s^2 + 2s + 5}{(s+2)(s+4)}\right|_{s=-6} = \frac{89}{8}$$

Therefore,

$$F_2(s) = \frac{3s^2 + 2s + 5}{s^3 + 12s^2 + 44s + 48} = \frac{9/8}{(s+2)} + \frac{-37/4}{(s+4)} + \frac{89/8}{(s+6)}$$

Case II: Complex Poles

Quite often, the poles of a proper rational function F(s) are complex, and since complex poles occur in complex conjugate pairs, the number of complex poles is even. Thus if p_k is a complex pole, then its complex conjugate $p_k{}^*$ is also a pole. The partial fraction expansion method can also be used in this case, as illustrated by the following example.

Example 12.3

Use partial fraction expansion to simplify $F_3(s)$ of (12.11) below.

$$F_3(s) = \frac{s+3}{s^3 + 5s^2 + 12s + 8} \tag{12.11}$$

Solution:

As a first step, we express the denominator in factored form to identify the poles of $F_3(s)$. Using the MATLAB script

syms s; factor(s^3 + 5*s^2 + 12*s + 8)

we obtain

```
ans =
(s+1)*(s^2+4*s+8)
```

Since the **factor(s)** function did not factor the quadratic term[*], we will use the **roots(p)** function to find its roots by treating it as a polynomial.

p=[1 4 8]; roots_p=roots(p)

```
roots_p =
  -2.0000+2.0000i
  -2.0000-2.0000i
```

Then,

$$F_3(s) = \frac{s+3}{s^3 + 5s^2 + 12s + 8} = \frac{s+3}{(s+1)(s+2+j2)(s+2-j2)}$$

$$= \frac{s+3}{s^3 + 5s^2 + 12s + 8} = \frac{r_1}{(s+1)} + \frac{r_2}{(s+2+j2)} + \frac{r_3}{(s+2-j2)} \tag{12.12}$$

and the residues are

---

[*] *For some undocumented reason, the factor(s) function does not seem to work with complex numbers.*

$$r_1 = \left.\frac{s+3}{s^2+4s+8}\right|_{s=-1} = \frac{2}{5}$$

$$r_2 = \left.\frac{s+3}{(s+1)(s+2-j2)}\right|_{s=-2-j2} = \frac{1-j2}{(-1-j2)(-j4)} = \frac{1-j2}{-8+j4}$$

$$= \frac{(1-j2)}{(-8+j4)}\frac{(-8-j4)}{(-8-j4)} = \frac{-16+j12}{80} = -\frac{1}{5}+j\frac{3}{20}$$

$$r_3 = \left.\frac{s+3}{(s+1)(s+2+j2)}\right|_{s=-2+j2} = \frac{1-j2}{(-1+j2)(j4)} = \frac{1-j2}{-8-j4}$$

$$= \frac{(1-j2)}{(-8-j4)}\frac{(-8+j4)}{(-8+j4)} = \frac{(-16)-j12}{80} = -\frac{1}{5}-j\frac{3}{20}$$

Of course, the last evaluation was not necessary since $r_3 = r_2{}^*$ or

$$r_3 = \left(-\frac{1}{5}+j\frac{3}{20}\right)^* = -\frac{1}{5}-j\frac{3}{20}$$

and this is always true since complex roots occur in conjugate pairs. Then, by substitution into (12.12), we obtain

$$F_3(s) = \frac{2/5}{(s+2)} + \frac{-1/5+j3/20}{(s+2+j2)} + \frac{-1/5-j3/20}{(s+2-j2)} \tag{12.13}$$

We can express (12.13) in a different form if we want to eliminate the complex presentation. This is done by combining the last two terms on the right side of (12.13) to form a single term and now is written as

$$F_3(s) = \frac{2/5}{(s+2)} - \frac{1}{5}\cdot\frac{(2s+1)}{(s^2+4s+8)} \tag{12.14}$$

Case III: Multiple (Repeated) Poles

In this case, $F(s)$ has simple poles but one of the poles, say p_1, has a multiplicity m. Then,

$$F(s) = \frac{N(s)}{(s-p_1)^m(s-p_2)\ldots(s-p_{n-1})(s-p_n)} \tag{12.15}$$

and denoting the m residues corresponding to multiple pole p_1 as $r_{11}, r_{12}, \ldots r_{1m}$, the partial fraction expansion of (12.15) can be expressed as

$$F(s) = \frac{r_{11}}{(s-p_1)^m} + \frac{r_{12}}{(s-p_1)^{m-1}} + \frac{r_{13}}{(s-p_1)^{m-2}} + \ldots + \frac{r_{1m}}{(s-p_1)}$$

$$+ \frac{r_2}{(s-p_2)} + \frac{r_3}{(s-p_3)} + \frac{r_n}{(s-p_n)} \qquad (12.16)$$

For the simple poles p_1, p_2, ... p_n we proceed as before, that is,

$$r_k = \lim_{s \to p_k} (s-p_k)F(s) = (s-p_k)F(s)\Big|_{s = p_k}$$

To find the first residue r_{11} of the repeated pole, we multiply both sides of (12.16) by $(s-p_1)^m$. Then,

$$(s-p_1)^m F(s) = r_{11} + (s-p_1)r_{12} + (s-p_1)^2 r_{13} + \ldots + (s-p_1)^{m-1} r_{1m}$$

$$+ (s-p_1)^m\left(\frac{r_2}{(s-p_2)} + \frac{r_3}{(s-p_3)} + \ldots + \frac{r_n}{(s-p_n)}\right) \qquad (12.17)$$

Next, taking the limit as $s \to p_1$ on both sides of (12.17), we obtain

$$\lim_{s \to p_1} (s-p_1)^m F(s)$$

$$= r_{11} + \lim_{s \to p_1} [(s-p_1)r_{12} + (s-p_1)^2 r_{13} + \ldots + (s-p_1)^{m-1} r_{1m}] \qquad (12.18)$$

$$+ \lim_{s \to p_1}\left[(s-p_1)^m\left(\frac{r_2}{(s-p_2)} + \frac{r_3}{(s-p_3)} + \ldots + \frac{r_n}{(s-p_n)}\right)\right]$$

or

$$r_{11} = \lim_{s \to p_1} (s-p_1)^m F(s) \qquad (12.19)$$

and thus (12.19) yields the residue of the first repeated pole.

To find the second residue r_{12} of the second repeated pole p_1, we first differentiate the relation of (12.18) with respect to s; then, we let $s \to p_1$, that is,

$$r_{12} = \lim_{s \to p_1} \frac{d}{ds}[(s-p_1)^m F(s)] \qquad (12.20)$$

To find the third residue r_{13} of the repeated pole p_1, we differentiate (12.18) twice with respect to s; then, we let $s \to p_1$, that is,

$$r_{13} = \lim_{s \to p_1} \frac{d^2}{ds^2}[(s-p_1)^m F(s)] \qquad (12.21)$$

This process is continued until all residues of the repeated poles have been found.

In general, for repeated poles the residue r_{1k} can be derived from the relation

$$(s - p_1)^m F(s) = r_{11} + r_{12}(s - p_1) + r_{13}(s - p_1)^2 + \ldots \tag{12.22}$$

whose $(m - 1)$th derivative of both sides is

$$(k - 1)! r_{1k} = \lim_{s \to p_1} \frac{d^{k-1}}{ds^{k-1}}[(s - p_1)^m F(s)] \tag{12.23}$$

or

$$r_{1k} = \lim_{s \to p_1} \frac{1}{(k - 1)!} \frac{d^{k-1}}{ds^{k-1}}[(s - p_1)^m F(s)] \tag{12.24}$$

Example 12.4

Use partial fraction expansion to simplify $F_4(s)$ of (12.25) below.

$$F_4(s) = \frac{s + 3}{(s + 2)(s + 1)^2} \tag{12.25}$$

Solution:

We observe that there is a pole of multiplicity 2 at $s = -1$ and thus, (12.25) in partial fraction expansion form is

$$F_4(s) = \frac{s + 3}{(s + 2)(s + 1)^2} = \frac{r_1}{(s + 2)} + \frac{r_{21}}{(s + 1)^2} + \frac{r_{22}}{(s + 1)} \tag{12.26}$$

The residues are

$$r_1 = \frac{s + 3}{(s + 1)^2}\bigg|_{s = -2} = 1$$

$$r_{21} = \frac{s + 3}{(s + 2)}\bigg|_{s = -1} = 2$$

$$r_{22} = \frac{d}{ds}\left(\frac{s + 3}{s + 2}\right)\bigg|_{s = -1} = \frac{(s + 2) - (s + 3)}{(s + 2)^2}\bigg|_{s = -1} = -1$$

Then, by substitution into (12.26),

$$F_4(s) = \frac{s + 3}{(s + 2)(s + 1)^2} = \frac{1}{(s + 2)} + \frac{2}{(s + 1)^2} + \frac{-1}{(s + 1)}$$

Instead of differentiation, the residue r_{22} could be found by substitution of the already known values of r_1 and r_{21} into (12.26), and letting $s = 0$ [*], that is,

$$\left.\frac{s+3}{(s+1)^2(s+2)}\right|_{s=0} = \left.\frac{1}{(s+2)}\right|_{s=0} + \left.\frac{2}{(s+1)^2}\right|_{s=0} + \left.\frac{r_{22}}{(s+1)}\right|_{s=0}$$

or $3/2 = 1/2 + 2 + r_{22}$ from which $r_{22} = -1$ as before.

To check our answers with MATLAB, we will use the **expand(s)** function. Like the **factor(s)** function, **expand(s)** is used with symbolic expressions. Its description can be displayed with the **help expand** command.

Check with MATLAB:

```
syms s                  % Create symbolic variable s
expand((s + 1)^2)       % Express it as a polynomial

ans =
s^2+2*s+1

Ns = [1  3];     % Coefficients of the numerator N(s)
d1 = [1  2  1];  % Coefficients of (s + 1)^2 = s^2 + 2*s + 1 term in D(s)
d2 = [0  1  2];  % Coefficients of (s + 2) term in D(s)
Ds=conv(d1,d2); % Multiplies polynomials d1 and d2 to express denominator D(s) as polynomial
[r,p,k]=residue(Ns,Ds)

r =
     1.0000
    -1.0000
     2.0000
p =
    -2.0000
    -1.0000
    -1.0000
k =
    []
```

Example 12.5

Use partial fraction expansion to simplify $F_5(s)$ of (12.27) below.

---

[*] *We must remember that (2.45) is an identity, and as such, it is true for any value of s.*

$$F_5(s) = \frac{s^2 + 3s + 1}{(s+1)^3(s+2)^2} \tag{12.27}$$

Solution:

We observe that there is a pole of multiplicity 3 at $s = -1$, and a pole of multiplicity 2 at $s = -2$. Then, in partial fraction expansion form

$$F_5(s) = \frac{r_{11}}{(s+1)^3} + \frac{r_{12}}{(s+1)^2} + \frac{r_{13}}{(s+1)} + \frac{r_{21}}{(s+2)^2} + \frac{r_{22}}{(s+2)} \tag{12.28}$$

We find the residue r_{11} by evaluating $F_5(s)$ at as $s = -1$

$$r_{11} = \frac{s^2 + 3s + 1}{(s+2)^2}\bigg|_{s=-1} = -1 \tag{12.29}$$

The residue r_{12} is found by first taking the first derivative of $F_5(s)$, and evaluating it at $s = -1$. Thus,

$$\begin{aligned}
r_{12} &= \frac{d}{ds}\left(\frac{s^2 + 3s + 1}{(s+2)^2}\right)\bigg|_{s=-1} \\
&= \frac{(s+2)^2(2s+3) - 2(s+2)(s^2+3s+1)}{(s+2)^4}\bigg|_{s=-1} = \frac{s+4}{(s+2)^3} = 3
\end{aligned} \tag{12.30}$$

The residue r_{13} is found by taking the second derivative of $F_5(s)$ and evaluating it at $s = -1$. Then,

$$\begin{aligned}
r_{13} &= \frac{1}{2!}\frac{d^2}{ds^2}\left(\frac{s^2+3s+1}{(s+2)^2}\right)\bigg|_{s=-1} = \frac{1}{2}\frac{d}{ds}\left[\frac{d}{ds}\left(\frac{s^2+3s+1}{(s+2)^2}\right)\right]\bigg|_{s=-1} \\
&= \frac{1}{2}\frac{d}{ds}\left(\frac{s+4}{(s+2)^3}\right)\bigg|_{s=-1} = \frac{1}{2}\left[\frac{(s+2)^3 - 3(s+2)^2(s+4)}{(s+2)^6}\right] \\
&= \frac{1}{2}\left(\frac{s+2-3s-12}{(s+2)^4}\right)\bigg|_{s=-1} = \frac{-s-5}{(s+2)^4}\bigg|_{s=-1} = -4
\end{aligned} \tag{12.31}$$

Similarly, the residue r_{21} if found by evaluating $F_5(s)$ at $s = -2$, and the residue r_{22} is found by first taking the first derivative of $F_5(s)$ and evaluating it at $s = -2$. Therefore,

$$r_{21} = \left.\frac{s^2 + 3s + 1}{(s+1)^3}\right|_{s=-2} = 1$$

$$r_{22} = \left.\frac{d}{ds}\left(\frac{s^2 + 3s + 1}{(s+1)^3}\right)\right|_{s=-2}$$

$$= \left.\frac{(s+1)^3(2s+3) - 3(s+1)^2(s^2+3s+1)}{(s+1)^6}\right|_{s=-2}$$

$$r_{22} = \left.\frac{(s+1)(2s+3) - 3(s^2+3s+1)}{(s+1)^4}\right|_{s=-2} = \left.\frac{-s^2 - 4s}{(s+1)^4}\right|_{s=-2} = 4$$

By substitution of these residues into (12.28), we obtain $F_5(s)$ in partial fraction expansion as

$$F_5(s) = \frac{-1}{(s+1)^3} + \frac{3}{(s+1)^2} + \frac{-4}{(s+1)} + \frac{1}{(s+2)^2} + \frac{4}{(s+2)} \tag{12.32}$$

We will now verify the values of these residues with MATLAB. Before we do this, we introduce the **collect(s)** function that we can use to multiply two or more symbolic expressions to obtain the result in a polynomial form. Its description can be displayed with the **help collect** command. We must remember that the **conv(p,q)** function is used with numeric expressions, i.e., polynomial coefficients only.

The MATLAB script for this example is as follows.

```
syms s;        % We must first define the variable s in symbolic form
               % The function "collect" below multiplies (s+1)^3 by (s+2)^2
Ds=collect(((s+1)^3)*((s+2)^2))

Ds =
s^5+7*s^4+19*s^3+25*s^2+16*s+4

% We now use this result to express the denominator D(s) as a
% polynomial so we can use its coefficients with the "residue" function
%
Ns=[1 3 1]; Ds=[1 7 19 25 16 4]; [r,p,k]=residue(Ns,Ds)

r =
    4.0000
    1.0000
   -4.0000
    3.0000
   -1.0000
```

```
p =
     -2.0000
     -2.0000
     -1.0000
     -1.0000
     -1.0000
k =
      []
```

Case for $m \geq n$

Our discussion thus far, was based on the condition that $F(s)$ is a proper rational function, that is, the highest power m of the numerator is less than the highest power n of the denominator, i.e., $m < n$. If $m \geq n$, $F(s)$ is an improper rational function, and before we apply the partial fraction expansion, we must divide the numeraror $N(s)$ by the denominator $D(s)$ to obtain an expression of the form

$$F(s) = k_0 + k_1 s + k_2 s^2 + \ldots + k_{m-n} s^{m-n} + \frac{N(s)}{D(s)} \tag{12.33}$$

so that $m < n$.

Example 12.6

Express $F_6(s)$ of (12.34) below in partial expansion form.

$$F_6(s) = \frac{s^2 + 2s + 2}{s + 1} \tag{12.34}$$

Solution:

In (12.34), $m > n$ and thus we need to express $F_6(s)$ in the form of (12.33). By long division,

$$F_6(s) = \frac{s^2 + 2s + 2}{s + 1} = \frac{1}{s + 1} + s + 1 \tag{12.35}$$

Check with MATLAB:

Ns = [1 2 2]; Ds = [1 1]; [r, p, k] = residue(Ns, Ds)

```
r =
      1
p =
     -1
k =
      1      1
```

The direct terms $k = \begin{bmatrix} 1 & 1 \end{bmatrix}$ are the coefficients of the s term and the constant in (2.54).

12.2 Alternate Method of Partial Fraction Expansion

The partial fraction expansion method can also be performed by the *equating the numerators procedure* thereby making the denominators of both sides the same, and then equating the numerators. We assume that the degree on the numerator $N(s)$ is less than the degree of the denominator. If not, we first perform a long division and then work with the quotient and the remainder as before.

We also assume that the denominator $D(s)$ can be expressed as a product of real linear and quadratic factors. If these assumptions prevail, we let $s - a$ be a linear factor of $D(s)$ and we suppose that $(s - a)^m$ is the highest power of $s - a$ that divides $D(s)$. Then, we can express $F(s)$ as

$$F(s) = \frac{N(s)}{D(s)} = \frac{r_1}{s - a} + \frac{r_2}{(s - a)^2} + \cdots \frac{r_m}{(s - a)^m} \tag{12.36}$$

Next, let $s^2 + \alpha s + \beta$ be a quadratic factor of $D(s)$ and suppose that $(s^2 + \alpha s + \beta)^n$ is the highest power of this factor that divides $F(s)$. Now, we perform the following steps:

1. To this factor, we assign the sum of n partial fractions as shown below.

$$\frac{r_1 s + k_1}{s^2 + \alpha s + \beta} + \frac{r_2 s + k_2}{(s^2 + \alpha s + \beta)^2} + \cdots + \frac{r_n s + k_n}{(s^2 + \alpha s + \beta)^n} \tag{12.37}$$

2. We repeat Step 1 for each of the distinct linear and quadratic factors of $D(s)$.

3. We set the given $F(s)$ equal to the sum of these partial fractions.

4. We multiply each term of the right side by the appropriate factor to make the denominators of both sides equal.

5. We arrange the terms of both sides in decreasing powers of s.

6. We equate the coefficients of corresponding powers of s.

7. We solve the resulting equations for the residues.

Example 12.7

Express $F_7(s)$ of (12.38) below as a sum of partial fractions using the equating the numerators procedure.

$$F_7(s) = \frac{-2s+4}{(s^2+1)(s-1)^2} \tag{12.38}$$

Solution:

By Steps 1 through 3 above,

$$F_7(s) = \frac{-2s+4}{(s^2+1)(s-1)^2} = \frac{r_1 s + A}{(s^2+1)} + \frac{r_{21}}{(s-1)^2} + \frac{r_{22}}{(s-1)} \tag{12.39}$$

By Step 4,

$$-2s+4 = (r_1 s + A)(s-1)^2 + r_{21}(s^2+1) + r_{22}(s-1)(s^2+1) \tag{12.40}$$

and by Steps 5, 6, and 7,

$$-2s+4 = (r_1 + r_{22})s^3 + (-2r_1 + A - r_{22} + r_{21})s^2$$
$$+ (r_1 - 2A + r_{22})s + (A - r_{22} + r_{21}) \tag{12.41}$$

Relation (12.41) is an identity in s; therefore, the coefficients of each power of s on the left and right sides are equal. Accordingly, by equating like powers of s, we obtain

$$0 = r_1 + r_{22}$$
$$0 = -2r_1 + A - r_{22} + r_{21}$$
$$-2 = r_1 - 2A + r_{22} \tag{12.42}$$
$$4 = A - r_{22} + r_{21}$$

Subtracting the second equation from the fourth in (12.42), we obtain

$$4 = 2r_1 \quad \text{or} \quad r_1 = 2 \tag{12.43}$$

and by substitution into the first equation of (12.42), we obtain

$$0 = 2 + r_{22} \quad \text{or} \quad r_{22} = -2 \tag{12.44}$$

Next, substitution of (12.43) and (12.44) into the third equation of (12.42), yields

$$-2 = 2 - 2A - 2 \quad \text{or} \quad A = 1 \tag{12.45}$$

and using the fourth equation of (12.42, we obtain:

$$4 = 1 + 2 + r_{21} \quad \text{or} \quad r_{21} = 1 \tag{12.46}$$

Therefore $F_7(s)$ in partial fraction expansion form becomes

$$F_7(s) = \frac{-2s+4}{(s^2+1)(s-1)^2} = \frac{2s+1}{(s^2+1)} + \frac{1}{(s-1)^2} - \frac{2}{(s-1)}$$

Example 12.8

Use the equating the numerators procedure to obtain the partial fraction expansion of $F_8(s)$ in (12.47) below.

$$F_8(s) = \frac{s+3}{s^3 + 5s^2 + 12s + 8} \tag{12.47}$$

Solution:

This is the same rational function as that of Example 12.3, where we found that the denominator can be expressed in factored form of a linear and a quadratic factor, that is,

$$F_7(s) = \frac{s+3}{(s+1)(s^2+4s+8)} \tag{12.48}$$

and in partial fraction expansion form,

$$F_7(s) = \frac{s+3}{(s+1)(s^2+4s+8)} = \frac{r_1}{s+1} + \frac{r_2 s + r_3}{s^2 + 4s + 8} \tag{12.49}$$

As in Example 12.3, we first find the residue of the linear factor as

$$r_1 = \left. \frac{s+3}{s^2+4s+8} \right|_{s=-1} = \frac{2}{5} \tag{12.50}$$

To compute r_2 and r_3, we use the equating the numerators procedure and we obtain

$$(s+3) = r_1(s^2 + 4s + 8) + (r_2 s + r_3)(s+1) \tag{12.51}$$

Since r_1 is already known, we only need two equations in r_2 and r_3. Equating the coefficient of s^2 on the left side, which is zero, with the coefficients of s^2 on the right side of (12.51), we obtain

$$0 = r_1 + r_2 \tag{12.52}$$

With $r_1 = 2/5$, (12.52) yields $r_2 = -2/5$. To find the third residue r_3, we equate the constant terms of (12.51), that is, $3 = 8r_1 + r_3$, and with $r_1 = 2/5$, we obtain $r_3 = -1/5$. Then, by substitution into (12.49), we obtain

$$F_7(s) = \frac{2/5}{(s+2)} - \frac{1}{5}\frac{(2s+1)}{(s^2+4s+8)}$$

as before. The remaining steps are the same as in Example 12.3.

We will conclude the partial fraction expansion topic with a few more examples, using the **residue(r,p,k)** function.

Example 12.9

Use the **residue(r,p,k)** function to compute the poles and residues of the function

$$F_9(s) = \frac{8s + 2}{s^2 + 3s + 2} \tag{12.53}$$

Solution:

Let p_1 and p_2 be the poles (the denominator roots) and r_1 and r_2 be the residues. Then, $F_9(s)$ can be written as

$$F_9(s) = \frac{r_1}{s + p_1} + \frac{r_2}{s + p_2} \tag{12.54}$$

The MATLAB script for this example is as follows:

```
num=[0 8 2];    %  The semicolon suppress the display of the row vector typed
                %  and zero is typed to make the numerator have same number
                %  of elements as the denominator; not necessary, but recommended
den=[1 3 2]; [r,p,k]=residue(num,den)

r =
      14
      -6
p =
      -2
      -1
k =
      []
```

Therefore, $F_9(s)$ in partial fraction expansion form is written as

$$F_9(s) = \frac{r_1}{s + p_1} + \frac{r_2}{s + p_2} = \frac{14}{s + 2} + \frac{-6}{s + 1} \tag{12.55}$$

Example 12.10

Use the **residue(r,p,k)** function to compute the poles and residues of $F_{10}(s)$ in (12.56) below.

$$F_{10}(s) = \frac{s + 3}{(s + 1)(s^2 + 4s + 8)} \tag{12.56}$$

Solution:

Let p_1, p_2, and p_3 be the poles (the denominator roots) and r_1, r_2, and r_3 be the residues of $F_{10}(s)$. Then, it can be written as

$$F_{10}(s) = \frac{r_1}{s + p_1} + \frac{r_2}{s + p_2} + \frac{r_3}{s + p_3} \tag{12.57}$$

The poles and the residues can be found with the statement **[r,p,k]=residue(num, den)**. Before we use this statement, we need to express the denominator as a polynomial. We will use the function **conv(a,b)** to multiply the two factors of the denominator of (12.56).

We recall that we can write two or more statements on one line if we separate them by commas or semicolons. We also recall that commas will display the results, whereas semicolons will suppress the display. Then,

a=[1 1]; b=[1 4 8]; c=conv(a,b); c, num=[1,3]; den=c;
[r,p,k]=residue(num,den)

```
c =
     1     5     12     8
r =
  -0.2000-  0.1500i
  -0.2000+  0.1500i
   0.4000
p =
  -2.0000+  2.0000i
  -2.0000-  2.0000i
  -1.0000
k =
     []
```

Therefore, $F_{10}(s)$ in partial fraction expansion form is

$$F_{10}(s) = \frac{r_1}{s + p_1} + \frac{r_2}{s + p_2} + \frac{r_3}{s + p_3} = \frac{-0.2 - 0.15j}{s + 2 - 2j} + \frac{-0.2 + 0.15j}{s + 2 + 2j} + \frac{0.4}{s + 1} \tag{12.58}$$

By repeated use of the **deconv(num,den)** function, we can reduce a rational polynomial to simple terms of a polynomial, where the last term is a rational polynomial whose order of the numerator is less than that of the denominator as illustrated by the following example.

Example 12.11

Use the **deconv(num,den)** function to express the following rational polynomial as a polynomial with four terms.

$$f_1(x) = \frac{x^3 + 2x^2 + 1}{0.5x - 1} \tag{12.59}$$

Solution:

num=[1 2 0 1]; den=[0 0 0.5 −1]; [q,r]=deconv(num,den)

```
q =
      2        8       16
r =
      0        0        0       17
```

Therefore, $f_1(x)$ can now be written as

$$f_1(x) = 2x^2 + 8x + 16 + \frac{17}{0.5x - 1} \qquad (12.60)$$

It is important to remember that the function **roots(p)** is used with polynomials only. If we want to find the zeros of any function, such as the function $f_2(x)$ defined as

$$f_2(x) = \frac{3x^3 + 7x^2 + 9}{(12x^6 + 2x^4 + 13x^2 + 25)} + \frac{0.5x^5 + 6.3x^2 + 4.35}{(23x^6 + 16x^3 + 7.5x)} + \frac{1}{4.11x + 2.75} \qquad (12.61)$$

we must use the function **fzero('function',x_0)**, where **function** is a pre–defined string, and x_0 is a required initial value. We can approximate this value by first plotting $f_2(x)$ to find out where it crosses the x –axis. This is discussed in Chapter 1, Page 1–27.

12.3 Summary

- The function

$$F(s) = \frac{N(s)}{D(s)} = \frac{b_m s^m + b_{m-1} s^{m-1} + b_{m-2} s^{m-2} + \ldots + b_1 s + b_0}{a_n s^n + a_{n-1} s^{n-1} + a_{n-2} s^{n-2} + \ldots + a_1 s + a_0}$$

where the coefficients a_k and b_k for $k = 0, 1, 2, \ldots, n$ are real numbers, is a proper rational function if the highest power of the numerator $N(s)$ is less than the highest power of of the denominator $D(s)$, i.e., $m < n$. If $m \geq n$, $F(s)$ is an improper rational function.

- Partial fraction expansion applies only to proper rational functions. If $F(s)$ is an improper rational function we divide the numeraror $N(s)$ by the denominator $D(s)$ to obtain an expression of the form

$$F(s) = k_0 + k_1 s + k_2 s^2 + \ldots + k_{m-n} s^{m-n} + \frac{N(s)}{D(s)}$$

so that $m < n$.

- If the function

$$F(s) = \frac{N(s)}{D(s)} = \frac{b_m s^m + b_{m-1} s^{m-1} + b_{m-2} s^{m-2} + \ldots + b_1 s + b_0}{a_n s^n + a_{n-1} s^{n-1} + a_{n-2} s^{n-2} + \ldots + a_1 s + a_0}$$

is a proper rational function where a_n is a non–zero integer other than unity, we rewrite this function as

$$F(s) = \frac{N(s)}{D(s)} = \frac{\dfrac{1}{a_n}(b_m s^m + b_{m-1} s^{m-1} + b_{m-2} s^{m-2} + \ldots + b_1 s + b_0)}{s^n + \dfrac{a_{n-1}}{a_n} s^{n-1} + \dfrac{a_{n-2}}{a_n} s^{n-2} + \ldots + \dfrac{a_1}{a_n} s + \dfrac{a_0}{a_n}}$$

to make a_n unity.

- The roots of the numerator are called the *zeros* of $F(s)$, and are found by letting $N(s) = 0$, and the roots of the denominator are called the *poles* of $F(s)$ and are found by letting $D(s) = 0$.

- The zeros and poles can be real and distinct, or repeated, or complex conjugates, or combinations of real and complex conjugates. In most engineering applications we are interested in the nature of the poles.

- If all the poles p_1, p_2, p_3, $\ldots$, p_n of $F(s)$ are distinct we can factor the denominator of $F(s)$ in the form

$$F(s) = \frac{N(s)}{(s-p_1) \cdot (s-p_2) \cdot (s-p_3) \cdot \ldots \cdot (s-p_n)}$$

where p_k is distinct from all other poles. Then, the partial fraction expansion method allows us to write the above expression as

$$F(s) = \frac{r_1}{(s-p_1)} + \frac{r_2}{(s-p_2)} + \frac{r_3}{(s-p_3)} + \ldots + \frac{r_n}{(s-p_n)}$$

where r_1, r_2, r_3, ..., r_n are the residues of $F(s)$. To evaluate the residue r_k, we multiply both sides of (12.5) by $(s-p_k)$; then, we let $s \to p_k$, that is,

$$r_k = \lim_{s \to p_k} (s-p_k)F(s) = (s-p_k)F(s)\Big|_{s=p_k}$$

- We can use the MATLAB **residue(r,p,k)** function to verify our answers. This function returns the residues, their associated poles, and a direct term. For proper rational functions there is no direct term.

- The partial fraction expansion can also be used if the poles are complex. Since complex poles occur in conjugate pairs, if p_k is a complex pole, then its complex conjugate p_k^* is also a pole.

- If a rational function $F(s)$ has simple poles but one of the poles, say p_1, has a multiplicity m, the function is expressed as

$$F(s) = \frac{N(s)}{(s-p_1)^m(s-p_2)\ldots(s-p_{n-1})(s-p_n)}$$

and denoting the m residues corresponding to multiple pole p_1 as r_{11}, r_{12}, ... r_{1m}, the partial fraction expansion can be expressed as

$$F(s) = \frac{r_{11}}{(s-p_1)^m} + \frac{r_{12}}{(s-p_1)^{m-1}} + \frac{r_{13}}{(s-p_1)^{m-2}} + \ldots + \frac{r_{1m}}{(s-p_1)}$$

$$+ \frac{r_2}{(s-p_2)} + \frac{r_3}{(s-p_3)} + \frac{r_n}{(s-p_n)}$$

- If a rational function $F(s)$ has simple poles but one of the poles, say p_1, has a multiplicity m, for the simple poles we use the same procedure as for distinct poles. The first residue of a repeated pole is found from

$$r_{11} = \lim_{s \to p_1} (s-p_1)^m F(s)$$

The second repeated pole is found from

$$r_{12} = \lim_{s \to p_1} \frac{d}{ds}[(s-p_1)^m F(s)]$$

the third from

$$r_{13} = \lim_{s \to p_1} \frac{d^2}{ds^2}[(s-p_1)^m F(s)]$$

and this process is continued until all residues of the repeated poles have been found.

- With the alternate method of partial fraction expansion we use the equating the numerators procedure thereby making the denominators of both sides the same, and then equating the numerators. We assume that the denominator $D(s)$ can be expressed as a product of real linear and quadratic factors.

12.4 Exercises

Perform partial fraction expansion for the following. Use MATLAB to simplify and to verify your results.

1. $\dfrac{1}{1-s^2}$

2. $\dfrac{1}{s^2 + 4s - 5}$

3. $\dfrac{s}{s^2 - 2s - 3}$

4. $\dfrac{5s - 3}{s^2 - 2s - 3}$

5. $\dfrac{s^2}{s^2 + 2s + 1}$

6. $\dfrac{1}{s(s + 1)^2}$

7. $\dfrac{1}{(s + 1)(s^2 + 1)}$

8. $\dfrac{1}{s(s^2 + s + 1)}$

12.5 Solutions to End-of-Chapter Exercises

1.

$$\frac{1}{1-s^2} = \frac{-1}{s^2-1} = \frac{-1}{(s+1)(s-1)} = \frac{r_1}{s+1} + \frac{r_2}{s-1}$$

$$r_1 = \left.\frac{-1}{s+1}\right|_{s=1} = -1/2 \qquad r_2 = \left.\frac{-1}{s-1}\right|_{s=-1} = 1/2$$

Then,

$$\frac{-1}{s^2-1} = -\frac{1/2}{s+1} + \frac{1/2}{s-1}$$

Ns = [0, 0, –1]; Ds = [1, 0, –1]; [r, p, k] = residue(Ns, Ds)

```
r =
    0.5000
   -0.5000
p =
   -1
    1
k =
    []
```

2.

$$\frac{1}{s^2+4s-5} = \frac{1}{(s-1)(s+5)} = \frac{r_1}{s-1} + \frac{r_2}{s+5}$$

$$r_1 = \left.\frac{1}{s+5}\right|_{s=1} = 1/6 \qquad r_2 = \left.\frac{1}{s-1}\right|_{s=-5} = -1/6$$

Then,

$$\frac{1}{s^2+4s-5} = \frac{1/6}{s-1} - \frac{1/6}{s+5}$$

format rat; Ns = [0, 0, 1]; Ds = [1, 4, –5]; [r, p, k] = residue(Ns, Ds)

```
r =
    -1/6
     1/6
p =
    -5
     1
k =
    []
```

3.

$$\frac{s}{s^2-2s-3} = \frac{s}{(s+1)(s-3)} = \frac{r_1}{s+1} + \frac{r_2}{s-3}$$

$$r_1 = \left.\frac{s}{s-3}\right|_{s=-1} = 1/4 \qquad r_2 = \left.\frac{s}{s+1}\right|_{s=3} = 3/4$$

Then,

$$\frac{s}{s^2-2s-3} = \frac{1/4}{s+1} + \frac{3/4}{s-3}$$

format rat; Ns = [0, 1, 0]; Ds = [1, −2, −3]; [r, p, k] = residue(Ns, Ds)

```
r =
      3/4
      1/4
p =
       3
      -1
k =
      []
```

4.

$$\frac{5s-3}{s^2-2s-3} = \frac{5s-3}{(s+1)(s-3)} = \frac{r_1}{s+1} + \frac{r_2}{s-3}$$

$$r_1 = \left.\frac{5s-3}{s-3}\right|_{s=-1} = 2 \qquad r_2 = \left.\frac{5s-3}{s+1}\right|_{s=3} = 3$$

Then,

$$\frac{5s-3}{s^2-2s-3} = \frac{2}{s+1} + \frac{3}{s-3}$$

Ns = [0, 5, −3]; Ds = [1, −2, −3]; [r, p, k] = residue(Ns, Ds)

```
r =
       3
       2
p =
       3
      -1
k =
      []
```

5.

This is an improper rational function, and before we apply the partial fraction expansion, we must divide the numeraror N(s) by the denominator D(s) to obtain an expression of the form

$$F(s) = k_0 + k_1 s + k_2 s^2 + \ldots + k_{m-n} s^{m-n} + \frac{N(s)}{D(s)}$$

We could perform long division but we will use the MATLAB **deconv(num,den)** function to express the following rational polynomial as a polynomial with four terms.

num=[1 0 0]; den=[1 2 1]; [q,r]=deconv(num,den)

q =

    1

r =

    0   -2   -1

and thus

$$\frac{s^2}{s^2 + 2s + 1} = 1 + \frac{-2s - 1}{s^2 + 2s + 1} = 1 - \frac{2s + 1}{(s + 1)^2} = 1 - \frac{r_1}{(s + 1)^2} + \frac{r_2}{(s + 1)}$$

$$\frac{-2s - 1}{(s + 1)^2} = \frac{r_1}{(s + 1)^2} + \frac{r_2}{(s + 1)} \qquad -2s - 1 = r_1 + r_2(s + 1)$$

$$r_2 = -2 \qquad r_1 + r_2 = -1 \qquad r_1 = 1$$

Then,

$$\frac{s^2}{s^2 + 2s + 1} = 1 + \frac{-2s - 1}{s^2 + 2s + 1} = 1 + \frac{1}{(s + 1)^2} + \frac{-2}{(s + 1)}$$

Ns = [1, 0, 0]; Ds = [1, 2, 1]; [r, p, k] = residue(Ns, Ds)

r =

    -2
     1

p =

    -1
    -1

k =

    1

6.

$$\frac{1}{s(s + 1)^2} = \frac{r_1}{s} + \frac{r_{21}}{(s + 1)^2} + \frac{r_{22}}{(s + 1)}$$

$$r_1 = \left.\frac{1}{s+1}\right|_{s=0} = 1 \qquad r_{21} = \left.\frac{1}{s}\right|_{s=-1} = -1 \qquad r_{22} = \left.\frac{d}{ds}\left(\frac{1}{s}\right)\right|_{s=-1} = \left.-\frac{1}{s^2}\right|_{s=-1} = -1$$

Then,

$$\frac{1}{s(s+1)^2} = \frac{1}{s} + \frac{-1}{(s+1)^2} + \frac{-1}{(s+1)}$$

```
syms s; expand(s*(s+1)^2)

ans =
   s^3+2*s^2+s
```

```
Ns = [0, 0, 0, 1]; Ds = [1, 2, 1, 0]; [r, p, k] = residue(Ns, Ds)

r =
      -1
      -1
       1
p =
      -1
      -1
       0
k =
      []
```

7.

$$\frac{1}{(s+1)(s^2+1)} = \frac{r_1}{s+1} + \frac{r_2 s + r_3}{s^2+1} = \frac{r_1(s^2+1)}{(s+1)(s^2+1)} + \frac{(r_2 s + r_3)(s+1)}{(s+1)(s^2+1)}$$

Equating numerators and like terms we obtain

$$1 = r_1 s^2 + r_1 + r_2 s^2 + r_2 s + r_3 s + r_3$$

$$r_1 + r_2 = 0 \qquad r_2 + r_3 = 0 \qquad r_1 + r_3 = 1$$

```
syms r1  r2  r3
eq1=r1+r2−0
eq2=r2+r3−0
eq3=r1+r3−1
S=solve(eq1, eq2, eq3)
eq1 =
 r1+r2

eq2 =
```

```
    r2+r3

eq3 =
  r1+r3-1

S =
    r1: [1x1 sym]
    r2: [1x1 sym]
    r3: [1x1 sym]
```

S.r1
```
ans =
  1/2
```

S.r2
```
ans =
-1/2
```

S.r3
```
ans =
1/2
```

The statement **S=solve(eq1, eq2, eq3, ...eqN)** returns the solutions in the structure S whose named fields hold hold the solution for each variable. Thus, $r_1 = 1/2$, $r_2 = -1/2$, and $r_3 = 1/2$. Then,

$$\frac{1}{(s+1)(s^2+1)} = \frac{1/2}{s+1} + \frac{(-1/2)s + 1/2}{s^2+1}$$

syms s; expand((s+1)*(s^2+1))

```
ans =
  s^3+s+s^2+1
```

Ns = [0, 0, 0, 1]; Ds = [1, 1, 1, 1]; [r, p, k] = residue(Ns, Ds)

```
r =
     1/2
    -1/4 - 1/4i
    -1/4 + 1/4i
p =
    -1
    -1/6004799503160662 + 1i
    -1/6004799503160662 - 1i
k =
    []
```

These values are inconsistent with those we've found. The MATLAB **help residue** command displays the following:

```
Warning: Numerically, the partial fraction expansion of a ratio of
polynomials represents an ill-posed problem.  If the denominator
polynomial, A(s), is near a polynomial with multiple roots, then
small changes in the data, including roundoff errors, can make arbi-
trarily large changes in the resulting poles and residues. Problem
formulations making use of state-space or zero-pole representations
are preferable.
```

8.

$$\frac{1}{s(s^2+s+1)} = \frac{r_1}{s} + \frac{r_2 s + r_3}{s^2+s+1} = \frac{r_1(s^2+s+1)}{s(s^2+s+1)} + \frac{(r_2 s + r_3)s}{s(s^2+s+1)}$$

Equating numerators and like terms we obtain

$$1 = r_1 s^2 + r_1 s + r_1 + r_2 s^2 + r_3 s$$

$$r_1 + r_2 = 0 \qquad r_2 + r_3 = 0 \qquad r_1 = 1$$

By inspection, $r_1 = 1$, $r_2 = -1$, and $r_3 = 1$. Then,

$$\frac{1}{(s+1)(s^2+1)} = \frac{1}{s} + \frac{-s+3}{s^2+s+1}$$

syms s; expand(s*(s^2+s+1))

```
ans =
   s^3+s^2+s
```

Ns = [0, 0, 0, 1]; Ds = [1, 1, 1, 0]; [r, p, k] = residue(Ns, Ds)

```
r =
    -1/2    +   390/1351i
    -1/2    -   390/1351i
      1
p =
    -1/2 + 1170/1351i
    -1/2 - 1170/1351i
     0
k =
    []
```

As in Exercise 7, these values are inconsistent with those we've found.

Chapter 13

The Gamma and Beta Functions and Distributions

his chapter is an introduction to the gamma and beta functions and their distributions used with many applications in science and engineering. They are also used in probability, and in the computation of certain integrals.

13.1 The Gamma Function

The *gamma function*, denoted as $\Gamma(n)$, is also known as *generalized factorial function*. It is defined as

$$\Gamma(n) = \int_0^\infty x^{n-1} e^{-x} dx \qquad (13.1)$$

and this improper[*] integral converges (approaches a limit) for all $n > 0$.

We will derive the basic properties of the gamma function and its relation to the well known factorial function

$$n! = n(n-1)(n-2)...3 \cdot 2 \cdot 1 \qquad (13.2)$$

We will evaluate the integral of (13.1) by performing integration by parts using the relation

$$\int u\, dv = uv - \int v\, du \qquad (13.3)$$

Letting

$$u = e^{-x} \quad \text{and} \quad dv = x^{n-1} \qquad (13.4)$$

we obtain

$$du = -e^{-x} dx \quad \text{and} \quad v = \frac{x^n}{n} \qquad (13.5)$$

Then, with (13.3), we write (13.1) as

---

[*] *Improper integrals are two types and these are:*

a. $\int_a^b f(x)dx$ *where the limits of integration a or b or both are infinite*

b. $\int_a^b f(x)dx$ *where f(x) becomes infinite at a value x between the lower and upper limits of integration inclusive.*

$$\Gamma(n) = \frac{x^n e^{-x}}{n}\Bigg|_{x=0}^{\infty} + \frac{1}{n}\int_0^{\infty} x^n e^{-x}dx \qquad (13.6)$$

With the condition that $n > 0$, the first term on the right side of (13.6) vanishes at the lower limit, that is, for $x = 0$. It also vanishes at the upper limit as $x \to \infty$. This can be proved with *L' Hôpital's rule*[*] by differentiating both numerator and denominator m times, where $m \geq n$. Then,

$$\lim_{x \to \infty} \frac{x^n e^{-x}}{n} = \lim_{x \to \infty} \frac{x^n}{ne^x} = \lim_{x \to \infty} \frac{\dfrac{d^m}{dx^m}x^n}{\dfrac{d^m}{dx^m}ne^x} = \lim_{x \to \infty} \frac{\dfrac{d^{m-1}}{dx^{m-1}}nx^{n-1}}{\dfrac{d^{m-1}}{dx^{m-1}}ne^x} = \dots$$

$$= \lim_{x \to \infty} \frac{n(n-1)(n-2)\dots(n-m+1)x^{n-m}}{ne^x} \qquad (13.7)$$

$$= \lim_{x \to \infty} \frac{(n-1)(n-2)\dots(n-m+1)}{x^{m-n}e^x} = 0$$

Therefore, (13.6) reduces to

$$\Gamma(n) = \frac{1}{n}\int_0^{\infty} x^n e^{-x}dx \qquad (13.8)$$

and with (13.1) we have

$$\Gamma(n) = \int_0^{\infty} x^{n-1}e^{-x}dx = \frac{1}{n}\int_0^{\infty} x^n e^{-x}dx \qquad (13.9)$$

By comparing the two integrals of (13.9), we observe that

$$\boxed{\Gamma(n) = \frac{\Gamma(n+1)}{n}} \qquad (13.10)$$

or

$$\boxed{n\Gamma(n) = \Gamma(n+1)} \qquad (13.11)$$

---

[*] *Quite often, the ratio of two functions, such as $\dfrac{f(x)}{g(x)}$, for some value of x, say a, results in the indeterminate form $\dfrac{f(a)}{g(a)} = \dfrac{0}{0}$. To work around this problem, we consider the limit $\lim\limits_{x \to a} \dfrac{f(x)}{g(x)}$, and we wish to find this limit, if it exists. L'Hôpital's rule states that if $f(a) = g(a) = 0$, and if the limit $\dfrac{d}{dx}f(x) / \dfrac{d}{dx}g(x)$ as x approaches a exists, then,*

$$\lim_{x \to a} \frac{f(x)}{g(x)} = \lim_{x \to a} \left(\frac{d}{dx}f(x) / \frac{d}{dx}g(x)\right)$$

It is convenient to use (13.10) for $n < 0$, and (13.11) for $n > 0$.

From (13.10), we see that $\Gamma(n)$ becomes infinite as $n \to 0$.

For $n = 1$, (13.1) yields

$$\Gamma(1) = \int_0^\infty e^{-x} dx = -e^{-x}\Big|_0^\infty = 1 \tag{13.12}$$

Thus, we have derived the important relation,

$$\Gamma(1) = 1 \tag{13.13}$$

From the recurring relation of (13.11), we obtain

$$\begin{aligned}
\Gamma(2) &= 1 \cdot \Gamma(1) = 1 \\
\Gamma(3) &= 2 \cdot \Gamma(2) = 2 \cdot 1 = 2! \\
\Gamma(4) &= 3 \cdot \Gamma(3) = 3 \cdot 2 = 3!
\end{aligned} \tag{13.14}$$

and in general

$$\boxed{\Gamma(n+1) = n! \quad \text{for } n = 1, 2, 3, \ldots} \tag{13.15}$$

The formula of (13.15) is a very useful relation; it establishes the relationship between the $\Gamma(n)$ function and the factorial $n!$.

We must remember that, whereas the factorial $n!$ is defined only for zero (recall that $0! = 1$) and positive integer values, the gamma function exists (is continuous) everywhere *except* at 0 and *negative integer numbers*, that is, $-1, -2, -3$, and so on. For instance, when $n = -0.5$, we can find $\Gamma(-0.5)$ in terms of $\Gamma(0.5)$, but if we substitute the numbers $0, -1, -2, -3$ and so on in (13.11), we obtain values which are not consistent with the definition of the $\Gamma(n)$ function, as defined in that relation.

Stated in other words, *the $\Gamma(n)$ function is defined for all positive integers and positive fractional values, and for all negative fractional, but not negative integer values.*

We can use MATLAB's **gamma(n)** function to plot $\Gamma(n)$ versus n. This is done with the script below which produces the plot shown in Figure 13.1.

```
n=-4: 0.05: 4; g=gamma(n); plot(n,g); axis([-4 4 -6 6]); grid;
title('The Gamma Function'); xlabel('n'); ylabel('Gamma(n)')
```

Figure 13.1 shows the plot of the function $\Gamma(n)$ versus n.

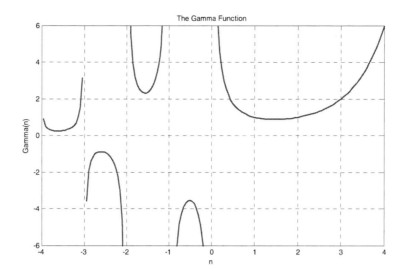

Figure 13.1. Plot of the gamma function

Numerical values of $\Gamma(n)$ for $1 \le n \le 2$, can be found in math tables, but we can use (13.10) or (13.11) to compute values outside this range. Of course, we can use MATLAB to find any valid values of n.

Example 13.1

Compute:

a. $\Gamma(3.6)$ b. $\Gamma(0.5)$ c. $\Gamma(-0.5)$

Solution:

a. From (13.11)

$$\Gamma(n+1) = n\Gamma(n)$$

Then,

$$\Gamma(3.6) = 2.6\Gamma(2.6) = (2.6)(1.6)\Gamma(1.6)$$

and from math tables

$$\Gamma(1.6) = 0.8953$$

Therefore,

$$\Gamma(3.6) = (2.6)(1.6)(0.8953) = 3.717$$

b. From (13.10)

$$\Gamma(n) = \frac{\Gamma(n+1)}{n}$$

Then,

$$\Gamma(0.5) = \frac{\Gamma(0.5+1)}{0.5} = \frac{\Gamma(1.5)}{0.5}$$

and from math tables

$$\Gamma(1.5) = 0.8862$$

Therefore,

$$\Gamma(0.5) = (2)(0.8862) = 1.772$$

c. From (13.10)

$$\Gamma(n) = \frac{\Gamma(n+1)}{n}$$

Then,

$$\Gamma(-0.5) = \frac{\Gamma(-0.5+1)}{-0.5} = \frac{\Gamma(0.5)}{-0.5} = -2\Gamma(0.5)$$

and using the result of (b),

$$\Gamma(-0.5) = -2\Gamma(0.5) = (-2)(1.772) = -3.544$$

We can verify these answers with MATLAB as follows:

a=gamma(3.6), b=gamma(0.5), c=gamma(-0.5)

```
a =
    3.7170
b =
    1.7725
c =
   -3.5449
```

Excel does not have a function which evaluates $\Gamma(n)$ directly. It does, however, have the **GAM-MALN(x)** function. Therefore, we can use the **EXP(GAMMALN(n))** function to evaluate $\Gamma(n)$ at some positive value of n. But because it first computes the natural log, it does not produce an answer if n is negative as shown in Figure 13.2.

x	gammaln(x)	exp(gammaln(x))= gamma(x)
3.6	1.3129	3.7170
0.5	0.5724	1.7725
-0.5	#NUM!	#NUM!

Figure 13.2. Using Excel to find $\Gamma(n)$

Example 13.2

Prove that when n is a positive integer, the relation

$$\boxed{\Gamma(n) = (n-1)!} \qquad (13.16)$$

is true.

Proof:

From (13.11),

$$\Gamma(n+1) = n\Gamma(n) \qquad (13.17)$$

Then,

$$\Gamma(n) = (n-1)\Gamma(n-1) \qquad (13.18)$$

Next, replacing n with $n-1$ on the left side of (13.18), we obtain

$$\Gamma(n-1) = (n-2)\Gamma(n-2) \qquad (13.19)$$

Substitution of (13.19) into (13.18) yields

$$\Gamma(n) = (n-1)(n-2)\Gamma(n-2) \qquad (13.20)$$

By n repeated substitutions, we obtain

$$\Gamma(n) = (n-1)(n-2)(n-3)\ldots1\Gamma(1) \qquad (13.21)$$

and since $\Gamma(1) = 1$, we have

$$\Gamma(n) = (n-1)(n-2)(n-3)\ldots1 \qquad (13.22)$$

or

$$\Gamma(n) = (n-1)! \qquad (13.23)$$

Example 13.3

Use the definition of the $\Gamma(n)$ function to compute the exact value of $\Gamma(1/2)$

Solution:

From (13.1),

$$\Gamma(n) = \int_0^\infty x^{n-1}e^{-x}dx \qquad (13.24)$$

Then,

$$\Gamma\left(\frac{1}{2}\right) = \int_0^\infty x^{0.5-1}e^{-x}dx = \int_0^\infty x^{-0.5}e^{-x}dx \qquad (13.25)$$

Letting

$$x = y^2$$

we obtain

$$\frac{dx}{dy} = 2y$$

or

$$dx = 2ydy$$

By substitution of the last three relations into (13.25), we obtain

$$\Gamma\left(\frac{1}{2}\right) = \int_0^\infty y^{2(-0.5)}e^{-y^2}2ydy = 2\int_0^\infty y^{-1}ye^{-y^2}dy = 2\int_0^\infty e^{-y^2}dy \qquad (13.26)$$

Next, we define $\Gamma(1/2)$ as a function of both x and y, that is, we let

$$\Gamma\left(\frac{1}{2}\right) = 2\int_0^\infty e^{-x^2}dx \qquad (13.27)$$

$$\Gamma\left(\frac{1}{2}\right) = 2\int_0^\infty e^{-y^2}dy \qquad (13.28)$$

Multiplication of (13.27) by (13.28) yields

$$\left[\Gamma\left(\frac{1}{2}\right)\right]^2 = 4\int_0^\infty e^{-x^2}dx\int_0^\infty e^{-y^2}dy = 4\int_0^\infty\int_0^\infty e^{-(x^2+y^2)}dxdy \qquad (13.29)$$

Now, we convert (13.29) to polar coordinates by making the substitution

$$\rho^2 = x^2 + y^2 \qquad (13.30)$$

and by recalling that:

1. the total area of a region is found by either one of the double integrals

$$A = \int\int dxdy = \int\int rdrd\theta \qquad (13.31)$$

2. from differential calculus

$$\frac{d}{du}e^{u^2} = e^{u^2}\frac{d}{du}u^2 = 2ue^{u^2} \qquad (13.32)$$

Then,

$$\int_{\rho_1}^{\rho_2}\rho e^{-\rho^2}d\rho = -\frac{1}{2}e^{-\rho^2} \qquad (13.33)$$

We observe that as $x \rightarrow \infty$ and $y \rightarrow \infty$,

$$\rho \to \infty \quad \text{and} \quad \theta \to \pi/2 \tag{13.34}$$

Substitution of (13.30), (13.33) and (13.34) into (13.29) yields

$$\left[\Gamma\left(\frac{1}{2}\right)\right]^2 = -2\int_0^{\pi/2}\left(e^{-\rho^2}\Big|_{\rho=0}^{\infty}\right)d\theta = -2\int_0^{\pi/2}(0-1)d\theta = 2\int_0^{\pi/2}d\theta = 2\theta\Big|_0^{\pi/2} = \pi$$

and thus, we have obtained the exact value

$$\boxed{\Gamma\left(\frac{1}{2}\right) = \sqrt{\pi}} \tag{13.35}$$

Example 13.4

Compute:

$$\text{a. } \Gamma(-0.5) \quad \text{b. } \Gamma(-1.5) \quad \text{c. } \Gamma(-2.5)$$

Solution:

Using the relations

$$\Gamma(n) = \frac{\Gamma(n+1)}{n} \quad \text{and} \quad \Gamma(0.5) = \sqrt{\pi}$$

we obtain:

a. for $n = -0.5$,

$$\Gamma(-0.5) = \frac{\Gamma(0.5)}{-0.5} = \frac{\sqrt{\pi}}{-0.5} = -2\sqrt{\pi}$$

b. for $n = -1.5$,

$$\Gamma(-1.5) = \frac{\Gamma(-1.5+1)}{-1.5} = \frac{\Gamma(-0.5)}{-1.5} = \frac{-2\sqrt{\pi}}{-1.5} = \frac{4}{3}\sqrt{\pi}$$

c. for $n = -2.5$,

$$\Gamma(-2.5) = \frac{\Gamma(-2.5+1)}{-2.5} = \frac{\Gamma(-1.5)}{-2.5} = \frac{\frac{4}{3}\sqrt{\pi}}{-2.5} = -\frac{8}{15}\sqrt{\pi}$$

Other interesting relations involving the $\Gamma(n)$ function are:

$$\boxed{\begin{array}{c}\Gamma(n)\Gamma(1-n) = \dfrac{\pi}{\sin n\pi} \\[6pt] \text{for} \quad 0 < n < 1\end{array}} \tag{13.36}$$

$$\boxed{\begin{array}{c}2^{2n-1}\Gamma(n)\Gamma\left(n+\dfrac{1}{2}\right) = \sqrt{\pi}\,\Gamma(2n) \\[6pt] \text{for any } n \neq \text{negative integer}\end{array}} \tag{13.37}$$

$$\Gamma(n+1) = n!$$

$$= \sqrt{2\pi n}\, n^n e^{-n}\left\{1 + \frac{1}{12n} + \frac{1}{288n^2} - \frac{139}{51840n^3} - \frac{571}{2488320n^4} + \ldots\right\} \qquad (13.38)$$

Relation (13.38) is referred to as *Stirling's asymptotic series for the* $\Gamma(n)$ *function*. If n is a positive integer, the factorial n! can be approximated as

$$\boxed{n! \approx \sqrt{2\pi n}\, n^n e^{-n}} \qquad (13.39)$$

Example 13.5

Use (13.36) to prove that

$$\Gamma\left(\frac{1}{2}\right) = \sqrt{\pi}$$

Proof:

$$\Gamma\left(\frac{1}{2}\right)\Gamma\left(1 - \frac{1}{2}\right) = \Gamma\left(\frac{1}{2}\right)\Gamma\left(\frac{1}{2}\right) = \frac{\pi}{\sin\frac{\pi}{2}}$$

or

$$\left[\Gamma\left(\frac{1}{2}\right)\right]^2 = \pi$$

Therefore,

$$\Gamma\left(\frac{1}{2}\right) = \sqrt{\pi}$$

Example 13.6

Compute the product

$$\Gamma\left(\frac{1}{3}\right)\Gamma\left(\frac{2}{3}\right)$$

Solution:

Using (13.36), we obtain

$$\Gamma\left(\frac{1}{3}\right)\Gamma\left(1 - \frac{1}{3}\right) = \frac{\pi}{\sin\frac{\pi}{3}}$$

or

$$\Gamma\left(\frac{1}{3}\right)\Gamma\left(\frac{2}{3}\right) = \frac{\pi}{\sqrt{3}/2} = \frac{2\pi}{\sqrt{3}} = \frac{2\sqrt{3}\pi}{3}$$

Example 13.7

Use (13.37) to find

$$\Gamma\left(\frac{3}{2}\right)$$

Solution:

$$2^{3-1}\Gamma\left(\frac{3}{2}\right)\Gamma\left(\frac{3}{2}+\frac{1}{2}\right) = \sqrt{\pi}\Gamma\left(2\cdot\frac{3}{2}\right)$$

$$2^2\Gamma\left(\frac{3}{2}\right)\Gamma(2) = \sqrt{\pi}\Gamma(3)$$

$$\Gamma\left(\frac{3}{2}\right) = \frac{\sqrt{\pi}\Gamma(3)}{4\Gamma(2)} = \frac{2!\sqrt{\pi}}{4\cdot 1} = \frac{\sqrt{\pi}}{2}$$

Example 13.8

Use (13.39) to compute 50!

Solution:

$$50! \approx \sqrt{2\pi\times 50}\times 50^{50}\times e^{-50}$$

We use MATLAB as a calculator, that is, we type and execute the expression

sqrt(2*pi*50)*50^50*exp(−50)

ans =
  3.0363e+064

This is an approximation. To find the exact value, we use the relation $\Gamma(n+1) = n!$ and the MATLAB **gamma(n)** function. Then,

gamma(50+1)

ans =
  3.0414e+064

We can check this answer with the Excel **FACT(n)** function, that is,

=FACT(50) and Excel displays **3.04141E+64**

The $\Gamma(n)$ function is very useful in integrating some improper integrals. Some examples follow.

Example 13.9

Using the definition of the $\Gamma(n)$ function, evaluate the integrals

$$\text{a.} \quad \int_0^\infty x^4 e^{-x} dx \quad \text{b.} \quad \int_0^\infty x^5 e^{-2x} dx$$

Solution:

By definition,

$$\int_0^\infty x^{n-1} e^{-x} dx = \Gamma(n)$$

Then,

a.

$$\int_0^\infty x^4 e^{-x} dx = \Gamma(5) = 4! = 24$$

b.

Let $2x = y$; then, $dx = dy/2$, and by substitution,

$$\int_0^\infty x^5 e^{-2x} dx = \int_0^\infty \left(\frac{y}{2}\right)^5 e^{-y} \frac{dy}{2} = \frac{1}{2^6} \int_0^\infty y^5 e^{-y} dy$$

$$= \frac{\Gamma(6)}{64} = \frac{5!}{64} = \frac{120}{64} = \frac{15}{8}$$

Example 13.10

A negatively charged particle is α meters apart from the positively charged side of an electric field. It is initially at rest, and then moves towards the positively charged side with a force inversely proportional to its distance from it. Assuming that the particle moves towards the center of the positively charged side, considered to be the center of attraction 0, derive an expression for the time required the negatively charged particle to reach 0 in terms of the distance α and its mass m.

Solution:

Let the center of attraction 0 be the point zero on the x–*axis*, as indicated in Figure 13.3.

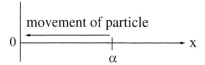

Figure 13.3. Sketch for Example 13.10

By Newton's law,

$$m\frac{dx^2}{dt^2} = -\frac{k}{x} \tag{13.40}$$

where

m = mass of particle

x = distance (varies with time)

k = positive constant of proportionality and the minus (–) sign indicates that the distance x decreases as time t increases.

At t = 0, the particle is assumed to be located on the x–axis at point x = α, and moves towards the origin at x = 0. Let the velocity of the particle be v. Then,

$$\frac{dx}{dt} = v \tag{13.41}$$

and

$$\frac{dx^2}{dt^2} = \frac{dv}{dt} = \frac{dv}{dx}\frac{dx}{dt} = v\frac{dv}{dx} \tag{13.42}$$

Substitution of (13.42) into (13.40) yields

$$mv\frac{dv}{dx} = -\frac{k}{x} \tag{13.43}$$

or

$$mvdv = -\frac{k}{x}(dx) \tag{13.44}$$

Integrating both sides of (13.44), we obtain

$$\frac{mv^2}{2} = -k\ln x + C \tag{13.45}$$

where C represents the constants of integration of both sides, and it is evaluated from the initial condition that v = 0 when x = α. Then,

$$C = k \ln \alpha \tag{13.46}$$

and by substitution into (13.45),

$$\frac{mv^2}{2} = k \ln \alpha - k \ln x = k \ln \frac{\alpha}{x} \tag{13.47}$$

Solving for v^2 and taking the square root of both sides we obtain

$$v = \frac{dx}{dt} = \pm \sqrt{\frac{2k}{m} \ln \frac{\alpha}{x}} \tag{13.48}$$

Since x decreases as t increases, we choose the negative sign, that is,

$$\frac{dx}{dt} = -\sqrt{\frac{2k}{m} \ln \frac{\alpha}{x}} \tag{13.49}$$

Solving (13.49) for dt we obtain

$$dt = -\sqrt{\frac{m}{2k}} \frac{dx}{\sqrt{\ln(\alpha/x)}} \tag{13.50}$$

We are interested in the time required for the particle to reach the origin 0. We denote this time as T; it is found from the relation below, noting that the integration on the right side is with respect to the distance x where at $t = 0$, $x = \alpha$, and at $\tau = t$, $x = 0$. Then,

$$T = \int_0^t d\tau = -\sqrt{\frac{m}{2k}} \int_\alpha^0 \frac{dx}{\sqrt{\ln(\alpha/x)}} \tag{13.51}$$

To simplify (13.51), we let

$$y = \ln\left(\frac{\alpha}{x}\right), \text{ then } e^y = \frac{\alpha}{x} \tag{13.52}$$

or

$$x = \alpha e^{-y}, \text{ and } dx = -\alpha e^{-y} dy \tag{13.53}$$

Also, since

$$\lim_{x \to \alpha} \ln\left(\frac{\alpha}{x}\right) = 0 \text{ and } \lim_{x \to 0} \ln\left(\frac{\alpha}{x}\right) = \infty$$

the lower and upper limits of integration in (13.51), are being replaced with 0 and ∞ respectively. Therefore, we express (13.51) as

$$T = -\sqrt{\frac{m}{2k}} \int_0^\infty \frac{-\alpha e^{-y} dy}{\sqrt{y}} = \alpha \sqrt{\frac{m}{2k}} \int_0^\infty y^{-1/2} e^{-y} dy$$

Finally, using the definition of the $\Gamma(n)$ function, we obtain

$$T = \alpha \Gamma\left(\frac{1}{2}\right) \sqrt{\frac{m}{2k}} = \alpha \sqrt{\pi} \sqrt{\frac{m}{2k}} = \alpha \sqrt{\frac{\pi m}{2k}} \qquad (13.54)$$

Example 13.11

Evaluate the integrals

$$\int_0^{\pi/2} \cos^n\theta d\theta \quad \text{and} \quad \int_0^{\pi/2} \sin^n\theta d\theta \qquad (13.55)$$

Solution:

From the definition of the $\Gamma(n)$ function,

$$\Gamma(n) = \int_0^\infty x^{n-1} e^{-x} dx \qquad (13.56)$$

Also,

$$\Gamma(m) = \int_0^\infty x^{m-1} e^{-x} dx \qquad (13.57)$$

For $m > 0$ and $n > 0$, multiplication of (13.56) by (13.57) yields

$$\Gamma(m)\Gamma(n) = \int_0^\infty u^{m-1} e^{-u} du \int_0^\infty v^{n-1} e^{-v} dv \qquad (13.58)$$

where u and v are dummy variables of integration. Next, letting $u = x^2$ and $v = y^2$, we obtain $du = 2xdx$ and $dv = 2ydy$. Then, with these substitutions, relation (13.58) it written as

$$\Gamma(m)\Gamma(n) = \int_0^\infty x^{2(m-1)} 2xe^{-x^2} dx \int_0^\infty y^{2(n-1)} 2ye^{-y^2} dy = 4\int_0^\infty x^{2m-2} xe^{-x^2} dx \int_0^\infty y^{2n-2} ye^{-y^2} dy$$

$$= 4\int_0^\infty \int_0^\infty x^{2m-1} y^{2n-1} e^{-(x^2+y^2)} dxdy \qquad (13.59)$$

Next, we convert (13.59) to polar coordinates by letting $x = \rho\cos\theta$ and $y = \rho\sin\theta$ Then,

$$\Gamma(m)\Gamma(n) = 4\int_0^{\pi/2}\int_0^{\infty}(\rho\cos\theta)^{2m-1}(\rho\sin\theta)^{2n-1}e^{-\rho^2}\rho\,d\rho\,d\theta$$

$$= 2\int_0^{\pi/2}\cos^{2m-1}\theta\cdot\sin^{2n-1}\theta\,d\theta\int_0^{\infty}\rho^{2m+2n-2}e^{-\rho^2}2\rho\,d\rho \qquad (13.60)$$

To simplify (13.60), we let $\rho^2 = w$; then, $dw = 2\rho\,d\rho$ and thus relation (13.60) is written as

$$\Gamma(m)\Gamma(n) = 2\int_0^{\pi/2}\cos^{2m-1}\theta\cdot\sin^{2n-1}\theta\,d\theta\int_0^{\infty}w^{m+n-1}e^{-w}dw$$

$$= 2\int_0^{\pi/2}\cos^{2m-1}\theta\cdot\sin^{2n-1}\theta\,d\theta\cdot\Gamma(m+n) \qquad (13.61)$$

Rearranging (13.61) we obtain

$$\int_0^{\pi/2}\cos^{2m-1}\theta\cdot\sin^{2n-1}\theta\,d\theta = \frac{\Gamma(m)\Gamma(n)}{2\Gamma(m+n)} \qquad (13.62)$$

and this expression can be simplified by replacing $2m-1$ with n, that is, $m = \dfrac{(n+1)}{2}$, and

$2n-1$ with 0, that is, $n = \dfrac{1}{2}$. Then, we obtain the special case of (13.62) as

$$\int_0^{\pi/2}\cos^n\theta\,d\theta = \frac{\Gamma\left(\dfrac{n+1}{2}\right)\Gamma\left(\dfrac{1}{2}\right)}{2\Gamma\left(\dfrac{n+1}{2}+\dfrac{1}{2}\right)} = \frac{\Gamma\left(\dfrac{n+1}{2}\right)}{\Gamma\left(\dfrac{n}{2}+1\right)}\frac{\sqrt{\pi}}{2} \qquad (13.63)$$

If, in (13.62), we replace $2m-1$ with 0 and $2n-1$ with m, we obtain the integral of the $\sin^n\theta$ function as

$$\int_0^{\pi/2}\sin^m\theta\,d\theta = \frac{\Gamma\left(\dfrac{1}{2}\right)\Gamma\left(\dfrac{m+1}{2}\right)}{2\Gamma\left(\dfrac{1}{2}+\dfrac{m+1}{2}\right)} = \frac{\Gamma\left(\dfrac{m+1}{2}\right)}{\Gamma\left(\dfrac{m}{2}+1\right)}\frac{\sqrt{\pi}}{2} \qquad (13.64)$$

We observe that (13.63) and (13.64) are equal since m and n can be interchanged. Therefore,

$$\boxed{\int_0^{\pi/2}\cos^n\theta\,d\theta = \int_0^{\pi/2}\sin^n\theta\,d\theta = \frac{\Gamma\left(\dfrac{n+1}{2}\right)}{\Gamma\left(\dfrac{n}{2}+1\right)}\frac{\sqrt{\pi}}{2} \qquad n > -1} \qquad (13.65)$$

The relations of (13.65) are known as *Wallis's formulas.*

13.2 The Gamma Distribution

One of the most common probability distributions[*] is the *gamma distribution* which is defined as

$$f(x, n, \beta) = \frac{x^{n-1} e^{-x/\beta}}{\beta^n \Gamma(n)} \quad x > 0, \quad n, \beta > 0 \tag{13.66}$$

A detailed discussion of this probability distribution is beyond the scope of this book; it will suffice to say that it is used in reliability and queuing theory. When n is a positive integer, it is referred to as *Erlang distribution.* Figure 13.4 shows the probability density function (pdf) of the gamma distribution for $n = 3$ and $\beta = 2$.

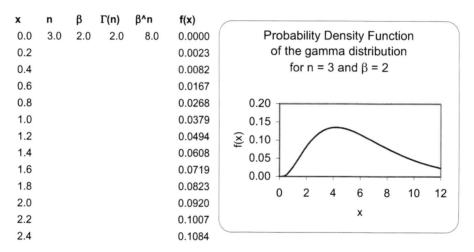

x	n	β	Γ(n)	β^n	f(x)
0.0	3.0	2.0	2.0	8.0	0.0000
0.2					0.0023
0.4					0.0082
0.6					0.0167
0.8					0.0268
1.0					0.0379
1.2					0.0494
1.4					0.0608
1.6					0.0719
1.8					0.0823
2.0					0.0920
2.2					0.1007
2.4					0.1084

Figure 13.4. The pdf for the gamma distribution.

We can evaluate the gamma distribution with the Excel **GAMMADIST** function whose syntax is

GAMMADIST(x,alpha,beta,cumulative)

where:

x = value at which the distribution is to be evaluated

alpha = the parameter n in (13.66)

beta = the parameter β in (13.66)

---

[*] *Several probability distributions are presented in Mathematics for Business, Science, and Technology, ISBN 0–9709511–0–8.*

cumulative = a TRUE / FALSE logical value; if TRUE, **GAMMADIST** returns the cumulative distribution function (cdf), and if FALSE, it returns the probability density function[*] (pdf).

Example 13.12

Use Excel's **GAMMADIST** function to evaluate $f(x)$, that is, the pdf of the *gamma distribution* if:

a. $x = 4$, $n = 3$, and $\beta = 2$

b. $x = 7$, $n = 3$, and $\beta = 2$

Solution:

Since we are interested in the probability density function (pdf) values, we specify the FALSE condition. Then,

a.

=GAMMADIST(4,3,2,FALSE) returns 0.1353

b.

=GAMMADIST(7,3,2,FALSE) returns 0.0925

We observe that these values are consistent with the plot of Figure 13.4.

13.3 The Beta Function

The *beta function*, denoted as $B(m, n)$, is defined as

$$B(m, n) = \int_0^1 x^{m-1}(1-x)^{n-1}dx \qquad (13.67)$$

where $m > 0$ and $n > 0$.

Example 13.13

Prove that

$$B(m, n) = B(n, m) \qquad (13.68)$$

Proof:

Let $x = 1 - y$; then, $dx = -dy$. We observe that as $x \to 0$, $y \to 1$ and as $x \to 1$, $y \to 0$. Therefore,

---

[*] *Several probability density functions are also presented on the text mentioned on the footnote of the previous page.*

$$B(m, n) = \int_0^1 x^{m-1}(1-x)^{n-1}dx = -\int_1^0 (1-y)^{m-1}[1-(1-y)]^{n-1}dy$$

$$= \int_0^1 (1-y)^{m-1}y^{n-1}dy = \int_0^1 y^{n-1}(1-y)^{m-1}dy = B(n, m)$$

and thus (13.68) is proved.

Example 13.14

Prove that

$$B(m, n) = 2\int_0^{\pi/2} \cos^{2m-1}\theta \cdot \sin^{2n-1}\theta\, d\theta \qquad (13.69)$$

Proof:

We let $x = \sin^2\theta$; then, $dx = 2\sin\theta\cos\theta\, d\theta$. We observe that as $x \to 0$, $\theta \to 0$ and as $x \to 1$, $\theta \to \pi/2$. Then,

$$B(m, n) = \int_0^1 x^{m-1}(1-x)^{n-1}dx$$

$$= 2\int_0^{\pi/2} (\sin^2\theta)^{m-1}(\cos^2\theta)^{n-1}\sin\theta\cos\theta\, d\theta \qquad (13.70)$$

$$= 2\int_0^{\pi/2} (\sin^{2m-1}\theta)(\cos^{2m-1}\theta)\, d\theta$$

Example 13.15

Prove that

$$\boxed{B(m, n) = \frac{\Gamma(m)\Gamma(n)}{\Gamma(m+n)}} \qquad (13.71)$$

Proof:

The proof is evident from (13.62) and (13.70).

The $B(m, n)$ function is also useful in evaluating certain integrals as illustrated by the following examples.

Example 13.16

Evaluate the integral

$$\int_0^1 x^4 (1-x)^3 \, dx \qquad (13.72)$$

Solution:

By definition

$$B(m, n) = \int_0^1 x^{m-1} (1-x)^{n-1} \, dx$$

and thus for this example,

$$\int_0^1 x^4 (1-x)^3 \, dx = B(5, 4)$$

Using (13.71) we obtain

$$B(5, 4) = \frac{\Gamma(5)\Gamma(4)}{\Gamma(9)} = \frac{4!3!}{8!} = \frac{24 \times 6}{40320} = \frac{144}{40320} = \frac{1}{280} \qquad (13.73)$$

We can also use MATLAB's **beta(m,n)** function. For this example,

```
format rat; % display answer in rational format
z=beta(5,4)

z =
    1/280
```

Excel does not have a function that computes the $B(m, n)$ function directly. However, we can use (13.71) for its computation as shown in Figure 13.5.

	$\Gamma(m)$ exp(gammaln(m))	$\Gamma(n)$ exp(gammaln(n))	$\Gamma(m+n)$ exp(gammaln(m+n))	Beta(m,n)= $\Gamma(m)$ x $\Gamma(n)$ / $\Gamma(m+n)$
m= 5				
	24.00	6.00	40320.00	1/280
n= 4				

Figure 13.5. Computation of the beta function with Excel.

Example 13.17

Evaluate the integral

$$\int_0^2 \frac{x^2}{\sqrt{2-x}} \, dx \qquad (13.74)$$

Solution:

Let $x = 2v$; then $x^2 = 4v^2$, and $dx = 2dv$. We observe that as $x \to 0$, $v \to 0$, and as $x \to 2$, $v \to 1$. Then, (13.74) becomes

$$\int_0^1 \frac{4v^2}{\sqrt{2-2v}} 2\,dv = \frac{8}{\sqrt{2}} \int_0^1 \frac{v^2}{\sqrt{1-v}} dv = 4\sqrt{2} \int_0^1 v^2 (1-v)^{-1/2} dv$$

$$= 4\sqrt{2} \cdot B\left(3, \frac{1}{2}\right) = 4\sqrt{2}\, \frac{\Gamma(3)\Gamma(1/2)}{\Gamma(7/2)} \tag{13.75}$$

where

$$\Gamma(3) = 2!$$
$$\Gamma(1/2) = \sqrt{\pi}$$
$$\Gamma(7/2) = (7/2 - 1)\Gamma(7/2 - 1) = (5/2)\Gamma(5/2) = (5/2)(5/2 - 1)\Gamma(5/2 - 1) \tag{13.76}$$
$$= 5/2 \cdot 3/2 \cdot (3/2 - 1)\Gamma(3/2 - 1) = (15/8)\Gamma(1/2) = 15\sqrt{\pi}/8$$

Then, from (13.74), (13.75) and (13.76) we obtain

$$\int_0^2 \frac{x^2}{\sqrt{2-x}} dx = \frac{4\sqrt{2} \cdot 2! \cdot \sqrt{\pi}}{15\sqrt{\pi}/8} = \frac{64\sqrt{2}}{15} \tag{13.77}$$

13.4 The Beta Distribution

The *beta distribution* is defined as

$$f(x, m, n) = \frac{x^{m-1}(1-x)^{n-1}}{B(m, n)} \quad x < 0 < 1, \quad m, n > 0 \tag{13.78}$$

A plot of the beta probability density function (pdf) for $m = 3$ and $n = 2$, is shown in Figure 13.6.

As with the gamma probability distribution, a detailed discussion of the beta probability distribution is beyond the scope of this book; it will suffice to say that it is used in computing variations in percentages of samples such as the percentage of the time in a day people spent at work, driving habits, eating times and places, etc.

Using (13.71) we can express the beta distribution as

$$f(x, m, n) = \frac{\Gamma(m+n)}{\Gamma(m)\Gamma(n)} \cdot x^{m-1}(1-x)^{n-1} \quad x < 0 < 1, \quad m, n > 0 \tag{13.79}$$

x	m	n	Γ(m)	Γ(n)	Γ(m+n)	$x^{(m-1)}$	$(1-x)^{(n-1)}$	f(x,m,n)
0.00	3.0	2.0	2.0	1.0	24.0	0.0000	1.0000	0.0000
0.02						0.0004	0.9800	0.0047
0.04								0.0184
0.06								0.0406
0.08								0.0707
0.10								0.1080
0.12								0.1521
0.14								0.2023
0.16								0.2580
0.18								0.3188
0.20								0.3840
0.22								0.4530
0.24								0.5253
0.26								0.6003
0.28								0.6774
0.30								0.7560
0.32								0.8356
0.34						0.1156	0.6600	0.9156

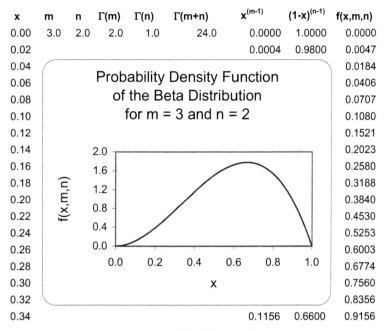

Figure 13.6. The pdf of the beta distribution

We can evaluate the beta *cumulative* distribution function (cdf) with Excels's **BETADIST** function whose syntax is

BETADIST(x,alpha,beta,A,B)

where:

x = value between **A** and **B** at which the distribution is to be evaluated

alpha = the parameter m in (13.79)

beta = the parameter n in (13.79)

A = the lower bound to the interval of **x**

B = the upper bound to the interval of **x**

From the plot of Figure 13.6, we see that when $x = 1$, f(x, m, n) which represents the probability density function, *is* zero. However, the cumulative distribution (the *area* under the curve) at this point is 100% or unity since this is the upper limit of the x –range. This value can be verified by

=BETADIST(1,3,2,0,1) which returns **1.0000**

13.5 Summary

- The gamma function, denoted as $\Gamma(n)$, is also known as generalized factorial function. It is defined as

$$\Gamma(n) = \int_0^\infty x^{n-1} e^{-x} dx$$

- It is convenient to use the relation

$$\Gamma(n) = \frac{\Gamma(n+1)}{n}$$

for $n < 0$ and the relation

$$n\Gamma(n) = \Gamma(n+1)$$

for $n > 0$.

- The $\Gamma(n)$ function is defined for all positive integers and positive fractional values, and for all negative fractional, but not negative integer values.

- The $\Gamma(n)$ function and the factorial n! are related as

$$\Gamma(n+1) = n! \quad \text{for } n = 1, 2, 3, \ldots$$

- We can use MATLAB's **gamma(n)** function to obtain values of $\Gamma(n)$.

- We can use the **EXP(GAMMALN(n))** function to evaluate $\Gamma(n)$ at some positive value of n.

- To evaluate $\Gamma(n)$ when n is a positive integer, we can use the relation

$$\Gamma(n) = (n-1)!$$

- Other useful relations are shown below.

$$\Gamma(1/2) = \sqrt{\pi}$$

$$\Gamma(n)\Gamma(1-n) = \frac{\pi}{\sin n\pi}$$

$$\text{for } 0 < n < 1$$

$$2^{2n-1}\Gamma(n)\sqrt{\pi} = \sqrt{\pi}\Gamma(2n)$$

for any $n \neq$ negative integer

- The relation

$$\Gamma(n+1) = n! = \sqrt{2\pi}n^n e^{-n}\left\{1 + \frac{1}{12n} + \frac{1}{288n^2} - \frac{139}{51840n^3} - \frac{571}{2488320n^4} + \ldots\right\}$$

is referred to as Stirling's asymptotic series for the $\Gamma(n)$ function. If n is a positive integer, the factorial n! can be approximated as

$$n! \approx \sqrt{2\pi n}\, n^n e^{-n}$$

- The $\Gamma(n)$ function is very useful in integrating some improper integrals.

- The relations

$$\int_0^{\pi/2} \cos^n\theta\, d\theta = \int_0^{\pi/2} \sin^n\theta\, d\theta = \frac{\Gamma\left(\dfrac{n+1}{2}\right)}{\Gamma\left(\dfrac{n}{2}+1\right)} \frac{\sqrt{\pi}}{2} \qquad n > -1$$

are known as *Wallis's formulas.*

- The gamma distribution which is defined as

$$f(x, n, \beta) = \frac{x^{n-1} e^{-x/\beta}}{\beta^n \Gamma(n)} \qquad x > 0, \quad n, \beta > 0$$

- The beta function, $B(m, n)$ where $m > 0$ and $n > 0$ is defined as

$$B(m, n) = \int_0^1 x^{m-1}(1-x)^{n-1} dx$$

- The beta function $B(m, n)$ and gamma function $\Gamma(n)$ are related by

$$.B(m, n) = \frac{\Gamma(m)\Gamma(n)}{\Gamma(m+n)}$$

- The beta $B(m, n)$ function is also useful in evaluating certain integrals.

- We can use MATLAB's **beta(m,n)** function to evaluate the beta $B(m, n)$ function.

- The beta distribution is defined as

$$f(x, m, n) = \frac{x^{m-1}(1-x)^{n-1}}{B(m, n)} \qquad x < 0 < 1, \quad m, n > 0$$

13.6 Exercises

1. Given that $m = 2.5$ and $n = -1.25$, compute

$$\frac{\Gamma(m + n)}{\Gamma(m)\Gamma(n)}$$

   Verify your answer with MATLAB and Excel

2. Given that $m = 10$ and $n = 8$, compute $B(m, n)$

   Verify your answer with MATLAB and Excel

3. Evaluate the following integrals

   a. $\displaystyle\int_0^\infty e^{-x^3}dx$

   b. $\displaystyle\int_0^\infty xe^{-x^3}dx$

   c. $\displaystyle\int_0^1 \frac{dx}{\sqrt{1 - x^4}}$

   d. $\displaystyle\int_0^{\pi/2} \sqrt{\tan\theta}\,d\theta$

   e. $\displaystyle\int_0^3 \frac{dx}{\sqrt{3x - x^2}}$

13.7 Solutions to End–of–Chapter Exercises

1.

By repeated use of the relations $n\Gamma(n) = \Gamma(n+1)$ for $n > 1$ and $\Gamma(n) = \dfrac{\Gamma(n+1)}{n}$ for $n < 1$, we obtain

$$\frac{\Gamma(m+n)}{\Gamma(m)\Gamma(n)} = \frac{\Gamma(2.5 + (-1.25))}{\Gamma(2.5)\Gamma(-1.25)} = \frac{\Gamma(1.25)}{\Gamma(2.5)\Gamma(-1.25)} = \frac{\Gamma(5/4)}{\Gamma(5/2)\cdot\Gamma(-5/4)}$$

$$= \frac{\Gamma(1/4 + 1)}{\Gamma(3/2 + 1)\cdot\dfrac{\Gamma(-5/4 + 1)}{(-5/4)}} = \frac{1/4 \cdot \Gamma(1/4)}{3/2 \cdot \Gamma(3/2)\cdot(-4/5)\cdot\Gamma(-1/4)} \quad (1)$$

$$= \frac{1/4 \cdot \Gamma(1/4)}{3/2 \cdot 1/2 \cdot \Gamma(1/2)\cdot(-4/5)\cdot\dfrac{\Gamma(3/4)}{-1/4}} = \frac{5}{48}\cdot\frac{\Gamma(1/4)}{\sqrt{\pi}\cdot\Gamma(3/4)}$$

There are no exact values for $\Gamma(1/4)$ and $\Gamma(3/4)$; therefore, we obtain their approximate values from tables, where we find that $\Gamma(1/4) = 3.6256$ and $\Gamma(3/4) = 1.2254$. Then, by substitution into (1) we obtain:

$$\frac{\Gamma(m+n)}{\Gamma(m)\Gamma(n)} = \frac{5}{48}\cdot\frac{3.6256}{\sqrt{\pi}\cdot 1.2254} = 0.1739$$

Check with MATLAB:

```
m=2.5; n=-1.25; gamma(m+n)/(gamma(m)*gamma(n))
```

```
ans =
    0.1739
```

We cannot check the answer with Excel because it cannot compute negative values.

2.

$$B(m, n) = \frac{\Gamma(m)\Gamma(n)}{\Gamma(m+n)} = \frac{\Gamma(10)\cdot\Gamma(8)}{\Gamma(18)} = \frac{(9!)\times(7!)}{17!}$$

$$= \frac{9\cdot 8\cdot 7\cdot 6\cdot 5\cdot 4\cdot 3\cdot 2\cdot 7\cdot 6\cdot 5\cdot 4\cdot 3\cdot 2}{17\cdot 16\cdot 15\cdot 14\cdot 13\cdot 12\cdot 11\cdot 10\cdot 9\cdot 8\cdot 7\cdot 6\cdot 5\cdot 4\cdot 3\cdot 2}$$

$$= \frac{7\cdot 6\cdot 5\cdot 4\cdot 3\cdot 2}{17\cdot 16\cdot 15\cdot 14\cdot 13\cdot 12\cdot 11\cdot 10} = \frac{5040}{980179200} = 5.1419\times 10^{-6}$$

Check with MATLAB:

```
beta(10,8)
```

```
ans =
   5.1419e-006
```

3.

a.

Let $x^3 = y$, then $x = y^{1/3}$, $dx = \frac{1}{3} \cdot y^{(-2/3)}dy$, so

$$\int_0^\infty e^{-x^3}dx = \int_0^\infty e^{-y} \cdot \frac{1}{3} \cdot y^{1/3-1}dy = \frac{1}{3}\Gamma\left(\frac{1}{3}\right)$$

b.

$$\int_0^\infty xe^{-x^3}dx = \int_0^\infty x^{2-1}e^{-x^3}dx = \Gamma(2) = 1! = 1$$

c.

We let $x^4 = y$ or $x = y^{1/4}$. Then, $dx = (1/4)y^{(-3/4)}dy$ and thus

$$\int_0^1 \frac{dx}{\sqrt{1-x^4}} = \frac{1}{4}\int_0^1 \frac{1}{\sqrt{1-y}} \cdot y^{(-3/4)}dy = \frac{1}{4}\int_0^1 (1-y)^{(-1/2)} \cdot y^{(-3/4)}dy$$

$$= \frac{1}{4}\int_0^1 (1-y)^{1/2-1} \cdot y^{1/4-1}dy = \frac{1}{4} \cdot \frac{\Gamma(1/4) \cdot \Gamma(1/2)}{\Gamma(3/4)} \qquad (1)$$

Also,

$$\Gamma(3/4) \cdot \Gamma(1-3/4) = \frac{\pi}{\sin(3\pi/4)} = \frac{\pi}{\sqrt{2}/2}$$

or

$$\Gamma(3/4) \cdot \Gamma(1/4) = \sqrt{2}\pi, \quad \Gamma(3/4) = \frac{\sqrt{2}\pi}{\Gamma(1/4)}$$

and by substitution into (1)

$$\int_0^1 \frac{dx}{\sqrt{1-x^4}} = \frac{1}{4} \cdot \frac{\Gamma(1/4) \cdot \Gamma(1/2)}{\Gamma(3/4)} = \frac{1}{4} \cdot \frac{\Gamma(1/4) \cdot \sqrt{\pi}}{\sqrt{2}\pi/\Gamma(1/4)}$$

$$= \frac{\sqrt{\pi}}{4\sqrt{2}\pi} \cdot \{\Gamma(1/4)\}^2 = \frac{\{\Gamma(1/4)\}^2}{4\sqrt{2}\pi}$$

d.

$$\int_0^{\pi/2} \sqrt{\tan\theta}\,d\theta = \int_0^{\pi/2} \left(\frac{\sin\theta}{\cos\theta}\right)^{1/2} d\theta = \int_0^{\pi/2} (\sin\theta)^{1/2}(\cos\theta)^{-1/2}(d\theta)$$

From (13.62),

$$\int_0^{\pi/2} \cos^{2m-1}\theta \cdot \sin^{2n-1}\theta\,d\theta = \frac{\Gamma(m)\Gamma(n)}{2\Gamma(m+n)}$$

Letting $2m - 1 = 1/2$ and $2n - 1 = -1/2$ we obtain $m = 3/4$ and $n = 1/4$. Then,

$$\int_0^{\pi/2} \sqrt{\tan\theta}\, d\theta = \frac{\Gamma(3/4)\Gamma(1/4)}{2\Gamma(3/4 + 1/4)} = \frac{\sqrt{2}\pi}{2\Gamma(1)} = \frac{\sqrt{2}\pi}{2}$$

e.

$$\int_0^3 \frac{dx}{\sqrt{3x - x^2}} = \int_0^3 \frac{dx}{\sqrt{x(3 - x)}} = \int_0^3 x^{-1/2} \cdot (3 - x)^{-1/2} dx \quad (1)$$

Let $x = 3y$, then $dx = 3dy$, $x^{-1/2} = (3y)^{-1/2} = (\sqrt{3}/3)y^{-1/2}$. When $x = 0$, $y = 0$ and when $x = 3$, $y = 1$, and the integral of (1) becomes

$$\frac{\sqrt{3}}{3} \cdot 3 \int_0^1 y^{-1/2} \cdot (3 - 3y)^{-1/2} dy = \sqrt{3} \int_0^1 y^{-1/2} \cdot \frac{1}{\sqrt{3}} (1 - y)^{-1/2} dy$$

$$= \int_0^1 y^{-1/2} (1 - y)^{-1/2} dy \quad (2)$$

Recalling that

$$B(m, n) = \int_0^1 x^{m-1} (1 - x)^{n-1} dx = \frac{\Gamma(m) \cdot \Gamma(n)}{\Gamma(m + n)}$$

it follows that $m - 1 = -1/2$, $m = 1/2$, $n - 1 = -1/2$, $m = 1/2$ and thus

$$\int_0^3 \frac{dx}{\sqrt{3x - x^2}} = B\left(\frac{1}{2}, \frac{1}{2}\right) = \frac{\Gamma(1/2) \cdot \Gamma(1/2)}{\Gamma(1/2 + 1/2)} = \frac{\{\Gamma(1/2)\}^2}{\Gamma(1)} = \pi$$

NOTES:

Chapter 14

Orthogonal Functions and Matrix Factorizations

This chapter is an introduction to orthogonal functions. We begin with orthogonal lines and functions, orthogonal trajectories, orthogonal vectors, and we conclude with the factorization methods LU, Cholesky, QR, and Singular Value Decomposition.

14.1 Orthogonal Functions

Orthogonal functions are those which are perpendicular to each other. Mutually orthogonal systems of curves and vectors are of particular importance in physical problems. From analytic geometry and elementary calculus we know that two lines are orthogonal if the product of their slopes is equal to minus one. This is shown in Figure 14.1.

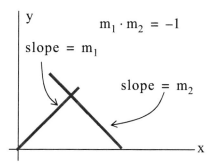

Figure 14.1. Orthogonal lines

Orthogonality applies also to curves. Figure 14.2 shows the angle between two curves C_1 and C_2.

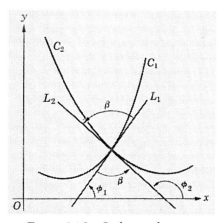

Figure 14.2. Orthogonal curves

By definition, in Figure 14.2, the angle between the curves C_1 and C_2 is the angle β between their tangent lines L_1 and L_2. If m_1 and m_2 are the slopes of these two lines, then, L_1 and L_2 are orthogonal if $m_2 = -1/m_1$.

Example 14.1

Prove that every curve of the family

$$xy = a \qquad a \neq 0 \tag{14.1}$$

is orthogonal to every curve of the family

$$x^2 - y^2 = b \qquad b \neq 0 \tag{14.2}$$

Proof:

At a point $P(x, y)$ on any curve of (14.1), the slope is

$$xdy + ydx = 0$$

or

$$\frac{dy}{dx} = -\frac{y}{x} \tag{14.3}$$

On any curve of (14,2) the slope is

$$2xdx - 2ydy = 0$$

or

$$\frac{dy}{dx} = \frac{x}{y} \tag{14.4}$$

From (14.3) and (14.4) we see that these two curves are orthogonal since their slopes are negative reciprocals of each other. The cases where $x = 0$ or $y = 0$ cannot occur because we defined $a \neq 0$ and $b \neq 0$.

Other orthogonal functions are the $\cos x$ and $\sin x$ functions as we've learned in Chapter 6.

14.2 Orthogonal Trajectories

Two families of curves with the property that each member of either family cuts every member of the other family at right angles are said to be *orthogonal trajectories* of each other. Thus, the curves of (14.2) are orthogonal trajectories of the curves of (14.1). The two families of these curves are shown in Figure 14.3.

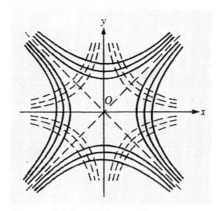

Figure 14.3. Orthogonal trajectories

Example 14.2

Find the orthogonal trajectories of the family of parabolas

$$y = cx^2 \qquad c \neq 0 \tag{14.5}$$

Solution:

The slope of (14.5) is

$$\frac{dy}{dx} = 2cx \tag{14.6}$$

From (14.5), $c = y/x^2$ and thus we rewrite (14.6) as

$$\frac{dy}{dx} = 2\frac{y}{x^2}x = \frac{2y}{x} \tag{14.7}$$

Therefore, the slope of the orthogonal family we are seeking must be

$$\frac{dy}{dx} = -\frac{x}{2y} \tag{14.8}$$

or

$$2ydy + xdx = 0$$

$$2\int ydy + \int xdx = 0$$

$$2\frac{y^2}{2} + \frac{x^2}{2} = k \text{ (constant)}$$

$$x^2 + 2y^2 = C \text{ (constant)} \tag{14.9}$$

Relation (14.9) represents a family of ellipses and the trajectories are shown in Figure14.4.

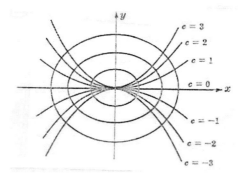

Figure 14.4. Orthogonal trajectories for families of parabolas and ellipses.

14.3 Orthogonal Vectors

Let $X = [x_1 \ x_2 \ x_3 \ ... \ x_n]$ and $Y = [y_1 \ y_2 \ y_3 \ ... \ y_n]$ be two vectors of the same length. Their *inner (dot)* product is defined as

$$X \cdot Y = x_1 y_1 + x_2 y_2 + x_3 y_3 + ... + x_n y_n \ (\text{a scalar}) \tag{14.10}$$

Example 14.3

Given that

$$X = [1 \ 1 \ 1] \text{ and } Y = [2 \ 1 \ 2]$$

find the dot product $X \cdot Y$

Solution:

$$X \cdot Y = (1) \cdot (2) + (1) \cdot (1) + (1) \cdot (2) = 5$$

Definition: Two vectors X_1 and X_2 are said to be orthogonal if their dot product is zero.

Example 14.4

Test the vectors

$$X_1 = [1 \ 1 \ 1] \text{ and } X_2 = [1 \ -2 \ 1]$$

for orthogonality.

Solution:

$$X_1 \cdot X_2 = (1) \cdot (1) + (1) \cdot (-2) + (1) \cdot (1) = 0$$

Therefore, the vectors X_1 and X_2 are orthogonal to each other.

With any vector $X \neq 0$ we may associate a unique unit vector U which is obtained by dividing each component of X by each magnitude $|X|$ defined as

$$|X| = \sqrt{x_1^2 + x_2^2 + \ldots + x_n^2}$$

where x_i represents an element of the vector X. This process is called *normalization*.

Example 14.5

Given that

$$X = [2 \quad 4 \quad 4]$$

compute the unit vector U_X.

Solution:

First, we compute the magnitude $|X|$. For this example,

$$|X| = \sqrt{2^2 + 4^2 + 4^2} = 6$$

To compute the unit vector U_X we divide each element of X by the magnitude $|X|$. Thus,

$$U_X = \begin{bmatrix} \dfrac{2}{6} & \dfrac{4}{6} & \dfrac{4}{6} \end{bmatrix} = \begin{bmatrix} \dfrac{1}{2} & \dfrac{2}{3} & \dfrac{2}{3} \end{bmatrix}$$

A basis that consists of mutually orthogonal vectors is referred to as an *orthogonal basis*. If these vectors are also unit vectors, the basis is called *orthonormal basis*.

If the column (or row) vectors of a square matrix A are mutually orthogonal unit vectors, the matrix A is orthogonal and

$$A \cdot A^T = I \tag{14.11}$$

where A^T is the transpose of A and I is the identity matrix.

Example 14.6

Given that

$$A = \begin{bmatrix} 1/2 & 1/4 \\ 1/2 & 1/2 \end{bmatrix}$$

find an orthonormal set of eigenvectors[*] and verify that the result satisfies (14.11).

Solution:

First, we find the eigenvalues of the matrix A from the relation $\det(A - \lambda I) = 0$ where for this example

$$\det\left(\begin{vmatrix} 1/2 & 1/4 \\ 1/2 & 1/2 \end{vmatrix} - \lambda \begin{vmatrix} 1 & 0 \\ 0 & 1 \end{vmatrix} \right) = 0$$

$$\det \begin{vmatrix} 1/2 - \lambda & 1/4 \\ 1/2 & 1/2 - \lambda \end{vmatrix} = 0$$

$$\lambda^2 - \lambda + 3/16 = 0$$

from which $\lambda_1 = 1/4$ and $\lambda_2 = 3/4$ and as we've learned in Chapter 5, with these eigenvalues we can obtain an infinite number of eigenvectors. To find a 2×2 square matrix Z such that

$$Z \cdot Z^T = I$$

we begin with

$$\begin{bmatrix} z_1 & z_2 \\ -z_1 & z_2 \end{bmatrix} \cdot \begin{bmatrix} z_1 & -z_1 \\ z_2 & z_2 \end{bmatrix} = \begin{bmatrix} 1 & 0 \\ 0 & 1 \end{bmatrix} \qquad (14.12)$$

or

$$\begin{bmatrix} z_1^2 + z_2^2 & -z_1^2 + z_2^2 \\ -z_1^2 + z_2^2 & z_1^2 + z_2^2 \end{bmatrix} = \begin{bmatrix} 1 & 0 \\ 0 & 1 \end{bmatrix} \qquad (14.13)$$

Equating like terms we obtain

$$z_1^2 + z_2^2 = 1 \qquad -z_1^2 + z_2^2 = 0$$

From the second equation we obtain $z_1^2 = z_2^2$ and by substitution into the first we obtain $2z_1^2 = 1$ or

---

[*] It is strongly suggested that the reader reviews the definitions of eigenvalues and eigenvectors in Chapter 5 at this time.

$$z_1 = z_2 = 1/(\pm\sqrt{2})$$

This result indicates that we can choose either $1/\sqrt{2}$ or $1/(-\sqrt{2})$ for the values of z_1 and z_2. We choose the value $1/\sqrt{2}$ and then the first (left most) matrix in (14.12) is

$$Z = \begin{bmatrix} 1/\sqrt{2} & 1/\sqrt{2} \\ 1/(-\sqrt{2}) & 1/\sqrt{2} \end{bmatrix}$$

and as a check,

$$\begin{bmatrix} 1/\sqrt{2} & 1/\sqrt{2} \\ 1/(-\sqrt{2}) & 1/\sqrt{2} \end{bmatrix} \cdot \begin{bmatrix} 1/\sqrt{2} & 1/(-\sqrt{2}) \\ 1/\sqrt{2} & 1/\sqrt{2} \end{bmatrix} = \begin{bmatrix} 1 & 0 \\ 0 & 1 \end{bmatrix}$$

The computations for finding orthonormal sets of eigenvectors for larger size (3×3 or higher) matrices using the above procedure becomes quite involved. A simpler procedure is the *Gram-Schmidt orthogonalization procedure* which is discussed on the next section.

14.4 The Gram-Schmidt Orthogonalization Procedure

Let $X_1, X_2, \ldots X_m$ be some column vectors. We can find an orthogonal basis $Y_1, Y_2, \ldots Y_m$ using the following relations. We must remember that the products in (14.14) below are the inner (dot) products and if $X = [x_1 \ x_2 \ x_3 \ \ldots \ x_n]$ and $Y = [y_1 \ y_2 \ y_3 \ \ldots \ y_n]$ are two vectors of the same length their dot product is defined as $X \cdot Y = x_1 y_1 + x_2 y_2 + x_3 y_3 + \ldots + x_n y_n$ (a scalar). Thus in the second equation in (14.14) the dot products on the numerator and denominator must be found first and the result must be from the dot product of it and Y_1

$$Y_1 = X_1$$

$$Y_2 = X_2 - \frac{Y_1 \cdot X_2}{Y_1 \cdot Y_1} \cdot Y_1$$

$$Y_3 = X_3 - \frac{Y_2 \cdot X_3}{Y_2 \cdot Y_2} \cdot Y_2 - \frac{Y_1 \cdot X_3}{Y_1 \cdot Y_1} \cdot Y_1 \qquad (14.14)$$

$$\ldots$$

$$Y_m = X_m - \frac{Y_{m-1} \cdot X_m}{Y_{m-1} \cdot Y_{m-1}} \cdot Y_{m-1} - \ldots - \frac{Y_1 \cdot X_m}{Y_1 \cdot Y_1} \cdot Y_1$$

Also, the unit vectors

$$U_i = \frac{Y_i}{|Y_i|} \qquad i = 1, 2, \ldots, m \qquad (14.15)$$

are mutually orthogonal and form an orthonormal basis.

In our subsequent discussion the column vectors will be denoted as row vectors transposed.

Example 14.7

Given that $X_1 = [1 \ 1 \ 1]^T$, $X_2 = [1 \ -2 \ 1]^T$, and $X_3 = [1 \ 2 \ 3]^T$, find an orthonormal basis.

Solution:

From (14.14)

$$Y_1 = X_1 = [1 \ 1 \ 1]^T$$

$$Y_2 = X_2 - \frac{Y_1 \cdot X_2}{Y_1 \cdot Y_1} \cdot Y_1 = [1 \ -2 \ 1]^T - \frac{0}{3} \cdot Y_1 = [1 \ -2 \ 1]^T$$

$$Y_3 = X_3 - \frac{Y_2 \cdot X_3}{Y_2 \cdot Y_2} \cdot Y_2 - \frac{Y_1 \cdot X_3}{Y_1 \cdot Y_1} \cdot Y_1 = [1 \ 2 \ 3]^T - \frac{0}{6} \cdot Y_2 - \frac{6}{3}[1 \ 1 \ 1]^T$$

$$= [1 \ 2 \ 3]^T - [2 \ 2 \ 2]^T = [-1 \ 0 \ 1]^T$$

and from (14.15)

$$U_1 = \frac{Y_1}{|Y_1|} = [1/\sqrt{3} \ \ 1/\sqrt{3} \ \ 1/\sqrt{3}]^T$$

$$U_2 = \frac{Y_2}{|Y_2|} = [1/\sqrt{6} \ \ -2/\sqrt{6} \ \ 1/\sqrt{6}]^T$$

$$U_3 = \frac{Y_3}{|Y_3|} = [-1/\sqrt{2} \ \ 0 \ \ 1/\sqrt{2}]^T$$

and denoting the matrix whose elements are the unit vectors as A, we have:

$$A = \begin{bmatrix} 1/\sqrt{3} & 1/\sqrt{6} & -1/\sqrt{2} \\ 1/\sqrt{3} & -2/\sqrt{6} & 0 \\ 1/\sqrt{3} & 1/\sqrt{6} & 1/\sqrt{2} \end{bmatrix}$$

We can verify that $A \cdot A^T = I$ with the MATLAB script below.

```
A=[1/sqrt(3) 1/sqrt(6) –1/sqrt(2); 1/sqrt(3) –2/sqrt(6) 0; 1/sqrt(3) 1/sqrt(6) 1/sqrt(2)];
I=A*A'
```

```
I =
    1.0000          0     0.0000
         0     1.0000          0
    0.0000          0     1.0000
```

We can also use the MATLAB function **orth(A)** to produce an orthonormal basis as shown below.

B=[1 1 1; 1 −2 1; 1 2 3]; C=orth(B)

```
C =
   -0.4027     0.0000     0.9153
    0.0000     1.0000     0.0000
   -0.9153     0.0000    -0.4027
```

We observe that the vectors of the C matrix produced by MATLAB are different from those we derived with the Gram-Schmidt orthogonalization procedure. The reason for this difference is that the orthogonalization process is not unique, that is, we may find different values depending on the process being used. As shown below, the vectors produced by MATLAB also satisfy the condition $C \cdot C^T = I$.

I=C*C'

```
I =
    1.0000    -0.0000     0.0000
   -0.0000     1.0000    -0.0000
    0.0000    -0.0000     1.0000
```

14.5 The LU Factorization

In matrix computations, computers use the so-called matrix factorization methods to decompose a matrix A into a product of other smaller matrices. The *LU factorization method* decomposes a matrix A into a lower triangular matrix L and an upper triangular matrix U so that $A = L \cdot U$. In Chapter 4 we saw how the method of Gaussian elimination proceeds by systematically removing the unknowns from a system of linear equations.

Consider the following 3×3 lower triangular case.

$$\begin{bmatrix} L_{11} & 0 & 0 \\ L_{21} & L_{22} & 0 \\ L_{31} & L_{32} & L_{33} \end{bmatrix} \cdot \begin{bmatrix} x_1 \\ x_2 \\ x_3 \end{bmatrix} = \begin{bmatrix} b_1 \\ b_2 \\ b_3 \end{bmatrix}$$

The unknowns are found from

$$x_1 = b_1/L_{11}$$
$$x_2 = (b_2 - L_{21}x_1)/L_{22} \quad\quad (14.16)$$
$$x_3 = (b_3 - L_{31}x_1 - L_{31}x_2)/L_{33}$$

provided that $L_{11} \cdot L_{22} \cdot L_{33} \neq 0$. The substitution order in (14.16) is referred to as *forward substitution.*

For the upper triangular case, the unknowns are written in reverse order. Thus, to solve

$$\begin{bmatrix} U_{11} & U_{12} & U_{13} \\ 0 & U_{22} & U_{23} \\ 0 & 0 & U_{33} \end{bmatrix} \cdot \begin{bmatrix} x_1 \\ x_2 \\ x_3 \end{bmatrix} = \begin{bmatrix} b_1 \\ b_2 \\ b_3 \end{bmatrix} \quad\quad (14.17)$$

we start from the bottom to the top as shown below.

$$x_3 = b_3/U_{33}$$
$$x_2 = (b_2 - U_{23}x_3)/U_{22} \quad\quad (14.18)$$
$$x_1 = (b_1 - U_{12}x_2 - U_{13}x_3)/U_{11}$$

provided that $U_{11} \cdot U_{22} \cdot U_{33} \neq 0$. The substitution order in (14.18) is referred to as *backward substitution.*

Example 14.8

Let us review the example given in Chapter 4 which consists of the following equations.

$$2v_1 - v_2 + 3v_3 = 5$$
$$-4v_1 - 3v_2 - 2v_3 = 8 \quad\quad (14.19)$$
$$3v_1 + v_2 - v_3 = 4$$

To find the three unknowns, we begin by multiplying the first equation by -2 and subtracting it from the second equation. This removes v_1 from the second equation. Likewise, we multiply the first equation by $3/2$ and we subtract it from the third equation. With these two reductions we obtain

$$2v_1 - v_2 + 3v_3 = 5$$
$$-5v_2 + 4v_3 = 18 \qu\quad (14.20)$$
$$2.5v_2 - 5.5v_3 = -3.5$$

Next we multiply the second equation of (14.20) by $-1/2$ and we subtract it from the third equation of (14.20) and we obtain the system of equations below.

$$\begin{aligned} 2v_1 - v_2 + 3v_3 &= 5 \\ -5v_2 + 4v_3 &= 18 \\ -3.5v_3 &= 5.5 \end{aligned} \tag{14.21}$$

We see that the eliminations have transformed the given square system into an equivalent upper triangular system that gives the same solution which is obtained as follows:

$$v_3 = -11/7$$

$$v_2 = (18 - 4v_3)/(-5) = -34/7$$

$$v_1 = (5 + v_2 - 3v_3)/2 = 17/7$$

The elements of the upper triangular matrix U are the coefficients of the unknowns in (14.21). Thus,

$$\begin{bmatrix} U_{11} & U_{12} & U_{13} \\ 0 & U_{22} & U_{23} \\ 0 & 0 & U_{33} \end{bmatrix} = \begin{bmatrix} 2 & -1 & 3 \\ 0 & -5 & 4 \\ 0 & 0 & -3.5 \end{bmatrix}$$

Now, let us use the relations of (14.16) and (14.18) to find the lower and upper triangular matrices of our example where

$$A = \begin{bmatrix} 2 & -1 & 3 \\ -4 & -3 & -2 \\ 3 & 1 & -1 \end{bmatrix}$$

We want to find L_{ij} and U_{ij} such that

$$\begin{bmatrix} L_{11} & 0 & 0 \\ L_{21} & L_{22} & 0 \\ L_{31} & L_{32} & L_{33} \end{bmatrix} \cdot \begin{bmatrix} U_{11} & U_{12} & U_{13} \\ 0 & U_{22} & U_{23} \\ 0 & 0 & U_{33} \end{bmatrix} = A = \begin{bmatrix} 2 & -1 & 3 \\ -4 & -3 & -2 \\ 3 & 1 & -1 \end{bmatrix} \tag{14.22}$$

where the first matrix on the left side is the lower triangular matrix L and the second is the upper triangular matrix U. The elements of matrix U are the coefficients of v_1, v_2, and v_3 in (14.20). To find the elements of matrix L we use MATLAB to multiply matrix A by the inverse of matrix U. Thus,

U=[2 −1 3; 0 −5 4; 0 0 −3.5]; A=[2 −1 3; −4 −3 −2; 3 1 −1]; L=A*inv(U)

L =

$$\begin{matrix} 1 & 0 & 0 \\ -2 & 1 & 0 \\ 3/2 & -1/2 & 1 \end{matrix}$$

Therefore, the matrix A has been decomposed to a lower triangular matrix L and an upper matrix U as shown below.

$$\begin{bmatrix} 2 & -1 & 3 \\ -4 & -3 & -2 \\ 3 & 1 & -1 \end{bmatrix} = \begin{bmatrix} 1 & 0 & 0 \\ -2 & 1 & 0 \\ 3/2 & -1/2 & 1 \end{bmatrix} \cdot \begin{bmatrix} 2 & -1 & 3 \\ 0 & -5 & 4 \\ 0 & 0 & -3.5 \end{bmatrix}$$

Check with MATLAB:

L=[1 0 0; –2 1 0; 3/2 –1/2 1]; U=[2 –1 3; 0 –5 4; 0 0 –3.5]; A=L*U

A =

$$\begin{matrix} 2 & -1 & 3 \\ -4 & -3 & -2 \\ 3 & 1 & -1 \end{matrix}$$

In the example above, we found the elements of the lower triangular matrix L by first computing the inverse of the upper triangular matrix U and performing the matrix multiplication $L = A \cdot U^{-1}$ but not $L = U^{-1} \cdot A$. Was this necessary? The answer is no. For a square matrix where none of the diagonal elements are zero, the lower triangular matrix has the form

$$L = \begin{bmatrix} 1 & 0 & 0 \\ L_{21} & 1 & 0 \\ L_{31} & L_{32} & 1 \end{bmatrix}$$

and in our example we found that the values of the subdiagonal elements are $L_{21} = -2$, $L_{31} = 3/5$, and $L_{32} = -1/2$. These values are the multipliers that we've used in the elimination process in succession.

Example 14.9

Use the MATLAB function **[L,U]=lu(A)** to decompose the matrix

$$A = \begin{bmatrix} 2 & -3 & 1 \\ -1 & 5 & -2 \\ 3 & -8 & 4 \end{bmatrix}$$

into a lower and an upper triangular.

Solution:

format rat; A=[2 –3 1; –1 5 –2; 3 –8 4]; [L,U]=lu(A)

```
L  =
        2/3              1              1
       -1/3              1              0
          1              0              0
U  =
          3             -8              4

          0            7/3           -2/3

          0              0             -1
```

We observe that while the upper triangular matrix U has the proper structure, the lower triangular matrix L lacks structure. When a matrix lacks structure we say that it is *permuted*. To put L in the proper structure, let us interchange the first and third rows. Then,

$$L' = \begin{bmatrix} 1 & 0 & 0 \\ -1/3 & 1 & 0 \\ 2/3 & 1 & 1 \end{bmatrix} \tag{14.23}$$

The new matrix L' has now the proper structure. Let us now use MATLAB to see if $L' \cdot U = A$.

L1=[1 0 0; –1/3 1 0; 2/3 1 1]; U=[3 –8 4; 0 7/3 –2/3; 0 0 –1]; A1=L1*U

```
A1  =
          3             -8              4

         -1              5             -2

          2             -3              1
```

We observe that matrix A is now permuted. To put it in the given form we need to make the same interchanges in rows as with the lower triangular matrix, that is, we must interchange the first and third rows.

To find out how MATLAB performs LU factorization, let us invoke the **help lu** command. Only part of the display is shown below.

```
help lu
  LU    LU factorization.
     [L,U] = lu(X) stores an upper triangular matrix in U and a
     "psychologically lower triangular matrix" (i.e. a product
```

```
of lower triangular and permutation matrices) in L, so
that X = L*U. X can be rectangular.

[L,U,P] = lu(X) returns unit lower triangular matrix L, upper
triangular matrix U, and permutation matrix P so that
P*X = L*U.
```

The permutation matrix P is an identity matrix that is permuted so that the rows of this matrix indicate the interchanges. Consider, for example, the identity matrix

$$I = \begin{bmatrix} 1 & 0 & 0 \\ 0 & 1 & 0 \\ 0 & 0 & 1 \end{bmatrix}$$

If we interchange the first and third rows of the identity matrix I above, we obtain the permutation matrix P below.

$$P = \begin{bmatrix} 0 & 0 & 1 \\ 0 & 1 & 0 \\ 1 & 0 & 0 \end{bmatrix} \tag{14.24}$$

and matrix P indicates the same interchanges as with the lower triangular matrix in Example 14.9.

Example 14.10

Use the MATLAB function **[L,U,P]=lu(A)** to decompose the matrix

$$A = \begin{bmatrix} 2 & -3 & 1 \\ -1 & 5 & -2 \\ 3 & -8 & 4 \end{bmatrix}$$

into a lower and an upper triangular and show that $P \cdot A = L \cdot U$.

Solution:

This is the same matrix as in Example 14.9. Thus,

A=[2 –3 1; –1 5 –2; 3 –8 4]; [L,U,P]=lu(A)

```
L =
        1           0           0
     -1/3           1           0
      2/3           1           1
```

```
U =
        3               -8              4
        0               7/3            -2/3
        0                0             -1
P =
        0                0              1
        0                1              0
        1                0              0
```

We observe that the lower triangular matrix has now the proper structure and the P matrix displayed by MATLAB is the same as in (14.24). Also,

PA=P*A, LU=L*U

```
PA =
        3               -8              4
       -1                5             -2
        2               -3              1
LU =
        3               -8              4
       -1                5             -2
        2               -3              1
```

We observe that $P \cdot A = L \cdot U$ with the first and second rows interchanged when compare with the given matrix A.

The MATLAB matrix left division operator $x = A \backslash b$ uses the $L \cdot U$ factorization approach.

The user−defined function **ExchRows** below, interchanges rows i and jj of a vector or matrix X.

```
% The function ExchRows interchanges rows i and j
% of a matrix or vector X
%
function X = ExchRows(X,i,j)
%
temp = X(i,:);
X(i,:) = X(j,:);
X(j,:) = temp;

% This file is saved as ExchRows.m
% To run this program, define the matrix or vector
% X and the indices i and j in MATLAB's Command Window
% as X=[....], i = {first row # to be interchanged},
% j = {row # to be interchanged with row i}, and
```

% then type ExchRows(X,i,j)at the command prompt.

Example 14.11

Given that the matrix X is defined as

$$X = \begin{bmatrix} -2 & 5 & -4 & 9 \\ -3 & -6 & 8 & 1 \\ 7 & -5 & 3 & 2 \\ 4 & -9 & -8 & -1 \end{bmatrix}$$

use the **ExchRows.m** user–defined function above to interchange rows 1 and 3.

Solution:

At the MATLAB command prompt we enter

X=[-2 5 -4 9; -3 -6 8 1; 7 -5 3 2; 4 -9 -8 -1]; i = 1; j = 3;
ExchRows(X,i,j)

and MATLAB outputs

```
X =
     7    -5     3     2
    -3    -6     8     1
    -2     5    -4     9
     4    -9    -8    -1
```

The user–defined function **GaussElimPivot** below, performs Gauss elimination with row pivoting. First, let us explain the use of MATLAB's built-in function **max(v)** where **v** is a row or a column vector, and for matrices is a row vector containing the maximum element from each column. As an example, let

v=[2 -1 3 -5 7 -9 -12]'; max(v)

```
ans =
     7
```

[Amax,m]=max(v)

```
Amax =
     7
m =
     5
```

[Amax,m]=max(abs(v))

```
Amax =
      12

m =
      7
```

% This user–defined function file solves A*x=b by
% the Gauss elimination with row pivoting method.
% A is a matrix that contains the coefficients of
% the system of equations, x is a column vector that
% will display the computed unknown values, and b
% is a column vector that contains the known values
% on the right hand side.

```
function x = GaussElimPivot(A,b)

if size(b,2) > 1; b=b';
end

n = length(b); z = zeros(n,1);

%  Set up scale factor array

for i = 1:n; z(i) = max(abs(A(i,1:n)));
end

% The statements below exchange rows if required

for k = 1:n–1
   [Amag,m] = max(abs(A(k:n,k))./z(k:n));
   m = m + k – 1;
if Amag < eps; error('Matrix is singular');
end

if m ~= k

b = ExchRows(b,k,m);
z = ExchRows(z,k,m);
A = ExchRows(A,k,m);
end

%  Elimination steps

for i = k+1:n
   if A(i,k) ~= 0
      alpha = A(i,k)/A(k,k);
      A(i,k+1:n) = A(i,k+1:n) – alpha*A(k,k+1:n)
      b(i) = b(i) – alpha*b(k);
   end
 end
end

%  Back substitution phase

for k = n:–1:1
   b(k) = (b(k) – A(k,k+1:n)*b(k+1:n))/A(k,k);
end
```

```
%  Enter the values of A and b at the MATLAB's
%  command window and type GaussElimPivot(A,b), x
```

Example 14.12

Given that

$$A = \begin{bmatrix} -2 & 5 & -4 & 9 \\ -3 & -6 & 8 & 1 \\ 7 & -5 & 3 & 2 \\ 4 & -9 & -8 & -1 \end{bmatrix}, \quad b = \begin{bmatrix} -3 \\ 2 \\ 8 \\ 5 \end{bmatrix}, \quad x = \begin{bmatrix} x_1 \\ x_2 \\ x_3 \\ x_4 \end{bmatrix}$$

use the **GaussElimPivot** user–defined function above to compute the values of the vector x.

Solution:

At the MATLAB command prompt we enter

A=[-2 5 -4 9; -3 -6 8 1; 7 -5 3 2; 4 -9 -8 -1]; b = [-3 2 8 5]';
GaussElimPivot(A,b), x

and MATLAB outputs the following:

```
A =
       7.0000      -5.0000       3.0000       2.0000
      -3.0000      -8.1429       9.2857       1.8571
      -2.0000       5.0000      -4.0000       9.0000
       4.0000      -9.0000      -8.0000      -1.0000
A =
       7.0000      -5.0000       3.0000       2.0000
      -3.0000      -8.1429       9.2857       1.8571
      -2.0000       3.5714      -3.1429       9.5714
       4.0000      -9.0000      -8.0000      -1.0000
A =
       7.0000      -5.0000       3.0000       2.0000
      -3.0000      -8.1429       9.2857       1.8571
      -2.0000       3.5714      -3.1429       9.5714
       4.0000      -6.1429      -9.7143      -2.1429
A =
       7.0000      -5.0000       3.0000       2.0000
      -3.0000      -8.1429       9.2857       1.8571
      -2.0000       3.5714       0.9298      10.3860
       4.0000      -6.1429      -9.7143      -2.1429
```

```
A =
      7.0000     -5.0000      3.0000      2.0000
     -3.0000     -8.1429      9.2857      1.8571
     -2.0000      3.5714      0.9298     10.3860
      4.0000     -6.1429    -16.7193     -3.5439
A =
      7.0000     -5.0000      3.0000      2.0000
     -3.0000     -8.1429      9.2857      1.8571
      4.0000     -6.1429    -16.7193     -3.5439
     -2.0000      3.5714      0.9298     10.1889
x =
      0.7451     -1.0980      0.1176     -0.1373
```

Check with MATLAB's left division:

x=b\A

```
x =
      0.7451     -1.0980      0.1176     -0.1373
```

The user–defined function **LUdecomp** below, performs LU decomposition, and returns matrix A as $A=L*U$ and the row permutation vector **permut**.

```
function [A,permut] = LUdecomp(A)

% LU decomposition of matrix A; returns A = L*U
% and the row permutation vector permut

n = size(A,1); z = zeros(n,1);
permut = (1:n)';

for i = 1:n; z(i) = max(abs(A(i,1:n)));
end

% Exchange rows if necessary

for k = 1:n-1
[Amag,m] = max(abs(A(k:n,k))./z(k:n));
m = m + k - 1;
if Amag < eps
   error('Matrix is singular')
end

if m ~= k
 z = ExchRows(z,k,m);
 A = ExchRows(A,k,m);
```

```
    permut = ExchRows(permut,k,m);
end

% Elimination pass

for i = k+1:n
   if A(i,k)~=0
     alpha = A(i,k)/A(k,k);
     A(i,k+1:n) = A(i,k+1:n) – alpha*A(k,k+1:n);
     A(i,k) = alpha;

   end
  end
end
```

Example 14.13

Given that

$$A = \begin{bmatrix} -2 & 5 & -4 & 9 \\ -3 & -6 & 8 & 1 \\ 7 & -5 & 3 & 2 \\ 4 & -9 & -8 & -1 \end{bmatrix}$$

use the **LUdecomp** user–defined function above to decompose matrix A and show how the given rows were permuted.

Solution:

At the MATLAB command prompt we enter

```
A=[-2 5 -4 9; -3 -6 8 1; 7 -5 3 2; 4 -9 -8 -1];
[A,permut] = LUdecomp(A)

A =
    7.0000   -5.0000    3.0000    2.0000
   -0.4286   -8.1429    9.2857    1.8571
    0.5714    0.7544  -16.7193   -3.5439
   -0.2857   -0.4386   -0.0556   10.1889

permut =
     3
     2
     4
     1
```

Check:

```
[L,U,P]=lu(A)

L =
     1.0000          0          0          0
    -0.0612     1.0000          0          0
     0.0816    -0.1376     1.0000          0
    -0.0408     0.0761     0.0417     1.0000

U =
     7.0000    -5.0000     3.0000     2.0000
          0    -8.4490     9.4694     1.9796
          0          0   -15.6612    -3.4347
          0          0          0     10.2632

P =
   1   0   0   0
   0   1   0   0
   0   0   1   0
   0   0   0   1

L*U

ans =
     7.0000    -5.0000     3.0000     2.0000
    -0.4286    -8.1429     9.2857     1.8571
     0.5714     0.7544   -16.7193    -3.5439
    -0.2857    -0.4386    -0.0556    10.1889
```

The user–defined function **LUsolPivot** listed below, is saved as **LUsolPivot.m** and will be used in the user–defined function **matInvert** that follows.

```
% In this user–defined function matrix A and column
% vector b are entered in MATLAB's command window
% and "permut" holds the row permutation data.
%
function x = LUsolPivot(A,b,permut)
%
% The six statements below rearrange vector b and
% stores it in vector x.
%
if size(b) > 1; b = b';
end
n = size(A,1);
x = b;
for i = 1:n; x(i)= b(permut(i));
```

```
end
%
% The next six statements perform forward and
% backward substitution
%
for k = 2:n
x(k) = x(k)– A(k,1:k–1)*x(1:k–1);
end
for k = n:–1:1
x(k) = (x(k) – A(k,k+1:n)*x(k+1:n))/A(k,k);
end
```

The user–defined function **matInvert** below, inverts matrix A with LU decomposition.

```
% This user–defined function inverts a matrix A
% defined in MATLAB's command prompt using LU
% decomposition
%
function Ainv = matInvert(A)
%
n = size(A,1); % Assigns the size of A to n.
%
Ainv = eye(n); % Creates identity matrix of size n.
% The statement below performs LU decomposition
% using the user–defined function LUdecomp(A) saved
% previously
%
[A,permut] = LUdecomp(A);
%
% The last three statements solve for each vector
% on the right side, and store results in Ainv
% replacing the corresponding vector using the
% user–defined function LUsolPivot saved previously.
%
for i = 1:n
   Ainv(:,i) = LUsolPivot(A,Ainv(:,i),permut);
end
```

Example 14.14

Invert the matrix A below with the user–defined function **matInvert**.

$$A = \begin{bmatrix} -2 & 5 & -4 & 9 \\ -3 & -6 & 8 & 1 \\ 7 & -5 & 3 & 2 \\ 4 & -9 & -8 & -1 \end{bmatrix}$$

Solution:

In MATLAB's command prompt we enter

```
A=[-2 5 -4 9; -3 -6 8 1; 7 -5 3 2; 4 -9 -8 -1];
Ainv = matInvert(A)
```

and MATLAB returns

```
Ainv =
    -0.0201    -0.0783     0.1174    -0.0242
    -0.0013    -0.0719     0.0078    -0.0683
    -0.0208     0.0369     0.0447    -0.0610
     0.0981     0.0389     0.0416     0.0055
```

Check with MATLAB's built–in **inv(A)** function.

```
inv(A)
```

```
ans =
    -0.0201    -0.0783     0.1174    -0.0242
    -0.0013    -0.0719     0.0078    -0.0683
    -0.0208     0.0369     0.0447    -0.0610
     0.0981     0.0389     0.0416     0.0055
```

14.6 The Cholesky Factorization

A matrix is said to be *positive definite* if

$$\mathbf{x}^T \cdot A \cdot \mathbf{x} > 0 \tag{14.25}$$

for every $\mathbf{x} \neq 0$ and A is symmetric, that is, $A^T = A$. Under those conditions, there exists an upper triangular matrix G with positive diagonal elements such that

$$G^T \cdot G = A \tag{14.26}$$

Relation (14.26) is referred to as Cholesky factorization. It is a special case of LU factorization and requires fewer computations than the LU factorization method of the previous section. Let us invoke the MATLAB **help chol** command to see how MATLAB performs this factorization.

```
CHOL    Cholesky factorization.
    CHOL(X) uses only the diagonal and upper triangle of X.
    The lower triangular is assumed to be the (complex conjugate)
    transpose of the upper. If X is positive definite, then
    R = CHOL(X) produces an upper triangular R so that R'*R = X.
    If X is not positive definite, an error message is printed.

    [R,p] = CHOL(X), with two output arguments, never produces an
    error message. If X is positive definite, then p is 0 and R
    is the same as above. But if X is not positive definite,
    then p is a positive integer.
```

We will consider an example using the Cholesky factorization after we review the MATLAB functions **eye(n)** and **diag(v,k)** as defined by MATLAB.

help eye

```
  EYE Identity matrix.
    EYE(N) is the N-by-N identity matrix.

    EYE(M,N) or EYE([M,N]) is an M-by-N matrix with 1's on
    the diagonal and zeros elsewhere.

    EYE(SIZE(A)) is the same size as A.

    See also ONES, ZEROS, RAND, RANDN.
```

help diag

```
  DIAG Diagonal matrices and diagonals of a matrix.
  DIAG(V,K) when V is a vector with N components is a  square
  matrix of order N+ABS(K) with the elements of V on the K-th
  diagonal. K = 0 is the main diagonal, K > 0 is above the main
  diagonal and K < 0 is below the main diagonal.

  DIAG(V) is the same as DIAG(V,0) and puts V on the main diagonal.

    DIAG(X,K) when X is a matrix is a column vector formed from
    the elements of the K-th diagonal of X.

  DIAG(X) is the main diagonal of X. DIAG(DIAG(X)) is a diagonal
  matrix.

    Example
      m = 5;
      diag(-m:m) + diag(ones(2*m,1),1) + diag(ones(2*m,1),-1)
      produces a tridiagonal matrix of order 2*m+1.

    See also SPDIAGS, TRIU, TRIL.
```

Example 14.15

Use MATLAB to compute the Cholesky factorization of matrix A as defined below.

```
format bank; B=[−0.25  −0.50  −0.75  −1.00];
A=5*eye(5)+diag(B, −1)+diag(B, 1), G=chol(A), A1=G'*G
```

Solution:

Execution of the MATLAB script above displays the following:

```
A  =
     5.00           -0.25              0              0              0
    -0.25            5.00          -0.50              0              0
        0           -0.50           5.00          -0.75              0
        0               0          -0.75           5.00          -1.00
        0               0              0          -1.00           5.00

G  =
     2.24           -0.11              0              0              0
        0            2.23          -0.22              0              0
        0               0           2.22          -0.34              0
        0               0              0           2.21          -0.45
        0               0              0              0           2.19

A1  =
     5.00           -0.25              0              0              0
    -0.25            5.00          -0.50              0              0
        0           -0.50           5.00          -0.75              0
        0               0          -0.75           5.00          -1.00
        0               0              0          -1.00           5.00
```

We observe that $A1 = A$, that is, the matrix product $G^T \cdot G = A$ is satisfied.

14.7 The QR Factorization

The QR factorization decomposes a matrix A into the product of an orthonormal matrix and an upper triangular matrix. The MATLAB function **[Q,R]=qr(A)** produces an $n \times n$ matrix whose columns form an orthonormal or *unitary*[*] matrix Q and an upper triangular matrix R of the same size as matrix A. In other words, A can be factored as

$$A = QR \tag{14.27}$$

---

[*] *An $n \times n$ matrix A is called unitary if $(A*)^T = A^{-1}$ where $A*$ is the complex conjugate matrix of A.*

Then, a system described by $Ax = b$ becomes

$$QRx = b \qquad (14.28)$$

and multiplying both sides of (14.28) by $Q \cdot Q^T = I$ we obtain

$$Rx = Q^T b \qquad (14.29)$$

The MATLAB **[Q,R]=qr(A)** is described as follows:
help qr

```
 QR       Orthogonal-triangular decomposition.
  [Q,R] = QR(A) produces an upper triangular matrix R of the same
     dimension as A and a unitary matrix Q so that A = Q*R.

  [Q,R,E] = QR(A) produces a permutation matrix E, an upper
     triangular R and a unitary Q so that A*E = Q*R.  The column
     permutation E is chosen so that abs(diag(R)) is decreasing.

  [Q,R] = QR(A,0) produces the "economy size" decomposition.
  If A is m-by-n with m > n, then only the first n columns of Q
  are computed.

  Q,R,E] = QR(A,0) produces an "economy size" decomposition in
  which E is a permutation vector, so that Q*R = A(:,E). The col-
  umn permutation E is chosen so that abs(diag(R)) is decreasing.

  By itself, QR(A) is the output of LAPACK'S DGEQRF or ZGEQRF rou-
  tine. TRIU(QR(A)) is R.

  R = QR(A) returns only R.  Note that R = chol(A'*A).
  [Q,R] = QR(A) returns both Q and R, but Q is often nearly full.
  C,R] = QR(A,B), where B has as many rows as A, returns C = Q'*B.
  R = QR(A,0) and [C,R] = QR(A,B,0) produce economy size results.

  The full version of QR does not return C.

  The least squares approximate solution to A*x = b can be found
  with the Q-less QR decomposition and one step of iterative
  refinement:

        x = R\(R'\(A'*b))
        r = b - A*x
        e = R\(R'\(A'*r))
        x = x + e;
```

Example 14.16

Given that

$$A = \begin{bmatrix} 2 & -3 & 1 \\ -1 & 5 & -2 \\ 3 & -8 & 4 \end{bmatrix} \text{ and } b = \begin{bmatrix} 2 \\ 4 \\ 5 \end{bmatrix}$$

solve $Ax = b$ using the MATLAB function **[Q,R]=qr(A)** and $x = R\backslash Q^T b$.

Solution:

A=[2 −3 1; −1 5 −2; 3 −8 4]; b=[2 4 5]'; [Q,R]=qr(A), x=R\Q'*b

```
Q =
        -0.53        -0.62        -0.58
         0.27        -0.77         0.58
        -0.80         0.15         0.58

R =
        -3.74         9.35        -4.28
            0        -3.24         1.54
            0            0         0.58

x =
         4.14
         4.43
         7.00
```

Check=A\b

```
Check =
         4.14
         4.43
         7.00
```

Let us verify that the matrix **Q** is unitary. Of course, since the elements are real numbers, the complex conjugate of Q is also Q and thus we only need to show that $Q^T = Q^{-1}$ or $Q \cdot Q^T = I$.

Q*Q'

```
ans =
         1.00         0.00        -0.00
         0.00         1.00        -0.00
        -0.00        -0.00         1.00
```

QR factorization is normally used to solve overdetermined systems,[*] that is, systems with more equations than unknowns as in applications where we need to find the least square distance in linear regression. In an overdetermined system, there is no vector X which can satisfy the entire system of equations, so we select the vector X which produces the minimum error. MATLAB does this with either the left division operator (\ ) or with the non–negative least–squares function **lsqnonneg(A,b)**. This function returns the vector X that minimizes **norm(A*X–b)** subject to $X \geq 0$ provided that the elements of A and b are real numbers. For example,

A=[2 –3 1; –1 5 –2; 3 –8 4]; b=[2 4 5]'; X=lsqnonneg(A,b)

returns

```
X =
    4.1429
    4.4286
    7.0000
```

Underdetermined systems have infinite solutions and MATLAB selects one but no warning message is displayed.

As we've learned in Chapter 4, the MATLAB function **inv(A)** produces the inverse of the square matrix A and an error message is displayed if A is not a square matrix. The function **pinv(A)** displays the pseudoobtaininverse of a $m \times n$ (non–square) matrix A. Of course, if A is square, then pinv(A)=inv(A).

14.8 Singular Value Decomposition

The *Singular Value Decomposition* (SVD) method decomposes a matrix A into a diagonal matrix S, of the same dimension as A and with nonnegative diagonal elements in decreasing order, and unitary matrices U and V so that

$$A = U \cdot S \cdot V^T \qquad (14.30)$$

The matrices U, S, and V, decomposed from a given matrix A, can be found with the MATLAB function **[U, S, V]=svd(A)**.

Example 14.17

Decompose the matrix

---

[*] *We defined overdetermined and underdetermined systems in Chapter 8*

$$A = \begin{bmatrix} 2 & -3 & 1 \\ -1 & 5 & -2 \\ 3 & -8 & 4 \end{bmatrix}$$

into two unitary matrices and a diagonal matrix with non–negative elements.

Solution:

We will use the MATLAB [U, S, V]=svd(A) function.

A=[2 –3 1; –1 5 –2; 3 –8 4]; [U,S,V]=svd(A)

```
U =
    -0.3150    -0.8050    -0.5028
     0.4731    -0.5924     0.6521
    -0.8228    -0.0325     0.5675


S =
    11.4605         0         0
         0    1.1782         0
         0         0    0.5184
V =
    -0.3116    -0.9463     0.0863
     0.8632    -0.2440     0.4420
    -0.3972     0.2122     0.8929
```

As expected, the diagonal elements of the triangular S matrix are non–negative and in decreasing values. We also verify that the matrices U and V are unitary as shown below.

U*U'

```
ans =

     1.0000    -0.0000    -0.0000
    -0.0000     1.0000     0.0000
    -0.0000     0.0000     1.0000
```

V*V'

```
ans =
     1.0000    -0.0000     0.0000
    -0.0000     1.0000    -0.0000
     0.0000    -0.0000     1.0000
```

14.9 Summary

- Orthogonal functions are those which are perpendicular to each other.

- Two families of curves with the property that each member of either family cuts every member of the other family at right angles are said to be orthogonal trajectories of each other.

- The inner (dot) product of two vectors $X = [x_1 \ x_2 \ x_3 \ \dots \ x_n]$ and $Y = [y_1 \ y_2 \ y_3 \ \dots \ y_n]$ is a scalar defined as $X \cdot Y = x_1 y_1 + x_2 y_2 + x_3 y_3 + \dots + x_n y_n$

- If the dot product of two vectors X_1 and X_2 is zero, these vector are said to be orthogonal to each other.

- The magnitude of a vector X, denoted as X, is defined as $|X| = \sqrt{x_1^2 + x_2^2 + \dots + x_n^2}$. A unique unit vector U is obtained by dividing each component of X by the magnitude $|X|$ and this process is referred to as normalization.

- A basis that consists of mutually orthogonal vectors is referred to as an orthogonal basis. If these vectors are also unit vectors, the basis is called orthonormal basis.

- If the column (or row) vectors of a square matrix A are mutually orthogonal unit vectors, the matrix A is said to be orthogonal and $A \cdot A^T = I$ where A^T is the transpose of A and I is the identity matrix.

- We can find an orthonormal set of eigenvectors in a 2×2 matrix easily from the eigenvalues but the computations for finding orthonormal sets of eigenvectors for larger size (3×3 or higher) matrices using the above procedure becomes quite involved. A simpler procedure is the Gram–Schmidt orthogonalization procedure which we will discuss on the next section.

- The LU factorization method decomposes a matrix A into a lower triangular matrix L and an upper triangular matrix U so that $A = L \cdot U$. The MATLAB function [L,U]=lu(A) decomposes the matrix A into a lower triangular matrix L and an upper triangular matrix U.

- A matrix is said to be positive definite if $x^T \cdot A \cdot x > 0$ for every $x \neq 0$ and A is symmetric, that is, $A^T = A$. Under those conditions, there exists an upper triangular matrix G with positive diagonal elements such that $G^T \cdot G = A$. This process is referred to as the Cholesky factorization.

- The QR factorization decomposes a matrix A into the product of an orthonormal matrix and an upper triangular matrix. The MATLAB function **[Q,R]=qr(A)** produces an $n \times n$ matrix whose columns form an orthonormal or unitary matrix Q and an upper triangular matrix R of the same size as matrix A.

- The Singular Value Decomposition (SVD) method decomposes a matrix A into a diagonal matrix S, of the same dimension as A and with nonnegative diagonal elements in decreasing order, and unitary matrices U and V so that $A = U \cdot S \cdot V^T$. The matrices U, S, and V, decomposed from a given matrix A, can be found with the MATLAB function [U, S, V]=svd(A).

14.10 Exercises

1. Show that the curve $x^2 + 3y^2 = k_1$ and the curve $3y = k_2x^3$ where k_1 and k_2 are constants, are orthogonal to each other.

2. Find the orthogonal trajectories of the curves of the family $2x^2 + y^2 = kx$

3. Given the vectors $X_1 = [2 \ -1]^T$ and $X_2 = [1 \ -3]^T$, use the Gram–Schmidt orthogonalization procedure to find two vectors Y_1 and Y_2 to form an orthonormal basis.

4. Use MATLAB to find another set of an orthonormal basis with the vectors given in Exercise 3.

5. Use the Gaussian elimination method as in Example 14.8 to decompose the system of equations

$$x_1 + 2x_2 + 3x_3 = 14$$
$$-2x_1 + 3x_2 + 2x_3 = 10$$
$$5x_1 - 8x_2 + 6x_3 = 7$$

into an upper triangular matrix U and a lower triangular matrix L. Verify your answers with MATLAB.

6. Using the MATLAB functions **eye(n)** and **diag(v,k)** to define and display the matrix A shown below.

$$A = \begin{bmatrix} 4.00 & -0.80 & 0 & 0 \\ -0.80 & 4.00 & -1.00 & 0 \\ 0 & -1.00 & 4.00 & -1.20 \\ 0 & 0 & -1.20 & 4.00 \end{bmatrix}$$

Then, use the MATLAB Cholesky factorization function to obtain the matrix G and verify that $G^T \cdot G = A$.

7. Use the appropriate MATLAB function to decompose the system of equations

$$A = \begin{bmatrix} 1 & 0 & -1 \\ 0 & 1 & 2 \\ -2 & -3 & 4 \end{bmatrix} \text{ and } b = \begin{bmatrix} 3 \\ 5 \\ 9 \end{bmatrix}$$

into an upper triangular matrix R of the same dimension as A and a unitary matrix Q so that $Q \cdot R = A$. Use a suitable function to verify your results.

8. Use the appropriate MATLAB function to decompose the matrix A given as

$$A = \begin{bmatrix} 1 & 0 & -1 \\ 0 & 1 & 2 \\ -2 & -3 & 4 \end{bmatrix}$$

into a diagonal matrix S of the same dimension as A and with non–negative diagonal elements in decreasing order and unitary matrices U and V so that $U \cdot S \cdot V^T = A$.

14.11 Solutions to End–of–Chapter Exercises

1.

$$x^2 + 3y^2 = k_1 \quad (1)$$

$$3y = k_2 x^3 \quad (2)$$

Implicit differentiation of (1) yields

$$2x + 6y\frac{dy}{dx} = 0$$

or

$$\frac{dy}{dx} = -\frac{1}{3} \cdot \frac{x}{y} \quad (3)$$

Differentiation of (2) yields

$$\frac{dy}{dx} = \frac{3k_2 x^2}{3} = k_2 x^2 \quad (4)$$

From (2),

$$k_2 = \frac{3y}{x^3}$$

and by substitution into (4) we obtain

$$\frac{dy}{dx} = \frac{3y}{x^3} \cdot x^2 = 3\frac{y}{x} \quad (5)$$

We observe that (5) is the negative reciprocal of (3) and thus the given curves are orthogonal to each other.

2.

$$2x^2 + y^2 = kx \quad (1)$$

Implicit differentiation of (1) yields

$$\frac{d}{dx}(2x^2) + \frac{d}{dx}y^2 = \frac{d}{dx}(kx)$$

$$4x + 2y\frac{dy}{dx} = k$$

and solving for dy/dx,

$$\frac{dy}{dx} = \frac{k - 4x}{2y} \quad (2)$$

From (1),

$$k = \frac{2x^2 + y^2}{x} \quad (3)$$

and by substitution into (2)

$$\frac{dy}{dx} = \frac{\dfrac{2x^2 + y^2}{x} - 4x}{2y} = \frac{-2x^2 + y^2}{2xy} \quad (4)$$

Now, we need to find the curves whose slopes are given by the negative reciprocal of (4), that is, we need to find the family of the curves of

$$\frac{dy}{dx} = \frac{2xy}{2x^2 - y^2} \quad (5)$$

We rewrite (5) as

$$(2x^2 - y^2)dy = 2xy\,dx \quad (6)$$

and we let $y = ux$. Then, $dy = u\,dx + x\,du$ and by substitution into (6)

$$(2x^2 - u^2x^2)(u\,dx + x\,du) = 2x^2u\,dx$$

Division of both sides of the above by x^2 yields

$$(2 - u^2)(u\,dx + x\,du) = 2u\,dx$$

Collecting like terms and simplifying we obtain

$$x(2 - u^2)du = u^3\,dx$$

Separating the variables we obtain

$$\frac{dx}{x} = \frac{(2 - u^2)}{u^3}du = \frac{2}{u^3}du - \frac{du}{u}$$

or

$$\frac{dx}{x} + \frac{du}{u} = \frac{2}{u^3}du$$

and by integrating these terms we find

$$\ln|x| + \ln|u| = \frac{-1}{u^2} + C$$

By substitution of $u = y/x$ we obtain

$$\ln|x| + \ln\left|\frac{y}{x}\right| + \ln|C| = \ln\left|\left(x \cdot \frac{y}{x} \cdot C\right)\right| = \ln|Cy| = \frac{-x^2}{y^2}$$

and thus the family of curves orthogonal to the given family is

$$x^2 = -y^2 \ln|Cy|$$

3.

From (14.14)

$$Y_1 = X_1 = [2 \quad -1]^T$$

$$Y_2 = X_2 - \frac{Y_1 \cdot X_2}{Y_1 \cdot Y_1} \cdot Y_1 = [1 \quad -3]^T - \frac{[2 \quad -1]^T \cdot [1 \quad -3]^T}{[2 \quad -1]^T \cdot [2 \quad -1]^T} \cdot [2 \quad -1]^T$$

$$= [1 \quad -3]^T - \frac{[2+3]}{[4+1]} \cdot [2 \quad -1]^T = [1 \quad -3]^T - \frac{5}{5} \cdot [2 \quad -1]^T =$$

$$= [1 \quad -3]^T - [2 \quad -1]^T = [-1 \quad 2]^T$$

and from (14.15)

$$U_1 = \frac{Y_1}{|Y_1|} = [2/\sqrt{5} \quad -1/\sqrt{5}]^T$$

$$U_2 = \frac{Y_2}{|Y_2|} = [-1/\sqrt{5} \quad -2/\sqrt{5}]^T$$

and denoting the matrix whose elements are the unit vectors as A, we have:

$$A = \begin{bmatrix} 2/\sqrt{5} & -1/\sqrt{5} \\ -1/\sqrt{5} & -2/\sqrt{5} \end{bmatrix}$$

We verify that $A \cdot A^T = I$ as shown below.

$$\begin{bmatrix} 2/\sqrt{5} & -1/\sqrt{5} \\ -1/\sqrt{5} & -2/\sqrt{5} \end{bmatrix} \cdot \begin{bmatrix} 2/\sqrt{5} & -1/\sqrt{5} \\ -1/\sqrt{5} & -2/\sqrt{5} \end{bmatrix} = \begin{bmatrix} 4/5+1/5 & -2/5+2/5 \\ -2/5+2/5 & 1/5+4/5 \end{bmatrix} = \begin{bmatrix} 1 & 0 \\ 0 & 1 \end{bmatrix}$$

4.

A=[2 −1; 1 −3]; B=orth(A)

B =
    -0.5257   -0.8507
    -0.8507   0.5257

I=B*B'

```
I =
    1.0000   -0.0000
   -0.0000    1.0000
```

5.

$$x_1 + 2x_2 + 3x_3 = 14$$
$$-2x_1 + 3x_2 + 2x_3 = 10 \quad (1)$$
$$5x_1 - 8x_2 + 6x_3 = 7$$

Multiplying the first equation of (1) by -2 and subtracting it from the second in (1) we obtain the second equation in (2) and thus x_1 is eliminated. Likewise, we multiply the first equation by 5 and we subtract it from the third in (1). Then,

$$x_1 + 2x_2 + 3x_3 = 14$$
$$7x_2 + 8x_3 = 38 \quad (2)$$
$$-18x_2 - 9x_3 = -63$$

Next, we multiply the second equation in (2) by $-18/7$ and we subtract it from the third in (2). Then, after simplification

$$x_1 + 2x_2 + 3x_3 = 14$$
$$7x_2 + 8x_3 = 38 \quad (3)$$
$$(81/7)x_3 = 243/7$$

Thus,

$$x_3 = 243/81 = 3$$
$$x_2 = (38 - 8x_3)/7 = 2$$
$$x_1 = (14 - 3x_3 - 2x_2) = 1$$

The multipliers that we've used are -2, 5, and $-18/7$. These are the elements L_{21}, L_{31}, and L_{32} respectively. Therefore, the lower triangular matrix is

$$L = \begin{bmatrix} 1 & 0 & 0 \\ -2 & 1 & 0 \\ 5 & -18/7 & 1 \end{bmatrix}$$

The elements of the upper triangular matrix are the coefficients of the unknowns in (3) and thus

$$U = \begin{bmatrix} 1 & 2 & 3 \\ 0 & 7 & 8 \\ 0 & 0 & 81/7 \end{bmatrix}$$

Now, we use MATLAB to verify that $L \cdot U = A$

L=[1 0 0; –2 1 0; 5 –18/7 1]; U=[1 2 3; 0 7 8; 0 0 81/7]; A=L*U

```
A =
      1                2                3
     -2                3                2
      5               -8                6
```

6.

format bank; B=[–0.8 –1.0 –1.2];
A=4*eye(4)+diag(B, –1)+diag(B, 1), G=chol(A), A1=G'*G

```
A =
      4.00            -0.80                0                0
     -0.80             4.00            -1.00                0
         0            -1.00             4.00            -1.20
         0                0            -1.20             4.00

G =
      2.00            -0.40                0                0
         0             1.96            -0.51                0
         0                0             1.93            -0.62
         0                0                0             1.90

A1 =
      4.00            -0.80                0                0
     -0.80             4.00            -1.00                0
         0            -1.00             4.00            -1.20
         0                0            -1.20             4.00
```

7.

A=[1 0 –1; 0 1 2; –2 -3 4]; b=[3 5 9]';
[Q,R]=qr(A), QQT=Q*Q', x=R\Q'*b, Check=A\b

```
Q =
    -0.4472        0.7171        0.5345
         0        -0.5976        0.8018
     0.8944        0.3586        0.2673

R =
    -2.2361       -2.6833        4.0249
         0        -1.6733       -0.4781
         0             0         2.1381
```

```
QQT =
    1.0000    0.0000    0.0000
    0.0000    1.0000    0.0000
    0.0000    0.0000    1.0000

x =
    6.7500
   -2.5000
    3.7500

Check =
    6.7500
   -2.5000
    3.7500
```

8.

```
A=[1 0 -1; 0 1 2; -2 -3 4];
[U,S,V]=svd(A), UUT=U*U', VVT=V*V'
U =
   -0.2093   -0.2076   -0.9556
    0.1977    0.9480   -0.2493
    0.9577   -0.2410   -0.1574

S =
    5.5985         0         0
         0    2.0413         0
         0         0    0.7000

V =
   -0.3795    0.1345   -0.9154
   -0.4779    0.8187    0.3184
    0.7922    0.5583   -0.2464

UUT =
    1.0000    0.0000    0.0000
    0.0000    1.0000    0.0000
    0.0000    0.0000    1.0000

VVT =
    1.0000    0.0000   -0.0000
    0.0000    1.0000   -0.0000
   -0.0000   -0.0000    1.0000
```

NOTES:

Chapter 15

Bessel, Legendre, and Chebyshev Functions

T his chapter is an introduction to some very interesting functions. These are special functions that find wide applications in science and engineering. They are solutions of differential equations with variable coefficients and, under certain conditions, satisfy the orthogonality principle.

15.1 The Bessel Function

The *Bessel functions*, denoted as $J_n(x)$, are used in engineering, acoustics, aeronautics, thermodynamics, theory of elasticity and others. For instance, in the electrical engineering field, they are used in frequency modulation, transmission lines, and telephone equations.

Bessel functions are solutions of the differential equation

$$x^2 \frac{d^2 y}{dx^2} + x \frac{dy}{dx} + (x^2 - n^2)y = 0 \tag{15.1}$$

where n can be any number, positive or negative integer, fractional, or even a complex number. Then, the form of the general solution of (15.1) depends on the value of n.

Differential equations with variable coefficients, such as (15.1), cannot be solved in terms of familiar functions as those which we encountered in ordinary differential equations with constant coefficients. The usual procedure is to derive solutions in the form of infinite series, and the most common are the *Method of Frobenius* and the *Method of Picard*. It is beyond the scope of this book to derive the infinite series which are approximations to the solutions of these differential equations; these are discussed in advanced mathematics textbooks. Therefore, we will accept the solutions without proof.

Applying the method of Frobenius to (15.1), we obtain the infinite power series

$$J_n(x) = \sum_{k=0}^{\infty} (-1)^k \cdot \left(\frac{x}{2}\right)^{n+2k} \cdot \frac{1}{k! \cdot \Gamma(n+k+1)} \quad n \geq 0 \tag{15.2}$$

This series is referred to as *Bessel function of order* n where n is any positive real number or zero. If in (15.2), we replace n with $-n$, we obtain the relation

$$J_{-n}(x) = \sum_{k=0}^{\infty} (-1)^k \cdot \left(\frac{x}{2}\right)^{-n+2k} \cdot \frac{1}{k! \cdot \Gamma(-n+k+1)} \tag{15.3}$$

and the function $J_{-n}(x)$ is referred to as the *Bessel function of negative order* n.

For the special case where n is a positive integer or zero,

$$\Gamma(n+k+1) = (n+k)! \tag{15.4}$$

and (15.2) reduces to

$$J_n(x) = \sum_{k=0}^{\infty} (-1)^k \cdot \left(\frac{x}{2}\right)^{n+2k} \cdot \frac{1}{k! \cdot (n+k)!} \quad n = 0, 1, 2, \ldots \tag{15.5}$$

or

$$J_n(x) = \frac{x^n}{2^n \cdot n!} \left\{ 1 - \frac{x^2}{2^2 \cdot 1! \cdot (n+1)} + \frac{x^4}{2^4 \cdot 2! \cdot (n+1)(n+2)} \right.$$
$$\left. - \frac{x^6}{2^6 \cdot 3! \cdot (n+1)(n+2)(n+3)} + \ldots \right\} \tag{15.6}$$

For $n = 0, 1$ and 2, (15.6) reduces to the following series:

$$J_0(x) = 1 - \frac{x^2}{2^2 \cdot (1!)^2} + \frac{x^4}{2^4 \cdot (2!)^2} - \frac{x^6}{2^6 \cdot (3!)^2} + \frac{x^8}{2^8 \cdot (4!)^2} - \ldots \tag{15.7}$$

$$J_1(x) = \frac{x}{2} - \frac{x^3}{2^3 \cdot 1! \cdot 2!} + \frac{x^5}{2^5 \cdot 2! \cdot 3!} - \frac{x^7}{2^7 \cdot 3! \cdot 4!} + \frac{x^9}{2^9 \cdot 4! \cdot 5!} - \ldots \tag{15.8}$$

$$J_2(x) = \frac{x^2}{2^2 \cdot 2!} - \frac{x^4}{2^4 \cdot 1! \cdot 3!} + \frac{x^6}{2^6 \cdot 2! \cdot 4!} - \frac{x^8}{2^8 \cdot 3! \cdot 5!} + \frac{x^{10}}{2^{10} \cdot 4! \cdot 6!} - \ldots \tag{15.9}$$

We observe from (15.7) through (15.9), that when n is zero or even, $J_n(x)$ is an even function of x, and odd when n is odd.

If we differentiate the series of $J_0(x)$ in (15.7), and compare with the series of $J_1(x)$ in (15.8), we see that

$$\frac{d}{dx} J_0(x) = -J_1(x) \tag{15.10}$$

Also, if we multiply the series for $J_1(x)$ by x and differentiate it, we will find that

$$\boxed{\frac{d}{dx}\{xJ_1(x)\} = xJ_0(x)}$$ (15.11)

Example 15.1

Compute, correct to four decimal places, the values of

a. $J_0(2)$ b. $J_1(3)$ c. $J_2(1)$

Solution:

a.

From (15.7),

$$J_0(2) = 1 - \frac{4}{4} + \frac{16}{64} - \frac{64}{64 \times 36} + \frac{256}{256 \times 576} = \frac{1}{4} - \frac{1}{36} + \frac{1}{576} = \frac{43}{192} = 0.2240$$

or from math tables, $J_0(2) = 0.2239$

b.

From (15.8),

$$J_1(3) = \frac{3}{2} - \frac{27}{16} + \frac{243}{384} - \frac{2187}{18432} + \frac{19683}{1474560} = 0.3400$$

or from math tables, $J_1(3) = 0.3391$

c.

From (15.9),

$$J_2(1) = \frac{1}{8} - \frac{1}{96} + \frac{1}{1536} - \frac{1}{184320} + \frac{1}{17694720} = 0.1152$$

or from math tables, $J_2(1) = 0.1149$.

We can use the MATLAB **besselj(n,x)** function or the Excel BESSELJ(x,n) function for the above computations. With MATLAB, we obtain

```
besselj(0,2), besselj(1,3), besselj(2,1)

ans =
    0.2239
ans =
    0.3391
ans =
    0.1149
```

and with Excel,

besselj(2,0)= 0.2239 besselj(3,1)= 0.3391 besselj(1,2)= 0.1149

The MATLAB script below plots $J_0(x)$, $J_1(x)$, and $J_2(x)$.

```
x = 0.00: 0.05: 10.00; v = besselj(0,x); w = besselj(1,x); z = besselj(2,x);
plot(x,v,x,w,x,z); grid; title('Bessel Functions of the First Kind'); xlabel('x'); ylabel('Jn(x)');
text(0.95, 0.85, 'J0(x)'); text(2.20, 0.60, 'J1(x)'); text(4.25, 0.35, 'J2(x)')
```

The plots for $J_0(x)$, $J_1(x)$ and $J_2(x)$ are shown in Figure 15.1.[*]

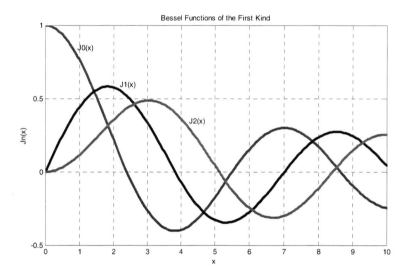

Figure 15.1. Plots of $J_0(x)$, $J_1(x)$ and $J_2(x)$ using MATLAB

We can also use Excel to plot these series as shown in Figure 15.2.

The definition of a Bessel function of the first kind will be explained shortly.

The x–axis crossings in the plot of Figures 15.1 and 15.2 show the first few roots of the $J_0(x)$, $J_1(x)$, and $J_2(x)$ series. However, all $J_n(x)$ are infinite series and thus, it is a very difficult and tedious task to compute all roots of these series. Fortunately, tables of some of the roots of $J_0(x)$ and $J_1(x)$ are shown in math tables.

---

[*] *In Frequency Modulation (FM), x is denoted as β and it is called **modulation index**. The functions $J_0(\beta)$, $J_1(\beta)$, $J_2(\beta)$ and so on, represent the carrier, first sideband, second sideband etc. respectively.*

Plot of Bessel Function Jn(x) for n = 0, 1 and 2

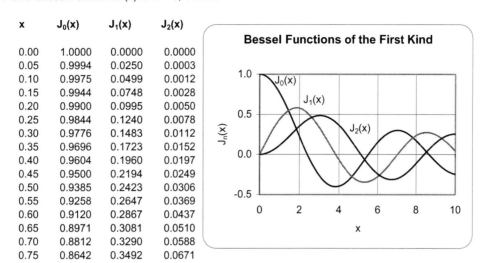

x	$J_0(x)$	$J_1(x)$	$J_2(x)$
0.00	1.0000	0.0000	0.0000
0.05	0.9994	0.0250	0.0003
0.10	0.9975	0.0499	0.0012
0.15	0.9944	0.0748	0.0028
0.20	0.9900	0.0995	0.0050
0.25	0.9844	0.1240	0.0078
0.30	0.9776	0.1483	0.0112
0.35	0.9696	0.1723	0.0152
0.40	0.9604	0.1960	0.0197
0.45	0.9500	0.2194	0.0249
0.50	0.9385	0.2423	0.0306
0.55	0.9258	0.2647	0.0369
0.60	0.9120	0.2867	0.0437
0.65	0.8971	0.3081	0.0510
0.70	0.8812	0.3290	0.0588
0.75	0.8642	0.3492	0.0671

Figure 15.2. Plots of $J_0(x)$, $J_1(x)$ and $J_2(x)$ using Excel

The equations $J_0(x) = 0$ and $J_1(x) = 0$ exhibit some interesting characteristics. The most noteworthy are:

1. They have no complex roots

2. Each has an infinite number of distinct real roots

3. Between two consecutive roots of one of these equations lies one and only one root of the other equation, that is, the roots of these equations separate each other. This is observed on Table 15.1 which shows the first 5 positive roots of these equations, and the differences between consecutive roots. For instance, we observe that the first root 3.8317 of $J_1(x)$ lies between the roots 2.4048 and 5.5201 of $J_0(x)$.

4. As the roots become larger and larger, the difference between consecutive roots approaches the value of π, that is, $J_0(x)$ and $J_1(x)$, are almost periodic with period almost 2π. In other words, these series behave like the $\cos x$ and $\sin x$ functions.

If n is half of an odd integer, such as $1/2$, $3/2$, $5/2$, and so on, then $J_n(x)$ can be expressed in a finite form of sines and cosines. Consider, for example, the so-called *half–order Bessel functions* $J_{1/2}(x)$ and $J_{(-1/2)}(x)$. If we let $n = 1/2$ in (15.2), we obtain

$$J_{1/2}(x) = \sqrt{\frac{2}{\pi x}} \sin x$$

(15.12)

TABLE 15.1 *The first few roots of* $J_0(x)$ *and* $J_1(x)$

$J_0(x) = 0$		$J_1(x) = 0$	
Roots	Differences	Roots	Differences
2.4048		3.8317	
	3.1153		3.1839
5.5201		7.0156	
	3.1336		3.1579
8.6537		10.1735	
	3.1378		3.1502
11.7915		13.3237	
	3.1394		3.1469
14.9309		16.4706	
	...		...
...		...	

Likewise, if we let n = 1/2 in (15.3), we obtain

$$J_{(-1/2)}(x) = \sqrt{\frac{2}{\pi x}} \cos x \qquad (15.13)$$

Example 15.2

Compute, correct to four decimal places, the values of

$$\text{a. } J_{1/2}\left(\frac{\pi}{4}\right) \qquad \text{b. } J_{(-1/2)}\left(\frac{\pi}{4}\right)$$

Solution:

a. Using (15.12),

$$J_{1/2}\left(\frac{\pi}{4}\right) = \sqrt{\frac{2}{\pi(\pi/4)}} \sin(\pi/4) = \frac{1}{\pi}\sqrt{8} \cdot \frac{\sqrt{2}}{2} = \frac{4}{2\pi} = \frac{2}{\pi} = 0.6366$$

b. Using (15.13).

$$J_{(-1/2)}\left(\frac{\pi}{4}\right) = \sqrt{\frac{2}{\pi(\pi/4)}} \cos(\pi/4) = \frac{1}{\pi}\sqrt{8} \cdot \frac{\sqrt{2}}{2} = \frac{2}{\pi} = 0.6366$$

Check with MATLAB:

besselj(0.5,pi/4), besselj(−0.5,pi/4)

```
ans =
     0.6366
ans =
     0.6366
```

The Bessel functions which we have discussed thus far, are referred to as *Bessel functions of the first kind*. Other Bessel functions, denoted as $Y_n(x)$ and referred to as *Bessel functions of the second kind*, or *Weber functions*, or *Neumann functions*. These are additional solutions of the Bessel's equation, and will be explained in the next paragraph. Also, certain differential equations resemble the Bessel equation, and thus their solutions are called *Modified Bessel functions*, or *Hankel functions*.

As mentioned earlier, a Bessel function $J_{-n}(x)$ for $n > 0$, can be obtained by replacing n with $-n$ in (15.2). If n is an integer, we will prove that

$$\boxed{\begin{array}{c} J_{-n}(x) = (-1)^n J_n(x) \\ \text{for } n = 1, 2, 3, \ldots \end{array}}$$

(15.14)

Proof:

From (15.3),

$$\begin{aligned} J_{-n}(x) &= \sum_{k=0}^{\infty} \frac{(-1)^k \cdot (x/2)^{-n+2k}}{k! \cdot \Gamma(-n+k+1)} \\ &= \sum_{k=0}^{n-1} \frac{(-1)^k \cdot (x/2)^{-n+2k}}{k! \cdot \Gamma(-n+k+1)} + \sum_{k=n}^{\infty} \frac{(-1)^k \cdot (x/2)^{-n+2k}}{k! \cdot \Gamma(-n+k+1)} \end{aligned}$$

(15.15)

Now, we recall from Chapter 13, that the numbers $n = 0, -1, -2, \ldots$ yield infinite values in $\Gamma(n)$; then, the first summation in the above relation is zero for $k = 0, 1, 2, \ldots, n-1$. Also, if we let $k = n + m$ in the second summation, after simplification and comparison with (15.5), we see that

$$\sum_{m=0}^{\infty} \frac{(-1)^{n+m} \cdot (x/2)^{-n+2n+2m}}{(n+m)! \cdot \Gamma(m+1)}$$

$$= (-1)^n \sum_{m=0}^{\infty} \left(\frac{x}{2}\right)^{n+2m} \frac{(-1)^m}{\Gamma(n+m+1)! \cdot m!} = (-1)^n J_n(x)$$

and thus (15.14) has been proved.

It is shown in advanced mathematics textbooks that, if n is not an integer, $J_n(x)$ and $J_{-n}(x)$ are linearly independent; for this case, the general solution of the Bessel equation is

$$\boxed{\begin{array}{c} y = AJ_n(x) + BJ_{-n}(x) \\ n \neq 0, 1, 2, 3, \ldots \end{array}}$$

(15.16)

For $n = 1, 2, 3, \ldots$ and so on, the functions $J_n(x)$ and $J_{-n}(x)$ are not linearly independent as we have seen in (15.14); therefore, (15.16) is not the general solution, that is, for this case, these two series produce only one solution, and for this reason, the Bessel functions of the second kind are introduced to obtain the general solution.

The following example illustrates the fact that when n is not an integer or zero, relation (15.16) is the general solution.

Example 15.3

Find the general solution of Bessel's equation of order $1/2$.

Solution:

By the substitution $n = 1/2$ in (15.1), we obtain

$$x^2 \frac{d^2 y}{dx^2} + x \frac{dy}{dx} + \left(x^2 - \frac{1}{4}\right) y = 0$$

(15.17)

We will show that the general solution of (15.17) is

$$y = AJ_{1/2}(x) + BJ_{-1/2}(x)$$

(15.18)

By substitution of (15.12) and (15.13) into (15.18), we obtain

$$y = A\sqrt{\frac{2}{\pi x}} \sin x + B\sqrt{\frac{2}{\pi x}} \cos x$$

(15.19)

and letting $C_1 = A\sqrt{2/\pi}$ and $C_2 = B\sqrt{2/\pi}$, (15.19) can be written as

$$y = C_1 \frac{\sin x}{\sqrt{x}} + C_2 \frac{\cos x}{\sqrt{x}}$$

(15.20)

Since the two terms on the right side of (15.20) are linearly independent, y represents the general solution of (15.17).

The Bessel functions of the second kind, third kind, and others, can be evaluated at specified values either with MATLAB or Excel. The descriptions, syntax, and examples for each can be found by invoking **help bessel** for MATLAB, and help for Excel.

One very important property of the Bessel's functions is that within certain limits, they constitute an *orthogonal system.*[*] For instance, if a and b are distinct roots of $J_0(x) = 0$, $J_0(a) = 0$ and $J_0(b) = 0$, then,

$$\int_0^1 x J_0(ax) J_0(bx) dx = 0 \tag{15.21}$$

and we say that $J_0(ax)$ and $J_0(bx)$ are orthogonal in the interval $0 \le x \le 1$. They are also orthogonal with the variable x.

The function

$$e^{\frac{x}{2}\left(t - \frac{1}{t}\right)} = \sum_{n=-\infty}^{\infty} J_n(x) t^n \tag{15.22}$$

is referred to as the *generating function for Bessel functions of the first kind of integer order.* Using this function, we can obtain several interesting properties for integer values of n. Some of these are given below without proof. More detailed discussion and proofs can be found in advanced mathematics textbooks.

$$\sum_{n=-\infty}^{\infty} J_n^2(x) = 1 \tag{15.23}$$

$$\cos(x \sin \phi) = J_0(x) + 2 \sum_{k=1}^{\infty} J_{2k}(x) \cos 2k\phi$$

$$\sin(x \sin \phi) = 2 \sum_{k=1}^{\infty} J_{2k-1}(x) \sin(2k-1)\phi \tag{15.24}$$

where the subscript 2k denotes that the first relation is valid for even values of k, whereas $2k - 1$ in the second, indicates that the second relation is valid for odd values of k. Also,

$$\int_0^\pi \cos n\phi \cdot \cos(x \sin \phi) d\phi = \begin{cases} \pi J_n(x) & n = \text{even} \\ 0 & n = \text{odd} \end{cases}$$

$$\int_0^\pi \sin n\phi \cdot \sin(x \sin \phi) d\phi = \begin{cases} 0 & n = \text{even} \\ \pi J_n(x) & n = \text{odd} \end{cases} \tag{15.25}$$

---

[*] *Two functions constitute an orthogonal system, when the average of their cross product is zero within some specified limits.*

and

$$J_n(x) = \frac{1}{\pi} \int_0^{\pi} \cos(n\phi - x\sin\phi)d\phi \qquad (15.26)$$

Relations (15.23) through (15.26) appear in frequency modulation. For example, the average power is shown to be

$$P_{ave} = \frac{1}{2}A_C^2 \sum_{n=-\infty}^{\infty} J_n^2(\beta)$$

and with (15.23), it reduces to

$$P_{ave} = \frac{1}{2}A_C^2$$

15.2 Legendre Functions

Another second–order differential equation with variable coefficients is the equation

$$\boxed{(1-x^2)\frac{d^2y}{dx^2} - 2x\frac{dy}{dx} + n(n+1)y = 0} \qquad (15.27)$$

known as *Legendre's equation*. Here, n is a constant, and if it is zero or a positive integer, then (15.27) has polynomial solutions of special interest.

Applying the method of Frobenius, as in the Bessel equation, we obtain two independent solutions y_1 and y_2 as follows.

$$y_1 = a_0\left[1 - \frac{n(n+1)}{2!}x^2 + \frac{(n-2)n(n+1)(n+3)}{4!}x^4 - \dots\right] \qquad (15.28)$$

$$y_2 = a_1\left[x - \frac{(n-1)(n+2)}{3!}x^3 + \frac{(n-3)(n-1)(n+2)(n+4)}{5!}x^5 - \dots\right] \qquad (15.29)$$

where a_0 and a_1 are constants. We observe that y_1 is an even function of x, while y_2 is an odd function. Then, the general solution of (15.27) is $y = y_1 + y_2$ or

$$y = a_0\left[1 - \frac{n(n+1)}{2!}x^2 + \frac{(n-2)n(n+1)(n+3)}{4!}x^4 - \dots\right] \qquad (15.30)$$
$$+ a_1\left[x - \frac{(n-1)(n+2)}{3!}x^3 + \frac{(n-3)(n-1)(n+2)(n+4)}{5!}x^5 - \dots\right]$$

and this series is *absolutely convergent*[*] for $|x| > 1$.

The parameter n is usually a positive integer. If n is zero, or an even positive integer, the first term on the right side of (15.30) contains only a finite number of terms; if it is odd, the second term contains only a finite number of terms. Therefore, whenever n is zero or a positive integer; the general solution of Legendre's equation contains a polynomial solution which is denoted as $P_n(x)$, and an infinite series solution which is denoted as $Q_n(x)$.

The *Legendre polynomials* are defined as

$$P_n(x) = (-1)^{n/2} \cdot \frac{1 \cdot 3 \cdot 5 \cdot \ldots \cdot (n-1)}{2 \cdot 4 \cdot 6 \cdot \ldots \cdot n}\left[1 - \frac{n(n+1)}{2!}x^2 + \ldots\right]$$

for $n = 0$ or $n =$ even integer

$$(15.31)$$

$$P_n(x) = (-1)^{(n-1)/2} \cdot \frac{1 \cdot 3 \cdot 5 \cdot \ldots \cdot n}{2 \cdot 4 \cdot 6 \cdot \ldots \cdot (n-1)}\left[x - \frac{(n-1)(n+2)}{3!}x^3 + \ldots\right]$$

for $n =$ odd integer

$$(15.32)$$

and these are also referred to as *surface zonal harmonics*. The infinite series solution $Q_n(x)$ is referred to as *Legendre functions of the second kind*. These become infinite as $x \rightarrow \pm 1$ and their applications to science and engineering problems are very limited. Accordingly, they will not be discussed in this text.

The even and odd functions of (15.31) and (15.32) can be combined to a single relation as

$$P_n(x) = \sum_{k=0}^{N} \frac{(-1)^k \cdot (2n-2k)!}{2^n k!(n-k)!(n-2k)!}x^{n-2k}$$

where $N = \frac{n}{2}$ for $n =$ even and $N = \frac{(n-1)}{2}$ for $n =$ odd

$$(15.33)$$

From (15.33), or (15.31) and (15.32), we obtain the following first 6 Legendre polynomials.

---

* Assume that the infinite series $\sum_{n=1}^{\infty} u_n(x_0) = u_1(x_0) + u_2(x_0) + \ldots$ converges, i.e., reaches a limit. If, when we replace the terms of this series by their absolute value, we find that the resulting series $\sum_{n=1}^{\infty} |u_n(x_0)| = |u_1(x_0)| + |u_2(x_0)| + \ldots$ also converges, this series is said to be absolutely convergent.

$$P_0(x) = 1 \qquad\qquad P_1(x) = x$$
$$P_2(x) = \frac{1}{2}(3x^2 - 1) \qquad\qquad P_3(x) = \frac{1}{2}(5x^3 - 3x)$$
$$P_4(x) = \frac{1}{8}(35x^4 - 30x^2 + 3) \qquad P_5(x) = \frac{1}{8}(63x^5 - 70x^3 + 15x)$$

(15.34)

The relation

$$P_n(x) = \frac{1}{2^n \cdot n!} \cdot \frac{d^n}{dx^n}(x^2 - 1)^n \qquad\qquad (15.35)$$

is known as *Rodrigues' formula*, and offers another method of expressing the Legendre polynomials. We prove (15.35) as follows.

From the binomial theorem,

$$(x^2 - 1)^n = \sum_{k=0}^{n} \frac{(-1)^k \cdot n!}{k! \cdot (n-k)!} x^{2n-2k} \qquad\qquad (15.36)$$

and differentiation of (15.36) with respect to x n times yields

$$\frac{d^n}{dx^n}(x^2 - 1)^n = \sum_{k=0}^{N} \frac{(-1)^k \cdot n!}{k!(n-k)!} \cdot \frac{(2n-2k)!}{(n-2k)!} x^{n-2k} \qquad\qquad (15.37)$$

Now, by comparison with (15.33), we recognize (15.37) as $2^n \cdot n! \cdot P_n(x)$ and thus (15.35) is proved.

Another important identity involving Legendre polynomials, is the *generating function for Legendre polynomials* which is defined as

$$\frac{1}{\sqrt{1 - 2xt + t^2}} = P_0(x) + P_1(x)t + P_2(x)t^2 + \ldots + P_n(x)t^n = \sum_{n=0}^{\infty} P_n(x)t^n \qquad\qquad (15.38)$$

We will illustrate the use of the Legendre polynomials with the following example.

Example 15.4

Find the potential difference (voltage) v at a point P developed by a nearby dipole[*] in terms of the distance between the point P and the dipole, and the angle which point P makes with the center of the dipole.

---

[*] *A dipole is a pair of electric charges or magnetic poles, of equal magnitude but of opposite sign or polarity, separated by a small distance. Alternately, a dipole is an antenna, usually fed from the center, consisting of two equal rods extending outward in a straight line.*

Solution:

Let the charges q and –q of the dipole be a distance *2d* apart with the origin *0* as the midpoint as shown in Figure 15.3.

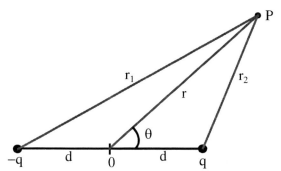

Figure 15.3. Figure for Example 15.4

Let the potential at point P be V_P. From electromagnetic field textbooks we find that

$$V_P = \frac{q}{4\pi\varepsilon_0}\left(\frac{1}{r_2} - \frac{1}{r_1}\right) \tag{15.39}$$

where ε_0 is the permittivity[*] of the vacuum. For simplicity, we will denote the quantity $1/(4\pi\varepsilon_0)$ as *k* and thus we rewrite (15.39) as

$$V_P = kq\left(\frac{1}{r_2} - \frac{1}{r_1}\right) = kq(r_2^{-1} - r_1^{-1}) \tag{15.40}$$

Next, we need to express r_1 and r_2 in terms or *d* and *r*. By the law of cosines,

$$r_1 = \sqrt{d^2 + r^2 - 2dr\cos(180° - \theta)} = \sqrt{d^2 + r^2 + 2dr\cos\theta} \tag{15.41}$$

and

$$r_2 = \sqrt{d^2 + r^2 - 2dr\cos\theta} \tag{15.42}$$

Dividing both sides of (15.42) by r, we obtain

$$\frac{r_2}{r} = \sqrt{\frac{d^2}{r^2} + 1 - \frac{2d\cos\theta}{r}} \tag{15.43}$$

or

---

[*] *Permittivity is a measure of the ability of a material to resist the formation of an electric field within it.*

$$r_2^{-1} = \frac{1}{r}\left(\frac{d^2}{r^2} + 1 - \frac{2d\cos\theta}{r}\right)^{-1/2} \tag{15.44}$$

In all practical applications, the point P is sufficiently far from the origin; thus, we assume that $r > d$. Now, we want to relate the terms inside the parentheses of (15.44), to a Legendre polynomial. We do this by expressing these terms in the form of the generating function of (15.38).

We let $x = \cos\theta$, and $y = d/r$; then, by substitution into (15.44) we obtain

$$\left(\frac{d^2}{r^2} + 1 - \frac{2d\cos\theta}{r}\right)^{-1/2} = (1 - 2xy + y^2)^{-1/2} = \sum_{n=0}^{\infty} P_n(x)y^n \tag{15.45}$$

We recall that (15.45) holds only if $|x| < 1$ and $|y| < 1$. This requirement is satisfied since x and y, as defined, are both less than unity.

To find a similar expression for r_1^{-1}, we simply replace x with $-x$ in (15.45), and thus

$$(1 + 2xy + y^2)^{-1/2} = \sum_{n=0}^{\infty} P_n(-x)y^n \tag{15.46}$$

By substitution of (15.45) and (15.46) into (15.40), we obtain

$$V_P = \frac{kq}{r} \sum_{n=0}^{\infty} [P_n(x)y^n - P_n(-x)y^n] \tag{15.47}$$

Since $x = \cos\theta$, and $y = d/r$, we can express (15.47) as

$$V_P = \frac{kq}{r} \sum_{n=0}^{\infty} [P_n(\cos\theta) - P_n(-\cos\theta)]\left(\frac{d}{r}\right)^n \tag{15.48}$$

However, if n is even in (15.48), $P_n(-\cos\theta) = P_n(\cos\theta)$, and therefore, all even powers vanish. But when n is odd, $P_n(-\cos\theta) = -P_n(\cos\theta)$ and the odd powers in (15.48) are duplicated. Then,

$$V_P = \frac{kq}{r} \sum_{n=0}^{\infty} 2P_{2n+1}(\cos\theta)\left(\frac{d}{r}\right)^{2n+1} \tag{15.49}$$

and for $r \gg d$, (15.49) can be approximated as

$$V_P \approx \frac{2kq}{r} P_1(\cos\theta)\left(\frac{d}{r}\right) = \frac{2kdq}{r^2}\cos\theta \; * \tag{15.50}$$

The term $2dq$ is the magnitude of the so–called *dipole moment*. It is a vector directed from the negative charge towards the positive charge. It is denoted with the letter p, that is,

$$\mathbf{p} = 2q\mathbf{d} \tag{15.51}$$

The relation of (15.50) can, of course, be derived without the use of Legendre polynomials as follows:

For $r \gg d$, the distances r_1, r, and r_2 can be approximated by parallel lines as shown in Figure 15.4. Then, the negative and positive charges look like a single point charge, and using (15.40) we obtain

$$V_P = kq\left(\frac{1}{r_2} - \frac{1}{r_1}\right) = \frac{kq}{r_1 \cdot r_2}(r_2 - r_1) = \frac{2kqd}{r^2}\cos\theta \tag{15.52}$$

We observe that (15.52) is the same as (15.50).

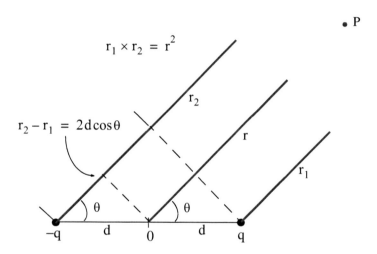

Figure 15.4. *Derivation of the voltage developed by a dipole*

Another interesting relation that can be used to find the Legendre polynomial series of a function $f(x)$ for $|x| < 1$, is

$$B_n = \frac{2n+1}{2}\int_{-1}^{1} f(x)P_n(x)dx \tag{15.53}$$

---

* *Here, we have used the identity $P_1(\cos\theta) = \cos\theta$. This will be seen shortly in (15.57) when we discuss the trigonometric form of the Legendre polynomials.*

Then, a function $f(x)$ can be expanded as

$$f(x) = B_0 P_0(x) + B_1 P_1(x) + B_2 P_2(x) + \ldots + B_n P_n(x) = \sum_{n=0}^{\infty} B_n P_n(x) \qquad (15.54)$$

The example below illustrates how this relation is being used.

Example 15.5

Compute the Legendre polynomial series representing the waveform of Figure 15.5.

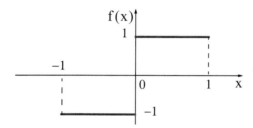

Figure 15.5. Waveform for Example 15.5

Solution:

We will first compute the coefficients B_n from (15.53); then, we will substitute these into (15.54). We will also use (15.34) for the polynomials of $P_n(x)$.

For this example,

$$B_0 = \frac{1}{2}\int_{-1}^{0}(-1)\cdot 1\cdot dx + \frac{1}{2}\int_{0}^{1}1\cdot 1\cdot dx = -\frac{1}{2}x\Big|_{-1}^{0} + \frac{1}{2}x\Big|_{0}^{1} = -\frac{1}{2} + \frac{1}{2} = 0$$

$$B_1 = \frac{3}{2}\int_{-1}^{0}(-1)\cdot x\cdot dx + \frac{3}{2}\int_{0}^{1}1\cdot x\cdot dx = -\frac{3}{4}x^2\Big|_{-1}^{0} + \frac{3}{4}x^2\Big|_{0}^{1} = \frac{3}{4} + \frac{3}{4} = \frac{3}{2}$$

$$B_2 = \frac{5}{2}\int_{-1}^{0}(-1)\cdot\frac{1}{2}(3x^2-1)dx + \frac{5}{2}\int_{0}^{1}1\cdot\frac{1}{2}(3x^2-1)dx = 0$$

$$B_3 = \frac{7}{2}\int_{-1}^{0}(-1)\cdot\frac{1}{2}(5x^3-3x)dx + \frac{7}{2}\int_{0}^{1}1\cdot\frac{1}{2}(5x^3-3x)dx = -\frac{7}{8}$$

$$B_4 = \frac{9}{2}\int_{-1}^{0}(-1)\cdot\frac{1}{8}(35x^4-30x^2+3)dx + \frac{9}{2}\int_{0}^{1}1\cdot\frac{1}{8}(35x^4-30x^2+3) = 0$$

$$B_5 = \frac{-11}{2}\int_{-1}^{0}\frac{1}{8}(63x^5 - 70x^3 + 15x)dx + \frac{11}{2}\int_{0}^{1}\frac{1}{8}(63x^5 - 70x^3 + 15x) = \frac{11}{16}$$

and so on. Therefore, using (15.54) and (15.34) we obtain

$$
\begin{aligned}
f(x) &= \frac{3}{2}P_1(x) - \frac{7}{8}P_3(x) + \frac{11}{16}P_5(x)\\[6pt]
&= \frac{3}{2}x - \frac{7}{8}\cdot\frac{1}{2}(5x^3 - 3x) + \frac{11}{16}\cdot\frac{1}{8}(63x^5 - 70x^3 + 15x)\\[6pt]
&= \frac{3}{2}x - \frac{7}{16}(5x^3 - 3x) + \frac{11}{128}\cdot(63x^5 - 70x^3 + 15x)\\[6pt]
&= \frac{525}{128}x - \frac{525}{64}x^3 + \frac{693}{128}x^5
\end{aligned}
\qquad (15.55)
$$

We observe that the waveform of $f(x)$ is an odd function and, as we found above, its expansion contains only odd Legendre polynomials.

In many applications, the algebraic form of the Legendre polynomials is usually the most useful. However, there are times when we want to express the polynomials in terms trigonometric functions, as we did in Example 15.4. Also, the trigonometric forms are most convenient with the cylindrical and spherical coordinate systems. It is shown in advanced mathematics textbooks that

$$
\begin{aligned}
P_n\cos\theta &= \frac{1\cdot 3\cdot\ldots\cdot(2n-1)}{2\cdot 4\cdot\ldots\cdot 2n}2\cos n\theta + \frac{1}{2}\cdot\frac{1\cdot 3\cdot\ldots\cdot(2n-3)}{2\cdot 4\cdot\ldots\cdot(2n-2)}2\cos(n-2)\theta\\[6pt]
&\quad + \frac{1\cdot 3}{2\cdot 4}\cdot\frac{1\cdot 3\cdot\ldots\cdot(2n-5)}{2\cdot 4\cdot\ldots\cdot(2n-4)}2\cos(n-4)\theta + \ldots
\end{aligned}
\qquad (15.56)
$$

From (15.56) we obtain the first 6 Legendre polynomials in trigonometric form listed below.

$$
\begin{aligned}
P_0\cos\theta &= 1\\[6pt]
P_1\cos\theta &= \cos\theta\\[6pt]
P_2\cos\theta &= \frac{3\cos 2\theta + 1}{4}\\[6pt]
P_3\cos\theta &= \frac{5\cos 3\theta + 3\cos\theta}{8}\\[6pt]
P_4\cos\theta &= \frac{35\cos 4\theta + 20\cos 2\theta + 9}{64}\\[6pt]
P_5\cos\theta &= \frac{63\cos 5\theta + 35\cos 3\theta + 30\cos\theta}{128}
\end{aligned}
\qquad (15.57)
$$

The Legendre polynomials in algebraic form, satisfy the orthogonality principle when $m \neq n$ as indicated by the following integral.

$$\int_{-1}^{1} P_m(x)P_n(x)dx = \begin{cases} 0 & m \neq n \\ \dfrac{2}{2n+1} & m = n \end{cases} \tag{15.58}$$

Similarly, the Legendre polynomials in trigonometric form satisfy the orthogonality principle when $m \neq n$ as indicated by the following integral.

$$\int_{0}^{\pi} P_m(\cos\theta)P_n(\cos\theta)\sin\theta d\theta = \begin{cases} 0 & m \neq n \\ \dfrac{2}{2n+1} & m = n \end{cases} \tag{15.59}$$

We must remember that all the Legendre polynomials we have discussed thus far are referred to as *surface zonal harmonics*, and math tables include values of these as computed from Rodrigues' formula of (15.35).

There is another class of Legendre functions which are solutions of the differential equation

$$\boxed{(1-x^2)\frac{d^2 y}{dx^2} - 2x\frac{dy}{dx} + \left[n(n+1) - \frac{m^2}{1-x^2}\right]y = 0} \tag{15.60}$$

and this is referred to as the *associated Legendre differential equation*. We observe that if $m = 0$, (15.60) reduces to (15.27).

The general solution of (15.60) is

$$y = C_1 P_n^m(x) + C_2 Q_n^m(x) \tag{15.61}$$

where C_1 and C_2 are arbitrary constants. The functions $P_n^m(x)$ and $Q_n^m(x)$ are referred to as *associated Legendre functions of the first and second kind* respectively. These are evaluated from

$$P_n^m(x) = (-1)^m (1-x^2)^{m/2} \cdot \frac{d^m}{dx^m} P_n(x) \tag{15.62}$$

and

$$Q_n^m(x) = (-1)^m (1-x^2)^{m/2} \cdot \frac{d^m}{dx^m} Q_n(x) \tag{15.63}$$

Relations (15.62) and (15.63) are also known as *spherical harmonics*.

We will restrict our subsequent discussion to the associated Legendre functions of the first kind, that is, the polynomials $P_n^m(x)$.

At present, Excel does not have any functions related to Legendre polynomials. MATLAB provides the **legendre(n,x)** function that computes the associated Legendre functions of the first kind of degree n, and order m = 0, 1, 2, ..., n evaluated for each element of x.

Example 15.6

Find the following associated Legendre functions and evaluate as indicated.

$$\text{a. } P_2^1(x)\Big|_{x=0.5} \quad \text{b.} P_3^2(x)\Big|_{x=-0.5} \quad \text{c. } P_2^3(x)\Big|_{x=0.25}$$

Solution:

For this example, we use the relation (15.62), that is,

$$P_n^m(x) = (-1)^m (1-x^2)^{m/2} \cdot \frac{d^m}{dx^m} P_n(x)$$

and the appropriate relations of (15.34). For this example,

a.

$$P_2^1(x)\Big|_{x=0.5} = (-1)^1 (1-x^2)^{1/2} \frac{d}{dx} P_2(x)\Big|_{x=0.5} = -(1-x^2)^{1/2} \frac{d}{dx}\left(\frac{3x^2-1}{2}\right)\Big|_{x=0.5}$$

$$= -(1-x^2)^{1/2}(3x) = -1.2990$$

For m = 0 in (15.62), we obtain

$$P_2^1(x)\Big|_{x=0.5} = \left(\frac{3x^2-1}{2}\right)\Big|_{x=0.5} = -0.125$$

As stated above, the MATLAB **legendre(n,x)** function computes the associated Legendre functions of the first kind of degree n and order m = 0, 1, 2, ..., n evaluated for each element of x. Here, n = 2 and thus MATLAB will return a matrix whose rows correspond to the values of m = 0, m = 1, and m = 2, for the first, second, and third rows respectively.

Check with MATLAB:

disp('The values for m = 0, m = 1 and m = 2 are:'); legendre(2,0.5)

```
The values for m = 0, m = 1 and m = 2 are:
```

```
ans =
   -0.1250
   -1.2990
    2.2500
```

or more elegantly,

```
m=0:2; y=zeros(3,2); y(:,1)=m'; y(:,2)=legendre(2,0.5);
fprintf('\n'); fprintf('m\t Legendre \n'); fprintf('%2.0f\t %7.4f \n',y')
```

```
   m   Legendre
   0   -0.1250
   1   -1.2990
   2    2.2500
```

b.

$$P_3^2(x)\Big|_{x=-0.5} = (-1)^2(1-x^2)^{2/2}\frac{d^2}{dx^2}P_3(x)\Big|_{x=-0.5} = (1-x^2)\frac{d^2}{dx^2}\Big(\frac{5x^3-3x}{2}\Big)\Big|_{x=-0.5}$$

$$= (1-x^2)\frac{d}{dx}\Big(\frac{15x^2-3}{2}\Big)\Big|_{x=-0.5} = (1-x^2)(15x)\Big|_{x=-0.5} = -5.6250$$

Here, $n=3$, and thus MATLAB will display a matrix whose rows correspond to the values of $m=0$, $m=1$, $m=2$, and $m=3$, for the first, second, third and fourth rows respectively.

Check with MATLAB:

```
m=0:3; y=zeros(4,2); y(:,1)=m'; y(:,2)=legendre(3,-0.5);
fprintf('\n'); fprintf('m\t Legendre \n'); fprintf('%2.0f\t %7.4f \n',y')
```

```
   m   Legendre
   0    0.4375
   1   -0.3248
   2   -5.6250
   3   -9.7428
```

c.

$$P_2^3(x)\Big|_{x=0.25} = (-1)^3(1-x^2)^{3/2}\frac{d^3}{dx^3}P_2(x)\Big|_{x=0.25} = -(1-x^2)^{3/2}\frac{d^3}{dx^3}(3x^2-1)\Big|_{x=0.25}$$

and since the third derivative of $3x^2-1$ is zero, it follows that $P_2^3(x)=0$.

In general, if $m>n$, then $P_n^m(x)=0$.

15.3 Laguerre Polynomials

Another class of polynomials that satisfy the orthogonality principle, are the *Laguerre polynomials* $L_n(x)$; these are solutions of the differential equation

$$x\frac{d^2y}{dx^2} + (1-x)\frac{dy}{dx} + ny = 0 \tag{15.64}$$

These polynomials are computed with the Rodrigues' formula

$$L_n(x) = e^x\frac{d^n}{dx^n}(x^n e^{-x}) \tag{15.65}$$

The orthogonality principle for these polynomials states that

$$\int_0^\infty e^{-x}L_m(x)L_n(x)dx = 0 \tag{15.66}$$

Example 15.7

Compute the Laguerre polynomials

a. $L_0(x)$ b. $L_1(x)$ c. $L_2(x)$ d. $L_3(x)$

Solution:

Using Rodrigues's formula of (15.65), we obtain

$$L_0(x) = e^x\frac{d^0}{dx^0}(x^0 e^{-x}) = e^x e^{-x} = e^0 = 1$$

$$L_1(x) = e^x\frac{d}{dx}(xe^{-x}) = e^x(e^{-x} - xe^{-x}) = 1 - x$$

$$L_2(x) = e^x\frac{d^2}{dx^2}(x^2 e^{-x}) = e^x e^{-x}(2 - 4x + x^2) = 2 - 4x + x^2 \tag{15.67}$$

$$L_3(x) = e^x\frac{d^3}{dx^3}(x^3 e^{-x}) = 6 - 18x + 9x^2 - x^3$$

The differentiation of the last two polynomials in (15.67) was performed with MATLAB as follows:

```
syms x y z
y=x^2*exp(−x); z=diff(y,2);% Differentiate y twice with respect to x
```

```
L2x=exp(x)*z; simplify(L2x)
```

```
ans =
2-4*x+x^2
```

```
syms x y z w ; y=x^3*exp(-x); z=diff(y,2);% Differentiate y twice
% we cannot differentiate three times at once
w=diff(z);% Differentiate one more time
L3x=exp(x)*w; simplify(L3x)
```

```
ans =
6-18*x+9*x^2-x^3
```

15.4 Chebyshev Polynomials

The *Chebyshev polynomials* are solutions of the differential equations

$$(1-x^2)\frac{d^2y}{dx^2} - x\frac{dy}{dx} + n^2y = 0 \qquad (15.68)$$

and

$$(1-x^2)\frac{d^2y}{dx^2} - 3x\frac{dy}{dx} + n(n+2)y = 0 \qquad (15.69)$$

The solutions of (15.68) are referred to as *Chebyshev polynomials of the first kind* and are denoted as $y = T_n(x)$.[*] The solutions of (15.69) are the *Chebyshev polynomials of the second kind*; these are denoted as $y = U_n(x)$. Both kinds comprise a set of orthogonal functions.

We will restrict our discussion to the $T_n(x)$ polynomials. We will plot some of these later in this section.

Two interesting properties of the $T_n(x)$ polynomials are:

1. They exhibit equiripple amplitute characteristics over the range $-1 \le x \le 1$, that is, within this range, they oscillate with the same ripple. This property is the basis for the Chebyshev approximation in the design of Chebyshev type electric filters.

2. For $|x| > 1$ they increase or decrease more rapidly than any other polynomial of order n.

---

[*] *Some books use the notation $C_k(x)$ for these polynomials. However, another class of orthogonal functions known as* **Genenbauer** *or* **Ultraspherical functions** *use the notation $C_n^{(a)}(x)$ and for this reason, we will avoid notation $C_k(x)$ for the Chebyshev polynomials.*

These polynomials are tabulated in reference books which contain mathematical functions. A good reference is the *Handbook of Mathematical Functions*, Dover Publications. They can also be derived from the following relations.

$$T_n(x) = \cos(n\cos^{-1}x) \quad \text{for } |x| \le 1 \tag{15.70}$$

$$T_n(x) = \cosh(n\cosh^{-1}x) \quad \text{for } |x| > 1 \tag{15.71}$$

Using (15.70) or (15.71), we can express $T_n(x)$ as polynomials in powers of x. Some are shown in Table 15.2.

TABLE 15.2 *Chebyshev polynomials expressed in powers of x*

n	$T_n(x)$
0	1
1	x
2	$2x^2 - 1$
3	$4x^3 - 3x$
4	$8x^4 - 8x^2 + 1$
5	$16x^5 - 20x^3 + 5x$
6	$32x^6 - 48x^4 + 18x^2 - 1$

To show that the relation of (15.70) can be expressed as a polynomial, we let

$$x = \cos y \tag{15.72}$$

and

$$T_n(y) = \cos ny \tag{15.73}$$

Next, in (15.73), we replace n with $n + 1$ and we obtain

$$T_{n+1}(y) = \cos(n+1)y = \cos ny \cos y - \sin ny \sin y \tag{15.74}$$

Similarly, replacing n with $n - 1$, we obtain

$$T_{n-1}(y) = \cos(n-1)y = \cos ny \cos y + \sin ny \sin y \tag{15.75}$$

Now, we add (15.74) with (15.75), and making use of (15.73) and (15.72), we obtain

$$T_{n+1}(y) + T_{n-1}(y) = 2\cos ny \cos y = 2T_n(y)x = 2xT_n(y) \tag{15.76}$$

or

$$T_{n+1}(y) = 2xT_n(y) - T_{n-1}(y)$$

Then, we can replace y with x to obtain

$$\boxed{\begin{array}{c} T_{n+1}(x) = 2xT_n(x) - T_{n-1}(x) \\ \text{Recurrence Relation} \end{array}} \tag{15.77}$$

The polynomials in Table 15.2, can now be verified by using a combination of the above relations. Thus, for $n = 0$, (15.73) yields

$$T_0(y) = T_0(x) = 1 \tag{15.78}$$

For $n = 1$, from (15.73) and (15.72), we obtain

$$T_1(y) = T_1(x) = x \tag{15.79}$$

To derive the algebraic expressions corresponding to $n = 2, 3, 4$ and so on, we use the recurrence formula of (15.77). For instance, when $n = 2$,

$$T_2(x) = 2xT_1(x) - T_0(x) = 2x^2 - 1 \tag{15.80}$$

and when $n = 3$,

$$T_3(x) = 2xT_2(x) - T_1(x) = 4x^3 - 2x - x = 4x^3 - 3x \tag{15.81}$$

Alternately, we can prove the first 3 entries of Table 15.2 with (15.70) by letting $y = \cos^{-1}x$. Thus, for $n = 0$,

$$T_0(x) = \cos(0 \cdot \cos^{-1}x) = \cos(0 \cdot y) = 1$$

For $n = 1$,

$$T_1(x) = \cos(1 \cdot \cos^{-1}x) = \cos(1 \cdot y) = \cos y = x$$

and for $n = 2$,

$$T_2(x) = \cos(2 \cdot \cos^{-1}x) = \cos(2 \cdot y) = \cos 2y = 2\cos^2 y - 1$$

$$= 2\cos^2(\cos^{-1}x) - 1 = 2\left[\underbrace{\cos(\cos^{-1}x)}_{x} \cdot \underbrace{\cos(\cos^{-1}x)}_{x}\right] - 1$$

or

$$T_2(x) = 2x^2 - 1$$

Relation (15.71) can be derived from (15.70) as follows:

We recall that

$$\cos\alpha = \frac{e^{j\alpha} + e^{-j\alpha}}{2} \tag{15.82}$$

and

$$\cosh\alpha = \frac{e^{\alpha} + e^{-\alpha}}{2} \tag{15.83}$$

Then,

$$\cos\alpha = \cosh j\alpha \tag{15.84}$$

and when $|x| > 1$,

$$\cos^{-1}x = -j\cosh^{-1}x \, ^* \tag{15.85}$$

By substitution into (15.70), making use of (15.85), and that $\cosh(-t) = \cosh t$, we obtain

$$T_n(x) = \cos[n(-j\cosh^{-1}x)] = \cos(-jn\cosh^{-1}x) = \cosh(jnj\cosh^{-1}x)$$

$$= \cosh[j(-jn\cosh^{-1}x)] = \cosh(n\cosh^{-1}x)$$

and this is the same as (15.71).

We can also use MATLAB to convert (15.70) and (15.71) to polynomials. For example, if $n = 3$,

```
syms x;
expand(cos(3*acos(x))), expand(cosh(3*acosh(x)))
```

```
ans =
4*x^3-3*x

ans =
4*x^3-3*x
```

The MATLAB script below plots the $T_n(x)$ for $n = 0$ through $n = 6$.

```
% Chebyshev polynomials
%
x=-1.2:0.01:1.2; Tnx0=cos(0*acos(x));
Tnx1=cos(1*acos(x)); Tnx2=cos(2*acos(x)); Tnx3=cos(3*acos(x)); Tnx4=cos(4*acos(x));
Tnx5=cos(5*acos(x)); Tnx6=cos(6*acos(x));
plot(x, Tnx0, x, Tnx1, x, Tnx2, x, Tnx3, x, Tnx4, x, Tnx5, x, Tnx6);....
axis([-1.2 1.2 -1.5 1.5]); grid; title('Chebyshev Polynomials of the First Kind');
xlabel('x'); ylabel('Tn(x)')
% We could have used the gtext function to label the curves but it is easier with the Figure text
% tool
```

---

* Let $\cos\alpha = \cosh j\alpha = v$; *then* $\alpha = \cos^{-1}v$, $j\alpha = \cosh^{-1}v$, $j\cos^{-1}v = \cosh^{-1}v$ *and (15.85) follows.*

Figure 15.6 shows the plot of the Chebyshev polynomials of the first kind $T_n(x)$ for $n = 0$ through $n = 6$.

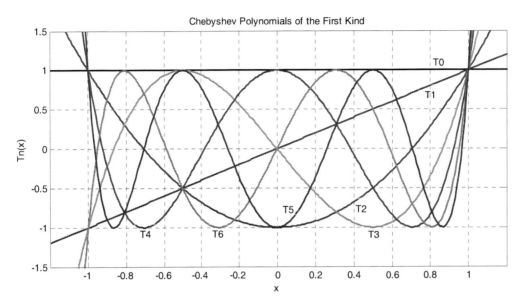

Figure 15.6. Plot of Chebyshev polynomials with MATLAB

As mentioned earlier, Chebyshev polynomials, among other applications, are used in the design of electric filters.* The filters are described in terms of rational polynomials that approximate the behavior of ideal filters. The basic Chebyshev low–pass filter approximation is defined as

$$A^2(\omega) = \frac{\alpha}{1 + \varepsilon^2 T_n^2(\omega/\omega_C)} \qquad (15.86)$$

where ω is the operating radian frequency, ω_C is the cutoff frequency, and α and ε are other parameters that are used to specify the order and type of the electric filter.

For example, if we want to design a second order Chebyshev low–pass filter, we use the Chebyshev polynomial

$$T_2(x) = 2x^2 - 1$$

and (15.86) becomes

$$A^2(\omega) = \frac{\alpha}{1 + \varepsilon^2[2(\omega/\omega_C)^2 - 1]} \qquad (15.87)$$

---

* For a thorough discussion on the design of analog and digital filters refer to Signals and systems with MATLAB Applications, Orchard Publications, ISBN 0–9744239–9–8.

15.5 Summary

- Differential equations with variable coefficients cannot be solved in terms of familiar functions as those which we encountered in ordinary differential equations with constant coefficients. The usual procedure is to derive solutions in the form of infinite series, and the most common are the Method of Frobenius and the Method of Picard.

- Bessel functions are solutions of the differential equation

$$x^2 \frac{d^2 y}{dx^2} + x \frac{dy}{dx} + (x^2 - n^2)y = 0$$

where n can be any number, positive or negative integer, fractional, or even a complex number. The general solution depends on the value of n.

- The series

$$J_n(x) = \sum_{k=0}^{\infty} (-1)^k \cdot \left(\frac{x}{2}\right)^{n+2k} \cdot \frac{1}{k! \cdot \Gamma(n+k+1)} \quad n \geq 0$$

where n is any positive real number or zero is referred to as Bessel function of order n.

- The series

$$J_{-n}(x) = \sum_{k=0}^{\infty} (-1)^k \cdot \left(\frac{x}{2}\right)^{-n+2k} \cdot \frac{1}{k! \cdot \Gamma(-n+k+1)}$$

is referred to as the Bessel function of negative order n.

- For $n = 0, 1$ and 2 the series reduce to

$$J_0(x) = 1 - \frac{x^2}{2^2 \cdot (1!)^2} + \frac{x^4}{2^4 \cdot (2!)^2} - \frac{x^6}{2^6 \cdot (3!)^2} + \frac{x^8}{2^8 \cdot (4!)^2} - \cdots$$

$$J_1(x) = \frac{x}{2} - \frac{x^3}{2^3 \cdot 1! \cdot 2!} + \frac{x^5}{2^5 \cdot 2! \cdot 3!} - \frac{x^7}{2^7 \cdot 3! \cdot 4!} + \frac{x^9}{2^9 \cdot 4! \cdot 5!} - \cdots$$

$$J_2(x) = \frac{x^2}{2^2 \cdot 2!} - \frac{x^4}{2^4 \cdot 1! \cdot 3!} + \frac{x^6}{2^6 \cdot 2! \cdot 4!} - \frac{x^8}{2^8 \cdot 3! \cdot 5!} + \frac{x^{10}}{2^{10} \cdot 4! \cdot 6!} - \cdots$$

- Two more useful relations are

$$\frac{d}{dx} J_0(x) = -J_1(x)$$

$$\frac{d}{dx}\{xJ_1(x)\} = xJ_0(x)$$

- Values of $J_n(x)$ can be calculated using the appropriate series given above. They also can be found in math table books, and can also be found with the MATLAB **besselj(n,x)** function or the Excel **BESSELJ(x,n)** function.

- The Bessel functions

$$J_{1/2}(x) = \sqrt{\frac{2}{\pi x}}\sin x \text{ and } J_{(-1/2)}(x) = \sqrt{\frac{2}{\pi x}}\cos x$$

are known as half–order Bessel functions.

- Besides the above functions known as Bessel functions of the first kind, other Bessel functions, denoted as $Y_n(x)$ and referred to as Bessel functions of the second kind, or Weber functions, or Neumann functions exist. Also, certain differential equations resemble the Bessel equation, and thus their solutions are called Modified Bessel functions, or Hankel functions.

- If n is not an integer, $J_n(x)$ and $J_{-n}(x)$ are linearly independent; for this case, the general solution of the Bessel equation is

$$y = AJ_n(x) + BJ_{-n}(x)$$

$$n \ne 0, 1, 2, 3, \dots$$

- If a and b are distinct roots of $J_0(x) = 0$, $J_0(a) = 0$ and $J_0(b) = 0$, then,

$$\int_0^1 xJ_0(ax)J_0(bx)dx = 0$$

and thus we say that $J_0(ax)$ and $J_0(bx)$ are orthogonal in the interval $0 \le x \le 1$.

- The differential equation

$$(1-x^2)\frac{d^2y}{dx^2} - 2x\frac{dy}{dx} + n(n+1)y = 0$$

where n is a constant, is known as Legendre's equation.

- The infinite series solution of the Legendre functions, denoted as $Q_n(x)$, is referred to as Legendre functions of the second kind.

- The Legendre polynomials are defined as

$$P_n(x) = \sum_{k=0}^{N} \frac{(-1)^k \cdot (2n-2k)!}{2^n k!(n-k)!(n-2k)!} x^{n-2k}$$

  where $N = \dfrac{n}{2}$ for $n = $ even and $N = \dfrac{(n-1)}{2}$ for $n = $ odd

and the first 6 Legendre polynomials are

$$P_0(x) = 1 \qquad\qquad P_1(x) = x$$

$$P_2(x) = \frac{1}{2}(3x^2 - 1) \qquad\qquad P_3(x) = \frac{1}{2}(5x^3 - 3x)$$

$$P_4(x) = \frac{1}{8}(35x^4 - 30x^2 + 3) \qquad P_5(x) = \frac{1}{8}(63x^5 - 70x^3 + 15x)$$

- The relation

$$P_n(x) = \frac{1}{2^n \cdot n!} \cdot \frac{d^n}{dx^n}(x^2 - 1)^n$$

is known as Rodrigues' formula, and offers another method of expressing the Legendre polynomials.

- The Legendre polynomial series of a function $f(x)$ for $|x| < 1$, is

$$B_n = \frac{2n+1}{2} \int_{-1}^{1} f(x)P_n(x)dx$$

and with this relation we can find a polynomial $f(x)$ defined as

$$f(x) = B_0 P_0(x) + B_1 P_1(x) + B_2 P_2(x) + \ldots + B_n P_n(x) = \sum_{n=0}^{\infty} B_n P_n(x)$$

- The trigonometric form of the Legendre polynomials is

$$P_n \cos\theta = \frac{1 \cdot 3 \cdot \ldots \cdot (2n-1)}{2 \cdot 4 \cdot \ldots \cdot 2n} 2\cos n\theta + \frac{1}{2} \cdot \frac{1 \cdot 3 \cdot \ldots \cdot (2n-3)}{2 \cdot 4 \cdot \ldots \cdot (2n-2)} 2\cos(n-2)\theta$$

$$+ \frac{1 \cdot 3}{2 \cdot 4} \cdot \frac{1 \cdot 3 \cdot \ldots \cdot (2n-5)}{2 \cdot 4 \cdot \ldots \cdot (2n-4)} 2\cos(n-4)\theta + \ldots$$

and the first 6 Legendre polynomials in trigonometric form listed below.

$$P_0 \cos\theta = 1$$

$$P_1 \cos\theta = \cos\theta$$

$$P_2 \cos\theta = \frac{3\cos 2\theta + 1}{4}$$

$$P_3 \cos\theta = \frac{5\cos 3\theta + 3\cos\theta}{8}$$

$$P_4 \cos\theta = \frac{35\cos 4\theta + 20\cos 2\theta + 9}{64}$$

$$P_5 \cos\theta = \frac{63\cos 5\theta + 35\cos 3\theta + 30\cos\theta}{128}$$

- The Legendre polynomials in algebraic form, satisfy the orthogonality principle when $m \neq n$ as indicated by the integral

$$\int_{-1}^{1} P_m(x)P_n(x)dx = \begin{cases} 0 & m \neq n \\ \dfrac{2}{2n+1} & m = n \end{cases}$$

- The Legendre polynomials in trigonometric form satisfy the orthogonality principle when $m \neq n$ as indicated by the integral

$$\int_{0}^{\pi} P_m(\cos\theta)P_n(\cos\theta)\sin\theta d\theta = \begin{cases} 0 & m \neq n \\ \dfrac{2}{2n+1} & m = n \end{cases}$$

- The differential equation

$$(1-x^2)\frac{d^2 y}{dx^2} - 2x\frac{dy}{dx} + \left[n(n+1) - \frac{m^2}{1-x^2}\right]y = 0$$

is referred to as the associated Legendre differential equation. The general solution of this equation is

$$y = C_1 P_n^m(x) + C_2 Q_n^m(x)$$

where C_1 and C_2 are arbitrary constants. The functions $P_n^m(x)$ and $Q_n^m(x)$ are referred to as associated Legendre functions of the first and second kind respectively. These functions, also known as spherical harmonics, are evaluated from the relations

$$P_n^m(x) = (-1)^m(1-x^2)^{m/2} \cdot \frac{d^m}{dx^m}P_n(x)$$

$$Q_n^m(x) = (-1)^m (1-x^2)^{m/2} \cdot \frac{d^m}{dx^m} Q_n(x)$$

• The MATLAB **legendre(n,x)** function computes the associated Legendre functions of the first kind of degree n, and order $m = 0, 1, 2, ..., n$ evaluated for each element of x.

• The solutions of the differential equation

$$x \frac{d^2 y}{dx^2} + (1-x) \frac{dy}{dx} + ny = 0$$

are known as Laguerre polynomials and are denoted as $L_n(x)$. These polynomials are satisfy the orthogonality principle. They are computed with the Rodrigues' formula

$$L_n(x) = e^x \frac{d^n}{dx^n} (x^n e^{-x})$$

• The Chebyshev polynomials are solutions of the differential equations

$$(1-x^2) \frac{d^2 y}{dx^2} - x \frac{dy}{dx} + n^2 y = 0$$

and

$$(1-x^2) \frac{d^2 y}{dx^2} - 3x \frac{dy}{dx} + n(n+2)y = 0$$

The solutions of the first differential equation are referred to as Chebyshev polynomials of the first kind and are denoted as $y = T_n(x)$. The solutions of the second are the Chebyshev polynomials of the second kind; these are denoted as $y = U_n(x)$. Both kinds comprise a set of orthogonal functions.

• The $T_n(x)$ polynomials are derived from the relations

$$T_n(x) = \cos(n \cos^{-1} x) \quad \text{for } |x| \le 1$$

$$T_n(x) = \cosh(n \cosh^{-1} x) \quad \text{for } |x| > 1$$

These polynomials exhibit equiripple amplitute characteristics over the range $-1 \le x \le 1$, that is, within this range, they oscillate with the same ripple as shown in Figure 15.6. This property is the basis for the Chebyshev approximation in the design of Chebyshev type electric filters.

15.6 Exercises

1. Use the appropriate series of the Bessel functions $J_n(x)$ to compute the following values using the first 4 terms of the series and check your answers with MATLAB or Excel.

   a. $J_0(3)$ b. $J_1(2)$ c. $J_{1/2}(\pi/6)$ d. $J_{-1/2}(\pi/3)$

2. Use the appropriate Legendre polynomials $P_n(x)$ or $P_n^m(x)$ and Rodrigues's formulas to compute the following, and check your answers with MATLAB.

   a. $P_1(x)|_{x = 0.5}$ b. $P_2(x)|_{x = 0.75}$ c. $P_3(x)|_{x = 0.25}$

   d. $P_1^2(x)|_{x = 0.5}$ e. $P_2^3(x)|_{x = -0.5}$ f. $P_3^2(x)|_{x = 0.25}$

3. Compute the Legendre polynomial $g(x)$ representing the waveform $f(x)$ of the figure below. The first 5 terms of $P_n(x)$, i.e., $P_0(x)$ through $P_4(x)$ will be sufficient. Then, use MATLAB or Excel to plot $g(x)$ and compare with $f(x)$.

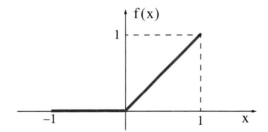

Numerical Analysis Using MATLAB® and Excel®, Third Edition
Copyright © Orchard Publications

15.7 Solutions to End–of–Chapter Exercises

1.

a.

$$J_0(x) = 1 - \frac{x^2}{2^2 \cdot (1!)^2} + \frac{x^4}{2^4 \cdot (2!)^2} - \frac{x^6}{2^6 \cdot (3!)^2}\bigg|_{x=3} = 1 - \frac{9}{4} + \frac{81}{64} - \frac{729}{2304} = -0.3008$$

b.

$$J_1(x) = \frac{x}{2} - \frac{x^3}{2^2 \cdot 4} + \frac{x^5}{2^2 \cdot 4^2 \cdot 6} - \frac{x^7}{2^2 \cdot 4^2 \cdot 6^2 \cdot 8}\bigg|_{x=2} = 1 - \frac{8}{16} + \frac{32}{384} - \frac{128}{18432} = 0.5764$$

c.

$$J_{1/2}(x) = \sqrt{\frac{2}{\pi x}}\sin x\bigg|_{x=\pi/6} = \sqrt{\frac{2}{\pi(\pi/6)}}\sin(\pi/6) = \frac{\sqrt{12}}{\pi}\cdot\frac{1}{2} = \frac{2\sqrt{3}}{2\pi} = \frac{\sqrt{3}}{\pi} = 0.5513$$

d.

$$J_{-1/2}(x) = \sqrt{\frac{2}{\pi x}}\cos x\bigg|_{x=\pi/3} = \sqrt{\frac{2}{\pi(\pi/3)}}\cos(\pi/3) = \frac{\sqrt{6}}{\pi}\cdot\frac{1}{2} = \frac{\sqrt{6}}{2\pi} = 0.3898$$

Check with MATLAB:

besselj(0,3), besselj(1,2), besselj(0.5,pi/6), besselj(–0.5,pi/3)

```
ans =
    -0.2601

ans =
    0.5767

ans =
    0.5513

ans =
    0.3898
```

We observe that the first value returned by MATLAB above is significantly different from that we obtained from the series. This is because our computation was based on the first 4 terms of the series. Had we taken also the fifth term our answer would have been –0.2563 and this is much closer to the value obtained with MATLAB.

2.

We will use the relations of (15.34) and (15.62). They are repeated below for convenience.

$$P_0(x) = 1 \qquad\qquad P_1(x) = x$$

$$P_2(x) = \frac{1}{2}(3x^2 - 1) \qquad\qquad P_3(x) = \frac{1}{2}(5x^3 - 3x)$$

$$P_4(x) = \frac{1}{8}(35x^4 - 30x^2 + 3) \qquad P_5(x) = \frac{1}{8}(63x^5 - 70x^3 + 15x)$$

$$P_n^m(x) = (-1)^m (1 - x^2)^{m/2} \cdot \frac{d^m}{dx^m} P_n(x)$$

a.

$$P_1(x)\big|_{x = 0.5} = x = 0.5$$

b.

$$P_2(x)\big|_{x = 0.75} = \frac{1}{2}(3x^2 - 1)\bigg|_{x = 0.75} = 0.5(3 \times (0.75)^2 - 1) = 0.3438$$

c.

$$P_3(x)\big|_{x = 0.25} = \frac{1}{2}(5x^3 - 3x)\bigg|_{x = 0.25} = 0.5(5 \times (0.25)^3 - 3 \times 0.25) = -0.3359$$

d.

$$P_1^2(x)\big|_{x = 0.5} = (-1)^2 (1 - x^2)^{2/2} \frac{d^2}{dx^2} P_1(x)\bigg|_{x = 0.5} = (1 - x^2) \cdot \frac{d^2}{dx^2}(x)\bigg|_{x = 0.5} = 0$$

We recall that if $m > n$, then $P_n^m(x) = 0$.

e.

$$P_2^3(x)\big|_{x = -0.5} = 0$$

This is because $m > n$. In other words,

$$\frac{d^3}{dx^3} P_2(x) = \frac{d^3}{dx^3}\left\{ \frac{1}{2}(3x^2 - 1) \right\} = 0$$

f.

$$P_3^2(x)\big|_{x = 0.25} = (-1)^2 (1 - x^2)^{2/2} \frac{d^2}{dx^2} P_3(x)\bigg|_{x = 0.25} = (1 - x^2) \frac{d^2}{dx^2}\left\{ \frac{1}{2}(5x^3 - 3x) \right\}\bigg|_{x = 0.25}$$

$$= (1 - 0.25^2) \cdot \frac{1}{2} \cdot \frac{d}{dx}(15x^2 - 3)\bigg|_{x = 0.25} = 0.4688(30x)\big|_{x = 0.25} = 3.5160$$

Check with MATLAB:

a.

```
m=0:1; y=zeros(2,2); y(:,1)=m'; y(:,2)=legendre(1,0.5);
fprintf('\n'); fprintf('m\t Legendre \n'); fprintf('%2.0f\t %7.4f \n',y')
```

```
 m Legendre
 0  0.5000
 1 -0.8660
```

b.

```
m=0:2; y=zeros(3,2); y(:,1)=m'; y(:,2)=legendre(2,0.75);
fprintf('\n'); fprintf('m\t Legendre \n'); fprintf('%2.0f\t %7.4f \n',y')
```

```
 m Legendre
 0  0.3438
 1 -1.4882
 2  1.3125
```

c.

```
m=0:3; y=zeros(4,2); y(:,1)=m'; y(:,2)=legendre(3,0.25);
fprintf('\n'); fprintf('m\t Legendre \n'); fprintf('%2.0f\t %7.4f \n',y')
```

```
 m Legendre
 0 -0.3359
 1  0.9985
 2  3.5156
 3 -13.6160
```

d.

```
m=0:1; y=zeros(2,2); y(:,1)=m'; y(:,2)=legendre(1,0.5);
fprintf('\n'); fprintf('m\t Legendre \n'); fprintf('%2.0f\t %7.4f \n',y')
```

```
m Legendre
 0  0.5000
 1 -0.8660
```

Here, the **legendre(n,x)** function computes the associated Legendre functions of degree **n** and order $m = 0, 1, ..., n$, evaluated for each element of **x**. For this example, $m > n$, that is, $m = 2$ and $n = 1$ and the statement **m=0:2** is not accepted. For this reason we've used **m=0:1**.

e.

```
m=0:2; y=zeros(3,2); y(:,1)=m'; y(:,2)=legendre(2,–0.5);
fprintf('\n'); fprintf('m\t Legendre \n'); fprintf('%2.0f\t %7.4f \n',y')
```

As in (d) above $m > n$ and MATLAB returns

```
m Legendre
0 -0.1250
1  1.2990
2  2.2500
```

f.

```
m=0:3; y=zeros(4,2); y(:,1)=m'; y(:,2)=legendre(3,0.25);
fprintf('\n'); fprintf('m\t Legendre \n'); fprintf('%2.0f\t %7.4f \n',y')
```

```
m Legendre
0  -0.3359
1   0.9985
2   3.5156
3 -13.6160
```

3.

$$B_n = \frac{2n+1}{2} \int_{-1}^{1} f(x)P_n(x)dx$$

For this exercise $f(x) = 0$ for $-1 \le x \le 0$ and thus

$$B_n = \frac{2n+1}{2} \int_{0}^{1} f(x)P_n(x)dx$$

Then,

$$B_0 = \frac{1}{2}\int_{0}^{1} x \cdot P_0(x) \cdot dx = \frac{1}{2} \cdot \frac{x^2}{2}\Big|_{0}^{1} = \frac{1}{4}$$

$$B_1 = \frac{3}{2}\int_{0}^{1} x \cdot P_1(x) \cdot dx = \frac{3}{2} \cdot \frac{x^3}{3}\Big|_{0}^{1} = \frac{1}{2}$$

$$B_2 = \frac{5}{2}\int_{0}^{1} x \cdot \frac{1}{2}(3x^2 - 1)dx = \frac{5}{4}\int_{0}^{1} (3x^3 - x)dx = \frac{5}{4}\left(\frac{3x^4}{4} - \frac{x^2}{2}\right)\Big|_{0}^{1} = \frac{5}{16}$$

$$B_3 = \frac{7}{2}\int_{0}^{1} x \cdot \frac{1}{2}(5x^3 - 3x)dx = \frac{7}{4}\int_{0}^{1} (5x^4 - 3x^2)dx = \frac{7}{4}\left(\frac{5x^5}{5} - \frac{3x^3}{3}\right)\Big|_{0}^{1} = 0$$

$$B_4 = \frac{9}{2}\int_{0}^{1} x \cdot \frac{1}{8}(35x^4 - 30x^2 + 3)dx = \frac{9}{16}\int_{0}^{1} (35x^5 - 30x^3 + 3x)dx$$

$$= \frac{9}{16}\left(\frac{35x^6}{6} - \frac{30x^4}{4} + \frac{3x^2}{2}\right)\Big|_{0}^{1} = -\frac{3}{32}$$

and by substitution into

$$f(x) = B_0 P_0(x) + B_1 P_1(x) + B_2 P_2(x) + \ldots + B_n P_n(x) = \sum_{n=0}^{\infty} B_n P_n(x)$$

we obtain

$$f(x) = \frac{1}{4} P_0(x) + \frac{1}{2} P_1(x) + \frac{5}{16} P_2(x) + 0 \cdot P_3(x) - \frac{3}{32} P_4(x)$$

$$= \frac{1}{4} \cdot 1 + \frac{1}{2} \cdot x + \frac{5}{16} \cdot \frac{1}{2}(3x^2 - 1) - \frac{3}{32} \cdot \frac{1}{8}(35x^4 - 30x^2 + 3)$$

$$= \frac{1}{4} + \frac{x}{2} + \frac{15x^2}{32} - \frac{5}{32} - \frac{105x^4}{256} + \frac{90x^2}{256} - \frac{9}{256}$$

or

$$f(x) = \frac{1}{256}(15 + 128x + 210x^2 - 105x^4)$$

we note that $f(-1) = -8/256$ and $f(0) = 15/256$. These values are close to zero. Also, $f(1) = 248/256$ and this value is close to unity.

We plot $f(x)$ with the MATLAB script below.

```
x=0:0.01:1; fx=(15+128.*x+210.*x.^2-105.*x.^4)./256; plot(x,fx); xlabel('x'); ylabel('f(x)'); grid
```

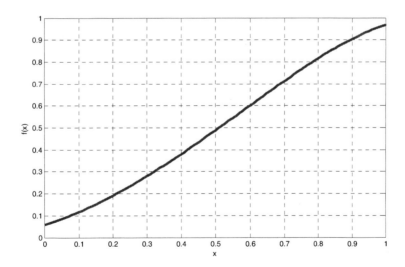

NOTES:

Chapter 16

T his chapter introduces three methods for maximizing or minimizing some function in order to achieve the optimum solution. These methods are topics discussed in detail in a branch of mathematics called *operations research* and it is concerned with financial and engineering economic problems. Our intent here is to introduce these methods with the basic ideas. We will discuss linear programming, dynamic programming, and network analysis, and we will illustrate these with some simple but practical examples.

16.1 Linear Programming

In linear* programming we seek to maximize or minimize a particular quantity, referred to as the *objective*, which is dependent on a finite number of variables. These variables may or may not be independent of each another, and in most cases are subject to certain conditions or limitations referred to as *constraints*.

Example 16.1

The ABC Semiconductor Corporation produces microprocessors (μPs) and memory (RAM) chips. The material types, A and B, required to manufacture the μPs and RAMs and the profits for each are shown in Table 16.1.

TABLE 16.1 Data for Example 16.1

	Parts of Material Types	
	μPs	RAMs (1000s)
Semiconductor Material A	3	2
Semiconductor Material B	5	10
Profit	$25.00 per unit	$20.00 per 1000

Due to limited supplies of silicon, phosphorus and boron, its product mix at times of high consumer demand, is subject to limited supplies. Thus, ABC Semiconductor can only buy 450 parts of Material A, and 1000 parts of Material B. This corporation needs to know what combination of μPs and RAMs will maximize the overall profit.

---

* A linear program is one in which the variables form a linear combination,i.e., are linearly related. All other programs are considered non−linear.

Solution:

Since with Material A we can produce 3 μPs and 2 RAMs, and with Material B 5 μPs and 10 RAMs, the corporation is confronted with the following constraints:

$$3x + 2y \leq 450$$
$$5x + 10y \leq 1000$$

We now can state the problem as

$$\text{Maximize} \quad z = 25 \times \mu P + 20 \times \text{RAMs} \qquad (16.1)$$

subject to the constraints

$$3x + 2y \leq 450$$
$$5x + 10y \leq 1000 \qquad (16.2)$$

Two additional constraints are $x \geq 0$, $y \geq 0$, and x and y must be integers.

For this example, there are only two variables, x and y; therefore, a graphical solution is possible. We will solve this example graphically.

The x and y intercept corresponding to the above equations is shown in the plot of Figure 16.1 where the cross–hatched area indicates the feasible region.[*]

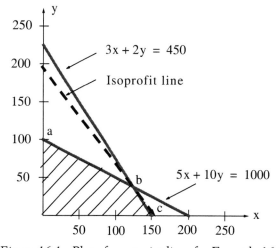

Figure 16.1. Plot of constraint lines for Example 16.1

The equation of the straight line of the maximum profit is referred to as *isoprofit line*. This line will pass through one of the three corners denoted as a, b, and c.

---

[*] *The feasible region is the area which includes all points (x, y) satisfying all constrains.*

The isoprofit line that we are interested is described by the equation

$$z = 25 \times \mu P + 20 \times RAM = \text{constant}$$
$$= 25x + 20y = C \tag{16.3}$$

We can express this equation in $y = mx + b$ form, that is,

$$y = -\frac{25}{20}x + \frac{C}{20} = -1.25x + k \tag{16.4}$$

where k is the y–intercept. Therefore, all possible isoprofit lines have the same slope, that is, are parallel to each another, and the highest isoprofit line passes through point b.

The coordinates of point b in Figure 16.1 are found by simultaneous solution of

$$3x + 2y = 450$$
$$5x + 10y = 1000 \tag{16.5}$$

Using MATLAB for the solution of (16.5) we obtain

```
syms x y
[x y]=solve(3*x+2*y−450, 5*x+10*y−1000)

x =
125
y =
75/2
```

Of course, these values must be integers, so we accept the values $x = 125$, and $y = 37$. Then, by substitution into (16.1),

$$z_{max} = 25 \times 125 + 20 \times 37 = \$3865 \tag{16.6}$$

and the isoprofit line can be drawn from the equation

$$25x + 20y = 3865 \tag{16.7}$$

by first letting $x = 0$, then, $y = 0$. Then, we obtain the points

$$x = \frac{3865}{25} = 154.6$$

and

$$y = \frac{3865}{20} = 193.25$$

This is shown as a dotted line on the plot of Figure 16.1.

It was possible to solve this problem graphically because it is relatively simple. In most cases, however, we cannot obtain the solution by graphical methods and therefore, we must resort to algebraic methods such as the *simplex method*. This and other methods are described in operations research textbooks.

We can find the optimum solution to this type of problems with Excel's *Solver* feature. The procedure is included in the spreadsheet of Figure 16.2

	A	B	C	D	E	F
1	**Optimization - Maximum Profit for Example 16.1**					
2	1. Enter zeros in B12 and B13		2. In B15 enter =25*B12+20*B13			
3	3. In B17 enter =3*B12+2*B13 and in B18 =5*B12+10*B13					
4	4. From the *Tools* drop menu select *Solver*. Use *Add-Ins* if necessary to add it.					
5	5. On the Solver Parameters screen enter the following:					
6	*Set Target Cell:* B15					
7	*Equal to:* Max					
8	*By Changing Cells:* B12:B13					
9	*Click on Add and enter Constraints:*					
10	*B12=Integer, Add B13=Integer, Add B12>=0, Add B13>=0,*					
11	*Add* B17<=450, *Add* B18<=1000, *OK*, *Solve*					
12	x(µPs)=	124				
13	y(RAMs)=	38				
14						
15	Maximum Profit=	$3,860				
16						
17	Semiconductor Material A=	448				
18	Semiconductor Material B=	1000				
19						
20	Note: Contents of A12:A18 are typed-in for information only					

Figure 16.2. Spreadsheet for solution of Example 16.1 with Excel's solver

16.2 Dynamic Programming

Dynamic Programming is based on R. Bellman's *Principle of Optimality* which states that:

An optimum policy has the property that whatever the initial state and the initial decisions are, the remaining decisions must constitute an optimum policy with regard to the state resulting from the first decision.

Figure 16.3 represents a *line graph*, where the nodes a through h represent the *states*, and the choice of alternative paths when leaving a given state, is called a *decision*. The alternative paths are represented by the line segments ab, ac, bd, and so on.

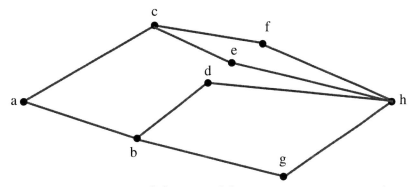

Figure 16.3. Line graph for a typical dynamic programming example

We assume that all segments are directed from left to right, and each has a value assigned to it which we will refer to as the *cost*. Thus, there is a cost associated with each segment, and it is usually denoted with the letter J. For example, for the path a, c, f, and h, the cost is

$$J_{ah} = J_{ac} + J_{cf} + J_{fh} \tag{16.8}$$

The costs for the other possible paths are defined similarly.

For the line graph of Figure 16.3, the objective is to go from state a to state h with minimum cost. Accordingly, we say that the *optimum path policy* for this line graph is

$$J_{min} = \min\{(J_{ac} + J_{cf} + J_{fh}), (J_{ac} + J_{ce} + J_{eh}), \\ (J_{ab} + J_{bd} + J_{dh}), (J_{ab} + J_{bg} + J_{gh})\} \tag{16.9}$$

Now, let us suppose that the initial state is a, and the initial decision has been made to go to state b. Then, the path from b to h must be selected optimally, if the entire path from a to h is to be optimum (minimum in this case).

Let the minimum cost from state b to h be denoted as g_b. Then,

$$g_b = \min\{(J_{bd} + J_{dh}), (J_{bg} + J_{gh})\} \tag{16.10}$$

Likewise, if the initial decision is to go from state a to c, the path from state c to h must be optimum, that is,

$$g_c = \min\{(J_{cf} + J_{fh}), (J_{ce} + J_{eh})\} \tag{16.11}$$

The optimum path policy of (16.9) can now be expressed in terms of (16.10) and (16.11) as

$$g_a = J_{min} = \min\{(J_{ab} + g_b), (J_{ac} + g_c)\} \tag{16.12}$$

This relation indicates that to obtain the minimum cost we must minimize:

1. The part which is related to the present decision, in this case, costs J_{ab} and J_{ac}.

2. The part which represents the minimum value of all future costs starting with the state which results from the first decision.

Example 16.2

Find the minimum cost route from state a to state m for the line graph of Figure 16.4. The line segments are directed from left to right and the costs are indicated beside each line segment.

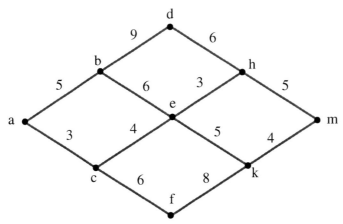

Figure 16.4. Line graph for Example 16.2

Solution:

We observe that at states h, k, d, and f have no alternative paths since the lines are directed from left to right. Therefore, we make the first decision at state e. Then,

$$g_e = \min\{(3 + g_h), (5 + g_k)\} = \min\{(3 + 5), (5 + 4)\} = 8$$

$$(e \rightarrow h \rightarrow m)$$

(16.13)

Next, we make decisions at states b and c.

$$g_b = \min\{(9 + g_d), (6 + g_e)\} = \min\{(9 + 6), (6 + 8)\} = 14$$

$$(b \rightarrow e)$$

(16.14)

$$g_c = \min\{(4 + g_e), (6 + g_f)\} = \min\{(4 + 8), (6 + 8)\} = 12$$

$$(c \rightarrow e)$$

(16.15)

The final decision is at state a and thus

$$g_a = \min\{(5 + g_b), (3 + g_c)\} = \min\{(5 + 14), (3 + 12)\} = 15$$

$$(a \rightarrow c)$$

(16.16)

Therefore, the minimum cost is 15 and it is achieved through path $a \rightarrow c \rightarrow e \rightarrow h \rightarrow m$, as shown in Figure 16.5

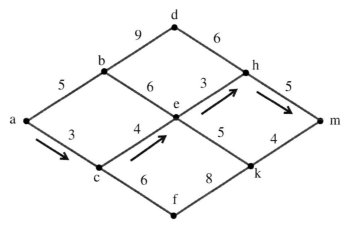

Figure 16.5. Line graph showing the minimum cost for Example 16.2

Example 16.3

On the line graph of Figure 16.6, node A represents an airport in New York City and nodes B through L several airports throughout Europe and Asia. All flights originate at A and fly eastward. A salesman must leave New York City and be in one of the airports H, J, K, or L at the shortest possible time. The encircled numbers represent waiting times in hours at each airport. The numbers in squares show the hours he must travel by an automobile to reach his destination, and the numbers beside the line segments indicated the flight times, also in hours. Which airport should he choose (H, J, K, or L) to minimize his total travel time, and in how many hours after departure from A will he reach his destination?

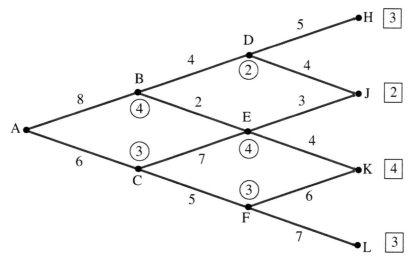

Figure 16.6. Line graph for Example 16.3

Solution:

The hours that the salesman must travel by automobile to reach his destination are

$$g_H = 3, \; g_J = 2, \; g_K = 4, \text{ and } g_L = 3 \tag{16.17}$$

The first decisions are made at D, E, and F. Then,

$$g_D = 2 + \min\{(5 + g_H), (4 + g_J)\} = 2 + \min\{(5 + 3), (4 + 2)\} = 2 + 6 = 8$$
$$D \rightarrow J \tag{16.18}$$

$$g_E = 4 + \min\{(3 + g_J), (4 + g_K)\} = 4 + \min\{(3 + 2), (4 + 4)\} = 4 + 5 = 9$$
$$E \rightarrow J \tag{16.19}$$

$$g_F = 3 + \min\{(6 + g_K), (7 + g_L)\} = 3 + \min\{(6 + 4), (7 + 3)\} = 3 + 10 = 13$$
$$F \rightarrow K \quad \text{or} \quad F \rightarrow L \tag{16.20}$$

The next decisions are made at B and C where we find that

$$g_B = 4 + \min\{(4 + g_D), (2 + g_E)\} = 4 + \min\{(4 + 8), (2 + 9)\} = 4 + 11 = 15$$
$$B \rightarrow E \tag{16.21}$$

$$g_C = 3 + \min\{(7 + g_E), (5 + g_F)\} = 3 + \min\{(7 + 9), (5 + 13)\} = 3 + 16 = 19$$
$$C \rightarrow E \tag{16.22}$$

The final decision is made at A, where we find

$$g_A = \min\{(8 + g_B), (6 + g_C)\} = \min\{(8 + 15), (6 + 19)\} = 23$$
$$A \rightarrow B \tag{16.23}$$

Therefore, the minimum cost (minimum time from departure to arrival at destination) is 23 hours and it is achieved through path $A \rightarrow B \rightarrow E \rightarrow J$, as shown in Figure 16.7.

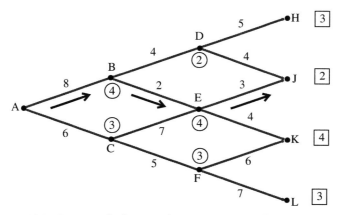

Figure 16.7. Line graph showing the minimum cost for Example 16.3

Example 16.4

A start–up, high–technology company, has $4,000,000 to invest in three different products A, B and C. Investments in each product are assumed to be multiples of $1,000,000 and the company may allocate all the money to just one product or split it between these three products. The expected profits are shown in Table 16.2.

TABLE 16.2 Amounts invested and return on investment for each product

Investments	Amount Invested				
	0	$1,000,000	$2,000,000	$3,000,000	$4,000,000
	Return on Investment				
Product A	0	$2,000,000	$5,000,000	$6,000,000	$7,000,000
Product B	0	$1,000,000	$3,000,000	$6,000,000	$7,000,000
Product C	0	$1,000,000	$4,000,000	$5,000,000	$8,000,000

How should the money be allocated so that company will realize the maximum profit?

Solution:

This problem can also be solved with linear programming methods but we will use the so called tabular form of solution. Let

$$p_i(x) \quad i = A, B, C \tag{16.24}$$

denote the profits in millions from product i, when x units of dollars are invested in it. For simplicity, we express the profits in millions, and we enter these in Table 16.3.

TABLE 16.3 Modified Table 16.2

x $p(x)$	0	1	2	3	4
$p_A(x)$	0	2	5	6	7
$p_B(x)$	0	1	3	6	7
$p_C(x)$	0	1	4	5	8

Our objective is to maximize the total profit z that represents the sum of the profits from each product, subject to the constraint that the amount invested does not exceed four million dollars. In other words is, we want to

$$\text{maximize } z = p_A(x) + p_B(x) + p_C(x) \tag{16.25}$$

subject to the constraint

$$x_A + x_B + x_C \leq 4 \qquad (16.26)$$

where x_A, x_B, and x_C, are the amounts to be invested in products A, B and C respectively.

The computations are done in three stages, one per product. We start by allocating units (millions) to Product C (*Stage C*), but since we do not know what units were allocated to the previous products A and B, we must consider all possibilities.

We let $v_j(u)$ denote the value of the optimum profit that can be achieved, where the subscript *j* indicates the number or stage assigned to the product, i.e., A for Product A, B for Product B, and C for Product C, and u represents the number of money units. Also, we let $d_j(u)$ be the decision that is being made to achieve the optimum value from $v_j(u)$.

At Stage C, j = C, and u = 4, i.e., 4 millions assumed to be allocated to Product C.

The possibilities that we allocate 0 or 1 or 2 or 3 or 4 units (millions) to Product C, and the corresponding returns are, from Table 16.3,

$$v_C(4) = \max\{p_C(0), p_C(1), p_C(2), p_C(3), p_C(4)\} = \max\{0, 1, 4, 5, 8\} = 8 \qquad (16.27)$$

with decision

$$d_C(4) = 8 \qquad (16.28)$$

that is, the maximum appears in the fourth position since the left most is the zero position.

The next possibility is that one unit was invested in either Product A or Product B, by a previous decision. In this case, do not have 4 units to invest in Product C; we have three or less.

If we invest the remaining three units in Product C, the optimum value $v_C(3)$ is found from

$$v_C(3) = \max\{p_C(0), p_C(1), p_C(2), p_C(3)\} = \max\{0, 1, 4, 5\} = 5 \qquad (16.29)$$

with decision

$$d_C(3) = 5 \qquad (16.30)$$

If we have only two units left, and we invest them in Product C, we obtain the maximum from

$$v_C(2) = \max\{p_C(0), p_C(1), p_C(2)\} = \max\{0, 1, 4\} = 4 \qquad (16.31)$$

with decision

$$d_C(2) = 4 \qquad (16.32)$$

With only one unit left to invest, we have

$$v_C(1) = \max\{p_C(0), p_C(1)\} = \max\{0, 1\} = 1 \qquad (16.33)$$

with decision

$$d_C(1) = 1 \qquad (16.34)$$

Finally, with no units left to invest in Product C,

$$v_C(0) = \max\{p_C(0)\} = \max\{0\} = 0 \qquad (16.35)$$

with decision

$$d_C(0) = 0 \qquad (16.36)$$

With these values, we construct Table 16.4.

TABLE 16.4 Optimum profit and decisions made for Stage C

		u				
		0	1	2	3	4
Stage C	$v_C(u)$	0	1	4	5	8
	$d_C(u)$	0	1	2	3	4
Stage B	$v_B(u)$	...	...	...	...	...
	$d_B(u)$	...	...	...	...	...
Stage A	$v_A(u)$	...	...	...	...	...
	$d_A(u)$	...	...	...	...	...

Next, we consider *Stage B*, and since we do not know what units were allocated to Product A (*Stage A*), again we must consider all possibilities.

With $j = B$ and $u = 4, 3, 2, 1$ and 0, we have

$$v_B(4) = \max\{p_B(0) + v_C(4-0), p_B(1) + v_C(4-1), p_B(2) \qquad (16.37)$$
$$+v_C(4-2), p_B(3) + v_C(4-3), p_B(4) + v_C(4-4)\}$$

This expression says that if zero units were invested in Product B, it is possible that all four units were invested in Product C, or if one unit was invested in Product B, it is possible that 3 units were invested in Product C, and so on. Inserting the appropriate values, we obtain

$$v_B(4) = \max\{0+8, 1+5, 3+4, 6+1, 7+0\} = 8 \qquad (16.38)$$

with decision

$$d_B(4) = 0 \qquad (16.39)$$

since the maximum value is the zero position term.

Using a similar reasoning, we have

$$v_B(3) = \max\{p_B(0) + v_C(3-0), p_B(1) + v_C(3-1), p_B(2) \\ + v_C(3-2), p_B(3) + v_C(3-3)\} \tag{16.40}$$

or

$$v_B(3) = \max\{0+5, 1+4, 3+1, 6+0\} = 6 \tag{16.41}$$

with decision

$$d_B(3) = 3 \tag{16.42}$$

Also,

$$v_B(2) = \max\{p_B(0) + v_C(2-0), p_B(1) + v_C(2-1), p_B(2) + v_C(2-2)\} \tag{16.43}$$

or

$$v_B(2) = \max\{0+4, 1+1, 3+0\} = 4 \tag{16.44}$$

with decision

$$d_B(2) = 0 \tag{16.45}$$

$$v_B(1) = \max\{p_B(0) + v_C(1-0), p_B(1) + v_C(1-1)\} \tag{16.46}$$

or

$$v_B(1) = \max\{0+1, 1+0\} = 1 \tag{16.47}$$

with decision

$$d_B(1) = 0 \tag{16.48}$$

if we consider the zero position term, or

$$d_B(1) = 1 \tag{16.49}$$

if we consider the first position term.

Also,

$$v_B(0) = \max\{p_B(0) + v_C(0-0)\} \tag{16.50}$$

or

$$v_B(0) = \max\{0+0\} = 0 \tag{16.51}$$

with decision

$$d_B(0) = 0 \tag{16.52}$$

Next, we update the previous table to include the *Stage B* values. These are shown in Table 16.5.

TABLE 16.5 Updated table to include Stage B values

		u				
		0	1	2	3	4
Stage C	$v_C(u)$	0	1	4	5	8
	$d_C(u)$	0	1	2	3	4
Stage B	$v_B(u)$	0	1	4	6	8
	$d_B(u)$	0	1	0	3	0
Stage A	$v_A(u)$	...	...	...	...	...
	$d_A(u)$	...	...	...	...	...

Finally, with $j = A$ and $u = 4$ *

$$v_A(4) = \max\{p_A(0) + v_B(4-0), p_A(1) + v_B(4-1), p_A(2) \atop + v_B(4-2), p_A(3) + v_B(4-3), p_A(4) + v_B(4-4)\} \qquad (16.53)$$

or

$$v_A(4) = \max\{0 + 8, 2 + 6, 5 + 4, 6 + 1, 7 + 0\} = 9 \qquad (16.54)$$

with decision

$$d_A(4) = 2 \qquad (16.55)$$

We complete the table by entering the values of Stage A in the last two rows as shown in Table 16.6. The only entries are in the last column, and this is always the case since in deriving $v_A(4)$ and $d_A(4)$, all possibilities have been considered.

Table 16.6 indicates that the maximum profit is realized with $v_A(4) = 9$, that is, 9 units, and thus the maximum profit is \$9,000,000.

To determine the investment allocations to achieve this profit, we start with $d_A(4) = 2$; this tells us that we should allocate 2 units to *Product* A, and the given table shows that 2 units (\$2,000,000) invested in this product will return \$5,000,000.

---

* Since this is the first stage, all 4 units can be allocated to the Product A or some of these can be allocated to Products B and C. Therefore, $v_A(4)$ considers all possibilities.

TABLE 16.6 Updated table to include values for all stages

		u				
		0	1	2	3	4
Stage C	$v_C(u)$	0	1	4	5	8
	$d_C(u)$	0	1	2	3	4
Stage B	$v_B(u)$	0	1	4	6	8
	$d_B(u)$	0	1	0	3	0
Stage A	$v_A(u)$	...	...	...	...	9
	$d_A(u)$	...	...	...	...	2

We now have two units left to invest in *Products* B and C. To find out where we should invest these units, we consider the decision at *Stage B*. Since two out of the four units have already been invested, we have $d_B(4 - 2) = d_B(2)$, and by reference to the Table 16.6, we see that $d_2(2) = 0$. This tells us that we should not invest any units in Product B if only two units are left. The decision at *Stage C* yields $d_C(4 - 0 - 2) = d_C(2)$, and from Table 16.6, $d_C(2) = 2$. This indicates that we should invest the remaining two units to *Product C* where we can obtain a return of $4,000,000.

In summary, to obtain the maximum profit of $9,000,000, we should allocate:

1. two units to *Product A* to earn $5,000,000

2. zero units to *Product B* to earn $0

3. two units to *Product C* to earn $4,000,000

16.3 Network Analysis

A *network*, as defined here, is a set of points referred to as *nodes* and a set of lines referred to as *branches*. Thus, Figure 16.8 is a network with 5 nodes A, B, C, D and E, and 6 branches AB, AC, AD, BD, BE, and CD.

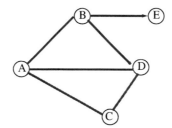

Figure 16.8. A typical network

Branches can be either *directed* (or *oriented*), if they have a direction assigned to them, that is, one–way, or two–way. If no direction is assigned, they are considered to be two–way. Thus, the branches BD and BE in Figure 16.8, are directed but the others are not.

A network is said to be *connected*, if there is a path (branch) connecting each pair of nodes. Thus, the network shown in Figure 16.8 is connected.

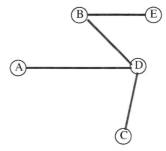

Figure 16.9. A network which is connected

The network of Figure 16.9 is also connected. However, the network of Figure 16.10 is not connected since the branch CD is removed.

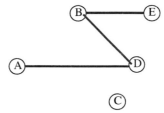

Figure 16.10. A network which is not connected

A *tree* is a *connected network* which has n branches and n + 1 nodes. For example, the network of Figure 16.11 is a tree network.

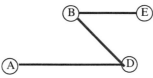

Figure 16.11. A tree network

Network analysis is a method that is used to solve *minimum span problems*. In such problems, we seek to find a tree which contains all nodes, and the sum of the costs (shortest total distance) is a minimum.

Example 16.5

Figure 16.12 represents a network for a project that requires telephone cable be installed to link 7 towns. The towns are the nodes, the branches indicate possible paths, and the numbers beside the branches, show the distance (not to scale) between towns in kilometers. Find the minimal spanning tree, that is, the least amount of telephone cable required to link each town.

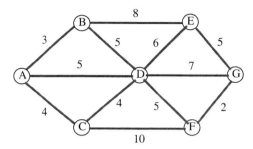

Figure 16.12. Network for Example 16.5

Solution:

For convenience, we redraw the given network with dotted lines as shown in Figure 16.13, and we arbitrarily choose A as the starting node.

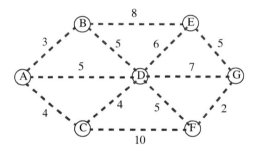

Figure 16.13. Network of Example 16.5 with no connections

We observe that there are 3 branches associated with node A, i.e., AB, AD, and AC. By inspection, or from the expression

$$\min\{AB = 3, AD = 5, AC = 4\} = AB = 3 \tag{16.56}$$

we find that branch AB is the shortest. We accept this branch as the first branch of the minimum span tree and we draw a solid line from Node A to Node B as shown in Figure 16.14.

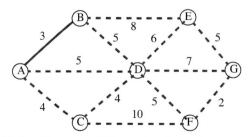

Figure 16.14. Network of Example 16.5 with first connection

Next, we consider all branches associated with Nodes A and B. We find that the minimum of these is

$$\min\{AD = 5, AC = 4, BD = 5, BE = 8\} = AC = 4 \tag{16.57}$$

and thus, AC is connected to the network as shown in Figure 16.15.

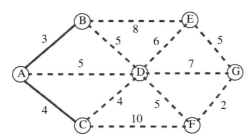

Figure 16.15. Network of Example 16.5 with the second connection

We continue by considering all branches associated with Nodes A, B and C, and we find that the shortest is

$$\min\{AD = 5, BE = 8, BD = 5, CD = 4, CF = 10\} = CD = 4 \tag{16.58}$$

and we add branch CD to the network shown in Figure 16.16. The dotted lines AD and BD have been removed since we no longer need to consider branch AD and BD, because Nodes B and D are already connected; otherwise, we will not have a tree network.

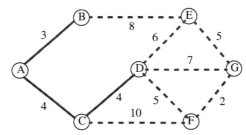

Figure 16.16. Network of Example 16.5 with the third connection

Next, considering all branches associated with Nodes B, C, and D and we find that the shortest is

$$\min\{BE = 8, DE = 6, DG = 7, DF = 5, CF = 10\} = DF = 5 \tag{16.59}$$

and the network now is connected as shown in Figure 16.17.

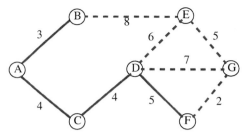

Figure 16.17. Network of Example 16.5 with the fourth connection

Continuing, we obtain

$$\min\{BE = 8, DE = 6, DG = 7\} = DE = 6 \tag{16.60}$$

and the network is connected as shown in Figure 16.18

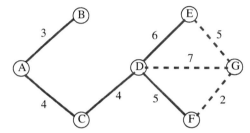

Figure 16.18. Network of Example 16.5 with the fifth connection

The last step is to determine the shortest branch to Node G. We find that

$$\min\{EG = 5, DG = 7, FG = 2\} = FG = 2 \tag{16.61}$$

and the complete minimum span tree is shown in Figure 16.19.

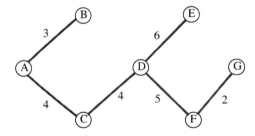

Figure 16.19. Network of Example 16.5 with all connections

Figure 16.19 shows that the minimum distance is $3 + 4 + 4 + 6 + 5 + 2 = 24$ kilometers.

16.4 Summary

- Linear programming is a procedure we follow to maximize or minimize a particular quantity, referred to as the objective, which is dependent on a finite number of variables. These variables may or may not be independent of each another, and in most cases are subject to certain conditions or limitations referred to as constraints.

- Dynamic Programming is based on R. Bellman's Principle of Optimality which states that an optimum policy has the property that whatever the initial state and the initial decisions are, the remaining decisions must constitute an optimum policy with regard to the state resulting from the first decision.

- A network, as defined in this chapter, is a set of points referred to as nodes and a set of lines referred to as branches.

- A tree is a connected network which has n branches and n + 1 nodes.

- Network analysis is a method that is used to solve *minimum span problems*. In such problems, we seek to find a tree which contains all nodes, and the sum of the costs (shortest total distance) is a minimum.

16.5 Exercises

1. A large oil distributor can buy *Grade A* oil which contains 7% lead for *$25.00* per barrel from one oil refinery company. He can also buy *Grade B* oil which contains 15% lead for *$20.00* per barrel from another oil refinery company. The Environmental Protection Agency (EPA) requires that all oil sold must not contain more than 10% lead. How many barrels of each grade of oil should he buy so that after mixing the two grades can minimize his cost while at the same time meeting EPA's requirement? Solve this problem graphically and check your answers with Excel's *Solver*.

2. Use dynamic programming to find the minimum cost route from state a to state m for the line graph shown below. The line segments are directed from left to right and the costs are indicated beside each line segment.

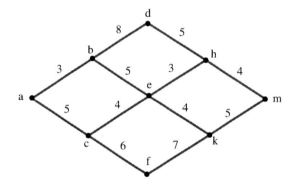

3. Repeat Example 16.3 for the line graph shown below.

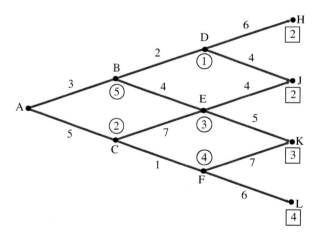

4. A salesman has 4 hours available to visit 4 of his customers. He will earn the commissions shown on the table below for various visiting times. Compute the optimal allocation of time that he should spent with his customers so that he will maximize the sum of his commissions. Consider only integer number of visiting hours, and ignore travel time from customer to customer. The third row (zero hours) indicates the commission that he will receive if he just calls instead of visiting them.

Visit Time (Hours)	Customer			
	1	2	3	4
0	$20	$40	$40	$80
1	$45	$45	$52	$91
2	$65	$57	$62	$95
3	$75	$61	$71	$97
4	$83	$69	$78	$98

5. Repeat Example 16.5 for the network shown below.

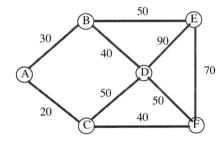

16.6 Solutions to End-of-Chapter Exercises

1.

Let x be the number of barrels of Grade A oil and y be the number of barrels of Grade B oil. The objective is to minimize $z = \$25.00x + \$20.00y$ or, for simplicity,

$$z = 25x + 20y \quad (1)$$

We want to minimize (1) because it represents a cost, not a profit.

Each barrel to be sold must not contain more than 10% lead and since Grade A contains 7% and Grade B 15%, we must have

$$0.07x + 0.15y \leq 0.10 \quad (2)$$

The oil of Grade A and Grade B used in each barrel to be sold must be equal to unity. Thus,

$$x + y = 1 \quad (3)$$

Moreover, x and y cannot be negative numbers, therefore

$$x \geq 0 \qquad y \geq 0 \quad (4)$$

The problem then can be stated as:

Minimize (1) subject to the constraints of (2), (3), and (4). To determine the feasible region we plot (2) and (3) where the x and y crossings are found by first setting $x = 0$ and then $y = 0$. Thus from (2), if $x = 0$, $y = 0.10/0.15 = 2/3$ and if $y = 0$, $x = 0.10/0.07 = 10/7$. Likewise, from (3), if $x = 0$, $y = 1$, and if $y = 0$, $x = 1$.

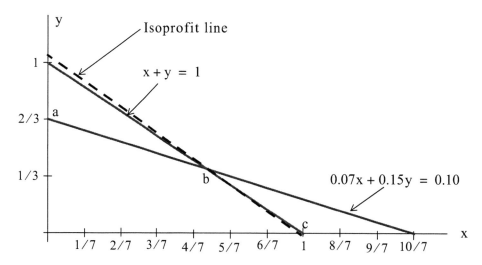

The isoprofit line passes through point b and its coordinates are found by simultaneous solution of (2) and (3). For convenience, we use the following MATLAB script:

syms x y; [x,y]=solve(0.07*x+0.15*y–0.10, x+y–1)

x =
  5/8

y =
  3/8

Therefore, the distributor should buy Grade A oil at the ratio $x = 5/8$ and Grade B at the ratio $y = 3/8$ and by substitution into (1)

$$z = 25 \times \frac{5}{8} + 20 \times \frac{3}{8} = \frac{185}{8} = \$23.125$$

and this represents his cost per barrel. The isoprofit line is

$$z = 25x + 20y = 23.125$$

and the y–intercept is found by setting x in the equation above to zero and we find that

$$y - intercept = 23.125/20 = 1.1563$$

Check with Excel's Solver:

	A	B	C	D	E	F
1	**Optimization - Minimum Cost for Exercise 16.1**					
2	1. Enter zeros in B12 and B13		2. In B15 enter =25*B12+20*B13			
3	3. In B17 enter =0.07*B12+0.15*B13 and in B18 =B12+B13					
4	4. From the *Tools* drop menu select *Solver*. Use *Add-Ins* if necessary to add it.					
5	5. On the Solver Parameters screen enter the following:					
6	*Set Target Cell:* B15					
7	*Equal to:* Min					
8	*By Changing Cells:* B12:B13					
9	*Click on Add and enter Constraints:*					
10	B12>=0, Add B13>=0,					
11	Add B17<=0.10, Add B18=1, OK, Solve					
12	Grade A=	0.625002				
13	Grade B=	0.374999				
14						
15	Minimum Cost=	$23.125				
16						
17	Lead Content=	0.10				
18	Grade A + Grade B=	1.000001				
19						
20	Note: Contents of A12:A18 are typed-in for information only					

2.

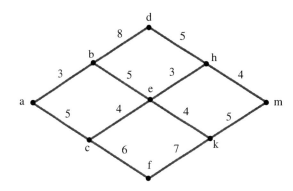

Since the segments are directed from left to right, state e is the first node where a decision must be made and thus

$$g_e = \min\{3 + g_h, 4 + g_k\} = \min\{3 + 4, 4 + 5\} = 7$$

Therefore, the best route from state e to state m passes through state h. Next,

$$g_b = \min\{8 + g_d, 5 + g_e\} = \min\{8 + 5, 5 + 7\} = 12$$

and

$$g_c = \min\{4 + g_e, 6 + g_f\} = \min\{4 + 7, 6 + 7\} = 11$$

Finally,

$$g_a = \min\{3 + g_b, 5 + g_c\} = \min\{3 + 12, 5 + 11\} = 15$$

Thus, the best (shortest) path is $a \rightarrow b \rightarrow e \rightarrow h \rightarrow m$

3.

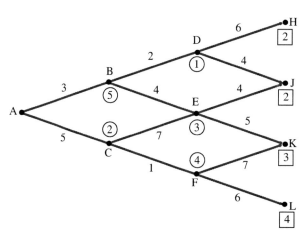

The numbers in circles represent the waiting time at these nodes.

Last stage: $g_H = 2$, $g_J = 2$, $g_K = 3$, and $g_L = 4$

Next stage to the left:

$$g_D = 1 + \min\{6 + g_H, 4 + g_J\} = 1 + 6 = 7 \qquad \text{from D to J}$$

$$g_E = 3 + \min\{4 + g_J, 5 + g_K\} = 3 + 6 = 9 \qquad \text{from E to J}$$

$$g_F = 4 + \min\{7 + g_K, 6 + g_L\} = 4 + 10 = 14 \qquad \text{from F to K or F to L}$$

Next to initial stage:

$$g_B = 5 + \min\{2 + g_D, 4 + g_E\} = 5 + 9 = 14 \qquad \text{from B to D}$$

$$g_C = 2 + \min\{7 + g_E, 1 + g_F\} = 2 + 15 = 17 \qquad \text{from C to F}$$

Initial stage:

$$g_A = 0 + \min\{3 + g_B, 5 + g_C\} = 17 \qquad \text{from A to B}$$

Therefore, minimum path is from $A \rightarrow B \rightarrow D \rightarrow J$ and the numerical minimum cost is 17.

4.

	A	B	C	D	E	F	G	H
1	Exercise 16.4 - Solution							
2	A salesman has 4 hours available to visit of his customers. He will earn the commissions shown on							
3	the table below for various visiting times. Compute the optimal allocation of time that he should spend							
4	with his customers so that he will maximize the sum of his commissions. Consider only integer number							
5	of visiting hours, and ignore travel time from customer to customer. The third row (zero hours)							
6	indicates the commission that he will receive if he just calls instead of visiting them.							
7								
8	Visit Time		Customer					
9	(Hours)	1	2	3	4			
10	0	$20	$40	$40	$80			
11	1	$45	$45	$52	$91			
12	2	$65	$57	$62	$95			
13	3	$75	$61	$71	$97			
14	4	$83	$69	$78	$98			
15								
16	Solution							
17	We will follow the same method as in Example 16.4							
18								
19	It is convenient to rearrange the table as shown below.							
20								

continued on the next page

	A	B	C	D	E	F	G	H	I
21				u					
22	f(x) \ x	0	1	2	3	4			
23	f1(x)	20	45	65	75	83			
24	f2(x)	40	45	57	61	69			
25	f3(x)	40	52	62	71	78			
26	f4(x)	80	91	95	97	98			
27									
28	m4(4)=	max(B26,C26,D26,E26,F26)					98	with d4(4)=	4
29	m4(3)=	max(B26,C26,D26,E26)					97	with d4(3)=	3
30	m4(2)=	max(B26,C26,D26)					95	with d4(2)=	2
31	m4(1)=	max(B26,C26)					91	with d4(1)=	1
32	m4(1)=	max(B26)					80	with d4(0)=	0
33									
34	The values of m4(u) and d4(u) are entered in the table below.								
35									
36				u					
37		0	1	2	3	4			
38	m4(u)	80	91	95	97	98			
39	d4(u)	0	1	2	3	4			
40	m3(u)	120	132	143	153	162			
41	d3(u)	0	1	1	2	3			
42	m2(u)	160	172	183	193	202			
43	d2(u)	0	0	0	0	0			
44	m1(u)					248			
45	d1(u)					2			
46									
47	**Next, we compute the values of m3(u) and d3(u)**								
48									
49	m3(4)=max[f3(0)+m4(4-0), f3(1)+m4(4-1), f3(2)+m4(4-2), f3(3)+m4(4-3), f3(4)+m4(4-4)]								
50		MAX(B25+F38,C25+E38,D25+D38,**E25+C38**,F25+B38)					162	with d3(4)=	3
51									
52	m3(3)=max[f3(0)+m4(3-0), f3(1)+m4(3-1), f3(2)+m4(3-2), f3(3)+m4(3-3)]								
53		MAX(B25+E38,C25+D38,**D25+C38**,E25+B38)					153	with d3(3)=	2
54									
55	m3(2)=max[f3(0)+m4(2-0), f3(1)+m4(2-1), f3(2)+m4(2-2)]								
56		MAX(B25+D38,**C25+C38**,D25+B38)					143	with d3(2)=	1
57									
58	m3(1)=max[f3(0)+m4(1-0), f3(1)+m4(1-1)]								
59		MAX(B25+C38,**C25+B38**)					132	with d3(1)=	1
60									
61	m3(0)=max[f3(0)+m4(0-0)]								
62		MAX(B25+B38)					120	with d3(0)=	0

continued on the next page.

	A	B	C	D	E	F	G	H	I
63									
64	These values are now added to the table above, Rows 40 and 41								
65									
66	Similarly, we compute the values of m2(u) and d2(u)								
67									
68	m2(4)=max[f2(0)+m3(4-0), f2(1)+m3(4-1), f2(2)+m3(4-2), f2(3)+m3(4-3), f2(4)+m3(4-4)]								
69		MAX(**B24+F40**,C24+E40,D24+D40,E24+C40,F25+B40)					202	with d2(4)=	0
70									
71	m2(3)=max[f2(0)+m3(3-0), f2(1)+m3(3-1), f2(2)+m3(3-2), f2(3)+m3(3-3)]								
72		MAX(**B24+E40**,C24+D40,D24+C40,E24+B40)					193	with d2(3)=	0
73									
74	m2(2)=max[f2(0)+m3(2-0), f2(1)+m3(2-1), f2(2)+m3(2-2)]								
75		MAX(**B24+D40**,C24+C40,D24+B40)					183	with d2(2)=	0
76									
77	m2(1)=max[f2(0)+m3(1-0), f2(1)+m3(1-1)]								
78		MAX(**B24+C40**,C24+B40)					172	with d2(1)=	0
79									
80	m2(0)=max[f2(0)+m3(0-0)]								
81		MAX(**B24+B40**)					160	with d2(0)=	0
82									
83	These values are added to the table above, Rows 42 and 43								
84									
85	Stage 1 is the last stage and there is only one state associated with it, u=4, and thus								
86									
87	m1(4)=max[f1(0)+m2(4-0), f1(1)+m2(4-1), f1(2)+m2(4-2), f1(3)+m2(4-3), f1(4)+m2(4-4)]								
88		MAX(B23+F42,C23+E42,**D23+D42**,E23+C42,F23+B42)					248	with d1(4)=	2
89									
90	These two values are the last entries into the table in Cells F44 and F45								
91									
92	The table (Rows 36 through 45) indicates that, to achieve the maximum sum of commissions,								
93	the salesman should spend 2 hours with Customer #1 (d1(4)=2), 0 hours with Customer #2								
94	(d2(4)=0), and for the remaining 2 hours he should spend 1 hour with Customer #3 and 1								
95	hour with Customer #4.								
96									
97	**Check:**								
98		Customer #1, 2 hours =			$65				
99		Customer #2, 0 hours =			$40				
100		Customer #3, 1 hour =			$52				
101		Customer #4, 1 hour =			$91				
102			Total =		**$248**				

5.

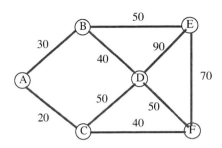

$$\min[A \rightarrow B, \ A \rightarrow C] = \min[30, 20] = 20 \qquad A \rightarrow C$$

$$\min[A \rightarrow B, \ C \rightarrow D, \ C \rightarrow F] = \min[30, 50, 40] = 30 \qquad A \rightarrow B$$

$$\min[B \rightarrow D, \ B \rightarrow E, \ C \rightarrow D, \ C \rightarrow F] = \min[40, 50, 50, 40] = 40 \qquad \text{choose } B \rightarrow D$$

$$\min[B \rightarrow E, \ D \rightarrow E, \ D \rightarrow F, \ C \rightarrow F] = \min[50, 90, 50, 40] = 40 \qquad C \rightarrow F$$

$$\min[B \rightarrow E, \ D \rightarrow E, \ F \rightarrow E] = \min[50, 90, 70] = 50 \qquad B \rightarrow E$$

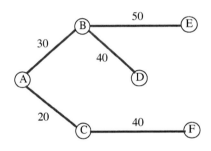

and thus the minimum distance is $20 + 30 + 40 + 40 + 50 = 180$ kilometers

Appendix A

Difference Equations in Discrete-Time Systems

This appendix is a treatment of linear difference equations with constant coefficients and it is confined to first– and second–order difference equations and their solution. Higher–order difference equations of this type and their solution is facilitated with the $\mathcal{Z}$–transform.[*]

A.1 Recursive Method for Solving Difference Equations

In mathematics, a *recursion* is an expression, such as a polynomial, each term of which is determined by application of a formula to preceding terms. The solution of a difference equation is often obtained by recursive methods. An example of a recursive method is Newton's method[†] for solving non–linear equations. While recursive methods yield a desired result, they do not provide a **closed–form** solution. If a closed–form solution is desired, we can solve difference equations using the Method of Undetermined Coefficients, and this method is similar to the classical method of solving linear differential equations with constant coefficients. This method is described in the next section.

A.2 Method of Undetermined Coefficients

A second–order difference equation has the form

$$y(n) + a_1 y(n-1) + a_2(n-2) = f(n) \qquad (A.1)$$

where a_1 and a_2 are constants and the right side is some function of n. This difference equation expresses the output $y(n)$ at time n as the linear combination of two previous outputs $y(n-1)$ and $y(n-2)$. The right side of relation (A.1) is referred to as the **forcing function**. The general (closed-form) solution of relation (A.1) is the same as that used for solving second–order differential equations. The three steps are as follows:

1. Obtain the **natural response** (complementary solution) $y_C(n)$ in terms of two arbitrary real constants k_1 and k_2, where a_1 and a_2 are also real constants, that is,

$$y_C(n) = k_1 a_1^n + k_2 a_2^n \qquad (A.2)$$

2. Obtain the forced response (particular solution) $y_P(n)$ in terms of an arbitrary real constant k_3,

---

[*] *For an introduction and applications of the $\mathcal{Z}$-transform please refer to Signals and Systems with MATLAB Computing and Simulink Modeling, Third Edition, ISBN 0-9744239-9-8.*

[†] *Newton's method is discussed in Chapter 2.*

that is,

$$y_P(n) = k_3 a_3^n \tag{A.3}$$

where the right side of (A.3) is chosen with reference to Table A.1.[*]

TABLE A.1 *Forms of the particular solution for different forms of the forcing function*

Form of forcing function	Form of particular solution[a]
Constant	k — a constant
an^k — a is a constant	$k_0 + k_1 n + k_2 n^2 + \ldots + k_k n^k$ — k_i is constant
$ab^{\pm n}$ — a and b are constants	Expression proportional to $b^{\pm n}$
$a\cos(n\omega)$ or $a\sin(n\omega)$	$k_1 \cos(n\omega) + k_2 \sin(n\omega)$

   a. As in the case with the solutions of ordinary differential equations with constant coefficients, we must remember that if $f(n)$ is the sum of several terms, the most general form of the particular solution $y_P(n)$ is the linear combination of these terms. Also, if a term in $y_P(n)$ is a duplicate of a term in the complementary solution $y_C(n)$, we must multiply $y_P(n)$ by the lowest power of n that will eliminate the duplication.

3. Add the natural response (complementary solution) $y_C(n)$ and the forced response (particular solution) $y_P(n)$ to obtain the total solution, that is,

$$y(n) = y_C(n) + y_P(n) = k_1 a_1^n + k_2 a_2^n + y_P(n) \tag{A.4}$$

4. Solve for k_1 and k_2 in (A.4) using the given initial conditions. It is important to remember that the constants k_1 and k_2 must be evaluated from the total solution of (A.4), not from the complementary solution $y_C(n)$.

It is best to illustrate the Method of Undetermined Coefficients via examples.

Example A.1

Find the total solution for the second–order difference equation

$$y(n) - \frac{5}{6}y(n-1) + \frac{1}{6}y(n-2) = 5^{-n} \qquad n \geq 0 \tag{A.5}$$

subject to the initial conditions $y(-2) = 25$ and $y(-1) = 6$.

---

[*] *Ordinary differential equations with constant coefficients are discussed in Chapter 5.*

Solution:

1. We assume that the complementary solution $y_C(n)$ has the form

$$y_C(n) = k_1 a_1^n + k_2 a_2^n \qquad \text{(A.6)}$$

The homogeneous equation of (A.5) is

$$y(n) - \frac{5}{6}y(n-1) + \frac{1}{6}y(n-2) = 0 \qquad n \geq 0 \qquad \text{(A.7)}$$

Substitution of $y(n) = a^n$ into (A.7) yields

$$a^n - \frac{5}{6}a^{n-1} + \frac{1}{6}a^{n-2} = 0 \qquad \text{(A.8)}$$

Division of (A.8) by a^{n-2} yields

$$a^2 - \frac{5}{6}a + \frac{1}{6} = 0 \qquad \text{(A.9)}$$

The roots of (A.9) are

$$a_1 = \frac{1}{2} \qquad a_2 = \frac{1}{3} \qquad \text{(A.10)}$$

and by substitution into (A.6) we obtain

$$y_C(n) = k_1 \left(\frac{1}{2}\right)^n + k_2 \left(\frac{1}{3}\right)^n = k_1 2^{-n} + k_2 3^{-n} \qquad \text{(A.11)}$$

2. Since the forcing function is 5^{-n}, we assume that the particular solution is

$$y_P(n) = k_3 5^{-n} \qquad \text{(A.12)}$$

and by substitution into (A.5),

$$k_3 5^{-n} - k_3 \left(\frac{5}{6}\right) 5^{-(n-1)} + k_3 \left(\frac{1}{6}\right) 5^{-(n-2)} = 5^{-n}$$

Division of both sides by 5^{-n} yields

$$k_3 \left[1 - \left(\frac{5}{6}\right) 5 + \left(\frac{1}{6}\right) 5^2 \right] = 1$$

or $k_3 = 1$ and thus

$$y_P(n) = 5^{-n} \qquad \text{(A.13)}$$

The total solution is the addition of (A.11) and (A.13), that is,

$$y(n) = y_C(n) + y_P(n) = k_1 2^{-n} + k_2 3^{-n} + 5^{-n} \qquad \text{(A.14)}$$

To evaluate the constants k_1 and k_2 we use the given initial conditions, i.e., s $y(-2) = 25$ and $y(-1) = 6$. For $n = -2$, (A.14) reduces to

$$y(-2) = k_1 2^2 + k_2 3^2 + 5^2 = 25$$

from which

$$4k_1 + 9k_2 = 0 \qquad \text{(A.15)}$$

For $n = -1$, (A.14) reduces to

$$y(-1) = k_1 2^1 + k_2 3^1 + 5^1 = 6$$

from which

$$2k_1 + 3k_2 = 1 \qquad \text{(A.16)}$$

Simultaneous solution of (A.15) and (A.16) yields

$$k_1 = \frac{3}{2} \qquad k_2 = -\frac{2}{3} \qquad \text{(A.17)}$$

and by substitution into (A.14) we obtain the total solution as

$$y(n) = y_C(n) + y_P(n) = \left(\frac{3}{2}\right)2^{-n} + \left(-\frac{2}{3}\right)3^{-n} + 5^{-n} \qquad n \geq 0 \qquad \text{(A.18)}$$

To plot this difference equation for the interval $0 \leq n \leq 10$, we use the following MATLAB script:

```
n=0:1:10; yn=1.5.*2.^(-n)-(2./3).*3.^(-n)+5.^(-n); stem(n,yn); grid
```

The plot is shown in Figure A.1.

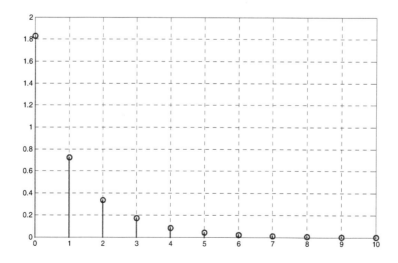

Figure A.1. Plot for the difference equation of Example A.1

Example A.2

Find the total solution for the second–order difference equation

$$y(n) - \frac{3}{2}y(n-1) + \frac{1}{2}y(n-2) = 1 + 3^{-n} \qquad n \geq 0 \qquad\qquad (A.19)$$

subject to the initial conditions $y(-2) = 0$ and $y(-1) = 2$

Solution:

1. We assume that the complementary solution $y_C(n)$ has the form

$$y_C(n) = k_1 a_1^n + k_2 a_2^n \qquad\qquad (A.20)$$

The homogeneous equation of (A.19) is

$$y(n) - \frac{3}{2}y(n-1) + \frac{1}{2}y(n-2) = 0 \qquad n \geq 0 \qquad\qquad (A.21)$$

Substitution of $y(n) = a^n$ into (A.21) yields

$$a^n - \frac{3}{2}a^{n-1} + \frac{1}{2}a^{n-2} = 0 \qquad\qquad (A.22)$$

Division of (A.22) by a^{n-2} yields

$$a^2 - \frac{3}{2}a + \frac{1}{2} = 0 \qquad\qquad (A.23)$$

The roots of (A.23) are

$$a_1 = \frac{1}{2} \qquad a_2 = 1 \qquad\qquad (A.24)$$

and by substitution into (A.20) we obtain

$$y_C(n) = k_1\left(\frac{1}{2}\right)^n + k_2(1)^n = k_1 2^{-n} + k_2 \qquad\qquad (A.25)$$

2. Since the forcing function is $1 + 3^{-n}$, in accordance with the first and third rows of Table A.1, we would assume that the particular solution is

$$y_P(n) = k_3 + k_4 3^{-n} \qquad\qquad (A.26)$$

However, we observe that both relations (A.25) and (A.26) contain common terms, that is, the constants k_2 and k_3. To avoid the duplication, we choose the particular solution as

$$y_P(n) = k_3 n + k_4 3^{-n} \qquad\qquad (A.27)$$

and by substitution of (A.27) into (A.19) we obtain

$$k_3 n + k_4 3^{-n} - \left(\frac{3}{2}\right) k_3 (n-1) - \left(\frac{3}{2}\right) k_4 3^{-(n-1)} + \frac{1}{2} k_3 (n-2) + \left(\frac{1}{2}\right) k_4 3^{-(n-2)} = 1 + 3^{-n}$$

$$k_3 n + k_4 3^{-n} - \left(\frac{3}{2}\right) k_3 n + \left(\frac{3}{2}\right) k_3 - \left(\frac{9}{2}\right) k_4 3^{-n} + \frac{1}{2} k_3 n - k_3 + \left(\frac{9}{2}\right) k_4 3^{-n} = 1 + 3^{-n}$$

$$k_4 3^{-n} + \left(\frac{3}{2}\right) k_3 - k_3 = 1 + 3^{-n}$$

Equating like terms, we obtain

$$\left(\frac{3}{2}\right) k_3 - k_3 = 1$$

$$k_4 3^{-n} = 3^{-n}$$

and after simplification,

$$k_3 = 2 \qquad k_4 = 1$$

By substitution into (A.27),

$$y_P(n) = 2n + 3^{-n} \tag{A.28}$$

The total solution is the addition of (A.25) and (A.28), that is,

$$y(n) = y_C(n) + y_P(n) = k_1 2^{-n} + k_2 + 2n + 3^{-n} \tag{A.29}$$

To evaluate the constants k_1 and k_2 we use the given initial conditions, i.e., s $y(-2) = 0$ and $y(-1) = 2$. For $n = -2$, (A.29) reduces to

$$y(-2) = k_1 2^2 + k_2 - 4 + 9 = 0$$

from which

$$4k_1 + k_2 = -5 \tag{A.30}$$

For $n = -1$, (A.29) reduces to

$$y(-1) = k_1 2^1 + k_2 - 2 + 3^1 = 2$$

from which

$$2k_1 + k_2 = 1 \tag{A.31}$$

Simultaneous solution of (A.30) and (A.31) yields

$$k_1 = -3 \qquad k_2 = 7 \tag{A.32}$$

and by substitution into (A.29) we obtain the total solution as

$$y(n) = y_C(n) + y_P(n) = (-3)2^{-n} + 7 + 2n + 3^{-n} \qquad n \geq 0 \qquad \text{(A.33)}$$

To plot this difference equation for the interval $0 \leq n \leq 10$, we use the following MATLAB script:

```
n=0:1:10; yn=(−3).*2.^(−n)+7+2.*n+3.^(−n); stem(n,yn); grid
```

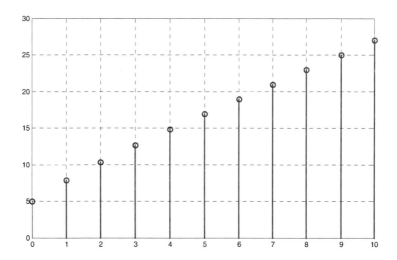

Figure A.2. Plot for the difference equation of Example A.2

Example A.3

Find the total solution for the first-order difference equation

$$y(n) - 0.9y(n-1) = 0.5 + (0.9)^{n-1} \qquad n \geq 0 \qquad \text{(A.34)}$$

subject to the initial condition $y(-1) = 5$

Solution:

1. We assume that the complementary solution $y_C(n)$ has the form

$$y_C(n) = k_1 a^n \qquad \text{(A.35)}$$

The homogeneous equation of (A.34) is

$$y(n) - 0.9y(n-1) = 0 \qquad n \geq 0 \qquad \text{(A.36)}$$

Substitution of $y(n) = a^n$ into (A.35) yields

$$a^n - 0.9a^{n-1} = 0 \qquad \text{(A.37)}$$

Division of (A.37) by a^{n-1} yields

$$a - 0.9 = 0$$

$$a = 0.9 \tag{A.38}$$

and by substitution into (A.35) we obtain

$$y_C(n) = k_1(0.9)^n \tag{A.39}$$

2. Since the forcing function is $0.5 + (0.9)^{n-1}$, in accordance with the first and third rows of Table A.1, we would assume that the particular solution is

$$y_P(n) = k_2 + k_3(0.9)^n \tag{A.40}$$

However, we observe that both relations (A.39) and (A.40) contain common terms, that is, the constants $k_1(0.9)^n$ and $k_3(0.9)^n$. To avoid the duplication, we choose the particular solution as

$$y_P(n) = k_2 + k_3 n(0.9)^n \tag{A.41}$$

and by substitution of (A.41) into (A.34) we obtain

$$k_2 + k_3 n(0.9)^n - 0.9 k_2 - 0.9 k_3(n-1)(0.9)^{(n-1)} = 0.5 + (0.9)^{n-1}$$

$$0.1 k_2 + k_3 n(0.9)^n - 0.9 k_3 n(0.9)^{(n-1)} + 0.9 k_3(0.9)^{(n-1)} = 0.5 + (0.9)^{n-1}$$

$$0.1 k_2 + k_3 n(0.9)^n - 0.9 k_3 n(0.9)^n 0.9^{-1} + 0.9 k_3(0.9)^n 0.9^{-1} = 0.5 + (0.9)^{n-1}$$

$$0.1 k_2 + k_3 n(0.9)^n - k_3 n(0.9)^n + k_3(0.9)^n = 0.5 + (0.9)^{n-1} = 0.5 + (0.9)^{-1}(0.9)^n$$

Equating like terms, we obtain

$$0.1 k_2 = 0.5$$

$$k_3(0.9)^n = (0.9)^{-1}(0.9)^n$$

and after simplification,

$$k_2 = 5 \qquad k_3 = \frac{10}{9}$$

By substitution into (A.41),

$$y_P(n) = 5 + \frac{10}{9} n(0.9)^n \tag{A.42}$$

The total solution is the addition of (A.39) and (A.42), that is,

$$y(n) = y_C(n) + y_P(n) = k_1(0.9)^n + \frac{10}{9} n(0.9)^n + 5 \tag{A.43}$$

To evaluate the constant k_1 we use the given initial condition, i.e., $y(-1) = 5$. For $n = -1$, (A.43) reduces to

$$y(-1) = k_1(0.9)^{-1} + \frac{10}{9}(-1)(0.9)^{-1} + 5 = 5$$

$$\frac{10}{9}k_1 - \frac{100}{81} = 0$$

from which

$$k_1 = \frac{10}{9} \qquad (A.44)$$

and by substitution into (A.43) we obtain the total solution as

$$y(n) = (0.9)^{n-1} + n(0.9)^{n-1} + 5$$

$$y(n) = (n+1)(0.9)^{n-1} + 5 \qquad n \geq 0 \qquad (A.45)$$

To plot this difference equation for the interval $0 \leq n \leq 10$, we use the following MATLAB script:

```
n=0:1:10; yn=(n+1).*(0.9).^(n-1)+5; stem(n,yn); grid
```

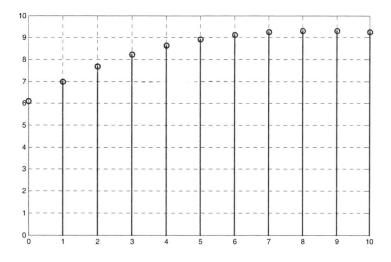

Figure A.3. Plot for the difference equation of Example A.3

Example A.4

Find the total solution for the second–order difference equation

$$y(n) - 1.8y(n-1) + 0.81y(n-2) = 2^{-n} \qquad n \geq 0 \qquad (A.46)$$

subject to the initial conditions $y(-2) = 25$ and $y(-1) = 6$

Solution:

No initial conditions are given and thus we will express the solution in terms of the unknown constants.

1. We assume that the complementary solution $y_C(n)$ has the form

$$y_C(n) = k_1 a_1^n + k_2 a_2^n \tag{A.47}$$

The homogeneous equation of (A.46) is

$$y(n) - 1.8y(n-1) + 0.81y(n-2) = 0 \qquad n \geq 0 \tag{A.48}$$

Substitution of $y(n) = a^n$ into (A.48) yields

$$a^n - 1.8a^{n-1} + 0.81a^{n-2} = 0 \tag{A.49}$$

Division of (A.49) by a^{n-2} yields

$$a^2 - 1.8a + 0.81 = 0 \tag{A.50}$$

The roots of (A.50) are repeated roots, that is,

$$a_1 = a_2 = 0.9 \tag{A.51}$$

and as in the case of ordinary differential equations, we accept the complementary solution to be of the form

$$y_C(n) = k_1(0.9)^n + k_2 n(0.9)^n \tag{A.52}$$

2. Since the forcing function is 2^{-n}, we assume that the particular solution is

$$y_P(n) = k_3 2^{-n} \tag{A.53}$$

and by substitution into (A.46),

$$k_3 2^{-n} - k_3(1.8)2^{-(n-1)} + k_3(0.81)2^{-(n-2)} = 2^{-n}$$

Division of both sides by 2^{-n} yields

$$k_3[1 - (1.8)2 + (0.81)2^2] = 1$$

$$k_3[1 - 3.6 + 3.24] = 1$$

$$k_3 = \frac{1}{0.64} = \frac{25}{16}$$

and thus

$$y_P(n) = \left(\frac{25}{16}\right)2^{-n} \tag{A.54}$$

The total solution is the addition of (A.52) and (A.54), that is,

$$y(n) = y_C(n) + y_P(n) = k_1(0.9)^n + k_2 n(0.9)^n + \left(\frac{25}{16}\right)2^{-n} \tag{A.55}$$

Example A.5

For the second–order difference equation

$$y(n) - 1.8y(n-1) + 0.81y(n-2) = (0.9)^n \qquad n \geq 0 \tag{A.56}$$

what would be the appropriate choice for the particular solution?

Solution:

This is the same difference equation as that of Example A.4 where the forcing function is $(0.9)^n$ instead of 2^{-n} where we found that the complementary solution is

$$y_C(n) = k_1(0.9)^n + k_2 n(0.9)^n \tag{A.57}$$

Row 3 in Table A.1 indicates that a good choice for the particular solution would be $k_3(0.9)^n$. But this is of the same form as the first term on the right side of (A.57). The next choice would be a term of the form $k_3 n(0.9)^n$ but this is of the same form as the second term on the right side of (A.57). Therefore, the proper choice would be

$$y_P(n) = k_3 n^2 (0.9)^n \tag{A.58}$$

Example A.6

Find the particular solution for the first-order difference equation

$$y(n) - 0.5y(n-1) = \sin\left(\frac{n\pi}{2}\right) \qquad n \geq 0 \tag{A.59}$$

Solution:

From Row 4 in Table A.1 we see that for a sinusoidal forcing function, the particular solution has the form

$$y_P(n) = k_1 \sin\left(\frac{n\pi}{2}\right) + k_2 \cos\left(\frac{n\pi}{2}\right) \tag{A.60}$$

---

and by substitution of (A.60) into (A.59)

$$k_1 \sin\left(\frac{n\pi}{2}\right) + k_2 \cos\left(\frac{n\pi}{2}\right) - 0.5 k_1 \sin\left[\frac{(n-1)\pi}{2}\right] - 0.5 k_2 \cos\left[\frac{(n-1)\pi}{2}\right] = \sin\left(\frac{n\pi}{2}\right)$$

$$k_1 \sin\left(\frac{n\pi}{2}\right) + k_2 \cos\left(\frac{n\pi}{2}\right) - 0.5 k_1 \sin\left[\frac{n\pi}{2} - \frac{\pi}{2}\right] - 0.5 k_2 \cos\left[\frac{n\pi}{2} - \frac{\pi}{2}\right] = \sin\left(\frac{n\pi}{2}\right) \tag{A.61}$$

From trigonometry,

$$\sin\left(\theta - \frac{\pi}{2}\right) = -\cos\theta$$

$$\cos\left(\theta - \frac{\pi}{2}\right) = \sin\theta$$

Then,

$$\sin\left[\frac{n\pi}{2} - \frac{\pi}{2}\right] = -\cos\left(\frac{n\pi}{2}\right)$$

$$\cos\left[\frac{n\pi}{2} - \frac{\pi}{2}\right] = \sin\left(\frac{n\pi}{2}\right)$$

and by substitution into (A.61)

$$k_1 \sin\left(\frac{n\pi}{2}\right) + k_2 \cos\left(\frac{n\pi}{2}\right) + 0.5 k_1 \cos\left(\frac{n\pi}{2}\right) - 0.5 k_2 \sin\left(\frac{n\pi}{2}\right) = \sin\left(\frac{n\pi}{2}\right) \tag{A.62}$$

Equating like terms, we obtain

$$k_1 - 0.5 k_2 = 1 \tag{A.63}$$

$$0.5 k_1 + k_2 = 0 \tag{A.64}$$

and simultaneous solution of (A.63) and (A.64) yields

$$k_1 = \frac{4}{5} \qquad k_2 = -\frac{2}{5}$$

Therefore, the particular solution of (A.59) is

$$y_P(n) = \frac{4}{5}\sin\left(\frac{n\pi}{2}\right) - \frac{2}{5}\cos\left(\frac{n\pi}{2}\right) \tag{A.65}$$

Appendix B

$\mathbf{T}$his appendix is a brief introduction to Simulink. This author feels that we can best introduce Simulink with a few examples. Some familiarity with MATLAB is essential in understanding Simulink, and for this purpose, it is highly recommended that the novice to MATLAB reader reviews Chapter 1 which serves as an introduction to MATLAB.

B.1 Simulink and its Relation to MATLAB

The MATLAB® and Simulink® environments are integrated into one entity, and thus we can analyze, simulate, and revise our models in either environment at any point. We invoke Simulink from within MATLAB. We will introduce Simulink with a few illustrated examples.

Example B.1

For the circuit of Figure B.1, the initial conditions are $i_L(0^-) = 0$, and $v_c(0^-) = 0.5 \text{ V}$. We will compute $v_c(t)$.

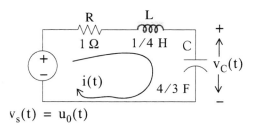

Figure B.1. Circuit for Example B.1

For this example,

$$i = i_L = i_C = C\frac{dv_C}{dt} \tag{B.1}$$

and by Kirchoff's voltage law (KVL),

$$Ri_L + L\frac{di_L}{dt} + v_C = u_0(t) \tag{B.2}$$

Substitution of (B.1) into (B.2) yields

$$RC\frac{dv_C}{dt} + LC\frac{d^2v_C}{dt^2} + v_C = u_0(t) \tag{B.3}$$

Substituting the values of the circuit constants and rearranging we obtain:

$$\frac{1}{3}\frac{d^2v_C}{dt^2} + \frac{4}{3}\frac{dv_C}{dt} + v_C = u_0(t)$$

$$\frac{d^2v_C}{dt^2} + 4\frac{dv_C}{dt} + 3v_C = 3u_0(t) \tag{B.4}$$

$$\frac{d^2v_C}{dt^2} + 4\frac{dv_C}{dt} + 3v_C = 3 \qquad t > 0 \tag{B.5}$$

To appreciate Simulink's capabilities, for comparison, three different methods of obtaining the solution are presented, and the solution using Simulink follows.

First Method – Assumed Solution

Equation (B.5) is a second–order, non–homogeneous differential equation with constant coefficients, and thus the complete solution will consist of the sum of the forced response and the natural response. It is obvious that the solution of this equation cannot be a constant since the derivatives of a constant are zero and thus the equation is not satisfied. Also, the solution cannot contain sinusoidal functions (sine and cosine) since the derivatives of these are also sinusoids.

However, decaying exponentials of the form ke^{-at} where k and a are constants, are possible candidates since their derivatives have the same form but alternate in sign.

It can be shown[*] that if $k_1 e^{-s_1 t}$ and $k_2 e^{-s_2 t}$ where k_1 and k_2 are constants and s_1 and s_2 are the roots of the characteristic equation of the homogeneous part of the given differential equation, the natural response is the sum of the terms $k_1 e^{-s_1 t}$ and $k_2 e^{-s_2 t}$. Therefore, the total solution will be

$$v_c(t) = \text{natural response} + \text{forced response} = v_{cn}(t) + v_{cf}(t) = k_1 e^{-s_1 t} + k_2 e^{-s_2 t} + v_{cf}(t) \tag{B.6}$$

The values of s_1 and s_2 are the roots of the characteristic equation

---

[*] *Please refer to Circuit Analysis II with MATLAB Applications, ISBN 0–9709511–5–9, Appendix B for a thorough discussion.*

$$s^2 + 4s + 3 = 0 \tag{B.7}$$

Solution of (B.7) yields of $s_1 = -1$ and $s_2 = -3$ and with these values (B.6) is written as

$$v_c(t) = k_1 e^{-t} + k_2 e^{-3t} + v_{cf}(t) \tag{B.8}$$

The forced component $v_{cf}(t)$ is found from (B.5), i.e.,

$$\frac{d^2 v_C}{dt^2} + 4\frac{dv_C}{dt} + 3v_C = 3 \qquad t > 0 \tag{B.9}$$

Since the right side of (B.9) is a constant, the forced response will also be a constant and we denote it as $v_{Cf} = k_3$. By substitution into (B.9) we obtain

$$0 + 0 + 3k_3 = 3$$

or

$$v_{Cf} = k_3 = 1 \tag{B.10}$$

Substitution of this value into (B.8), yields the total solution as

$$v_C(t) = v_{Cn}(t) + v_{Cf} = k_1 e^{-t} + k_2 e^{-3t} + 1 \tag{B.11}$$

The constants k_1 and k_2 will be evaluated from the initial conditions. First, using $v_C(0) = 0.5$ V and evaluating (B.11) at $t = 0$, we obtain

$$v_C(0) = k_1 e^0 + k_2 e^0 + 1 = 0.5$$

$$k_1 + k_2 = -0.5 \tag{B.12}$$

Also,

$$i_L = i_C = C\frac{dv_C}{dt}, \quad \frac{dv_C}{dt} = \frac{i_L}{C}$$

and

$$\left.\frac{dv_C}{dt}\right|_{t=0} = \frac{i_L(0)}{C} = \frac{0}{C} = 0 \tag{B.13}$$

Next, we differentiate (B.11), we evaluate it at $t = 0$, and equate it with (B.13). Thus,

$$\left.\frac{dv_C}{dt}\right|_{t=0} = -k_1 - 3k_2 \tag{B.14}$$

By equating the right sides of (B.13) and (B.14) we obtain

$$-k_1 - 3k_2 = 0 \tag{B.15}$$

Simultaneous solution of (B.12) and (B.15), gives $k_1 = -0.75$ and $k_2 = 0.25$. By substitution into (B.8), we obtain the total solution as

$$v_C(t) = (-0.75e^{-t} + 0.25e^{-3t} + 1)u_0(t) \tag{B.16}$$

Check with MATLAB:

```
syms t                              %  Define symbolic variable t
y0=-0.75*exp(-t)+0.25*exp(-3*t)+1;  %  The total solution y(t), for our example, vc(t)
y1=diff(y0)                         %  The first derivative of y(t)

y1 =
3/4*exp(-t)-3/4*exp(-3*t)

y2=diff(y0,2)                       %  The second derivative of y(t)

y2 =
-3/4*exp(-t)+9/4*exp(-3*t)

y=y2+4*y1+3*y0                      %  Summation of y and its derivatives

y =
3
```

Thus, the solution has been verified by MATLAB. Using the expression for $v_C(t)$ in (B.16), we find the expression for the current as

$$i = i_L = i_C = C\frac{dv_C}{dt} = \frac{4}{3}\left(\frac{3}{4}e^{-t} - \frac{3}{4}e^{-3t}\right) = e^{-t} - e^{-3t} \text{ A} \tag{B.17}$$

Second Method – Using the Laplace Transformation

The transformed circuit is shown in Figure B.2.

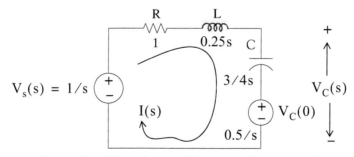

Figure B.2. Transformed Circuit for Example B.1

By the voltage division[*] expression,

$$V_C(s) = \frac{3/4s}{(1 + 0.25s + 3/4s)} \cdot \left(\frac{1}{s} - \frac{0.5}{s}\right) + \frac{0.5}{s} = \frac{1.5}{s(s^2 + 4s + 3)} + \frac{0.5}{s} = \frac{0.5s^2 + 2s + 3}{s(s + 1)(s + 3)}$$

Using partial fraction expansion,[†] we let

$$\frac{0.5s^2 + 2s + 3}{s(s + 1)(s + 3)} = \frac{r_1}{s} + \frac{r_2}{(s + 1)} + \frac{r_3}{(s + 3)} \tag{B.18}$$

$$r_1 = \left.\frac{0.5s^2 + 2s + 3}{(s + 1)(s + 3)}\right|_{s = 0} = 1$$

$$r_2 = \left.\frac{0.5s^2 + 2s + 3}{s(s + 3)}\right|_{s = -1} = -0.75$$

$$r_3 = \left.\frac{0.5s^2 + 2s + 3}{s(s + 1)}\right|_{s = -3} = 0.25$$

and by substitution into (B.18)

$$V_C(s) = \frac{0.5s^2 + 2s + 3}{s(s + 1)(s + 3)} = \frac{1}{s} + \frac{-0.75}{(s + 1)} + \frac{0.25}{(s + 3)}$$

Taking the Inverse Laplace transform[‡] we find that

$$v_C(t) = 1 - 0.75e^{-t} + 0.25e^{-3t}$$

Third Method – Using State Variables

$$Ri_L + L\frac{di_L}{dt} + v_C = u_0(t)^{**}$$

---

[*] *For derivation of the voltage division and current division expressions, please refer to Circuit Analysis I with MATLAB Applications, ISBN 0–9709511–2–4.*

[†] *Partial fraction expansion is discussed in Chapter 12, this text.*

[‡] *For an introduction to Laplace Transform and Inverse Laplace Transform, please refer to Chapters 2 and 3, Signals and Systems with MATLAB Computing and Simulinl Modeling, ISBN 0-9744239-9-8.*

[**] *Usually, in State–Space and State Variables Analysis, u(t) denotes any input. For distinction, we will denote the Unit Step Function as u_0(t). For a detailed discussion on State–Space and State Variables Analysis, please refer to Chapter 5, Signals and Systems with MATLAB Computing and Simulinl Modeling, ISBN 0-9744239-9-8.*

By substitution of given values and rearranging, we obtain

$$\frac{1}{4}\frac{di_L}{dt} = (-1)i_L - v_C + 1$$

or

$$\frac{di_L}{dt} = -4i_L - 4v_C + 4 \qquad (B.19)$$

Next, we define the state variables $x_1 = i_L$ and $x_2 = v_C$. Then,

$$\dot{x}_1 = \frac{di_L}{dt}* \qquad (B.20)$$

and

$$\dot{x}_2 = \frac{dv_C}{dt} \qquad (B.21)$$

Also,

$$i_L = C\frac{dv_C}{dt}$$

and thus,

$$x_1 = i_L = C\frac{dv_C}{dt} = C\dot{x}_2 = \frac{4}{3}\dot{x}_2$$

or

$$\dot{x}_2 = \frac{3}{4}x_1 \qquad (B.22)$$

Therefore, from (B.19), (B.20), and (B.22), we obtain the state equations

$$\dot{x}_1 = -4x_1 - 4x_2 + 4$$

$$\dot{x}_2 = \frac{3}{4}x_1$$

and in matrix form,

$$\begin{bmatrix} \dot{x}_1 \\ \dot{x}_2 \end{bmatrix} = \begin{bmatrix} -4 & -4 \\ 3/4 & 0 \end{bmatrix}\begin{bmatrix} x_1 \\ x_2 \end{bmatrix} + \begin{bmatrix} 4 \\ 0 \end{bmatrix}u_0(t) \qquad (B.23)$$

Solution[†] of (B.23) yields

---

* *The notation $\dot{x}$ (x dot) is often used to denote the first derivative of the function x, that is, $\dot{x} = dx/dt$.*

† *The detailed solution of (B.23) is given in Chapter 5, Signals and Systems with MATLAB Computing and Simulinl Modeling, ISBN 0-9744239-9-8.*

$$\begin{bmatrix} x_1 \\ x_2 \end{bmatrix} = \begin{bmatrix} e^{-t} - e^{-3t} \\ 1 - 0.75e^{-t} + 0.25e^{-3t} \end{bmatrix}$$

Then,

$$x_1 = i_L = e^{-t} - e^{-3t} \tag{B.24}$$

and

$$x_2 = v_C = 1 - 0.75e^{-t} + 0.25e^{-3t} \tag{B.25}$$

Modeling the Differential Equation of Example B.1 with Simulink

To run Simulink, we must first invoke MATLAB. Make sure that Simulink is installed in your system. In the MATLAB Command prompt, we type:

simulink

Alternately, we can click on the Simulink icon shown in Figure B.3. It appears on the top bar on MATLAB's Command prompt.

Figure B.3. The Simulink icon

Upon execution of the Simulink command, the **Commonly Used Blocks** appear as shown in Figure B.4.

In Figure B.4, the left side is referred to as the **Tree Pane** and displays all Simulink libraries installed. The right side is referred to as the **Contents Pane** and displays the blocks that reside in the library currently selected in the Tree Pane.

Let us express the differential equation of Example B.1 as

$$\frac{d^2 v_C}{dt^2} = -4\frac{dv_C}{dt} - 3v_C + 3u_0(t) \tag{B.26}$$

A block diagram representing relation (B.26) above is shown in Figure B.5. We will use Simulink to draw a similar block diagram.[*]

---

[*] *Henceforth, all Simulink block diagrams will be referred to as models.*

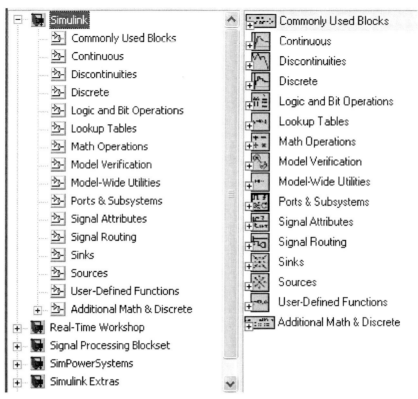

Figure B.4. The Simulink Library Browser

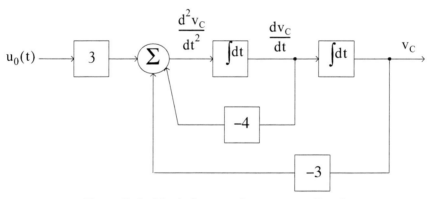

Figure B.5. Block diagram for equation (B.26)

To model the differential equation (B.26) using Simulink, we perform the following steps:

1. On the **Simulink Library Browser**, we click on the leftmost icon shown as a blank page on the top title bar. A new model window named **untitled** will appear as shown in Figure B.6.

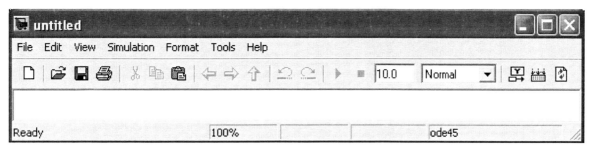

Figure B.6. The Untitled model window in Simulink.

The window of Figure B.6 is the model window where we enter our blocks to form a block diagram. We save this as model file name **Equation_1_26**. This is done from the File drop menu of Figure B.6 where we choose **Save as** and name the file as **Equation_1_26**. Simulink will add the extension **.mdl**. The new model window will now be shown as **Equation_1_26**, and all saved files will have this appearance. See Figure B.7.

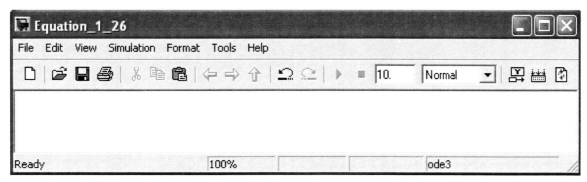

Figure B.7. Model window for Equation_1_26.mdl file

2. With the **Equation_1_26** model window and the **Simulink Library Browser** both visible, we click on the **Sources** appearing on the left side list, and on the right side we scroll down until we see the unit step function shown as **Step**. See Figure B.8. We select it, and we drag it into the **Equation_1_26** model window which now appears as shown in Figure B.8. We save file Equation_1_26 using the File drop menu on the **Equation_1_26** model window (right side of Figure B.8).

3. With reference to block diagram of Figure B.5, we observe that we need to connect an amplifier with Gain 3 to the unit step function block. The gain block in Simulink is under **Commonly Used Blocks** (first item under Simulink on the **Simulink Library Browser**). See Figure B.8. If the **Equation_1_26** model window is no longer visible, it can be recalled by clicking on the white page icon on the top bar of the **Simulink Library Browser**.

4. We choose the gain block and we drag it to the right of the unit step function. The triangle on the right side of the unit step function block and the > symbols on the left and right sides of the gain block are connection points. We point the mouse close to the connection point of the unit step function until is shows as a cross hair, and draw a straight line to connect the two

blocks.* We double–click on the gain block and on the **Function Block Parameters**, we change the gain from 1 to 3. See Figure B.9.

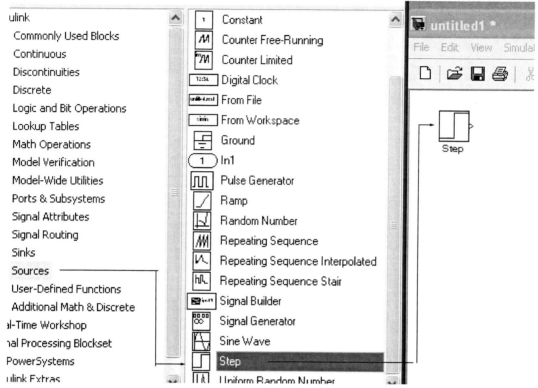

Figure B.8. Dragging the unit step function into File Equation_1_26

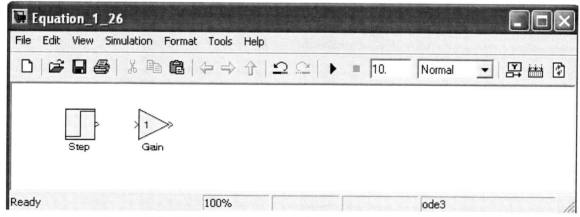

Figure B.9. File Equation_1_26 with added Step and Gain blocks

---

* *An easy method to interconnect two Simulink blocks is by clicking on the source block to select it, then holding down the **Ctrl** key, and left–clicking on the destination block.*

B–10 *Numerical Analysis Using MATLAB® and Excel®, Third Edition*
Copyright © Orchard Publications

5. Next, we need to add a thee–input adder. The adder block appears on the right side of the **Simulink Library Browser** under **Math Operations**. We select it, and we drag it into the Equation_1_26 model window. We double click it, and on the **Function Block Parameters** window which appears, we specify 3 inputs. We then connect the output of the of the gain block to the first input of the adder block as shown in Figure B.10.

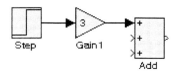

Figure B.10. File Equation_1_26 with added gain block

6. From the **Commonly Used Blocks** of the **Simulink Library Browser**, we choose the **Integrator** block, we drag it into the Equation_1_26 model window, and we connect it to the output of the **Add** block. We repeat this step and to add a second **Integrator** block. We click on the text "Integrator" under the first integrator block, and we change it to Integrator 1. Then, we change the text "Integrator 1" under the second Integrator to "Integrator 2" as shown in Figure B.11.

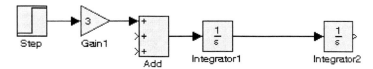

Figure B.11. File Equation_1_26 with the addition of two integrators

7. To complete the block diagram, we add the **Scope** block which is found in the **Commonly Used Blocks** on the **Simulink Library Browser**, we click on the Gain block, and we copy and paste it twice. We flip the pasted Gain blocks by using the **Flip Block** command from the Format drop menu, and we label these as Gain 2 and Gain 3. Finally, we double–click on these gain blocks and on the **Function Block Parameters** window, we change the gains from to –4 and –3 as shown in Figure B.12.

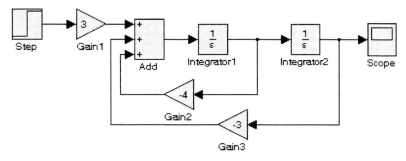

Figure B.12. File Equation_1_26 complete block diagram

8. The initial conditions $i_L(0^-) = C(dv_C/dt)\big|_{t=0} = 0$, and $v_c(0^-) = 0.5$ V are entered by double clicking the Integrator blocks and entering the values 0 for the first integrator, and 0.5 for the second integrator. We also need to specify the simulation time. This is done by specifying the simulation time to be 10 seconds on the **Configuration Parameters** from the **Simulation** drop menu. We can start the simulation on **Start** from the **Simulation** drop menu or by clicking on the ▶ icon.

9. To see the output waveform, we double click on the **Scope** block, and then clicking on the Autoscale 🔍 icon, we obtain the waveform shown in Figure B.13.

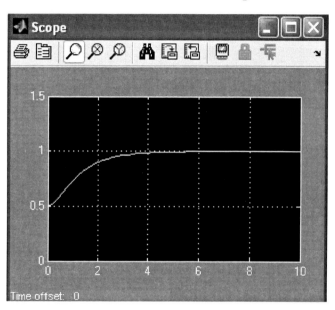

Figure B.13. The waveform for the function $v_C(t)$ for Example B.1

Another easier method to obtain and display the output $v_C(t)$ for Example B.1, is to use **State–Space** block from **Continuous** in the Simulink Library Browser, as shown in Figure B.14.

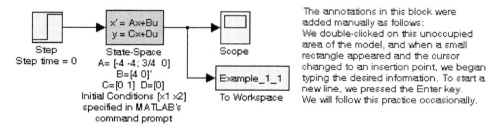

Figure B.14. Obtaining the function $v_C(t)$ for Example B.1 with the State–Space block.

The **simout To Workspace** block shown in Figure B.14 writes its input to the workspace. The data and variables created in the MATLAB Command window, reside in the MATLAB Workspace. This block writes its output to an array or structure that has the name specified by the block's Variable name parameter. This gives us the ability to delete or modify selected variables. We issue the command **who** to see those variables. From Equation B.23, Page B–6,

$$\begin{bmatrix} \dot{x}_1 \\ \dot{x}_2 \end{bmatrix} = \begin{bmatrix} -4 & -4 \\ 3/4 & 0 \end{bmatrix} \begin{bmatrix} x_1 \\ x_2 \end{bmatrix} + \begin{bmatrix} 4 \\ 0 \end{bmatrix} u_0(t)$$

The output equation is

$$y = Cx + du$$

or

$$y = [0 \ \ 1] \begin{bmatrix} x_1 \\ x_2 \end{bmatrix} + [0]u$$

We double–click on the **State–Space** block, and in the **Functions Block Parameters** window we enter the constants shown in Figure B.15.

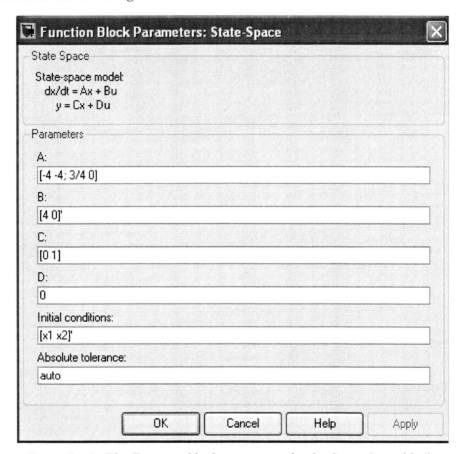

Figure B.15. The Function block parameters for the State–Space block.

The initials conditions [x1 x2]' are specified in MATLAB's Command prompt as

x1=0; x2=0.5;

As before, to start the simulation we click clicking on the ▶ icon, and to see the output wave-

form, we double click on the **Scope** block, and then clicking on the Autoscale 🔭 icon, we obtain the waveform shown in Figure B.16.

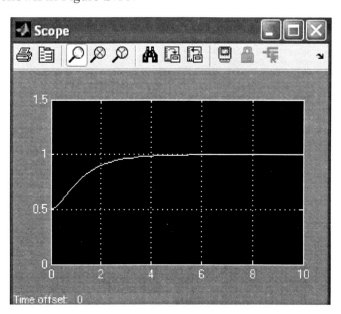

Figure B.16. The waveform for the function $v_C(t)$ for Example B.1 with the State–Space block.

The state–space block is the best choice when we need to display the output waveform of three or more variables as illustrated by the following example.

Example B.2

A fourth–order network is described by the differential equation

$$\frac{d^4 y}{dt^4} + a_3 \frac{d^3 y}{dt^3} + a_2 \frac{d^2 y}{dt^2} + a_1 \frac{dy}{dt} + a_0 y(t) = u(t) \tag{B.27}$$

where $y(t)$ is the output representing the voltage or current of the network, and $u(t)$ is any input, and the initial conditions are $y(0) = y'(0) = y''(0) = y'''(0) = 0$.

a. We will express (B.27) as a set of state equations

Numerical Analysis Using MATLAB® and Excel®, Third Edition

Copyright © Orchard Publications

b. It is known that the solution of the differential equation

$$\frac{d^4y}{dt^4} + 2\frac{d^2y}{dt^2} + y(t) = \sin t \qquad \text{(B.28)}$$

subject to the initial conditions $y(0) = y'(0) = y''(0) = y'''(0) = 0$, has the solution

$$y(t) = 0.125[(3 - t^2) - 3t\cos t] \qquad \text{(B.29)}$$

In our set of state equations, we will select appropriate values for the coefficients $a_3, a_2, a_1,$ and a_0 so that the new set of the state equations will represent the differential equation of (B.28), and using Simulink, we will display the waveform of the output $y(t)$.

1. The differential equation of (B.28) is of fourth–order; therefore, we must define four state variables that will be used with the four first–order state equations.

We denote the state variables as $x_1, x_2, x_3,$ and x_4, and we relate them to the terms of the given differential equation as

$$x_1 = y(t) \qquad x_2 = \frac{dy}{dt} \qquad x_3 = \frac{d^2y}{dt^2} \qquad x_4 = \frac{d^3y}{dt^3} \qquad \text{(B.30)}$$

We observe that

$$\begin{aligned}
\dot{x}_1 &= x_2 \\
\dot{x}_2 &= x_3 \\
\dot{x}_3 &= x_4
\end{aligned} \qquad \text{(B.31)}$$

$$\frac{d^4y}{dt^4} = \dot{x}_4 = -a_0x_1 - a_1x_2 - a_2x_3 - a_3x_4 + u(t)$$

and in matrix form

$$\begin{bmatrix} \dot{x}_1 \\ \dot{x}_2 \\ \dot{x}_3 \\ \dot{x}_4 \end{bmatrix} = \begin{bmatrix} 0 & 1 & 0 & 0 \\ 0 & 0 & 1 & 0 \\ 0 & 0 & 0 & 1 \\ -a_0 & -a_1 & -a_2 & -a_3 \end{bmatrix} \begin{bmatrix} x_1 \\ x_2 \\ x_3 \\ x_4 \end{bmatrix} + \begin{bmatrix} 0 \\ 0 \\ 0 \\ 1 \end{bmatrix} u(t) \qquad \text{(B.32)}$$

In compact form, (B.32) is written as

$$\dot{x} = Ax + bu \qquad \text{(B.33)}$$

Also, the output is

$$y = Cx + du \qquad \text{(B.34)}$$

where

$$\dot{x} = \begin{bmatrix} \dot{x}_1 \\ \dot{x}_2 \\ \dot{x}_3 \\ \dot{x}_4 \end{bmatrix}, \quad A = \begin{bmatrix} 0 & 1 & 0 & 0 \\ 0 & 0 & 1 & 0 \\ 0 & 0 & 0 & 1 \\ -a_0 & -a_1 & -a_2 & -a_3 \end{bmatrix}, \quad x = \begin{bmatrix} x_1 \\ x_2 \\ x_3 \\ x_4 \end{bmatrix}, \quad b = \begin{bmatrix} 0 \\ 0 \\ 0 \\ 1 \end{bmatrix}, \quad \text{and } u = u(t) \tag{B.35}$$

and since the output is defined as

$$y(t) = x_1$$

relation (B.34) is expressed as

$$y = \begin{bmatrix} 1 & 0 & 0 & 0 \end{bmatrix} \cdot \begin{bmatrix} x_1 \\ x_2 \\ x_3 \\ x_4 \end{bmatrix} + [0]u(t) \tag{B.36}$$

2. By inspection, the differential equation of (B.27) will be reduced to the differential equation of (B.28) if we let

$$a_3 = 0 \qquad a_2 = 2 \qquad a_1 = 0 \qquad a_0 = 1 \qquad u(t) = \sin t$$

and thus the differential equation of (B.28) can be expressed in state–space form as

$$\begin{bmatrix} \dot{x}_1 \\ \dot{x}_2 \\ \dot{x}_3 \\ \dot{x}_4 \end{bmatrix} = \begin{bmatrix} 0 & 1 & 0 & 0 \\ 0 & 0 & 1 & 0 \\ 0 & 0 & 0 & 1 \\ -a_0 & 0 & -2 & 0 \end{bmatrix} \begin{bmatrix} x_1 \\ x_2 \\ x_3 \\ x_4 \end{bmatrix} + \begin{bmatrix} 0 \\ 0 \\ 0 \\ 1 \end{bmatrix} \sin t \tag{B.37}$$

where

$$\dot{x} = \begin{bmatrix} \dot{x}_1 \\ \dot{x}_2 \\ \dot{x}_3 \\ \dot{x}_4 \end{bmatrix}, \quad A = \begin{bmatrix} 0 & 1 & 0 & 0 \\ 0 & 0 & 1 & 0 \\ 0 & 0 & 0 & 1 \\ -a_0 & 0 & -2 & 0 \end{bmatrix}, \quad x = \begin{bmatrix} x_1 \\ x_2 \\ x_3 \\ x_4 \end{bmatrix}, \quad b = \begin{bmatrix} 0 \\ 0 \\ 0 \\ 1 \end{bmatrix}, \quad \text{and } u = \sin t \tag{B.38}$$

Since the output is defined as

$$y(t) = x_1$$

in matrix form it is expressed as

$$y = [1 \ \ 0 \ \ 0 \ \ 0] \cdot \begin{bmatrix} x_1 \\ x_2 \\ x_3 \\ x_4 \end{bmatrix} + [0]\sin t \qquad (B.39)$$

We invoke MATLAB, we start Simulink by clicking on the Simulink icon, on the **Simulink Library Browser** we click on the **Create a new model** (blank page icon on the left of the top bar), and we save this model as Example_1_2. On the **Simulink Library Browser** we select **Sources**, we drag the **Signal Generator** block on the Example_1_2 model window, we click and drag the **State–Space** block from the **Continuous** on **Simulink Library Browser**, and we click and drag the **Scope** block from the **Commonly Used Blocks** on the **Simulink Library Browser**. We also add the **Display** block found under **Sinks** on the **Simulink Library Browser**. We connect these four blocks and the complete block diagram is as shown in Figure B.17.

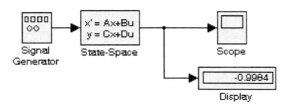

Figure B.17. Block diagram for Example B.2

We now double–click on the **Signal Generator** block and we enter the following in the **Function Block Parameters:**

Wave form: sine

Time (t): Use simulation time

Amplitude: 1

Frequency: 2

Units: Hertz

Next, we double–click on the **state–space** block and we enter the following parameter values in the **Function Block Parameters:**

A: [0 1 0 0; 0 0 1 0; 0 0 0 1; –a0 –a1 –a2 –a3]

B: [0 0 0 1]'

C: [1 0 0 0]

D: [0]

Initial conditions: x0

Absolute tolerance: auto

Now, we switch to the MATLAB Command prompt and we type the following:

>> a0=1; a1=0; a2=2; a3=0; x0=[0 0 0 0]';

We change the **Simulation Stop time** to 25, and we start the simulation by clicking on the ▶ icon. To see the output waveform, we double click on the **Scope** block, then clicking on the Autoscale 🔍 icon, we obtain the waveform shown in Figure B.18.

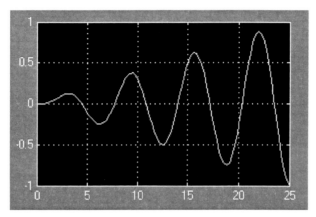

Figure B.18. Waveform for Example B.2

The **Display** block in Figure B.17 shows the value at the end of the simulation stop time.

Examples B.1 and B.2 have clearly illustrated that the State–Space is indeed a powerful block. We could have obtained the solution of Example B.2 using four Integrator blocks by this approach would have been more time consuming.

Example B.3

Using **Algebraic Constraint** blocks found in the **Math Operations** library, **Display** blocks found in the **Sinks** library, and **Gain** blocks found in the **Commonly Used Blocks** library, we will create a model that will produce the simultaneous solution of three equations with three unknowns.

The model will display the values for the unknowns z_1, z_2, and z_3 in the system of the equations

$$a_1 z_1 + a_2 z_2 + a_3 z_3 + k_1 = 0$$
$$a_4 z_1 + a_5 z_2 + a_6 z_3 + k_2 = 0 \qquad (B.40)$$
$$a_7 z_1 + a_8 z_2 + a_9 z_3 + k_3 = 0$$

The model is shown in Figure B.19.

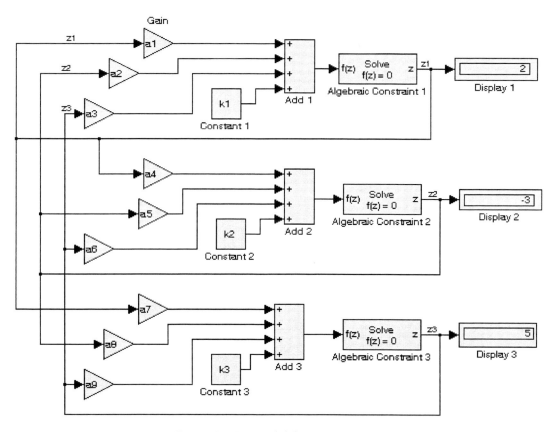

Figure B.19. Model for Example B.3

Next, we go to MATLAB's Command prompt and we enter the following values:

a1=2; a2=−3; a3=−1; a4=1; a5=5; a6=4; a7=−6; a8=1; a9=2;...
k1=−8; k2=−7; k3=5;

After clicking on the simulation icon, we observe the values of the unknowns as $z_1 = 2$, $z_2 = -3$, and $z_3 = 5$. These values are shown in the Display blocks of Figure B.19.

The **Algebraic Constraint** block constrains the input signal $f(z)$ to zero and outputs an algebraic state z. The block outputs the value necessary to produce a zero at the input. The output must affect the input through some feedback path. This enables us to specify algebraic equations for index 1 differential/algebraic systems (DAEs). By default, the Initial guess parameter is zero. We can improve the efficiency of the algebraic loop solver by providing an Initial guess for the algebraic state z that is close to the solution value.

An outstanding feature in Simulink is the representation of a large model consisting of many blocks and lines, to be shown as a single Subsystem block.[*] For instance, we can group all blocks and lines in the model of Figure B.19 except the display blocks, we choose **Create Subsystem** from the **Edit** menu, and this model will be shown as in Figure B.20[†] where in MATLAB's Command prompt we have entered:

a1=5; a2=–1; a3=4; a4=11; a5=6; a6=9; a7=–8; a8=4; a9=15;...
k1=14; k2=–6; k3=9;

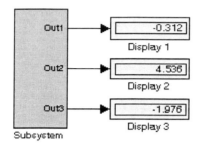

Figure B.20. The model of Figure B.19 represented as a subsystem

The Display blocks in Figure B.20 show the values of z_1, z_2, and z_3 for the values specified in MATLAB's Command prompt.

B.2 Simulink Demos

At this time, the reader with no prior knowledge of Simulink, should be ready to learn Simulink's additional capabilities. It is highly recommended that the reader becomes familiar with the block libraries found in the Simulink Library Browser. Then, the reader can follow the steps delineated in The MathWorks Simulink User's Manual to run the Demo Models beginning with the **thermo** model. This model can be seen by typing

thermo

in the MATLAB Command prompt.

---

[*] *The Subsystem block is described in detail in Chapter 2, Section 2.1, Page 2–2, Introduction to Simulink with Engineering Applications, ISBN 0–9744239–7–1.*

[†] *The contents of the Subsystem block are not lost. We can double–click on the Subsystem block to see its contents. The Subsystem block replaces the inputs and outputs of the model with Inport and Outport blocks. These blocks are described in Section 2.1, Chapter 2, Page 2–2, Introduction to Simulink with Engineering Applications, ISBN 0–9744239–7–1.*

Appendix C

T his appendix supplements Chapters 4 and 14 with concerns when the determinant of the coefficient matrix is small. We will introduce a reference against which the determinant can be measured to classify a matrix as a well– or ill–conditioned.

C.1 The Norm of a Matrix

A *norm* is a function which assigns a positive length or size to all vectors in a vector space, other than the zero vector. An example is the two–dimensional Euclidean space denoted as R^2. The elements of the Euclidean vector space (e.g., (2,5)) are usually drawn as arrows in a two–dimensional cartesian coordinate system starting at the origin (0,0). The Euclidean norm assigns to each vector the length of its arrow.

The *Euclidean norm* of a matrix A, denoted as $\|A\|$, is defined as

$$\|A\| = \sqrt{\sum_{i=1}^{n} \sum_{j=1}^{n} A_{ij}^2} \tag{C.1}$$

and it is computed with the MATLAB function **norm(A)**.

Example C.1

Using the MATLAB function **norm(A)**, compute the Euclidean *norm* of the matrix A, defined as

$$A = \begin{bmatrix} -2 & 5 & -4 & 9 \\ -3 & -6 & 8 & 1 \\ 7 & -5 & 3 & 2 \\ 4 & -9 & -8 & -1 \end{bmatrix}$$

Solution:

At the MATLAB command prompt, we enter

A=[–2 5 –4 9; –3 –6 8 1; 7 –5 3 2; 4 –9 –8 –1]; norm(A)

and MATLAB outputs

```
ans =
   14.5539
```

C.2 Condition Number of a Matrix

The *condition number of a matrix* A is defined as

$$k(A) = \|A\| \cdot \|A^{-1}\| \tag{C.2}$$

where $\|A\|$ is the *norm* of the matrix A defined in relation (C.1) above. Matrices with condition number close to unity are said to be *well–conditioned matrices*, and those with very large condition number are said to be *ill–conditioned matrices*.

The condition number of a matrix A is computed with the MATLAB function **cond(A)**.

Example C.2

Using the MATLAB function **cond(A)**, compute the condition number of the matrix A defined as

$$A = \begin{bmatrix} -2 & 5 & -4 & 9 \\ -3 & -6 & 8 & 1 \\ 7 & -5 & 3 & 2 \\ 4 & -9 & -8 & -1 \end{bmatrix}$$

Solution:

At the MATLAB command prompt, we enter

A=[–2 5 –4 9; –3 –6 8 1; 7 –5 3 2; 4 –9 –8 –1]; cond(A)

and MATLAB outputs

```
ans =
   2.3724
```

This condition number is relatively close to unity and thus we classify matrix A as a well-conditioned matrix.

We recall from Chapter 4 that if the determinant of a square matrix A is singular, that is, if $det(A) = 0$, the inverse of A is undefined. Please refer to Chapter 4, Page 4–22.

Now, let us consider that the coefficient matrix[*] is very small, i.e., almost singular. Accordingly, we classify such a matrix as ill–conditioned.

C.3 Hilbert Matrices

Let n be a positive integer. A *unit fraction* is the reciprocal of this integer, that is, $1/n$. Thus, $1/1, 1/2, 1/3, \ldots$ are unit fractions. A *Hilbert matrix* is a matrix with unit fraction elements

$$B_{ij} = 1/(i+j-1) \tag{C.3}$$

Shown below is an example of the 5×5 Hilbert matrix.

$$\begin{bmatrix} \frac{1}{1} & \frac{1}{2} & \frac{1}{3} & \frac{1}{4} & \frac{1}{5} \\ \frac{1}{2} & \frac{1}{3} & \frac{1}{4} & \frac{1}{5} & \frac{1}{6} \\ \frac{1}{3} & \frac{1}{4} & \frac{1}{5} & \frac{1}{6} & \frac{1}{7} \\ \frac{1}{4} & \frac{1}{5} & \frac{1}{6} & \frac{1}{7} & \frac{1}{8} \\ \frac{1}{5} & \frac{1}{6} & \frac{1}{7} & \frac{1}{8} & \frac{1}{9} \end{bmatrix} \tag{C.4}$$

---

* *In general, a system with m linear equations and n unknowns can be written as*

$$a_{11}x_1 + a_{12}x_2 + \ldots + a_{1n}x_n = b_1$$
$$a_{21}x_1 + a_{22}x_2 + \ldots + a_{2n}x_n = b_2$$
$$\ldots \quad \ldots \quad \ldots \quad \ldots \quad \ldots$$
$$a_{m1}x_1 + a_{m2}x_2 + \ldots + a_{mn}x_n = b_m$$

where $x_1, x_2, \ldots, x_n$ are the unknowns and the numbers $a_{11}, a_{12}, \ldots, a_{mn}$ are the coefficients of the system. The coefficient matrix is the $m \times n$ matrix with the coefficient a_{ij} as the (i,j)-th entry:

$$\begin{bmatrix} a_{11} & a_{12} & \cdots & a_{1n} \\ a_{21} & a_{22} & \cdots & a_{2n} \\ \cdots & \cdots & \cdots & \cdots \\ a_{m1} & a_{m1} & a_{m1} & a_{m1} \end{bmatrix}$$

MATLAB's function **hilb(n)** displays the Hilbert n × n matrix.

Example C.3

Compute the determinant and the condition number of the 6 × 6 Hilbert matrix using MATLAB.

Solution:

At the MATLAB command prompt, we enter

det(hilb(6))

and MATLAB outputs

ans =

  5.3673e-018

This is indeed a very small number and for all practical purposes this matrix is singular.

We can find the condition number of a matrix A with the **cond(A)** MATLAB function. Thus, for the 6 × 6 Hilbert matrix,

cond(hilb(6))

ans =
  1.4951e+007

This is a large number and if the coefficient matrix is multiplied by this number, seven decimal places might be lost.

Let us consider another example.

Example C.4

Let $Ax = b$ where $A = \begin{bmatrix} 0.585 & 0.379 \\ 0.728 & 0.464 \end{bmatrix}$ and $b = \begin{bmatrix} 0.187 \\ 0.256 \end{bmatrix}$

Compute the values of the vector x.

Solution:

Here, we are asked to find the values of x_1 and x_2 of the linear system

$$\begin{bmatrix} 0.585 & 0.379 \\ 0.728 & 0.464 \end{bmatrix} \cdot \begin{bmatrix} x_1 \\ x_2 \end{bmatrix} = \begin{bmatrix} 0.187 \\ 0.256 \end{bmatrix}$$

Using MATLAB, we define A and b, and we use the left division operation, i.e.,

A=[0.585 0.378; 0.728 0.464]; b=[0.187 0.256]'; x=b\A

```
x =
    2.9428    1.8852
```

Check:

A=[0.585 0.378; 0.728 0.464]; x=[2.9428 1.8852]'; b=A*x

```
b =
    2.4341
    3.0171
```

but these are not the given values of the vector b, so let us check the determinant and the condition number of the matrix A.

determinant = det(A)

```
determinant =
    -0.0037
```

condition=cond(A)

```
condition =
    328.6265
```

Therefore, we conclude that this system of equations is ill-conditioned and the solution is invalid.

Example C.4 above should serve as a reminder that when we solve systems of equations using matrices, we should check the determinants and the condition number to predict possible floating point and roundoff errors.

References and Suggestions for Further Study

A. The following publications by The MathWorks, are highly recommended for further study. They are available from The MathWorks, 3 Apple Hill Drive, Natick, MA, 01760, www.mathworks.com.

1. *Getting Started with MATLAB*

2. *Using MATLAB*

3. *Using MATLAB Graphics*

4. *Financial Toolbox*

5. *Statistics Toolbox*

B. Other references indicated in footnotes throughout this text, are listed below.

1. *Mathematics for Business, Science, and Technology with MATLAB and Excel Computations, Third Edition*, ISBN-13: 978–1–934404–01–2

2. *Circuit Analysis I with MATLAB Applications*, ISBN 0–9709511–2–4

3. *Circuit Analysis II with MATLAB Applications*, ISBN 0–9709511–5–9

4. *Introduction to Simulink with Engineering Applications*, ISBN 0-9744239-7-1

5. *Signals and Systems with MATLAB Computing and Simulink Modeling, Third Edition*, ISBN 0-9744239-9-8

6. *Handbook of Mathematical Functions*, ISBN 0-4866127-2-4

7. *CRC Standard Mathematical Tables*, ISBN 0-8493-0626-4

Index

Newton's divided difference interpolation method 7-15
Newton-Cotes 8-panel rule 10-10
non-homogeneous difference equation 11-2
non-homogeneous ODE 5-6
non-singular matrix - see matrix
norm C-1
norm(A) MATLAB function C-1
numeric expressions in MATLAB 12-4
numerical evaluation of Fourier coefficients 6-36

O

odd functions 6-8, 6-31
odd symmetry 6-7
ODE (Ordinary Differential Equation) 5-3
ode23 MATLAB function 9-9
ode45 MATLAB function 9-9
one-dimensional wave equation 5-3
optimum path policy 16-5
order of a differential equation 5-3
ordinary differential equation 5-3
oriented network 16-15
orthogonal basis 14-5
orthogonal functions 6-2, 14-1, 14-2
orthogonal system 15-9
orthogonal trajectories 14-2
orthogonal unit vectors 14-5
orthogonal vectors 5-39, 14-4
orthonormal basis 14-5
out-of-phase sinusoids 3-3
overdetermined system 8-3

P

parabolic curve 8-1
partial differential equation (PDE) 5-3
partial fraction expansion 12-1
PDE (Partial Differential Equation) 5-3
Pearson correlation coefficient 8-10
period 3-2, 3-3
periodic waveform 3-2
phasor 3-2
plot area in Excel 6
plot MATLAB command 1-9, 1-12, 1-15
plot3 MATLAB command 1-16
polar form of complex numbers 3-15
polar plot in MATLAB 1-24
polar(theta,r) MATLAB function 1-24
poles 12-2
poly MATLAB function 1-4
polyder MATLAB function 1-7
polyfit(x,y,n) MATLAB function 8-11
polynomial construction from known roots in MATLAB 1-4
polyval(p,x) MATLAB function 1-5, 8-11
power series 6-40
proper rational function 12-1

Q

QR factorization 14-25
qr(A) MATLAB function 14-25
quad MATLAB function 10-10
quad8 MATLAB function 10-10
quadratic curve 8-1

quadratic factor 1-9
quit MATLAB command 1-2

R

radian frequency 3-2
rational polynomial 1-8
rationalization of the quotient 3-13
real axis 3-10
real number 3-11
real(z) MATLAB function 1-24
recursion A-1
recursive method A-1
regression 8-1
regression analysis 8-7
relative cell in Excel 2-19
repeated poles 5-9, 12- 6
residue(r,p,k) MATLAB function 12-1
revolutions per second 3-5
Rodrigues' formula 15-12, 15-18
roots of polynomials 1-3
roots(p) MATLAB function 1-3, 1-8
rotating vector 3-5
round(n) MATLAB function 1-24
row vector 1-3, 1-19
Runge-Kutta method 5-24, 9-5

S

sawtooth waveform 6-10, 6-18
scalar matrix - see matrix
Scope block in Simulink B-12
script file in MATLAB 1-26
second divided difference 7-1
second harmonic 6-1
semicolons in MATLAB 1-7
semilogx MATLAB command 1-12
semilogy MATLAB command 1-12
simple differential equations 5-1
simplex method 16-4
Simpson's rule 10-6
Simulation drop menu in Simulink B-12
simulation start icon in Simulink B-12
Simulink icon B-7
Simulink Library Browser B-8
singular matrix - see matrix
Singular Value Decomposition 14-28
Sinks library B-18
sinusoids 3-2
size of a matrix - see matrix
skew-Hermitian matrix - see matrix
skew-symmetric matrix - see matrix
solution of the homogeneous ODE 5-8
solutions of ODEs 5-6
space equations 5-24
spectrum analyzer 33
spherical harmonics 15-18
sprintf MATLAB command 2-5
sqrt MATLAB function 10-12
square matrix - see matrix
square waveform 6-9, 6-14, 6-16, 6-48, 6-49
start simulation in Simulink B-12
state equations 5-24
state transition matrix 5-28
state variables 5-24
State-Space block in Simulink B-12

Stirling's asymptotic series for the G(n) function 13-9
string in MATLAB 1-17
subplot MATLAB command 1-18
surface zonal harmonics 15-11
svd(A) MATLAB function 14-28
sym, syms MATLAB symbolic expressions 12-4
symbolic expressions in MATLAB 12-3
Symbolic Math Toolbox in MATLAB 12-4
symmetric matrix - see matrix
symmetry 6-7, 6-14, 6-31

T

Taylor series 5-24, 6-41, 6-44
Taylor series expansion method 9-1
text MATLAB command 1-14, 1-17
third harmonic 6-1
title('string') in MATLAB 1-12
trace of a matrix - see matrix
transpose of a matrix - see matrix
trapezoidal rule 10-1
trapz(x,y) MATLAB function 10-3, 10-5
Tree Pane in Simulink B-7
Trendline Excel feature 8-9
triangular waveform 6-11, 6-19
trigonometric Fourier series 6-1
trigonometric relations 3-5
two-dimensional plots 7-32
type of a diferential equation 5-2

U

ultraspherical functions 15-22
undetermined system 8-3
unit fraction C-3
unitary matrix - see matrix
upper triangular matrix - see matrix

V

VLOOKUP Excel function 7-23

W

Wallis's formulas 13-16
Weber functions 15-7
well-conditioned matrix - see matrix
while end in MATLAB 2-4
Wronskian determinant 5-10, 11-2

X

xlabel MATLAB command 1-12
XY (Scatter) in Excel 8-6

Y

ylabel MATLAB command 1-12

Z

zero matrix - see matrix
zeros 12-2
zlabel MATLAB command 1-17